6. Bayes's Theorem If subsets H_i form a partition of S,

$$P(H_i|E) = P(H_i \cap E)/P(E) = P(H_i \cap E)/\sum P(H_i \cap E)$$
$$= P(H_i)P(E|H_i)/\sum P(H_i)P(E|H_i).$$

7. Odds If $P(A) = m/n$, the odds in favor of A are $m/(n - m)$. If the odds in favor of A are a/b, then $P(A) = a/(a + b)$.

D. POPULATIONS: MEANS, VARIANCES, COVARIANCES, AND CORRELATION

Notation. X and Y are random variables with values x_i, y_i; $X_1, X_2, \ldots, X_n$ are random variables; a and b are constants; $P(X = x_i) = f(x_i)$; $E(X)$ is the expected value of X; σ_i^2 is the variance of X_i.

1. Mean $\mu_X = E(X) = \sum x_i f(x_i)$; $\mu_{aX+b} = a\mu_X + b$.

2. Variance $\sigma_X^2 = E(X - \mu_X)^2 = E(X^2) - \mu_X^2 = \mathrm{Var}(X)$; $\sigma_{aX+b}^2 = a^2\sigma_X^2$.

Standard deviation. $\sigma_X = \sqrt{\mathrm{Var}(X)}$; $\sigma_{aX+b} = |a|\sigma_X$.

3. Standardized random variable *Z* If $Z = (X - \mu_X)/\sigma_X$, then $\mu_Z = 0$, $\sigma_Z = 1$.

4. Chebyshev's Theorem $P(|X - \mu| > h\sigma) < 1/h^2$; $P(|X - \mu| \leq h\sigma) > 1 - (1/h^2)$.

5. Covariance $\mathrm{Cov}(X, X) = \sigma_X^2$;

$$\mathrm{Cov}(X, Y) = E[(X - \mu_X)(Y - \mu_Y)] = E(XY) - \mu_X\mu_Y = \rho\sigma_X\sigma_Y.$$

6. Correlation $\rho = \mathrm{Cov}(X, Y)/\sigma_X\sigma_Y$.

7. Sums $E(X + Y) = E(X) + E(Y)$; $E(\sum X_i) = \sum E(X_i)$;

$$\sigma_{X+Y}^2 = \sigma_X^2 + \sigma_Y^2 + 2\rho\sigma_X\sigma_Y; \qquad \sigma_{\Sigma X_i}^2 = \sum\sigma_i^2 + 2\sum \rho_{ij}\sigma_i\sigma_j, \quad i < j.$$

$$E\left(\sum w_i X_i\right) = \sum w_i\mu_i, \qquad \sigma_{\Sigma w_i X_i}^2 = \sum w_i^2\sigma_i^2 + 2\sum_{i<j} w_i w_j \,\mathrm{Cov}(X_i, X_j).$$

Special cases. If all correlations $\rho_{ij} = \rho$ and all variances $\sigma_i^2 = \sigma^2$, then $\sigma_{\Sigma X_i}^2 = n\sigma^2[1 + (n - 1)\rho]$.
If X and Y are independent or uncorrelated ($\rho = 0$):

$$\sigma_{X+Y}^2 = \sigma_X^2 + \sigma_Y^2, \qquad \sigma_{aX+bY}^2 = a^2\sigma_X^2 + b^2\sigma_Y^2,$$
$$\sigma_{X-Y}^2 = \sigma_X^2 + \sigma_Y^2, \qquad E(XY) = E(X)E(Y) = \mu_X\mu_Y.$$

Continued inside back cover

Glossary

$\approx$	approximately equals
$>, <$	greater than, less than
$\geq, \leq$	greater than or equal to, less than or equal to
$\neq$	not equal to
$\|x\|$	absolute value of x
$\sum$	sum of
$\binom{n}{r}$	binomial coefficient; also ${}_nC_r$
Capital letter	usually denotes a set or random variable
S	sample space (or sum of squares), according to context
df	degrees of freedom
e_i	sample point
E_i	set consisting of sample point e_i
$\bar{A}$	complement of the set A
$P(A)$	probability of A
$A \cup B$	union of sets A and B
$A \cap B$	intersection of sets A and B
$\|$	given
$\{\ \}$	set consisting of
ϕ	empty set
$:$	such that
H_i	hypothesis
$P(H_i)$	*a priori* probability of H_i
$P(H_i \| E)$	*a posteriori* probability of H_i, given E
$E(X)$	expected value of X
μ	population mean, often with subscript
$\bar{x}$	sample average
σ	population standard deviation, often with subscript
s	sample standard deviation, often with subscript
σ^2	population variance
s^2	sample variance
ρ	population correlation coefficient
r	sample correlation coefficient
t	Student's statistic
Z	often standard normal random variable ($\mu = 0, \sigma = 1$)
$\hat{a}, \hat{b}, \hat{m}, \hat{g}, \hat{h}, \hat{S}$	least squares values
$A(x)$	area from 0 to x of the standard normal distribution
N	finite population size
$\square$	end of proof

Probability with statistical applications

Second edition

Probability with statistical applications

Second edition

Frederick Mosteller
Harvard University

Robert E. K. Rourke
St. Stephen's School, Rome

George B. Thomas, Jr.
Massachusetts Institute of Technology

Addison-Wesley Publishing Company
Reading, Massachusetts / Menlo Park, California
London / Amsterdam / Don Mills, Ontario / Sydney

This book is in the
ADDISON-WESLEY SERIES IN STATISTICS

 Printed in the United States of America. Published simultaneously in Canada. Library of Congress Catalog Card No. 74-87039.

ISBN 0-201-04857-4
KLMNOPQRS-MA-89876543210

To A. W. Tucker and S. S. Wilks

Preface

In writing this text, the authors have kept before them two basic objectives: the first is to acquaint the reader with the theory of probability—the mathematics of uncertainty; the second is to illustrate some applications of probability to statistical theory. Chapter 1 elaborates these objectives, explains current interpretations of probability, illustrates how probability and statistical theory are applied to important practical and scientific problems, and exhibits some applications of probability where intuition offers a strong guide.

The reader may expect to gain four things from this book: first, an understanding of the kinds of regularity that occur amid random fluctuations; second, experience in associating probabilistic mathematical models with phenomena in the real world; third, skill in using these mathematical models to interpret such phenomena and in predicting, with appropriate measures of uncertainty, the outcomes of related experiments; and fourth, some insight into statistical inference, both classical and Bayesian.

The level of mathematics required for an understanding of the material is that of a second course in high school algebra. No knowledge of calculus is assumed, but readers having a background in calculus will recognize that some topics (for example, probabilities as areas, and least squares) could be handled by methods different from those presented here.

As distinctive features, we think our treatment has the following:

1. The first four chapters give, in a readily accessible form, a brief course in elementary probability theory for finite sample spaces, and, for those who wish it, an extension to countably infinite sample spaces.

2. Chapters 5 and 6 offer an elementary introduction to random variables, their distributions, and properties of their distributions.

3. The key position of the normal distribution both in probability and in statistics has led to the inclusion in Chapter 7 of an intuitive introduction to continuous random variables. This, together with the study of joint distributions of two or more discrete random variables, lays a foundation for the study

of distributions of functions of several random variables. Without the normal distribution very little can be done analytically or numerically about computing approximate numerical probabilities for functions of several independent random variables. With it, the theory of sampling in Chapter 10 offers an immediate and major application whose results are routinely used in statistical inference in every field.

4. The properties of binomial probability distributions are studied in detail as a means of introducing, and applying, the central limit theorem. The development, given in Chapter 8, is new.

5. In addition to applications of classical statistical inference, we include some examples of modern Bayesian inference in Chapter 9. After introducing the ideas of confidence limits and of testing hypotheses through the binomial distribution, we study the comparison of two proportions. This is equivalent to testing for dependence in two-by-two contingency tables. We use the hypergeometric distribution in small samples and the normal distribution in large samples.

In Chapter 12, we follow statistical inference further into measured data, setting confidence limits on and testing hypotheses about population means and their differences.

6. Our treatment of statistical matters strongly emphasizes illustrative examples and exercises drawn from real life; and, where it does not impede our progress, we have not hesitated in our discussion to go a bit beyond the mathematics.

7. Our treatment of least squares is unusual because it depends only on high school algebra. A welcome by-product is that the calculation of the residual sum of squares seems easier with this approach than with that of traditional calculus. We also include a discussion of models of regression not usually found in elementary texts. In this spirit, we have extended the statistical discussion of regression lines through the origin. The new development shows that the good way to estimate the slope depends upon the appropriate model of variation in the population.

8. We offer a special set of projects for use with high-speed computers, for those who wish them.

The authors introduce each new concept through examples, and additional examples are given after each important definition or theorem. Readers who desire a faster pace may scan some of these examples rapidly and concentrate their attention on the numbered definitions, theorems, and corollaries. However, mastery of the theory will usually be increased by studying the illustrative examples, and by working exercises in the lists appearing at the ends of most sections.

The endpapers at the very front and back of the book contain a summary of most of the key formulas and a glossary of symbols.

The binomial tables at the back of the book include values of p less than, equal to, and greater than $\frac{1}{2}$. They also contain both individual terms and cumulatives. The consequent size of the tables is justified, we believe, by the added convenience.

By a suitable selection of topics, the text can be used for courses of varying lengths. The taste and experience of the instructor are the best guides for such selections.

A reader studying the book without the guidance of an instructor should generally follow the order of presentation in the text. A reasonable minimal goal is to complete Chapter 6; and, if pressed for time, the reader may omit Sections 3–5 through 3–9, 4–7 through 4–9, 6–2, and 6–4. This minimal goal leaves out many useful and important topics, and makes it difficult to understand those parts of Chapters 8, 9, and 10 that refer to the normal distribution. However, Chapter 11 is still accessible if preceded by Sections 6–2 and 7–1 and the paragraph in Section 10–4 defining the sample correlation coefficient. Chapter 11 does not depend on Chapters 8, 9, or 10 (except for a starred Supplement to Section 11–3).

A reader particularly interested in the statistical applications (acceptance sampling, hypothesis testing, estimation, regression) should include Section 7–4, Chapter 8, Chapter 9 including Section 9–7, Sections 10–1, 10–2, and 10–3, and Chapters 11 and 12.

Theorems, corollaries, and important definitions are numbered sequentially, by chapters. For example, 7–1 Definition, 7–2 Definition, 7–3 Theorem, 7–4 Theorem, 7–5 Corollary are the first two definitions and the first two theorems and first corollary in Chapter 7. The third theorem of Chapter 7 is 7–6 Theorem. This numbering system, with the number on the left, set in boldface type, is intended to make it easier to look up a reference. In the body of the text, however, we refer to Definition 7–1, Corollary 7–5, and so on. We have starred sections (or even parts of sections such as Remarks or Comments) to suggest that they might be omitted on a first reading unless the reader has a special interest in the material. In Chapters 4 and 5, this device is used to guide those readers who wish to study finite sample spaces without the complications of countably infinite sample spaces. We have starred Section 10–5, which treats sampling without replacement from a finite population. The important findings of Section 10–5 are summarized in its opening paragraph, since these findings suggest that study time may be more profitably spent on sampling with replacement, as in Section 10–3.

A star on an exercise means that the authors believe it difficult.

The use of a single parenthesis rather than the customary pair of parentheses in numbered or lettered listings is dictated by typographical considerations. No difference in meaning is intended between "1)" and "(1)" or between "a)" and "(a)".

Changes in the revised edition The revised edition, like the first edition, is accessible to students who have completed two years of high school algebra. We have kept the same pace.

To heighten interest, the order of some topics has been changed and new material has been added. For example:

1) We have brought elementary probability forward into Chapter 1.
2) We have applied new counting methods to probability as soon as the theory of the new methods has been given.
3) We have brought in the binomial distribution much earlier.
4) We have introduced the geometric and the negative binomial distributions in extending the applications beyond finite sample spaces to countably infinite sample spaces. These distributions may be omitted without disturbing the continuity.

The chapter on counting closes with a new comparison of a variety of mathematical models to which the counting ideas apply.

Chapter 9 gives an improved and extended introduction to confidence intervals and the ideas of testing hypotheses through the binomial distribution. Section 9–5, in which we apply the hypergeometric and the normal distributions to contingency tables in testing for differences of proportions, is completely new. The applications are frequent in nearly every field of investigation—medicine, social science, business, and natural sciences. [*Note:* Since we are not treating contingency tables larger than 2×2, nor combining information from several tables, we do not need chi-square. Everything we can do with chi-square for one 2×2 table, we can do better either with the hypergeometric treatment (which faces the discreteness rigorously) or, when approximation is needed, with the normal (which preserves the directionality). Consequently, we prefer to defer treatment of chi-square until a more elaborate treatment of contingency tables, goodness of fit, or some other topics makes it more necessary.]

The treatment of Bayesian inference in Chapter 9 has been amplified because of its growing importance both in applied problems and in theory. Chart I, newly computed, exemplifies a meeting of confidence limits and posterior Bayesian probabilities in the standard problem of estimating a binomial probability.

The material in Chapter 12 on statistical inference for means is entirely new in this revision. It includes a discussion of the useful t-distribution and its applications to setting confidence limits and testing hypotheses. The methods discussed here, like those in Chapter 9, have very wide applicability. They are used in nearly every field where measurements are taken. We have included treatments of the most frequently needed one- and two-sample problems. This chapter, together with the material in Chapter 9, offers an extensive set of applications of statistical inference presented in a unified way. If one is eager to get to Chapter 12, he can skip from Section 10–3 directly to Chapter 12. Since there are many different cases, the introduction to Chapter 12 explains the attitude the reader may wish to adopt toward this material.

We have also added Table I–B, a short table of random normal deviates to aid in understanding statistics based on normally distributed measurements.

A new treatment of regression lines through the origin and new illustrative examples and exercises constitute the main changes in Chapter 11.

In many places, old examples have been replaced by newer, more vital ones; and we have added many new exercises, most of which have been taken directly from the scientific literature.

The ideas of elementary set theory are now so widespread that we decided to omit the Appendix on sets.

Some institutions have high-speed computing machines and will want to use them in this course. We have provided a special set of projects for use with computers. The introduction to that material suggests how it might be treated.

Those who have used the first edition may wish to know how the chapters of the original material have been redistributed: old Chapter 1 → new Chapter 1; 2 → 2; 3 → 1 and 3; 4 → 4; first half of 5 → 5; last half of 5 → 6; 6 → 7; Sections 7–1, 7–2, 7–4 → 4; Section 7–3 → 5; Sections 7–5, 7–6, 7–7 → 8; 8 → 9; 9 → 10; 10 → 11; Appendix I deleted; Appendix II → Appendix 1; Appendix III → Appendix 2. Note that the renumbering is primarily due to splitting the first edition's Chapter 5 into two chapters.

The authors are indebted to many for help, criticism, and encouragement in connection with the writing of this book. Their experiences as members of the Commission on Mathematics of the College Entrance Examination Board have been especially valuable. In particular, they have used the evaluations and experiences associated with the experimental text *Introductory Probability and Statistical Inference*, also known as "the grey book". The authors assisted in the writing of that text, and they recall with pleasure the collaboration of Edwin C. Douglas, Richard S. Pieters, Donald E. Richmond, and the late Samuel S. Wilks. Parts of the present work have been based on, or quoted from, material copyrighted by the College Entrance Examination Board, and this has been done by permission. This permission does not imply any responsibility for, or endorsement of, this work by the College Entrance Examination Board or the Commission on Mathematics.

Much of the material in this text was used as the basis of the nationally televised NBC Continental Classroom course in Probability and Statistics, first presented early in 1961.

We gratefully express our appreciation to W. G. Cochran, A. P. Dempster, G. E. Noether, and John W. Pratt for critically reading and discussing parts of the manuscript, and for many helpful conversations and suggestions.

For the revised edition, our thanks go to colleagues, teachers, and students who have been so generous and enthusiastic in their suggestions for amplification that the additions are more extensive than we had planned: Robert Berk, Miss Lois Chew, William G. Chinn, Paul C. Clifford, Shulamith Gross, Gus W. Haggstrom, William Kruskal, Ransom V. Lynch, and Mrs. Virginia C. Sakoda. We needed only the magic touch that shortens a book by adding to it.

Our thanks for the revised edition also go to Cleo Youtz, for typing, computations, tables and graphs, and efficient organization and management; to Laurence Herbst, for his work on the binomial tables; to Linda Alger, Keewhan

Choi, Robert M. Elashoff, Miles Davis, Farhad Mehran, and Joseph I. Naus, for their assistance with problem solutions; to William and Gale Mosteller, for the card-shuffling data of the first-ace example in Chapter 1, and for new calculations for tables and charts; and to Mrs. Holly Grano, Mrs. Nancy Larson, Miss Gale Mosteller, Miss Phyllis Ruby, and Miss Anita Siskind, for production of the manuscript.

In addition, we wish to thank Willard H. Longcor for kindly making available for publication the extensive data from his long-term dice-tossing experiments; The Free Press, Glencoe, Illinois, and The RAND Corporation for permission to publish random digits from the book *A Million Random Digits with 100,000 Normal Deviates*, copyright 1955 by The RAND Corporation; and The Macmillan Company of Canada for permission to use some problems from *An Advanced Course in Algebra* by Norman Miller and Robert E. K. Rourke.

July 1969

F. M.
R. E. K. R.
G. B. T., Jr.

Contents

1 | Probability and statistics: the study of variability

This chapter provides initial answers to questions such as: What is probability? What is statistics? How are they used today? The aim is to introduce some background ideas in probability and statistics, and to transmit a feeling for the questions posed and the answers given in these subjects. Algebraic skills are not involved here. Consequently:

1. We discuss the nature and role of mathematical models, and the relation of these models to the real world.
2. We exhibit and discuss probability models, which are special kinds of mathematical models.
3. We offer opportunities for personal experience with the fluctuations and regularities of experiments involving chance.

At a first reading, one should not expect full understanding of all the ideas in this chapter. Something less will set the stage for the mathematical work to come.

1–1 PROBABILITY AND STATISTICS

The wealth and variety of applications of the theory of probability attract many students. Some find beauty in the extensive mathematical structure that emerges from a few assumptions and definitions; others, both the practical and the philosophical, enjoy discussing the meanings that may be attached to probabilistic statements. Still others admire the order that emerges from seeming chaos—toss a penny once, and no one knows whether it will fall heads or tails; toss two tons of pennies, and we all know that one ton will fall heads, the other tails.

We are used to the notion that the idealized triangles of plane geometry can be used as mathematical representations, or *mathematical models*, of physical triangles in the real world. In a similar way, we build mathematical models for probabilistic problems and develop consequences of them. For example, a tossed coin has probability $\frac{1}{2}$ of coming up "heads"; and we shall develop for coin tosses a probability model that gives the probability that when n coins are tossed, exactly x fall heads and $n - x$ fall tails. In particular, the probability that all land heads is $(\frac{1}{2})^n$. The theory and its consequences apply to idealized coins, and we hope it applies to real coins when they are tossed. In Section 1-3, we give brief descriptions of a variety of problems in the real world that are studied by probabilistic models.

The field of statistical inference leans heavily upon the theory of probability, but supplements it. When data are gathered, we may use statistical theory to help choose among alternative mathematical models. For a given town, consider drawing a sample of families to estimate the fraction of homes with color television sets. The theory of probability tells us, for a given fraction owning sets in the community, what the ownership fraction in the sample is likely to be. But statistical inference uses the sample result to estimate the fraction in the town who own sets. In this example, probability theory deduces from the known content of the population the probable content of the sample, while statistical inference infers the content of the population from the observed content of the sample. More generally, the theory of probability deduces from a mathematical model the properties of a physical process, while statistical inference infers the properties of the model from observed data.

The field of statistics includes more than statistical inference. In general, statistics is the *art and science of gathering, analyzing, and making inferences from data.* Some parts of statistics are not mathematical, while other parts are. Although we have tried to separate probability from statistics, statisticians must work on problems in probability as well as in statistics.

1-2 INTERPRETATIONS OF PROBABILITY

At the mathematical level, there is hardly any disagreement about the foundations of probability or about its mathematical consequences. The foundation in set theory was laid in 1933 by the great Russian probabilist, A. Kolmogorov, still an active research worker in 1969. At the level of interpretation and use, there are two extreme positions that are often adopted and, of course, many positions in between.

1) *Objective.* This position holds that probability is applicable only to events that can be repeated over and over under much the same conditions. Thus the objectivist is happy to talk about probabilities in connection with the tossing of a coin or the manufacture of a mass-produced item. He can readily think of many light bulbs being produced, and of the probability of a good light bulb as the long-run ratio of the number of good bulbs to the total number produced. But he draws the line at unique events. For example, he would not care to talk about the probability that Romulus founded Rome, or that Chile and Argentina would unite to become a single country in the next ten years. Thus a large class of problems is set aside by the objectivist as not appropriate for

the application of probability, because there is no long-run ratio in view. Furthermore, the objectivist likes to make interpretations only from repeated events.

2) *Personal.* The personalist regards probability as a measure of personal belief in a particular proposition, such as the proposition that it will rain tomorrow. This school of thought believes that different "reasonable" individuals may differ in their degrees of belief, even when offered the same evidence; and so their personal probabilities for the same event may differ. The personalist will apply probability to all the problems an objectivist studies, and to many more. For example, at least in principle, the personalist would take the Romulus question in his stride. The personalist also has available some additional techniques; in particular, he may have more use for Bayes's Theorem, treated in Chapter 4, than the objectivist. On the other hand, when the amount of data is large, the objectivist and the personalist usually get similar answers.

A beginner would be unwise to try to decide at once where he fits in with respect to these two views. Furthermore, the last word is never said on such matters because new schools of thought arise. But the distinction between probability as a long-run relative frequency and probability as a measure of degree of belief is one that he may wish to reconsider from time to time as he understands the issues better.

1-3 ILLUSTRATIONS OF PROBABILISTIC MODELS

Games of chance loom large in the early history of probability, and even today they provide instructive problems for both the beginner and the expert. About 1654, the Chevalier de Méré, an amateur mathematician, consulted Blaise Pascal (mathematician, scientist, and theologian) about the solution of generalizations of the following problem.*

Problem of points On each play of a game, one of two players scores a point, and the two players have equal chances of making the point. Three points are required to win. If the players must end the game when one has 2 points and the other 1, how should the stakes be divided? Pascal said the stakes should be split 3 to 1, in favor of the man who was ahead. What do you think?

Pascal engaged in a profitable correspondence with Fermat, another great mathematician, on this and other problems in probability, and between them they developed many results, some of which are presented in Chapters 2, 3, and 4.

Another early gambling problem was that of duration of play. We mention it because it has evolved and developed through the years, and in this evolution has become of value to both scientists and industrialists.

---

* You may enjoy reading Oystein Ore, "Pascal and the invention of probability theory", *American Mathematical Monthly*, Vol. 67, No. 5 (May, 1960), pp. 409–419.

Duration of play Two players in a fair game have as fortunes m and n units, and the stake on each play is one unit each. Each player has an equal chance of winning a play. If they play until one player is ruined, how long will they play, and what is the chance that the player starting with m units wins? The probability that the game lasts t plays is difficult to compute; the probability that the player with fortune m wins is $m/(m + n)$.

This problem is a forerunner of that of the random motion of a physical particle which is absorbed when it strikes a barrier—one of the many kinds of "random-walk" problems studied by physicists. To show the relation between the problems, suppose that a particle starts at the origin, O, and in each unit of time moves one unit to the right or one unit to the left, the direction being randomly determined. Erect barriers m units to the right and n units to the left of the origin, and suppose the particle stops when it strikes a barrier. The position of the particle after t units of time corresponds exactly to the amount of money won after t plays.

For each of the foregoing problems, the probability model consists, essentially, of a fair coin, a rule for assigning points for plays, and a rule for deciding the winner of the game. A few examples of problems requiring probability models for their solution may help give the flavor of applications that are made today. The answers to the questions raised are either beyond the scope of this book or require an extensive specialized development. Such models are developed by applied mathematicians, statisticians, physicists, biologists, or other scientists.

Queueing theory People arrive at random times at a counter to be served by an attendant, lining up in a queue if others are waiting. Given information about the rate of arrival and the length of time an attendant requires to serve each customer, how much of the time is the attendant idle? How much of the time is the queue more than 10 persons long? What would be the effect of adding another attendant? If people are not allowed to wait in line but must go elsewhere, what percentage of arrivals go unserved? Variations of this problem are of interest in the maintenance of a battery of machines, in deciding how many toll booths to provide at the entrance to a throughway, in considering equipment needed for telephone lines and for high-speed computers, and even in the construction and control of dams.

Inheritance in biology The Mendelian theory of heredity in its simplest form requires little more probability than that presented in this text; but, of course, the theory has gone far beyond Mendel. Suppose parents are classified on the basis of one pair of genes, and that d represents a dominant gene, and r represents a recessive gene. Then a parent with genes dd is pure dominant, dr is hybrid, and rr is pure recessive. The pure dominant and the hybrid are alike in appearance. Offspring receive one gene from each parent, and are classified the same way, dd, dr, or rr. The following table gives, for a simple case, the proportions of offspring of each type, for a given type of parents. Typical problems are: Knowing the proportions of the types of parents, what can we say of the composition of the population of offspring after $1, 2, \ldots, n$ generations?

If the model is modified so that dominant characteristics are favored in some way, do recessives die out?

Parents		Offspring		
		dd	*dr*	*rr*
dd	*dd*	1		
dd	*dr*	$\frac{1}{2}$	$\frac{1}{2}$	
dd	*rr*		1	
dr	*dr*	$\frac{1}{4}$	$\frac{1}{2}$	$\frac{1}{4}$
dr	*rr*		$\frac{1}{2}$	$\frac{1}{2}$
rr	*rr*			1

Theory of epidemics Suppose an infectious disease is spread by contact, that a susceptible person has a chance of catching it with each contact with an infected person, but that one becomes immune after having had the disease and can no longer transmit it. Then the mathematical theories of epidemics describe the progress of an epidemic in terms of the numbers susceptible, infected, and immune through time. Typical questions are: How many susceptibles will be left when the number of infected is zero (epidemic over)? How long will the epidemic last? For a city of given size, what is the probability that the disease will die out?

Naturally, in this book we cannot expect to study such difficult problems in full generality, but we can lay a foundation for their study.

1-4 APPLICATIONS OF STATISTICS

We have already indicated that statistics deals, in part, with the analysis of data stemming from probability models, and that statisticians may also develop probability models like those needed for the problems discussed in Section 1-3. A few examples of applications of statistics in other fields may interest the student.

Screening of drugs A pharmaceutical house tests hundreds of new medications, trying to find one that will be safe, and superior to the standard treatment of a disease. People vary in their responses to a medication, and so do the animals on which medications are initially tested. This variation introduces a probabilistic aspect to the problem. Usually, testing is done in stages: most medications are eliminated at an initial stage based on a small number of subjects. If a medication looks promising it is carried on to a later stage where a more elaborate and severe test is made. One problem is to choose the sizes and the severities of the experiments at the successive stages so that good new medications are unlikely to be discarded, but so that poorer medications do not receive expensive investigations.

Field tests The addition of fertilizer increases crop yield. The farmer's profit depends on yield, costs, and sale price. Agricultural experiment stations help the farmer by carrying out field trials designed to measure the additional yield, say of corn, for given amounts of nitrogen fertilizer. These trials produce curves that relate yield to amount of fertilizer, and farmers use these curves together with cost information and anticipated sales price to decide on the amount of fertilizer to use. The efficient design of the field trials is part of the work of the agricultural statistician.

Sample surveys The use of sample surveys is not restricted to opinion polls. Surveys are also used by large companies to assess their inventories or their book value. Surveys are taken to determine what and how much mathematics is available in colleges. You may have seen figures in newspapers estimating the number of unemployed; these come from a periodic government sample survey. To find and correct errors in the U.S. Census, the Census Bureau uses special sample surveys. That surveys are widely used instead of a complete census is partly a matter of cost in time and money, but partly a matter of quality. A more thorough and careful job can be done on a sample than on a large population.

Many problems can be treated only by sampling methods: breaking strength of steel rods, life testing of vacuum tubes, and, in general, destructive tests. Other problems have infinite populations; there is no end to the number of measurements the Bureau of Standards can take on its platinum-iridium standard meter bars.

Genetics and radiation The development and testing of atomic bombs has led to extensive experiments on the genetic effects of radiation in insects and mammals. Mutation, a suddenly produced variation in the character of offspring, is sometimes produced by radiation. For example, fruit flies are exposed to radiation of different kinds and in different doses, and mutations in offspring are observed. Several sites on the fly are possible places for mutations. Here are typical questions: Are the different sites equally likely to mutate? Is the frequency of mutation proportional to the dose? Do kinds of radiation differ in their effect? Statistical studies of the effects of radiation on humans are still carried on at Hiroshima and Nagasaki.

Geology Large boulders are left scattered by a glacier. From the distribution of the angles that the long axes of the boulders make with the North, it is desired to estimate the direction of the path of the glacier.

Reliability of components The increasingly complex mechanisms used both in space and on the ground require higher and higher reliability of their parts, especially of those parts whose failure destroys the operation of the whole mechanism. For a mechanism not to fail for, say, 10 years, suppose that 1000 parts must survive at least this long. Then the single parts must be reliable, on the average, for more years than we can conceivably test them. Alternatively, the mechanism can be designed so that the failure of one or of several parts does not stop its performance. A combination of engineering research with probability models and statistical methods is needed to achieve the high reliabilities required.

Other examples of statistical applications will be found throughout the text.

1-5 THE EMPIRICAL STUDY OF VARIABILITY

As we have seen, probability and statistics deal with the fluctuations and the regularities in processes that have random or chance elements. Although we all experience such variability every day in traffic flow, in time taken to brush our teeth, in our own changing weight, in our expenditures for necessities, in our time used for study, and in our games and races, we rarely study variability systematically. Thus we have impressions about variability, but usually no data.

Better personal experience of probability processes can be acquired by doing a few experiments of a simple sort, keeping records of the results, and analyzing them. Some of the results will be much as expected; others may be a bit surprising. In this chapter, we study the results of some simple experiments that you can do, and we suggest additional ones so that you can gain experience with probability models and variability. You are asked to *write down* your initial thoughtful guess about the outcome in each example without peeking ahead, so that you will gain experience in such estimates, and so that you will honestly know whether the result is as you expected or not. When you are seriously wrong, you should ask yourself what features of the problem you did not take into account. You should understand that even professional mathematicians cannot solve all the mathematical examples without hard work, and that some cannot be solved without empirical data.

First-ace problem An ordinary deck of 52 playing cards containing four aces is shuffled thoroughly, and we count from the top the number of cards down to and including the first ace, and record the count. The process is repeated.

a) What is the average count? (Without reading further, write down your thoughtful guess.)

b) What is the probability that the first ace is on card 1? 2? . . . 52? (A series of three dots, as used here, stands for all the whole numbers between the numbers immediately preceding and following the dots.)

c) Within what number of cards will we find the first ace half the time? (Write down your thoughtful guess.)

Discussion Parts (b) and (c) of this problem are treated theoretically in Chapter 3, but here we study the matter empirically.

In Table 1-1, we list in order in columns of 20 the results of 100 shuffles for this experiment. We observe that the counts vary considerably, from 1 to 32. They might have varied more—from 1 to 49—because all four aces could be clustered on the bottom of the deck. Furthermore, as you look down a column

Table 1-1 Count (number of cards to and including the first ace) for each of 100 shuffles

Shuffle number	First 20 counts	Second 20 counts	Third 20 counts	Fourth 20 counts	Fifth 20 counts
1	5	17	5	7	27
2	4	8	5	15	18
3	29	19	8	17	11
4	3	18	4	16	2
5	24	20	1	9	1
6	3	13	5	11	28
7	3	2	24	10	17
8	22	2	1	1	9
9	5	19	15	21	3
10	16	9	2	1	4
11	1	7	3	4	17
12	1	1	18	15	7
13	5	3	22	4	2
14	23	6	25	26	5
15	16	2	13	8	9
16	6	15	11	13	6
17	26	3	2	13	3
18	10	4	5	11	22
19	1	12	3	1	13
20	32	5	22	29	1
Total	235	185	194	232	205
Average	11.75	9.25	9.70	11.60	10.25

the numbers change without much rhyme or reason. We call such changes *sampling fluctuations* or *sampling variation*. If the shuffling is thoroughly done, knowledge of one count is no help in predicting what the next will yield.

That there is order in this chaos is suggested by the stability of the column totals and column averages. The averages vary only from 9.25 to 11.75. The changes in the average are much less than the changes from one count to the next.

We summarize these data in a *frequency distribution* in Table 1-2, by obtaining the number of times each count occurred in our 100 shuffles. For example, the count 1 occurs 11 times in Table 1-1. We also give the theoretical frequency, computed from a probability model for this problem. We defer the calculation of such theoretical frequencies to Chapter 3, but here we can compute the probability that the count is 1. There are 4 aces out of 52 cards that can be on top of the deck, and any one of these yields a count of 1. So it is natural to say that the probability that an ace is on top is $\frac{4}{52}$ or $\frac{1}{13}$. Since we had 100 shuffles, the theoretical frequency is $\frac{100}{13} \approx 7.7$. (The symbol $\approx$ means "is approximately equal to".) Of course we cannot have 7.7 counts of 1; that is the long-run rate per 100 counts if thousands of trials are made. We call it the theoretical or *expected* frequency, and discuss it in Chapter 5. The theoretical frequencies of

Table 1-2 Observed frequency distribution of counts and theoretical frequencies of counts for first-ace problem

Count	Number of times observed	Theoretical frequencies	Count	Number of times observed	Theoretical frequencies
1	11	7.7	21	1	1.7
2	7	7.2	22	4	1.5
3	9	6.8	23	1	1.3
4	6	6.4	24	2	1.2
5	9	6.0	25	1	1.1
6	3	5.6	26	2	1.0
7	3	5.2	27	1	.8
8	3	4.9	28	1	.7
9	4	4.6	29	2	.7
10	2	4.2	30	0	.6
11	4	3.9	31	0	.5
12	1	3.6	32	1	.4
13	5	3.4	33	0	.4
14	0	3.1	34	0	.3
15	4	2.9	35	0	.3
16	3	2.6	36	0	.2
17	4	2.4	37	0	.2
18	3	2.2	38	0	.1
19	2	2.0	39	0	.1
20	1	1.8	40–49	0	.3
				100	99.9

counts from 40 to 49 are each less than 0.05, so for our purposes it is sufficient to report their sum, 0.3.

We observe that the frequencies for theoretical counts do not match those of the observed counts exactly, but that they have the same general trend, i.e., they decrease as the size of the count increases. The discrepancies you see between the observed and theoretical frequencies are part of the experience this chapter can give. There are two sources for such discrepancies—sampling variation and failure of the theoretical probability model to fit the facts of real-life shuffling and counting. The counting was carefully checked, but it is harder to check the shuffling.

The authors have made a statistical study of these counts, and they find no evidence of disagreement between the theoretical model and the actual data. The study is not presented here, but part is shown in Section 11–6.

Let us return to our first question:

a) What is the average count?

Our total for the five sets is

$$235 + 185 + 194 + 232 + 205 = 1051,$$

Table 1-3 Average count

	Observed average for 20 shufflings
Part 1: cards before the first ace	9.75
Part 2: cards between the first ace and the second	12.55
Part 3: cards between the second ace and the third	6.65
Part 4: cards between the third ace and the fourth	9.40
Part 5: cards after the fourth ace	9.65

so the average for 100 counts is 10.51. This is close to the theoretical value, as we now show, using considerations of symmetry. But the reader needs to take the argument partly on faith.

The 4 aces break the rest of the pack into 5 parts, as shown in Table 1-3. Any part may have from 0 to 48 cards in it. It seems reasonable (and it can be proved) that all 5 parts have the same long-run average count. In Table 1-3, we show opposite each part its average count for 20 new shufflings.

If it is true that the 5 parts have the same long-run average, then $\frac{48}{5} = 9.6$ is the theoretical count for a part. When we counted to the first ace we included the ace, so the expected count including the first ace is $9.6 + 1 = 10.6$. This theoretical number is very near our average, 10.51, for 100 hands. However, agreement this close is unexpected. Application of large-sample theory shows that one-third of the repetitions of the 100-hand experiment would produce averages that deviate more than 0.85 from 10.6, and that 5% of the repetitions would be more than 1.7 from it.

The theoretical numbers of Table 1-2, when divided by 100, answer question (b):

b) What is the probability that the first ace is on card 1? 2? . . . 52?

Our final question was:

c) Within what number of cards will we find the first ace half the time?

Table 1-2 shows that 51 times in 100 trials we observed the first ace at a count of 8 or less. So we could use the number 8 as an estimate of the answer to (c). Since we are also given the expected frequencies, we can add them starting from a count of 1 and continuing until the sum of the frequencies is 50% or more. Doing so, we find the theoretical answer to be 9; the total theoretical frequency for counts of 9 or less is 54.4%. We call this count 9 the *median* count, to distinguish it from the *mean* count of 10.6.

We next consider an example of an entirely different nature.

Distribution of word length What is the average length of words, measured in letters, used in sports reporting? (Write down a thoughtful guess.)

Solution We show in Table 1-4 the results for a sample of 50 words from one newspaper article on baseball. Naturally, a more extensive sample would be needed for firm conclusions.

Table 1–4 Distribution of word length in sports article

Length in letters	Frequency	Length in letters	Frequency
1	1	7	5
2	6	8	2
3	12	9	3
4	7	10	0
5	7	11	1
6	5	12	1
		Total	50

The sum of the lengths for the 50 words can be obtained by multiplying each length by its frequency and adding these products to obtain a sum of 243. The average length is $243/50 \approx 4.9$. We observe that three-letter words are most frequent, and that about half the words are 1, 2, 3, or 4 letters long.

Last digits of phone numbers From a telephone book, find the frequency distribution of the last digits for 100 phone numbers. (Write down your guess for the frequency distribution.)

Table 1–5 Frequency distribution of last digits from 100 telephone numbers

Digit	Frequency
0	11
1	13
2	11
3	11
4	10
5	5
6	7
7	14
8	8
9	10
Total	100

Solution Many people expect the digits 0, 1, . . . , 9 to be about equally frequent. Table 1–5 gives the results for one sample of 100. We observe that the digits are about equally frequent, as people expect. Of course, this is only one sample of 100.

Distribution of first digits Find the frequency distribution of first (leftmost) digits in counts of votes for a given candidate by some unit of population, such as state, county, or precinct (or in physical measurements such as areas of

Table 1-6 Frequency distribution of first digits for voting statistics

Digit	Frequency
1	23
2	16
3	18
4	9
5	14
6	7
7	6
8	5
9	4
Total	102

states, heights of mountains, or the first significant digits in physical constants). In the number 345, the first significant digit is 3, as it is also in 0.00345.

Solution Most people guess that the digits 1, 2, . . . , 9 are about equally frequent. Write down your guess. Table 1-6 gives the first digits of counts of votes for Nixon in counties in Illinois in the 1968 presidential election. We observe that 1's are quite frequent, and that the low numbers are much more frequent than the high ones. Note that the digits 7, 8, and 9 together, instead of representing $\frac{1}{3}$ of the total or 34, show only 15. A number of scientific papers have set up probability models to explain this phenomenon—unexpected for most of us. The high frequency for the low numbers is said to have been first pointed out by a man who observed that the early pages of a well-used table of logarithms were much dirtier than the late pages. He decided on this evidence that first digits were most frequently small, and counts on a large variety of measures have borne him out.

Random walk Suppose a man stands facing north and tosses a coin to decide whether to take one step north or one step south. Suppose he continues tossing and stepping in this manner for 25 steps.

a) On the average, how far is he from his starting point? (Write down a thoughtful guess.)

b) On how many steps is he on the north side of his starting point; on how many is he on the south side?

c) How often does he return to the starting point during the walk? (Write down your guess.)

Discussion These are difficult mathematical problems, but we can simulate the experiment by tossing a coin 25 times and counting steps north and south. (Alternatively, we could use last digits from telephone numbers, using odd numbers to represent a step north and even numbers for south. Or we could use

Table 1–7 Results for 10 random walks of 25 steps (N and S indicate north and south)

Walk number	Final position	Times on north side	Times on south side	Times at origin
1	1S	16	5	4
2	7N	19	3	3
3	5N	20	2	3
4	9N	24	0	1
5	1N	4	15	6
6	5S	6	11	8
7	5S	7	12	6
8	3S	2	19	4
9	11N	25	0	0
10	3N	9	15	1
Sum of distances	50	Totals 132	82	36

the random digits given in Table I–A at the back of the book in the same manner.) Table 1–7 shows the results for 10 walks of 25 steps.

The column totals show that the average distance from the origin is $50/10 = 5$. A theoretical answer from advanced work is about 4. (For n steps, the theoretical mean distance is about $0.8\sqrt{n}$, for large n.)

We note that there is considerable imbalance between time spent on the north and on the south. But the symmetry of north and south and of heads and tails shows us that in the long run, over many walks, half the time will be spent on each side of the starting point. The imbalance of an average of 13.2 steps on the north versus 8.2 on the south must therefore be due to large sampling fluctuations. Note that on walk 9 all 25 steps were on the north; on walk 4, 24 out of 25 were on the north; and on walk 8, 19 were on the south. We seem to have discovered that instead of each walk being split about equally—about half on the north, and half on the south—a very substantial fraction of the time is likely to be spent on one side in any one walk. This surprising result is *not* a feature of the smallness of the total number of steps taken or of an unusual sample. It is a general feature of this kind of random-walk problem.

Finally, the average number of returns to the origin was observed to be $36/10 = 3.6$. Advanced theory gives about 3.0 as the theoretical mean.

Random digits You may like to see the magnitudes of departures from expected frequencies observed in a table of random numbers, entitled *A Million Random Digits*, made by The RAND Corporation and published by The Free Press, Glencoe, Illinois. The second column in Table 1–8 gives the frequencies of the digits 0, 1, . . . , 9 in the first block of 50,000 random digits in the table; the expected frequency for each digit is, of course, $50{,}000/10 = 5000$. The third column gives the frequencies for the million digits, the expected frequencies being each 100,000.

Table 1–8 Frequencies of random digits

Digit	Frequencies in first block of 50,000	Frequencies in a million digits
0	4923	99,802
1	5013	100,050
2	4916	100,641
3	4951	100,311
4	5109	100,094
5	4993	100,214
6	5055	99,942
7	5080	99,559
8	4986	100,107
9	4974	99,280

Large-sample theory suggests that about $\frac{2}{3}$ of the observed frequencies for the 50,000 blocks should be within 67 of the expected frequency, and that about $\frac{2}{3}$ of the observed frequencies for the million digits should be within 300 of the expected frequencies. In both instances, 6 digits have frequencies within the interval where $\frac{2}{3}$ (or 6.7 digits) are expected, so the agreement is close.

EXERCISES FOR SECTION 1–5

1. Obtain a deck of ordinary playing cards and, after thorough shuffling, count the number of cards down to and including the first ace; record the count for five shuffles. Get the average count for the five shuffles and compare it with the theoretical value of 10.6.
2. For each of five shuffles of an ordinary deck of playing cards, record the counts of the cards before the first ace, between the first ace and the second, and so on, as in Table 1–3. Then get the averages for each part as in the final column of Table 1–3, and compare the results with 9.6.
3. Obtain a frequency distribution for the lengths of the first 50 words in a sports article in a newspaper, and compare the mean word length with that obtained from Table 1–4.
4. Open a residential telephone book to any page, and obtain the frequency distribution of the last digit for 25 telephone numbers. Find the average value of the last digit and compare it with 4.5, the theoretical value if all digits are equally likely.
5. From an almanac, or other source, obtain the distribution of leftmost digits of areas of states of the United States of America (or populations) and compare the distribution of digits with that of Table 1–6.
6. From a chemical or physical handbook, obtain the frequency distribution of the first significant digits of 50 physical constants.
7. Use the random digits of Table I–A at the back of the book to carry out 4 random walks of size 25, like those described in the text. Use your data to answer the three questions in the text.

8. Use the random numbers of Table I–A at the back of the book to make 10 random walks of length 10 steps each, and use these results to answer the three random-walk questions in the text (for walks of length 10).

9. Obtain the frequency distribution of the digits 0, 1, . . . , 9 of the 50 random numbers in the first 5 columns and first 10 rows of the random digit Table I–A at the back of the book.

10. Split an ordinary pack of playing cards into two packs, the reds and the blacks. Lay out the reds in a row (in order A, 2, 3, . . . , 10, J, Q, K of diamonds, then A, 2, 3, . . . , K of hearts). Shuffle the blacks and lay them out beneath the reds. Then count the number of times the value of a black card matches that of the red. Repeat 5 times, and obtain the average number of matches for the 5 shuffles. Make a thoughtful guess at the theoretical average number of matches.

11. Open a novel to a page near the middle, and choose the first 10 full lines of text. Record the number of *e*'s in each line, and get the average number of *e*'s per line. Use the letter count from one line as a base, and estimate the percent of letters that are *e*'s.

12. Consider the duration-of-play problem, Section 1–3, with $m = 3$, $n = 2$. By flipping a coin (or using random numbers, Table I–A) and scoring a point for the player starting with m units when a head appears, and one for the other player when a tail appears, play 10 games, recording the total tosses required for each game, and the winner. (a) Find the average number of tosses per game. (b) Compute the fraction of games won by the player starting with m units, and compare it with 0.6.

13. Refer to the problem of points, Section 1–3. Use a coin toss (or random numbers, Table I–A) to simulate the finish of the game 20 times. Compare the number of times the player with 2 points won to the remaining number. Are the numbers approximately in the ratio 3 to 1, as Pascal thought?

14. Record the number of rolls of a die before a 6 appears. Repeat the experiment 10 times, and obtain the average number of rolls required.

15. Record the total number of rolls of a die required before every face has appeared. Repeat the experiment 5 times, and obtain the average number of rolls.

16. *Simplified epidemic.* An infectious disease has a one-day infectious period, and after that day the patient is immune. Six hermits (numbered 1, 2, 3, 4, 5, 6) live on an island, and if one has the disease he randomly visits another hermit for help during his infectious period. If the visited hermit has not had the disease, he catches it and is infectious the next day. Assume hermit 1 has the disease today, and the rest have not had it. Throw a die to choose which hermit he visits (ignore face 1). That hermit is infectious tomorrow. Then throw again to see whom he visits, and so on. Continue the process until a sick hermit visits an immune one and the disease dies out. Repeat the experiment 5 times and find the average number who get the disease.

17. *Server problem.* In a unit of time there is a 50:50 chance that a customer appears at a counter to be served. If others are ahead of him at the counter, he lines up in the queue; otherwise the server serves him and takes 2 units of time to complete the service. In the 10th unit, what is the average number in the queue if the process starts with no customers at the counter? Use a coin to carry out the experiment

5 times and get the average number. Also get the average number served at the close of the 10th unit of time. Example (a_i stands for a customer who arrived in the ith time interval):

Time unit	1	2	3	4	5	6	7	8	9	10
Arrivals	a_1	a_2	—	—	—	a_6	a_7	a_8	a_9	a_{10}
Being served	a_1	a_1	a_2	a_2	—	a_6	a_6	a_7	a_7	a_8
In line	a_1	a_1, a_2	a_2	a_2	—	a_6	a_6, a_7	a_7, a_8	a_7, a_8, a_9	a_8, a_9, a_{10}

Total served: 4
Number in queue in 10th unit: 3

1–6 FURTHER EXPERIMENTS

One of the earliest notions in probability was that of equally likely cases, or equally likely outcomes. The words "equally likely" are meant to convey the notion of "equally probable". The idea is essentially an intuitive one. For example, if a coin is tossed it seems reasonable to assume that the coin is just as likely to fall "heads" as to fall "tails": the two outcomes, heads and tails, are considered to be equally likely. In other words, we say that heads and tails have *equal chances*.

A "die" (plural, *dice*) is a homogeneous cube whose faces are marked with dots as follows:

If a die is thrown, most people find it reasonable to assume that when the die stops, one face is as likely to be on top as another: the six faces have equal chances. Thus, the throw of a die gives rise to six "equally likely" outcomes: the upper face shows one dot (an ace), or two dots (a deuce), or three dots (a trey), and so on. We say that each of these outcomes has "one chance in six" of occurring.

Cards in an ordinary bridge deck are made so as to be indistinguishable from one another when they are placed face down. Suppose that we shuffle such a deck and draw one card. We then say that each card has the same chance of being drawn as every other card: one chance in fifty-two.

What makes us feel, in the tossing of a coin, that the two sides are equally likely, that in the throwing of a die the six faces are equally likely, and that in the drawing of a card the fifty-two cards are equally likely? It is the symmetry and homogeneity of the coin and the die, and the similarity of the cards. If the die were heavily loaded on one side, then the opposite side would be more likely to appear on top, and the six faces would not be equally likely. Experience with ordinary coins, dice, and cards confirms the notion that heads and tails appear about equally often when coins are tossed, that one face of a die appears on top about as often as another when dice are thrown, and that one card appears about as often as another when cards are drawn. Thus both reasoning based on symmetry and similarity, and experience with actual physical objects, support the idea of equally likely outcomes for such experiments.

J. E. Kerrich* designed a coin-tossing experiment while he was interned during World War II. In ten sets of 1000 tosses, he found that the numbers of heads were

502, 511, 497, 529, 504, 476, 507, 528, 504, 529.

We see that these numbers cluster around 500, although none is exactly 500.

Willard H. Longcor obtained some unusual data by performing the following experiments. He tossed precision-made dice 2,000,000 times, using a new die for each 20,000 tosses. For each toss, the number of dots showing on the top face was recorded as even or odd. Longcor repeated this experiment with inexpensive dice. His results were as follows:

Type of dice	Number of tosses	Percentage even
Precision-made	2,000,000	50.045
Inexpensive	1,160,000	50.725

To check that theory fits the facts, it is important to examine the results of such experiments with physical objects and to compare these results with theory. It is even more important to experience at first hand the relation between mathematical theory and events in the real world. For this reason, we have suggested and we shall continue to suggest experiments for you to perform.

To appreciate the importance of experimental verification as well as clear reasoning, consider the experiment of tossing two coins. It is said that the mathematician D'Alembert reasoned that there are three possible ways the coins can fall: (a) both heads, (b) one head and one tail, and (c) both tails. He thought these three outcomes were equally likely. If that were true, we should expect each of them to occur about $\frac{1}{3}$ of the time in actual experiments. Before reading further, you are urged to do Exercise 1 at the end of this section.

Table 1–9 shows another analysis of the two-coin problem. This analysis leads to four possible outcomes instead of three. If these four cases were equally likely, we should expect one head and one tail to occur about twice as often as two heads (or two tails), because cases 2 and 3 both contribute to the "one head and one tail" count.

Would reason alone enable us to decide which analysis of the two-coin problem is correct? Possibly so. But intelligent people have disagreed, some choosing the "three equally likely cases" and others the "four equally likely

* J. E. Kerrich, *An Experimental Introduction to the Theory of Probability*. Belgisk Import Co., Copenhagen.

Table 1–9 Possible ways 2 coins can land

Case	First coin	Second coin
1	head	head
2	head	tail
3	tail	head
4	tail	tail

cases". As we learn more about the theory of probability, we shall discover that the "four equally likely cases" corresponds to the assumption that the outcome on either coin is independent of the outcome on the other—an assumption that most experts nowadays believe fits the facts for coins.

Four major uses of the "equally likely cases" approach in probability are:

1. To describe a physical experiment such as the tossing of a coin.

2. To offer a "baseline" for comparison with empirical results. We may not believe that a phenomenon can be described by "equally likely cases", but we see where such an assumption leads, and then we check the consequences experimentally. For instance, an experiment of tossing 2 coins 1000 times furnishes evidence for deciding in favor of three or of four equally likely cases.

3. To give a satisfactory approximation to a process where we know very well that the cases are not equally likely. For example, even though more boys than girls are born, we often assume that every new baby has an equal chance of being a boy or a girl, and for many purposes this assumption is adequate.

4. To achieve random sampling in sample surveys and other experimental investigations. The analysis of data from random samples is a relatively straightforward matter, but the analysis of data from samples not based on randomness requires difficult judgments.

The idea of an experiment We have used the word *experiment* to describe such things as the throwing of one or more coins, the tossing of a die, or the drawing of a card from a bridge deck. Many more serious experiments are associated with every medical and scientific research campaign, such as the search for a polio vaccine, the study of the cause and cure of cancer, the studies of genetics by Gregor Mendel,* the psychological experiments of Pavlov, and the synthesis of penicillin by Sheehan. In the experiments of such campaigns, it is difficult to imagine all possible outcomes and it is not to be expected that these outcomes would be equally likely.

Nearly all serious experiments produce observations or measurements whose interpretation may well use an application of probability or of statistical reasoning. We shall use the word "experiment" to describe any act that can be

---

* See "Mathematics of Heredity" by Gregor Mendel, in *World of Mathematics*, James R. Newman, ed., Vol. 2, pp. 937–949.

repeated under given conditions. Usually the exact result of the act cannot be predicted with certainty. We focus attention on experiments that have only a finite number of possible outcomes and, in this chapter, we usually deal with outcomes that are equally likely.

For example, selecting a sample of 3 persons from a population of 30 persons is an "experiment". In Chapter 2 we shall see that 3 persons can be selected from 30 persons in 4060 different ways. To select such a sample "at random" means that each of the possible 4060 selections has 1 chance in 4060 of being chosen.

Probability: a measure of chance Although we used the word loosely earlier, we now introduce the word "probability" as a technical word that we shall define and use when dealing with experiments involving "chance".

The experiments discussed to this point suggest a method of assigning numbers to the chances of certain events. Consider, for example, the throwing of a coin. In everyday language, we say that "the coin has one chance in two of falling heads"; in technical language, we say that "the probability of heads is $\frac{1}{2}$". In symbols, we write

$$P(\text{head}) = \tfrac{1}{2}.$$

Similarly, in the tossing of a die, the face with four dots has 1 chance in 6 of landing on top; thus, we say that "the probability of 4 dots on top is $\frac{1}{6}$". In symbols:

$$P(\text{4 dots on top}) = \tfrac{1}{6}.$$

Likewise, when we draw a card from a shuffled bridge deck, we have

$$P(\text{ace of hearts}) = \tfrac{1}{52};$$

and when we select at random a sample of 3 people from a population of 30, the probability of selecting three specified persons A, B, and C is

$$P(A,\ B, \text{ and } C) = \tfrac{1}{4060}.$$

In general, we use

$$\text{“}P(\underline{\qquad})\text{”}$$

to denote "the probability of ______", where the blank may be filled so as to name or indicate an outcome.

As a further example, consider a 10-ticket draw for a prize. A name is written on each of 10 tickets, the tickets are then thoroughly mixed in a bag, and one ticket is drawn at random. The person whose name appears on the ticket so drawn is the winner.

Now if Carter's name appears on just one ticket, his chance of winning the prize is 1 in 10, since all outcomes in the drawing are equally likely. Therefore we say

$$P(\text{Carter wins}) = \tfrac{1}{10}.$$

Similarly, if Carter's name appears on 7 tickets, his chances of winning are 7 in 10, and

$$P(\text{Carter wins}) = \tfrac{7}{10}.$$

The general idea is that of separating from the whole set of equally likely outcomes the special subset of "favorable" outcomes. Then we assign the probability of a favorable outcome by the following rule:

$$P(\text{favorable outcome}) = \frac{\text{number of favorable outcomes}}{\text{number of possible outcomes}}.$$

This method of assigning to a favorable outcome a measure, or number, called its probability has an immediate consequence; for if there are no favorable outcomes in the set of possible outcomes, then

$$P(\text{favorable outcome}) = 0;$$

and, if all possible outcomes are favorable, then

$$P(\text{favorable outcome}) = 1.$$

It follows that

$$0 \leq P(\text{favorable outcome}) \leq 1.$$

This is consistent with our intuitive feeling. If Carter's name is on no ticket, he has no chance of winning and $P(\text{Carter wins}) = 0$; if his name is on all the tickets, his winning is a sure thing and $P(\text{Carter wins}) = 1$.

The number assigned to the probability of a favorable outcome also measures what we feel would be the long-run proportion of occurrences of the outcome in many repetitions of the experiment.

Example 1 Four faces of a homogeneous cube are colored red, and the other two faces are colored green. If the die is rolled once, what is the probability that the top face is (a) red, (b) green?

Solution Since the six faces are equally likely, the probability that the top face is red is given by

$$\frac{\text{number of favorable outcomes}}{\text{number of possible outcomes}} = \frac{4}{6} = \frac{2}{3}.$$

Thus we have

$$P(\text{red}) = \tfrac{2}{3}.$$

Similarly,

$$P(\text{green}) = \tfrac{2}{6} = \tfrac{1}{3}.$$

Note that the answer is not $\frac{1}{2}$. The two outcomes "red" and "green" are not equally likely.

The idea of an event Sometimes it is convenient to regard a set of outcomes as corresponding to a single *event*. In Example 1, we may think of any one of the 4 red faces landing on top (4 outcomes) as corresponding to the event "top face red". The foregoing method can be used to assign probabilities to events.

1–1 Definition: Probability of an event

If an experiment can result in any one of n different, equally likely outcomes, and if exactly m of these outcomes correspond to event A, then the probability of event A is

$$P(A) = \frac{m}{n}. \tag{1}$$

Remark We denote the event not-A by $\overline{A}$. It follows at once from the foregoing definition that the probability of event not-A is given by

$$P(\overline{A}) = \frac{n-m}{n} = 1 - \frac{m}{n} = 1 - P(A). \tag{2}$$

1–2 Definition: Odds

The relative chances for A and $\overline{A}$ are often expressed in terms of the *odds in favor of A*:

$$\text{odds in favor of } A = \frac{P(A)}{P(\overline{A})} = \frac{m/n}{(n-m)/n} = \frac{m}{n-m}. \tag{3}$$

Thus for the die with 4 red faces and 2 green faces, the odds in favor of red are $\frac{4}{2}$ or $\frac{2}{1}$. If we know the probabilities of the events A and $\overline{A}$, Eq. (3) gives the odds in favor of A. Sometimes these are also called the odds *against* $\overline{A}$.

On the other hand, if we know that the odds in favor of A are a/b, then

$$P(A) = \frac{a}{a+b}, \qquad P(\overline{A}) = \frac{b}{a+b}. \tag{4}$$

The idea of odds can be extended to deal with more than two categories of events, but Definition 1–2 does not generalize in the obvious manner. In this book we shall restrict the application of the notion of odds to two events.

Example 2 A dealer decides to give away 10 radios because of defects. Six of the radios have marred finishes, 3 have electronic defects, and 1 has both kinds of defects. A radio is picked at random and given to you. Find the probability and the odds that this radio has a marred finish.

Solution Since the radio is picked at random, there are 10 equally likely outcomes. Of these, 7 outcomes correspond to the event "finish marred". Hence,

$$P(\text{finish marred}) = \tfrac{7}{10}.$$

The odds in favor of getting a marred finish are 7 to 3, because

$$P(\text{finish not marred}) = \frac{3}{10} \quad \text{and} \quad \frac{\tfrac{7}{10}}{\tfrac{3}{10}} = \frac{7}{3}.$$

Alternatively, the odds against "finish marred" are 3 to 7. We get the same results by observing that

$$\text{odds in favor of marred finish} = \frac{\text{number with marred finish}}{\text{number with unmarred finish}} = \frac{7}{3}$$

and then computing

$$P(\text{finish marred}) = \frac{7}{7+3} = \frac{7}{10}.$$

EXERCISES FOR SECTION 1–6

1. Perform the following experiment. Toss 2 coins 50 times, keeping a careful tally of the possible outcomes—two heads, one head and one tail, two tails. Compute the proportion of occurrence for each of the three outcomes. Do your results appear to support the reasoning of D'Alembert that the three outcomes are equally likely?
2. If a die is thrown, what is the probability that the upper face shows 3? More than 3? Less than 3? An even number? An odd number?
3. In a throw of two ordinary dice, what is the probability that the numbers on their upper faces add up to 3? to 4? to 11?
4. When two coins are tossed, what is the probability that both show heads? That they show one head and one tail? What are the odds in favor of both coins showing heads?
5. From a class of 40 students with 25 girls, one student is chosen by lot. What is the probability that a boy is chosen?
6. A bag contains 10 white marbles and 8 black marbles. If one marble is drawn from the bag at random, what is the probability that it is black? That it is white? What are the odds against drawing a black marble?
7. In a family of three children, what is the probability that all three are boys? Assume that boys and girls have an equal chance of being born.
8. Toss a coin 100 times, keeping a record of the number of heads; or use the results of Exercise 1 for 50 tosses of 2 coins. Compute the relative frequency of heads. How does your result compare with the probability of getting a head on one toss of a single coin?

9. One card is drawn at random from a well-shuffled bridge deck of 52 cards. What is the probability of drawing a heart? An ace? A black card? What are the odds in favor of drawing an ace?
10. What is wrong with the following procedure? To find the probability that an American citizen chosen at random was born in a given state, divide the number of favorable cases, 1, by the total number of states, 50, and obtain the answer $\frac{1}{50}$.
11. A letter is chosen at random from the word *equations*. What is the probability that the letter is a vowel? A consonant? The letter m?
12. A person holds a ticket in a lottery that offers 10 prizes and sells 120 tickets. What is the probability that the person will win a prize? That he will not win a prize? What are the odds in favor of his winning a prize?
13. Four balls, numbered 1 to 4, are placed in a bag and two of the balls are drawn at random. What is the probability that balls numbered 3 and 4 are drawn?
14. The numbers 1 to 9 inclusive are written on slips of paper, and the slips are placed in a bag and thoroughly mixed. One slip is drawn from the bag at random. What is the probability that the number on the slip is odd? Even? Prime? [*Note:* We do not count the number 1 as a prime.] A multiple of 3?
15. Three balls, numbered 1 to 3, are placed in a bag, mixed, and drawn out, one at a time. What is the probability that the balls are drawn in the order 1, 2, 3?

1–7 A SAMPLE SPACE OF AN EXPERIMENT

We have discussed the notion of an experiment and examined a number of simple experiments with equally likely outcomes. We now introduce an important related idea: "a sample space of an experiment". We approach this idea with the aid of an illustrative example.

Consider the experiment of tossing two coins: a dime and a quarter. How shall we list the possible outcomes of this experiment? It may be done in a number of ways, and the particular method preferred depends upon what our interest is centered on. Suppose, for example, that each coin falls heads (H) or tails (T). Then the set

$$S = \{HH, HT, TH, TT\} \tag{1}$$

provides a list that represents the possible outcomes of one toss, if we understand that the first letter in a pair designates the outcome for the dime, and the second letter that for the quarter. Thus HT means that the dime fell heads and the quarter fell tails. Every outcome of the experiment corresponds to exactly one element of the set (1).

Alternatively, we may be interested only in the *number* of heads or tails that appear. If we agree to denote a heads and b tails by the ordered pair (a, b), then the set

$$S_1 = \{(2, 0), (1, 1), (0, 2)\} \tag{2}$$

lists all possible outcomes of the experiment. And every outcome of the experiment corresponds to exactly one element of the set (2).

Again, we may be concerned only with whether the coins fall alike (A) or different (D). We could then list all possible outcomes with the set

$$S_2 = \{A, D\}. \tag{3}$$

As before, every outcome of the experiment corresponds to exactly one element of the set (3).

Hence each of the sets (1), (2), and (3) provides a list that includes all possible outcomes of the experiment. Each such set is called "a sample space of the experiment"; that's why we talk about "a" sample space of an experiment, rather than "the" sample space. More than one sample space can be used to list the possible outcomes of our probabilistic model.

Note S is a more fundamental sample space than S_1 or S_2 because it offers more information. If we know which element of S occurs, we can tell which outcomes occur in S_1 and S_2; but the reverse is not always true.

1-3 Definitions: Sample space; sample point

A sample space of an experiment is a set S of elements such that any outcome of the experiment corresponds to exactly one element in the set. An element in a sample space is called a sample point.

Example 1 *Three-child families.* To study the distribution of boys and girls in families having three children, a survey of such families is made. What is a sample space for the experiment of drawing one family from a population of three-child families?

Solution Let B stand for "boy", and G stand for "girl". If we use a triple of letters to represent the oldest, the second, and the youngest child, in that order, then the following set is a sample space for a single family:

$$\{BBB, BBG, BGB, GBB, BGG, GBG, GGB, GGG\}.$$

The triple GBB, for instance, represents the outcome "oldest child is a girl, second and third are boys". Another useful way of listing the possibilities is provided by a tree graph as shown in Fig. 1–1. If we follow all branches from a starting point, O, to the right-hand edge of the tree, we get the 8 possibilities listed in the column on the extreme right.

A third sample space is obtained by listing the number of girls in families with three children. The required numbers follow from the grouping of elements in the first sample space, thus:

$$\begin{array}{cccc} \{BBB, & (BBG, BGB, GBB), & (BGG, GBG, GGB), & GGG\} \\ \downarrow & \downarrow & \downarrow & \downarrow \\ \{\ 0, & 1, & 2, & 3\ \} \end{array}$$

Example 2 The numbers 1, 2, 3, and 4 are written separately on four slips of paper. The slips are then put into a hat and mixed. A blindfolded person

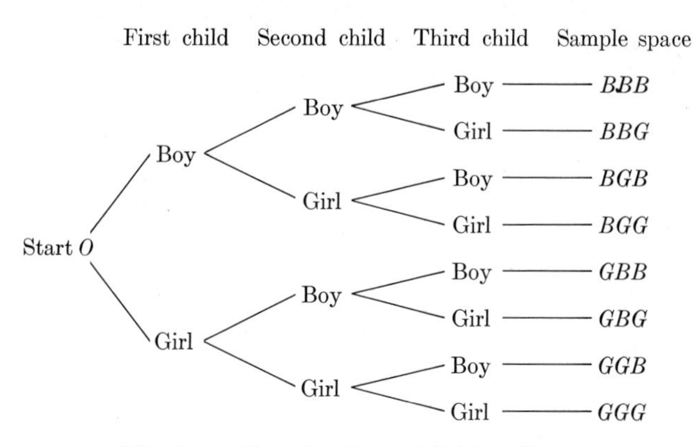

Fig. 1–1 Tree for three-child families.

draws two slips from the hat, one after the other, without replacement. Describe a sample space for the experiment.

Solution We may consider that each outcome of the experiment is represented by an ordered pair of numbers (x, y), where x is the number on the first slip and y is the number on the second. The restrictions on x and y are as follows:

$$1 \leq x \leq 4, \qquad 1 \leq y \leq 4, \qquad x \neq y.$$

Table 1–10 shows a sample space.

Table 1–10 Sample space for 2 numbered slips

y: number on second slip

x: number on first slip	$x \backslash y$	1	2	3	4
	1		(1, 2)	(1, 3)	(1, 4)
	2	(2, 1)		(2, 3)	(2, 4)
	3	(3, 1)	(3, 2)		(3, 4)
	4	(4, 1)	(4, 2)	(4, 3)	

EXERCISES FOR SECTION 1–7

1. A coin is tossed and then a die is thrown. List a sample space for this experiment. Illustrate with a tree graph.
2. Three coins are tossed. List two sample spaces for this experiment.
3. Two letters are randomly chosen, one after another, from the word *tack*. List a sample space.

4. A boy has in his pocket a penny, a nickel, a dime, and a quarter. He takes two coins out of his pocket, one after the other. List a sample space. Illustrate with a tree graph.

5. Suppose you plan to make a survey of families having two children. You want to record the sex of each child, in the order of their births. For example, if the first child is a boy and the second a girl, you record (boy, girl). This is one point in the sample space. List all the sample points.

6. If the survey in Exercise 5 is undertaken for families having four children, list an appropriate sample space. How many sample points does it have? How many of these points correspond to families having 3 boys and 1 girl? How many correspond to families in which the first child is a girl?

7. Two dice, one black and one red, are tossed and the numbers of dots on their upper faces are noted. List a sample space for the experiment. [*Note:* A tabular arrangement is convenient.]

8. An engineer's ruler has a cross section that is an equilateral triangle. Two such rulers, one red and one green, have their faces numbered 1, 2, and 3. The rulers are tossed onto the floor and the numbers on the bottom faces are read when they come to rest. Set up a table for the sample space of outcomes.

9. An experiment consists of selecting 3 radios from a lot of 25 and testing them. The test shows that a radio is defective (D), or nondefective (N). List a sample space for this experiment.

10. From five different books, A, B, C, D, and E, three are selected. List a suitable sample space of outcomes. In your sample space: (a) How many sample points correspond to a selection including A? (b) How many correspond to a selection without A? (c) How many correspond to a selection including both B and C? (d) How many correspond to a selection including either D or E?

11. A letter is chosen at random from the word *ground*. Which of the following sets are acceptable as sample spaces for the experiment and which are not?

    a) $\{g, r, o, u, n, d\}$;
    b) $\{\text{vowel}, g, r, n, d\}$;
    c) $\{r, o, u, n, d\}$;
    d) $\{\text{vowel}, \text{consonant}\}$;
    e) $\{\text{consonant}, u\}$.

12. A bag contains a number of marbles, identical in every way except that some are red, some white, and some blue, at least 2 of each color. Two marbles are drawn, one after the other, without replacement. What is a sample space for this experiment? How many points of the sample space correspond to drawing two marbles of the same color? How many points correspond to drawing marbles of different colors? How many correspond to drawing a red *and* a blue marble? How many correspond to drawing a red *or* a blue marble?

13. In the sample space of Exercise 7, how many sample points correspond to a total of more than 10 dots? To a total of less than 5 dots? To an even total? To the black die showing more than 5 dots? To the red die showing less than 3 *and* the black die showing more than 5? To the red die showing an even number *or* the black die showing 3?

14. There are 12 ordered pairs listed in Table 1-10. If these represent equally likely outcomes, what is the probability that

    a) $x + y = 5$?
    b) $x + y > 5$?
    c) x is even and y is odd?

1-8 PROBABILITIES IN A FINITE SAMPLE SPACE

When an experiment is performed, we may want to know the probabilities of various outcomes or events associated with the experiment. Often such probabilities can be computed by setting up a sample space of equally likely outcomes and counting the sample points. The following examples illustrate the method. Since we shall be making considerable use of these examples in this and in later chapters, they should be studied with special care.

Example 1 *Two dice.* An experiment consists of throwing two ordinary six-sided dice and observing the numbers of dots on their upper faces.

Discussion For the purposes of this experiment, we assume that the dice are distinguishable: one die is *red*, and the other die is *clear*. It would serve our purpose just as well to throw a single die twice, the first throw corresponding to the red die, the second throw corresponding to the clear die. Table 1-11 shows a sample space that lists the possible outcomes of the experiment.

In Chapter 3, Table 1-11 is referred to repeatedly, so inserting a bookmark here may speed your reading.

Each *row* of the table corresponds to a fixed value of r, the outcome for the *red* die; and each *column* corresponds to a fixed value of c, the outcome for the *clear* die. For instance, the entry (2, 4) in the second row and the fourth column represents the event "red die shows 2, clear die shows 4". The entire sample space, S, is the set of ordered pairs (r, c) with r and c each taking the values 1, 2, 3, 4, 5, or 6:

$$S = \{(r, c): 1 \leq r \leq 6,\ 1 \leq c \leq 6\}. \tag{1}$$

Since each of the 6 admissible values of r may be associated with 6 admissible values of c, we can see, without making a list, that there are 6×6, or 36, possible outcomes of the experiment. We assume that the dice are well balanced

Table 1-11 Sample space for two-dice experiment

		Clear die outcome (c)					
	$r \backslash c$	1	2	3	4	5	6
Red die outcome (r)	1	(1, 1)	(1, 2)	(1, 3)	(1, 4)	(1, 5)	(1, 6)
	2	(2, 1)	(2, 2)	(2, 3)	(2, 4)	(2, 5)	(2, 6)
	3	(3, 1)	(3, 2)	(3, 3)	(3, 4)	(3, 5)	(3, 6)
	4	(4, 1)	(4, 2)	(4, 3)	(4, 4)	(4, 5)	(4, 6)
	5	(5, 1)	(5, 2)	(5, 3)	(5, 4)	(5, 5)	(5, 6)
	6	(6, 1)	(6, 2)	(6, 3)	(6, 4)	(6, 5)	(6, 6)

Table 1–12 Events and probabilities for the two-dice experiment

Question number	Verbal description	Algebraic condition	Solution set (subset of S)	Probability
1	throwing a double	$c = r$	(1, 1), (2, 2), (3, 3), (4, 4), (5, 5), (6, 6)	$\frac{6}{36} = \frac{1}{6}$
2	clear score at least 3 greater than red score	$c \geq r + 3$	(1, 4), (1, 5), (1, 6), (2, 5), (2, 6), (3, 6)	$\frac{6}{36} = \frac{1}{6}$
3	sum equal to 10	$c + r = 10$	(4, 6), (5, 5), (6, 4)	$\frac{3}{36} = \frac{1}{12}$

and fairly thrown, so that these outcomes are equally likely to occur. We therefore attach probability $\frac{1}{36}$ to each point in the sample space.

Once the sample space of the experiment has been set up and probabilities assigned to the sample points, we can answer questions such as the following:

1. What is the probability of throwing a double?
2. What is the probability that the number on the clear die is at least 3 greater than the number on the red die?
3. What is the probability that the sum $r + c$ is 10?

When an event is described by a verbal expression (for example, "throwing a double"), we often find it helpful to translate the verbal expression into an algebraic condition such as "$c = r$". Then we focus attention on the subset of the sample space whose members satisfy this algebraic condition, which is usually an equation or an inequality. Counting does the rest. For example, consider the three questions just posed in connection with the two-dice experiment. We list, in Table 1–12, the verbal descriptions, the corresponding algebraic conditions, the solution sets, and the required probabilities.

Remark The foregoing procedure is useful for finding the probabilities of various outcomes of an experiment. We list the steps in the method:

1. *Set up a sample space S of all possible outcomes.* The sample space may be listed as in Table 1–11, or it may be indicated by the set-builder notation as in Eq. (1).

2. *Assign probabilities to the elements of the sample space (sample points).* In a sample space of n equally likely outcomes, we assign probability $1/n$ to each sample point. The sum of the probabilities of all the sample points in a given sample space must equal 1.

3. *To obtain the probability of an event E, add the probabilities assigned to the elements of the subset of S that corresponds to E.* Since the empty set has no elements, its probability is zero.

Example 2 For a chronic disease, there are five standard ameliorative treatments: a, b, c, d, and e. A doctor has resources for conducting a comparative study of three of these treatments. If he chooses the three treatments for study at random from the five, what is the probability that (a) treatment a will be chosen, (b) treatments a and b will be chosen, (c) at least one of a and b will be chosen?

Solution In Table 1–13, we list the 10 possible selections of the 5 treatments, taken 3 at a time. For reference, the sample points are numbered.

Table 1–13 S for study of treatments

1	2	3	4	5	6	7	8	9	10
abc	*abd*	*abe*	*acd*	*ace*	*ade*	*bcd*	*bce*	*bde*	*cde*

Next, we assign probability $\frac{1}{10}$ to each sample point, since we assume that all 10 selections are equally likely. Then the probability that treatment a is chosen is $\frac{6}{10}$, because there are 6 selections corresponding to the event "treatment a is chosen".

Similarly, the probability that treatments a and b are both chosen is $\frac{3}{10}$, because there are exactly 3 selections containing both a and b. Finally, the probability that at least one of the treatments a and b is chosen is $\frac{9}{10}$; only the tenth sample point contains neither a nor b.

EXERCISES FOR SECTION 1–8

Exercises 1 through 7 refer to the two-dice experiment. Consult the sample space in Table 1–11.

1. What is the probability of not throwing a double?
2. What is the probability that the number on one die is double the number on the other?
3. What is the probability that one die gives a 5 and the other die a number less than 5?
4. What is the probability that the clear die gives a number less than 3 and the red die a number greater than 3?
5. Evaluate:

   a) $P(r + c = 6)$ b) $P(r + c = 8)$ c) $P(r + c < 5)$
   d) $P(r + c > 9)$ e) $P(r \geq c + 4)$

6. Give algebraic descriptions of the following verbally described events: (a) not throwing a double, (b) red die shows two less than clear die, (c) clear die shows number at least 2 greater than red die, (d) number on red die twice that on clear die.

7. Give verbal descriptions of the following algebraically described events:

   a) $r = 3c$ b) $r - c = 1$ c) $r \neq c$
   d) $r + c > 8$ e) $c = r^2$ f) $r \geq c$

8. For the sample space of Exercise 4 at the end of Section 1-7, answer the following: (a) What is the probability that he takes out 35 cents? (b) What is the probability that the value of the coins selected is less than 20 cents? Less than 15 cents? More than 15 cents? A prime number? A number divisible by 10?

9. In Exercise 6, Section 1-7, assume that all points in the sample space have the same probability. What is the probability that in a family of four children the first two are girls? What is the probability that three are boys and one is a girl? That there are two boys and two girls?

10. In Exercise 8, Section 1-7, assume that all points of the sample space have equal probabilities. Let r denote the number on the red ruler, and g the number on the green. Evaluate:

    a) $P(r = g)$ b) $P(r + g > 3)$
    c) $P(r > g)$ d) $P(r \neq g)$
    e) $P(r = g^2)$

11. In the ancient Indian game of Tong, two players simultaneously show their right hands to each other, exhibiting either one or two or three extended fingers. If each player is equally likely to extend one, two, or three fingers, what is the probability that the total number of fingers extended is even? Odd? Greater than 4? Less than 2? Prime? [*Note:* Set up a sample space as a first step.]

12. Two rods, one black and one white, have square cross sections. Each rod has its faces numbered 1, 2, 3, and 4. The rods are rolled on the floor, and the numbers on their upper faces are read after they come to rest. Set up a table for a sample space of outcomes. If b is the number on the upper face of the black rod, and w that on the upper face of the white rod, evaluate:

    a) $P(b + w = 5)$ b) $P(b = w)$ c) $P(b > w + 1)$
    d) P(black 1 or 3 and white 2 or 4) e) P(sum of numbers even)
    f) P(larger number shown is a 4)

13. Suppose that you have a black rod from Exercise 12 and a red engineer's ruler from Exercise 8, Section 1-7. Rod and ruler are rolled on the floor, and the number on the top face of the rod and that on the bottom face of the ruler are noted. Set up a sample space and find the probability that the number on the black rod is greater than that on the red ruler. What is the probability that both numbers are the same? That the sum of the numbers is prime?

14. In the sample space of Exercise 10, Section 1-7, assume that all points are equally likely. What is the probability that B will be included in the selection? That both A and B will be included? That either A or B will be included? That the selection will be C, D, and E?

1-9 DO PROBABILITIES GROW?

Most people correctly believe that when a *fair* coin is tossed many times, the fraction of heads will be close to $\frac{1}{2}$. Some feel that a logical consequence is that, after 10 heads have appeared in a row, the probability of a tail is larger than before. This view stems from a misapprehension about the way the "law of averages" works for coins. Since the coin has neither memory, conscience, nor force of its own, it can scarcely change its probability. The great probabilist Feller puts the explanation succinctly. He says that the law of averages works by *swamping* rather than by compensation. Thus, if a set of tosses starts with 10 heads, the 10 will be largely swamped after 1000 tosses, and negligible after a million.

One reason for believing that probabilities grow is that in some problems they do. Can you think of such a problem? In the first-ace problem, if we have dealt 30 cards without an ace, the probability of an ace on the next card is large, $\frac{4}{22}$; and after 48 cards without an ace, the probability of an ace is 1. This growth happens because we draw *without* replacement from the pack, and the composition of the population has changed. But when a coin is tossed, there is no sense in which we have used up a head from a finite pool of heads. The model of drawing without replacement is the wrong one for coins.

Another reason for thinking that probabilities grow arises from a misunderstanding of the question. The probability of getting at least one head in the first 5 tosses of a coin is less than the probability of getting at least one head in the first 10 tosses; the probability of getting at least one head does grow with the number of tosses allowed. While true, this is irrelevant to the usual question about the probability of a head changing from one toss to another.

In some problems, superficially like the fair-coin problem, probabilities change from time to time. At the start of the season, a pitcher may not be in good physical condition, and his probability of throwing a strike may be low. But later he may improve. Still later a small injury may plague him. For this pitcher, one may well believe that the probability of throwing a strike will wax and wane with time. But simple forms of the law of averages are not readily applicable to such a complicated process.

REVIEW EXERCISES FOR SECTIONS 1-6, 1-7, AND 1-8

1. The numbers from 1 through 15 are painted on 15 balls, one number per ball. If one of these balls is drawn at random, what is the probability that the number on it is: (a) Divisible by 5? (b) Even? (c) Odd? (d) A perfect square? (e) A two-digit number? (f) A prime number? (g) A prime number that is 2 more than another prime?

2. A bag contains 5 times as many red marbles as black marbles (identical except for color). One marble is drawn at random. What is the probability that it is red?

3. A regular *icosahedron* is a symmetrical convex solid with 20 faces. Some of the faces of such a solid are painted red and the rest are painted blue. If, when the icosahedron is thrown onto the floor, the probability of a red face landing on the bottom is 4 times the probability of a blue face, how many faces are painted red?

4. A poll is taken among 70 residents of a suburb of Boston on the question of an ordinance to prohibit motorboats on the upper Mystic Lake. The results of the poll are tabulated as follows:

	Own motorboat only	Own sailboat only	Own motorboat and sailboat	Own neither	Totals
Favor	0	7	1	18	26
Oppose	20	2	3	5	30
No opinion	0	1	1	12	14
Totals	20	10	5	35	70

If one of the 70 persons is chosen at random, what is the probability that he: (a) Favors the ordinance? (b) Opposes the ordinance? (c) Favors the ordinance or has no opinion on it? (d) Owns a boat? (e) Owns a sailboat? (f) Owns a motorboat?

5. A committee of two persons is to be selected from three men (Archer, Baker, Connor) and two women (Davis and Eads). Describe two different sample spaces for the experiment.

6. You ask a friend to "think of a number". Describe a sample space for the experiment.

7. A teacher asks each member of his class to count the number of pencils that he (or she) has brought to class. Describe a sample space for the experiment.

8. Cards are dealt one after another from an ordinary bridge deck until the first ace appears. Describe two different sample spaces for the experiment.

9. A coin is tossed repeatedly until a head first appears, or until tails appear four times in succession. Describe a sample space for the experiment.

10. A die is thrown until a "2" appears. Describe a sample space for the experiment.

11. Twenty-five people tell you their birthdays. Describe a sample space for the experiment.

12. A plant breeder crosses two parent strains, each possessing a gene pair of type aA. Each parent contributes one-half of this gene pair (either a or A) to the offspring, where the two halves are combined. Describe a sample space for the genetic type of the offspring.

2 Permutations, combinations, and the binomial theorem

A basic mathematical model for many of our sample spaces is provided by the set of all possible lineups of n different objects in a row. Finding the number of ways of lining up objects in a row is a problem that has intrigued people for generations. . . . In how many ways can 12 different books be arranged on a shelf? . . . How many automobile license plates can be made using 2 letters followed by 3 digits? . . . How many even numbers can be formed from the digits 1, 2, 3, 4, 5, 6, 7? . . . In how many ways can a person take a walk of 9 blocks, if he always walks 5 blocks west and 4 blocks north? . . . Although such questions are fascinating and challenging in their own right, we shall consider them for an additional reason: we often need answers to such questions in the study of probability.

The question "How many?" occurred frequently in Chapter 1, and it will continue to occur frequently in the chapters that follow. The present chapter deals with some new methods of answering this question. At first glance, it may appear that new methods are not needed. Surely, the answer is a matter of counting! And are we not always able to count sets of objects by simply matching them with the natural numbers? In practice, unfortunately, we are *not* always able to do so: time runs out on us. Even a high-speed computer runs out of time in attempting to count, in the usual way, the number of elements in some of the sets that we shall meet.

Therefore, in this chapter:

1. We present a new, general principle that leads to a formula for counting the number of possible lineups of n different objects in a row.
2. We extend this new principle to deal with related counting problems involving sets that have some indistinguishable members, and lineups in which the order of objects is not considered.
3. We apply the new counting methods to finding probabilities.
4. We use the new methods to find the expansion of $(a + b)^n$, where n is a positive integer or zero.

2–1 PERMUTATIONS: THE MULTIPLICATION PRINCIPLE

If the outcomes of an experiment are equally likely, the method of finding probabilities of events, as set forth on pages 28–29, requires us to find two numbers:

1) the number of possible outcomes, and
2) the number of these outcomes that correspond to a given event, E.

To this point we have obtained these numbers by listing all possible outcomes and by counting, in the ordinary way, sample points in the resulting sample space. The foregoing discussion indicates that this method of counting has practical limitations, and the following example underlines this fact.

Example 1 Three persons designated A, B, and C line up at random to buy tickets for a game. Find the probability that A is beside B in the lineup.

Solution Our sample space of possible outcomes is

$$S = \{\underline{ABC},\ ACB,\ \underline{BAC},\ BCA,\ \underline{CAB},\ \underline{CBA}\}.$$

Assuming equally likely outcomes, we find by counting sample points that

$$P(A \text{ beside } B) = \tfrac{4}{6}.$$

Query If, in this example, there were five persons, say A, B, C, D, and E, in the random lineup, what would $P(A$ beside $B)$ be? . . . Try to list the possible lineups. . . . How many possible lineups are there? . . . In how many of these lineups is A beside B?

You soon see that, even with five people, the list-and-then-count method has gotten out of hand. We shall soon learn that there are 120 possible lineups, in 48 of which A is beside B. For ten people, the corresponding numbers would be 3,628,800 and 725,760. But even these rather well-fed results do not hint at the astronomical numbers of lineups possible with relatively small groups of people. Suppose that we program a high-speed computer to line up 17 people in a row in all possible ways and to count the number of these lineups. If the computer can construct and count 50,000 lineups per second, it would take more than a lifetime to obtain the count. And if we present the computer with the same problem for 26 people, it would take more than one hundred trillion years to complete the job! Clearly, we need help of another kind.

Let us seek to discover a general principle that will enable us to find, without resorting to our old counting methods, the number of possible lineups of n different objects in a row. To this end, consider the following example.

Example 2 In how many ways can 3 books, denoted by A, B, and C, be arranged in order on a shelf?

Solution 1 One way to solve this problem is to list the possible arrangements and count them, just as in Example 1.

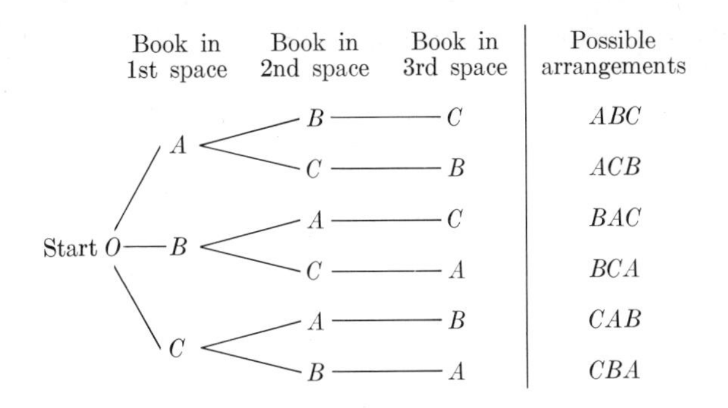

Fig. 2–1 Tree for arrangements of 3 books.

A tree graph (Fig. 2–1) provides an organized way of listing the arrangements so that none is missed. Note that the tree diagram takes *order* into account. Thus ABC and ACB count as different arrangements of the 3 books because they are in different orders. *Order is the essence of such arrangements; a change in order yields a different arrangement.*

Solution 2 A more convenient solution to this example is suggested by a further study of the tree diagram. The reasoning is as follows:

The problem requires us to fill 3 spaces, which can be represented as

In the first space, we can put A or B or C. Hence the first space can be filled in three ways:

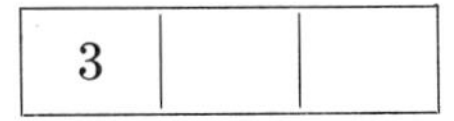

(This is indicated on the tree graph by the 3 branches from O that end at the column headed "Book in 1st space".) For each of the 3 ways of filling the first space, we have 2 ways of filling the second space, because either of the 2 remaining books can be used:

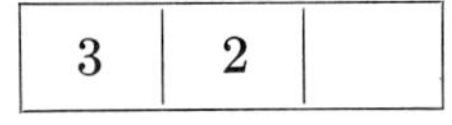

Thus, we can fill the first 2 spaces in 3×2, or 6, ways. (Note that 6 branches of the tree end at the column headed "Book in 2nd space".) For each of the 6 ways of filling the first 2 spaces, we have one way of filling the third space, because only one book remains. Therefore, we can fill the 3 spaces in 6×1, or 6, ways. (Note that 6 branches of the tree end at the column headed "Book in 3rd space".) We can indicate the number of ways of filling each of the 3 spaces thus:

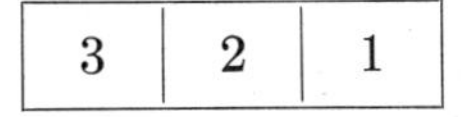

And, as indicated by the tree, we can obtain the total number of arrangements by *multiplication:*

$$3 \times 2 \times 1 = 6.$$

To this point, we have used the word "arrangements" to describe orderings of objects that result from operations such as that of placing books in a line. "Arrangement" is a common word that is informally descriptive. But we are dealing with special kinds of arrangements: we are concerned with arrangements, or orderings, of objects *in a line*, not with other kinds of arrangements, such as those of flowers in a vase. Since we refer to a special kind of arrangement, we need, for more precise description, a special word. This special word is *permutation.*

Each of the six arrangements in the foregoing example is called a permutation of the three books. We say that there are six permutations of the three books, taken three at a time, or all together.

2-1 Definition: Permutation

A permutation of a number of objects is any arrangement of these objects in a definite order.

To "permute" a collection of objects means to arrange them in a definite order.

Example 3 If at least 3 copies each of book A, book B, and book C are available, in how many distinguishable ways can we arrange 3 of the books on a shelf? (Regard the copies as indistinguishable.)

Solution With at least 3 copies of each book available, we can now have arrangements like AAA or ABA. Because the copies are indistinguishable in appearance, even though they are composed of different molecules, one arrangement of 3 copies of book A is indistinguishable from any other arrangement of those copies, or of any other 3 copies of book A. However, the arrangements ABA and AAB are distinguishable. By reasoning similar to that used in Example 2, we can show that each of the 3 spaces can now be filled in 3 ways. The choices are indicated thus:

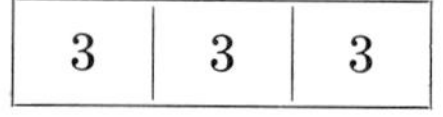

As before, the total number of permutations is found by multiplication:

$$3 \times 3 \times 3 = 27.$$

A shortcut in counting When the number of objects in a set is large, the number of permutations of the objects cannot, without great labor, be found by listing and counting. Fortunately, the method of reasoning suggested by the tree graph and used in Examples 2 and 3 can be extended, and used to provide a convenient general method for dealing with problems in permutations.

2-2 The multiplication principle

If an operation can be performed in n_1 ways and, after it is performed in any one of these ways, a second operation can be performed in n_2 ways and, after it is performed in any one of these ways, a third operation can be performed in n_3 ways, and so on for k operations, then the k operations can be performed together in

$$\boxed{n_1 \times n_2 \times n_3 \times \cdots \times n_k} \quad \text{ways.} \tag{1}$$

A note on notation We have used subscripts on the letter n, along with three dots, to indicate a set of variables of arbitrary length. This device may seem complicated, but some such method is necessary. All the letters in the English alphabet would denote only 26 variables, but the subscripts and the three dots enable us to denote any finite number.

Observe the special function of the three dots. They indicate that we are to begin with the factor n_1 and write additional factors until we reach the kth factor, n_k. The dots do *not* imply that k is greater than 3. If, for example, $k = 2$, then expression (1) becomes

$$n_1 \times n_2;$$

and if $k = 1$, the expression means simply n_1.

The need for subscripts becomes apparent if we try to get along without them when the number of variables is large, or indefinite. If we denote a set of variables of arbitrary number by

$$a, b, c, \ldots, h,$$

a little thought shows that this notation implies eight variables, not an arbitrary number. When we become familiar with the use of subscripts, we appreciate their convenience and usefulness. (See Appendix 1.)

A tree diagram to illustrate the multiplication principle The tree in Fig. 2-2 illustrates the multiplication principle for $n_1 = 2$, $n_2 = 3$, and $n_3 = 2$. The total number of paths along branches of the tree, from the origin O to the right-hand edge of the diagram, is

$$n_1 \times n_2 \times n_3 = 2 \times 3 \times 2.$$

Examples such as the foregoing make the multiplication principle intuitively evident. We shall in future accept its truth, and use it freely as a shortcut in counting the number of permutations of sets of objects. Note that the multiplication principle takes *order* into account.

Example 4 a) How many license plates can be made using 2 letters followed by a 3-digit number?

b) Among such license plates, what is the probability of getting one with 2 vowels followed by 3 identical digits?

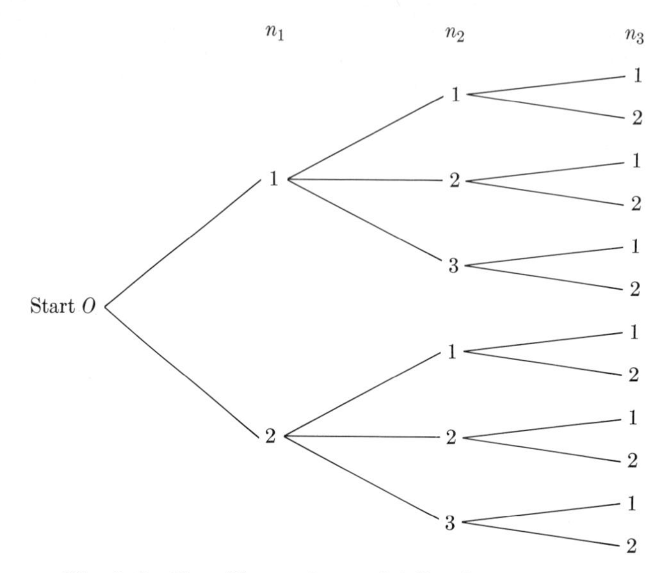

Fig. 2–2 Tree illustrating multiplication principle.

Solution a) There are 5 spaces to fill. The first space can be filled with any one of 26 letters, and so in 26 ways. After the first space has been filled in any one of these ways, the second space can be filled in 26 ways (repetitions of a letter allowed). Similarly, the third space can be filled in 9 ways (zero not allowed), the fourth space in 10 ways, and the fifth space in 10 ways (zero and repetitions of a digit allowed). By the multiplication principle, the answer is

$$26 \times 26 \times 9 \times 10 \times 10 = 608{,}400.$$

b) Similarly, by the multiplication principle, the number of license plates having 2 vowels followed by 3 identical digits is

$$5 \times 5 \times 9 \times 1 \times 1 = 225.$$

Assuming that the license plates are assigned at random, we get

$$P(\text{2 vowels, 3 identical digits}) = 225/608{,}400 = 1/2704.$$

Example 5 In planning a round trip from Chicago to Southampton by way of New York, a traveler decides to travel between Chicago and New York by air and between New York and Southampton by steamship. If there are 6 airlines operating between Chicago and New York, and 4 steamship lines operating between New York and Southampton, in how many ways can the round trip be made without traveling over any line twice?

Solution The trip from Chicago to New York can be made in 6 ways; after it has been made in any one of these ways, the trip from New York to Southampton can be made in 4 ways. Then the trip from Southampton to New York can be made in 3 ways, after which the trip from New York to Chicago can be made in 5 ways. By the multiplication principle, the number of possible ways of making the round trip is

$$6 \times 4 \times 3 \times 5 = 360.$$

Example 6 Given the digits 1, 2, 3, 4, and 5, find how many 4-digit numbers can be formed from them (a) if no digit may be repeated, (b) if repetitions of a digit are allowed, and (c) if the number must be odd, without any repeated digit.

Solution a) *No repetitions.* There are 4 places to fill. The first place can be filled with any one of the 5 digits, and so in 5 ways. Then, since no digit may be used more than once, the second place can be filled with any one of the remaining digits, and so in 4 ways. Similarly, the third place can be filled in 3 ways, and the fourth place in 2 ways. From the multiplication principle, it follows that the number of 4-digit numbers is

$$5 \times 4 \times 3 \times 2 = 120.$$

b) *Repetitions allowed.* If repetitions of a digit are allowed, each of the 4 places can be filled with any one of the given 5 digits, and so in 5 ways. The number of 4-digit numbers, with repetitions allowed, is therefore

$$5 \times 5 \times 5 \times 5 = 625.$$

c) *Odd, without repetitions.* If the number must be odd, the final digit has to be 1 or 3 or 5. Therefore the fourth place can be filled in 3 ways. After this has been done in any one of these ways, the remaining places can be filled in 4 ways, 3 ways, and 2 ways, respectively, since no digit may be used more than once. The number of odd, 4-digit numbers, without repeated digits, is

$$4 \times 3 \times 2 \times 3 = 72.$$

Note We filled the fourth place first. In such calculations, *if some operation must be performed in a special way, it is usually advisable to do it first.* However, for nonspecial operations, the *sequence* in which the spaces are filled is often arbitrary. Thus in part (c) of the foregoing example, once the fourth space is filled, it doesn't matter which of the three remaining spaces is filled next.

Similarly, in Example 2, it doesn't matter which space on the shelf is filled first. We can put a book in the middle space, then put a book to its left, and then one to its right. The multiplication principle still applies and gives the same answer as before. We think of the first operation as that of placing a book in the middle space; the second, as that of placing a book in the leftmost space; and the third, as that of placing a book in the rightmost space. It helps us to analyze the problem if we think of performing a definite sequence of operations one after another, even though our physical actions might be done in a different order. In fact, the three books can all be put on the shelf at the same time, rather than one after another; but such a way of looking at the problem provides no insight into its solution, whereas the one-book-after-another approach does.

Example 7 The digits 1, 2, 3, 4, and 5 are used, with repetitions of a digit allowed, to form 4-digit numbers in all possible ways. The 4-digit numbers are then printed on discs, which are placed in a box and thoroughly mixed. If a disc is drawn at random from the box, what is the probability that it has a number without repeated digits?

Solution Refer to Example 6. From Example 6(b), we see that the sample space has 625 sample points, corresponding to the 625 equally likely outcomes of the drawing. Of these outcomes, 120 are favorable; that is, there are 120 possible ways of drawing a disc that does not show repeated digits. Hence,

$$P(\text{no repeated digits}) = 120/625 = 24/125.$$

Example 8 Bill's Pizza Palace offers pepper, onion, sausage, mushrooms, and anchovies as toppings for the plain cheese base of the pizzas. How many different pizzas can be made?

Solution There are 5 ingredients. In adding a topping to the base, we deal with the available ingredients one at a time. The pepper can be dealt with in 2 ways—take it or leave it. After the pepper has been dealt with, we can dispose of the onion in 2 ways—take it or leave it. Similarly, each of the remaining 3 toppings can be dealt with in 2 ways. Therefore, there are 2^5, or 32, possible pizzas, including the plain pizza and the one with everything.

Note Example 8 is a special case of the general problem of finding the number of subsets in a set of n elements. This general problem is left as an exercise. (See Exercise 18, page 42.)

The addition principle Consider two operations, one of which can be performed in m ways and the other in n ways. Then the *multiplication principle* says: If, after the first operation is performed in any one of the m ways, the second operation can be performed in n ways, the two operations can be performed together in mn ways. In short, the multiplication principle is concerned with situations where we can perform the first operation *and then* perform the second.

A different situation is faced if we wish to perform the first operation *or* the second operation, not both. Consider the following example.

Example 9 Three different flags are available. In how many ways can a signal with at least 2 flags be arranged on a flagpole, if the order of the flags on the flagpole counts?

Solution As our first operation, let us arrange 2 flags on the flagpole (Fig. 2–3). By the multiplication principle, this can be done in 3×2, or 6, ways. As our second operation, let us arrange 3 flags on the flagpole. This can be done in $3 \times 2 \times 1$, or 6, ways.

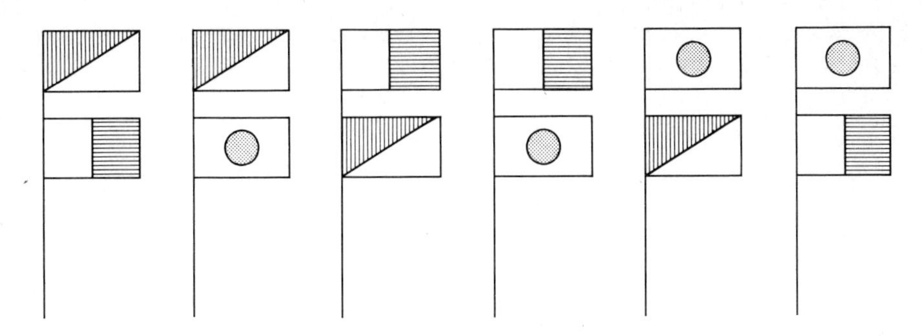

Fig. 2-3 Signals using 2 flags out of 3.

Now we have only one signal to arrange, and this signal may be a two-flag signal or a three-flag signal, but not both together. It is a question of performing the first operation *or* the second, not the first operation *and then* the second. The operations are mutually exclusive: they cannot both occur together. The total number of signals is therefore $6 + 6 = 12$.

2-3 The addition principle

> If two operations are mutually exclusive, and the first can be done in m ways and the second in n ways, then one operation or the other can be done in $m + n$ ways.

This principle is readily generalized to include any finite number of operations. The statement is left as an exercise.

EXERCISES FOR SECTION 2-1

Use the multiplication principle or the addition principle to solve the following exercises.

1. In how many ways can 8 people line up at a theater box office?
2. How many 5-digit numbers can be formed from the integers 1, 2, 4, 6, 7, 8, if no integer can be used more than once? How many of these numbers will be even? How many odd?
3. If the call letters of a broadcasting station must begin with the letter W, how many different stations could be designated by using only 3 letters, with repetitions of a letter allowed? How many by using 4 letters, without repetitions?
4. In how many ways can 3 letters be mailed in 6 mailboxes, if each letter must be mailed in a different box? If the letters are not necessarily mailed in different boxes, how many ways are there of posting them? If the letters are mailed at random, and not necessarily in different boxes, what is the probability that all the letters are put in the same mailbox?
5. There are 7 seats available in a sedan. In how many ways can 7 persons be seated for a journey if only 3 are able to drive? [*Hint:* See note following Example 6(c).]

6. A passenger train has 9 coaches. In how many ways can 4 people be assigned to coaches if they must ride in different coaches?
7. In how many ways can 6 students be seated in a classroom with 30 desks?
8. Twelve boys try out for the basketball team. Two can play only at center, four only as right or left guard, and the rest can play only as right or left forward. In how many ways could the coach assign a team?
9. How many numbers, each with at least 3 digits, can be formed from the 5 digits 1, 2, 3, 4, 5, if no digit may be used more than once?
10. In how many ways can 5 boys and 5 girls be seated alternately in a row of 10 chairs, numbered from 1 to 10, if a boy always occupies chair number one?
11. In how many ways can 3 different presents, A, B, and C, be given to any 3 of 15 persons? If a specified person must receive A, and if no person is to receive more than one present, in how many ways can the presents be distributed?
12. In how many ways can a selection of at least one book be made from 8 different books? [*Hint:* See Example 8 of the text.]
13. Given 4 flags of different colors, how many different signals can be made by arranging them on a vertical mast, if at least 2 flags must be used for each signal?
14. An encyclopedia consists of 9 volumes numbered 1 to 9. In how many ways can the 9 volumes be arranged together on a shelf so that some or all of the volumes are out of order? If the 9 volumes are placed on the shelf at random, what is the probability that they are not in correct order?
15. How many 5-digit numbers can be formed? How many of these begin with 2 and end with 4? How many do not contain the digit 5? How many are divisible by 5? If a 5-digit number is chosen at random from a table listing all possible 5-digit numbers, what is the probability that the chosen number does not contain the digit 5? What is the probability that the chosen number is divisible by 5?
16. How many different parties of 2 or more can be formed from 9 people?
17. Five men compete in a race. In how many ways can the first two places be taken?
18. (a) How many subsets, including the empty and universal sets, can be formed from a set of 10 different objects? (b) From a set of n different objects?
19. How many ordered pairs of symbols (x, y) can be formed if x can be replaced by a or b or c, and y can be replaced by 1 or 2 or 3 or 4? Draw a tree diagram exhibiting the set of possible ordered pairs (x, y).
20. How many permutations are there of n different objects, taken r at a time, with repetitions allowed? (It is assumed that there are at least r copies of each of the n objects available.)
21. On stepping off a train, a man finds that he has a nickel, a dime, a quarter, and a half-dollar in his pocket. In how many ways can he give the porter a tip?
22. Six people, designated A, B, C, D, E, and F, line up at random at a theater box office. What is the probability that A is beside B in the lineup?
23. A farm is divided into 64 plots of land. An agricultural experimenter wishes to compare, on this farm, the yield of 5 strains of beets using 3 kinds of insecticide and 4 different fertilizers. Will he have enough plots to compare all possible combinations of strain, insecticide, and fertilizer?

2-2 FORMULAS FOR PERMUTATIONS

The multiplication principle provides a general method for finding the number of permutations of sets of objects. For some important types of problems, this method can be shortened by means of some convenient symbols and formulas that we now introduce.

The factorial symbol As we saw in Section 2-1, the multiplication principle enables us to establish facts such as the following:

a) 7 people can be arranged in a line in

$$7 \times 6 \times 5 \times 4 \times 3 \times 2 \times 1 \quad \text{ways;}$$

b) 20 books can be arranged on a shelf in

$$20 \times 19 \times 18 \times \cdots \times 3 \times 2 \times 1 \quad \text{ways;}$$

c) n objects can be arranged in a line in

$$n(n-1)(n-2)\cdots 3 \times 2 \times 1 \quad \text{ways;}$$

and so on. Once again, note that the dots do not imply that n is greater than 3. The dots indicate that we are to begin with the integer n and continue to multiply factors, each of which is one less than its predecessor, until 1 is reached.

Problems such as the three foregoing may lead to very large numbers or very long sequences of factors. For convenience, therefore, we introduce a special symbol.

2-4 Definition: *n* factorial

The product of all whole numbers from 1 through n is called n factorial, and denoted by $n!$.

Thus,

$$\boxed{n! = n(n-1)(n-2)\cdots 3 \times 2 \times 1.}$$

In particular, we have

$$\begin{aligned}
1! &= 1,\\
2! &= 2 \times 1 = 2 \times 1! = 2,\\
3! &= 3 \times 2 \times 1 = 3 \times 2! = 6,\\
4! &= 4 \times 3 \times 2 \times 1 = 4 \times 3! = 24,\\
5! &= 5 \times 4 \times 3 \times 2 \times 1 = 5 \times 4! = 120.
\end{aligned}$$

Proceeding in this way, we can make the table of $n!$ shown in Table 2-1, or the more extensive Table II in the back of the book. Table II also gives log $n!$.

Table 2-1

n	1	2	3	4	5	6	7	8	9	10
$n!$	1	2	6	24	120	720	5040	40,320	362,880	3,628,800

The task soon becomes laborious, because the factorials increase in size at a fantastic rate. The number of permutations of the letters of the alphabet, 26!, is greater than 4×10^{26}.

The factorial symbol provides a useful notation for representing large numbers of the type encountered in the study of permutations and related topics. Note that

$$20! = 20 \times 19!,$$
$$100! = 100 \times 99!,$$
$$(n + 1)! = (n + 1) \times n!.$$

Example 1 From the multiplication principle, we can show (cf. Section 2-1) that 50 people can form a line in

$$50 \times 49 \times 48 \times \cdots \times 3 \times 2 \times 1 = 50! \quad \text{ways.}$$

2-5 Theorem: Permutations of *n* things, all together

The number of permutations of a set of n different objects, taken all together, is $n!$.

Proof The proof is a direct application of the multiplication principle. We have n spaces to fill. The first space can be filled with any one of the n objects, and so in n ways. After this has been done in any one of these ways, the second space can be filled with any one of the remaining objects, and so in $n - 1$ ways. Similarly, the third space can be filled in $n - 2$ ways, the fourth space in $n - 3$ ways, and so on. Therefore, by the multiplication principle, the number of ways of filling the n spaces is

$$n(n - 1)(n - 2) \cdots 3 \times 2 \times 1 = n!.$$

The number of permutations of n different objects, taken all together, is denoted by ${}_nP_n$. Therefore we have

$$\boxed{{}_nP_n = n!.} \quad \square$$

We now consider permutations of n different objects in which some, but not necessarily all, of the objects are used.

Example 2 In how many ways can 3 books be chosen from 7 different books and arranged in 3 spaces on a bookshelf?

Solution The first space can be filled with any one of the 7 books, and so in 7 ways. After this has been done in any one of these ways, the second space can be filled in 6 ways. Similarly, the third space can be filled in 5 ways. By the multiplication principle, the 3 spaces can be filled in

$$7 \times 6 \times 5 \quad \text{ways.}$$

Factorial symbols can also be used to denote the product $7 \times 6 \times 5$:

$$7 \times 6 \times 5 = \frac{7 \times 6 \times 5 \times 4 \times 3 \times 2 \times 1}{4 \times 3 \times 2 \times 1} = \frac{7!}{4!}.$$

The number of permutations of 7 objects, taken 3 at a time, is denoted by ${}_7P_3$, and its value is $7 \times 6 \times 5$. Thus,

$${}_7P_3 = 7 \times 6 \times 5 = \frac{7!}{4!}.$$

Note To evaluate ${}_7P_3$, we "begin with 7 and proceed for 3 factors".

2-6 Definition: ${}_nP_r$

An arrangement of r objects, taken from a set of n objects, is called a permutation of the n objects, taken r at a time. The total number of such permutations is denoted by ${}_nP_r$, $r \leq n$.

2-7 Theorem: Permutations of *n* things, *r* at a time

The number of permutations of a set of n different objects, taken r at a time, without repetitions, is

$$\boxed{{}_nP_r = \frac{n!}{(n-r)!}.}$$

Proof Once again, the proof is an application of the multiplication principle. Suppose that we have r spaces to fill and n objects from which to choose. The first space can be filled with any one of the n objects, and so in n ways. After this has been done in any one of these ways, there remain $n - 1$ objects, any one of which can be put in the second space. Thus the second space can be filled in $n - 1$ ways. Similarly, the third space can be filled in $n - 2$ ways, the fourth space in $n - 3$ ways, and so on. The pattern shows that the tenth space can be filled in $n - 9$ ways, the twenty-fifth space in $n - 24$ ways and, in general, the rth space in $n - (r - 1)$ ways. From the multiplication principle, the r spaces can be filled in

$$\boxed{{}_nP_r = n(n-1)(n-2)\cdots(n-r+1)} \quad \text{ways.} \tag{1}$$

The right-hand member of formula (1) consists of r factors. It takes another convenient form if we multiply by $(n-r)!/(n-r)!$, because then we can write

$$_nP_r = \frac{n!}{(n-r)!}\cdot \quad \square \tag{2}$$

Note that, in formula (2), the numerator is the number of arrangements of all the objects, and the denominator is the number of arrangements of the objects *not* selected:

$$_nP_r = \frac{\text{number of arrangements of } n \text{ objects}}{\text{number of arrangements of } n-r \text{ unselected objects}}.$$

In Exercise 17, page 49, you are asked to explain why this is so.

The meaning of 0! Formula (1) is defined for $r \leq n$. Formulas (1) and (2) agree if $r < n$; but if $r = n$, formula (2) gives

$$_nP_n = \frac{n!}{0!},$$

which is meaningless since 0! has not been defined. If we define 0! to be 1, then formula (2) would have a meaning for the case $r = n$; and formulas (1) and (2) would agree for $r \leq n$. We shall so define 0!.

2-8 Definition: Zero factorial

$0! = 1.$

This definition gives us an additional dividend; for, if $n = 1$, the formula

$$n! = n \times (n-1)! \tag{3}$$

becomes

$$1! = 1(0!),$$

which is true. Hence, formula (3) now holds for $n = 1$.

Example 3 A memory expert has found 8 compatible steps that improve the ability to recall lists of items. He also finds that (a) beginners can keep only 3 steps in mind, and (b) the order of using the steps is important. How many experiments must he perform in order to find the 3 ordered steps that together produce the best recall?

Solution The problem is that of finding the number of permutations of 8 things, 3 at a time. This number is

$$_8P_3 = 8 \times 7 \times 6 = 336,$$

which is the number of experiments required to compare the performances of all possible ordered 3-step rules.

The foregoing calculation is important to the memory expert. It gives him a basis for computing the cost of such an investigation, and, if sufficient funds are not available, it informs him that a more restricted set of experiments must be designed.

Example 4 How many permutations are there of 5 cards, taken from a bridge deck of 52 different cards?

Solution From Theorem 2–7, the number is

$$_{52}P_5 = 52 \times 51 \times 50 \times 49 \times 48 = 311{,}875{,}200.$$

Note Order counts here. Thus, 2, 3, 4, 5, 6 of hearts differs from 2, 4, 3, 5, 6 of hearts.

Example 5 How many words can be formed from the letters of the word *hyperbola*, taken all together? In how many of these words will the letters h and y occur together? In how many will the letters h and y not occur together? Supposing that the letters of the word *hyperbola* are arranged at random in a row, find (a) $P(h$ and y occur side by side) and (b) $P(h$ and y do not occur side by side). [*Note.* A "word", as used in such problems, means any arrangement of letters. It does not need to be a word in some language.]

Solution From Theorem 2–5, the number of ways of arranging 9 different letters, all together, is 9!, or 362,880. Therefore, the required number is 9!, if there are no restrictions. If the letters h and y must occur together, it is a good idea to consider them as one letter, hy. We now have 8 different letters to be arranged all together. This gives 8! arrangements. However, in each of these arrangements, the order hy may be changed to yh, so that each of the 8! arrangements gives rise to two arrangements that satisfy the given restriction. Hence, the total number of words in which the letters h and y occur together is $2(8!) = 80{,}640$.

The number of words in which the letters h and y do not occur together is the difference $362{,}880 - 80{,}640 = 282{,}240$.

To find the required probabilities, we imagine a sample space with 362,880 sample points corresponding to the possible arrangements of the letters of the word *hyperbola*. Since all arrangements are equally likely, we assign probability 1/362,880 to each sample point. Then,

$$P(h \text{ and } y \text{ side by side}) = \frac{80{,}640}{362{,}880} = \frac{2 \times 8!}{9!} = \frac{2}{9},$$

and

$$P(h \text{ and } y \text{ not side by side}) = \frac{282{,}240}{362{,}880} = \frac{7 \times 8!}{9!} = \frac{7}{9}.$$

For solving problems involving arrangements of objects in a set, the reader now has available the multiplication and addition principles and some formulas. He must not expect all problems to yield to the direct application of a formula. Flexibility is the key to the situation. Special problems may require formulas, the multiplication or addition principle, some special device, or a combination of these methods.

Example 6 Twenty-six persons $A, B, \ldots, Z$ are to be arranged in a line. If this is done randomly, what is the probability that A and B will be adjacent in the lineup?

Solution There are 26! ways of lining up the 26 persons. To compute the number of ways of having A and B together, we first pick two adjacent places (which can be done in 25 ways), then put A and B in these places (which can be done in 2 ways), and finally put the remaining 24 persons in the remaining 24 places (which can be done in 24! ways). Therefore, the desired probability is

$$P(A \text{ and } B \text{ together}) = \frac{25 \times 2 \times 24!}{26!} = \frac{2}{26} = \frac{1}{13}.$$

EXERCISES FOR SECTION 2–2

Note. A "word", as used in these exercises, means any arrangement of letters.

1. Evaluate the following: ${}_9P_3$, ${}_mP_1$, ${}_7P_7$, ${}_kP_2$.
2. Compute ${}_nP_0$ and interpret it.
3. How many words can be formed from the letters of the word *fragments* (a) taken all at a time, (b) taken 8 at a time, (c) taken 4 at a time?
4. A student has 4 examinations to write and there are 10 examination periods available. How many possible arrangements are there of his examination program?
5. (a) A musical concert is to consist of 3 songs and 2 violin selections. In how many ways can the program be arranged so that the concert begins and ends with a song, and neither violin selection follows immediately after the other? (b) If the 5 selections are arranged at random, what is the probability that the program follows the plan in part (a)?
6. Prove that the number of 3-letter words that can be formed from the letters of the word *background* is the same as the number of words that can be made by rearranging the letters of the word *ground*.
7. Use your knowledge of whole numbers to find the number of 4-digit whole numbers. (What is the first such whole number? What is the last?) Then check your answer by using the multiplication principle.
8. How many automobile license plates bearing 5-digit numbers can be made if no license number starts with 0? If letters of the alphabet are used in place of the first digit and the next digit is not 0, how many plates can be made?
9. A passenger train consists of 2 baggage cars, 4 day coaches, and 3 parlor cars. In how many ways can the train be made up if the 2 baggage cars must come in front, and the 3 parlor cars must come in the rear?
10. If the passenger train in Exercise 9 is assembled at random, what is the probability that the 2 baggage cars will be in front and the 3 parlor cars will be in the rear?

11. If there are 3 roads from town A to town B, and 4 roads from town B to town C, in how many ways can one make a trip from A to C by way of B, and return from C to A by way of B?
12. In the Hotel Superba, 7 rooms in a row are assigned at random to 7 guests, 2 of whom are from Minnesota. What is the probability that the guests from Minnesota are assigned rooms side by side? What are the odds against such an assignment?
13. In geometry, polygons are commonly labeled by placing letters at their vertices. How many ways are there of labeling a triangle with letters of the alphabet? A pentagon? A decagon? (Do not multiply out the answers.)
14. How many 5-letter words can be made from 10 different letters (a) if any letter may be repeated any number of times, (b) if repetitions of a letter are not allowed? (c) In how many of the words of (a) will repeated letters actually occur?
15. Ten different letters are marked on 10 identical discs, and the discs are then placed in a bag and mixed. A disc is drawn at random from the bag, its letter is noted, and the disc is returned to the bag. This operation is repeated until 5 letters are recorded. Find the probability that repeated letters occur among the 5 letters recorded.
16. (Refer to Example 3, page 46.) If the memory expert decides to use always the same first step in each ordered 3-step rule, how many experiments are needed to study all admissible 3-step rules? If each experiment costs $200, how much money is needed for the complete investigation?
17. Use the fact noted after formula (2) to show that (the number of permutations of n things taken r at a time) $\times$ (the number of permutations of the remaining $n - r$ things) $=$ (the number of permutations of n things). Explain why this should be so.

2-3 COMBINATIONS

In Section 2-2, we obtained some formulas for counting permutations. These formulas enable us to count the number of ways objects can be arranged in an *ordered* row. Sometimes a problem requires us to make a *selection* of objects without regard to order; in such cases a change in order does *not* yield a new selection. For example, in choosing a committee of 3, "Smith, Jones, and White" is the same committee as "White, Jones, and Smith", and these two triples are counted as one selection, not two. The total number of such selections is not obtained by the formulas of Section 2-2, as the following example shows.

Example 1 Four drugs, designated A, B, C, and D, are beneficial in the treatment of a certain chronic disease. A research scientist has funds sufficient for a study of the comparative efficacy of 3 of the drugs. How many different studies are possible?

Solution The scientist makes a selection of 3 drugs from A, B, C, and D without taking order into account. He has only 4 possible selections:

$$ABC, \quad ABD, \quad ACD, \quad BCD. \tag{1}$$

We do not list ACB, for example, because the selection ABC is the same selection and corresponds to the same study as ACB.

As we saw in Section 2–2, an entirely different problem arises if we require the number of permutations of 4 objects, taken 3 at a time. This number is

$$_4P_3 = 4 \times 3 \times 2 = 24.$$

In these permutations, or arrangements, the *order* of the objects is taken into account; in making selections for a 3-drug study, order is *not* taken into account.

The word "selection" is a good, everyday word that describes the outcome of an operation such as that performed by the scientist in choosing 3 drugs for a study in Example 1. However, since we are dealing with a special kind of selection that is not concerned with the order of objects, we need a special word. Each selection in the list (1) is called a *combination* of the 4 drugs, taken 3 at a time. The total number of such combinations is denoted by

$$_4C_3, \qquad \text{or by} \qquad \binom{4}{3},$$

each of which is read "number of combinations of 4 things, taken 3 at a time". The symbol $\binom{4}{3}$ has no bar in the middle; it is not a fraction. By counting items in list (1), we see that

$$_4C_3 = \binom{4}{3} = 4.$$

Example 1 underlines the difference between a permutation and a combination:

in a permutation, order is taken into account;
in a combination, order is not taken into account.

It is not always feasible to count combinations by listing them and counting in the usual way. The list may get much too long. Thus, in Example 1, if there had been 8 available drugs, with 4 to be selected, the list of possible selections would have had 70 items. We shall shortly develop a formula for counting combinations.

Practical considerations Ordinarily, we must decide from the nature of the problem whether permutations or combinations are involved. The decision hinges on the answer to the question: *Does order count or doesn't it?* For example, if we are *arranging* 3 books on a shelf, it is natural to regard ABC and ACB as different arrangements, and to take order into consideration; permutations are involved. But if we are *selecting* 3 books for weekend reading, ABC and ACB are regarded as the same selection; order does not count, and combinations are involved. Likewise two men, X and Y, can line up in 2 ways: XY or YX. But these two men can form a committee of two in only one way, because XY and YX yield the same committee. Order counts in a lineup; order does not count in a committee of the usual type, unless it matters which member is chairman.

Subsets of a given set The language of sets can be used in discussing Example 1. We talk about subsets of 3 elements that can be formed from the set

$$\{A, B, C, D\}.$$

For brevity, we sometimes call a subset of 3 elements a 3-subset. Thus we say that the number of 3-subsets in the given 4-set is 4.

2-9 Definition: Combinations

A combination is a selection of objects considered without regard to their order. A subset of r objects selected without regard to their order from a set of n different objects is called a combination of the n objects, taken r at a time. The total number of such combinations is denoted by

$$_nC_r, \quad \text{or by} \quad \binom{n}{r}, \quad \text{where} \quad r \leq n.$$

Alternatively, we say that the number of r-subsets in a given n-set is $_nC_r$, or $\binom{n}{r}$. We must now find out how to evaluate these symbols.

Evaluation of $\binom{n}{r}$, or $_nC_r$ Consider list (1), the list of possible selections of 3 drugs from 4. If the drugs are in different bottles and the bottles are arranged in a row, we get 6 permutations from each of the 4 selections

$$ABC, \quad ABD, \quad ACD, \quad BCD,$$

since each 3-subset can be arranged in 3! ways. This operation yields a total of 4(3!) or 24 permutations, as listed in Table 2–2.

Table 2–2 $_4C_3$ and $_4P_3$ (each combination has $3! = 6$ permutations)

Combinations	Permutations
ABC	$ABC,\ ACB,\ BAC,\ BCA,\ CAB,\ CBA$
ABD	$ABD,\ ADB,\ BAD,\ BDA,\ DAB,\ DBA$
ACD	$ACD,\ ADC,\ CAD,\ CDA,\ DAC,\ DCA$
BCD	$BCD,\ BDC,\ CBD,\ CDB,\ DBC,\ DCB$

It is evident that all 24 permutations of the 4 bottles, taken 3 at a time, are obtained by thus rearranging the combinations. In other words,

$$(\text{number of combinations}) \times 3! = (\text{number of permutations}).$$

Or, in symbols:

$$_4C_3 \times 3! = {}_4P_3,$$

$$\binom{4}{3} \times 3! = 4 \times 3 \times 2.$$

Thus

$$ {}_4C_3 = \binom{4}{3} = \frac{{}_4P_3}{3!} = 4. $$

A generalization of the foregoing reasoning enables us to evaluate ${}_nC_r$ or $\binom{n}{r}$.

2-10 Theorem: Combinations of *n* things, *r* at a time

The number of combinations of a set of n different objects, taken r at a time, is

$$ \boxed{\binom{n}{r} = {}_nC_r = \frac{n!}{r!(n-r)!}\cdot} \qquad (2) $$

Proof Each combination of r objects can be arranged in $r!$ ways, and therefore gives rise to $r!$ permutations. Hence, $r!$ permutations of each of the ${}_nC_r$ combinations yield ${}_nC_r \times r!$ permutations. Moreover, the number ${}_nC_r \times r!$ is the total number of permutations, since each permutation of r objects arises from some combination of r objects. Therefore,

$$ {}_nC_r \times r! = {}_nP_r = \frac{n!}{(n-r)!}\cdot $$

Or, dividing by $r!$, we get

$$ \binom{n}{r} = {}_nC_r = \frac{n!}{r!(n-r)!}\cdot \quad \square \qquad (2) $$

By direct application of formula (2), we obtain

$$ {}_{100}C_2 = \frac{100!}{2!98!} = \frac{100 \times 99}{1 \times 2} = 4950, $$

$$ \binom{n}{1} = \frac{n!}{1!(n-1)!} = n, $$

$$ \binom{n}{n} = \frac{n!}{n!(n-n)!} = \frac{1}{0!} = 1. $$

2-11 Corollary

The number of combinations of n things taken $n - r$ at a time is the same as the number taken r at a time:

$$ \boxed{\binom{n}{n-r} = \frac{n!}{(n-r)!r!} = \binom{n}{r}\cdot} \qquad (3) $$

Proof The denominator of the middle term of Eq. (3) can be rearranged to give $\binom{n}{r}$. $\square$

Discussion That the number of combinations of n objects taken $(n - r)$ at a time is the same as the number taken r at a time is not surprising; for whenever we select r objects from the n, we leave $(n - r)$ objects behind. Thus,

$$\binom{9}{5} = \binom{9}{4} \quad \text{and} \quad \binom{50}{5} = \binom{50}{45}.$$

Example 2 In how many ways can a hand of 13 cards be selected from a standard bridge deck of 52 cards? If a bridge deck is thoroughly shuffled, what are the chances of getting a hand with 13 cards of the same suit?

Solution The number of ways of selecting 13 cards from a deck of 52 different cards is given by formula (2):

$$\binom{52}{13} = \frac{52!}{13!39!} = 635{,}013{,}559{,}600.$$

The sample space of possible hands has 635,013,559,600 sample points. There are 4 suits: spades, hearts, diamonds, and clubs. Assuming that all hands are equally likely, we have:

$$P(\text{13 cards of same suit}) = 4/635{,}013{,}559{,}600 = 1/158{,}753{,}389{,}900.$$

Query A newspaper reported that a bridge player got two perfect hands (13 cards of the same suit) in one evening. What is your comment on this report?

Example 3 In how many ways can a committee of 3 be chosen from 4 married couples (a) if all are equally eligible, (b) if the committee must consist of 2 women and 1 man, (c) if a husband and wife cannot both serve on the same committee? (d) The names of each possible committee of 3 (chosen without restrictions) are written on cards, one committee to a card. The cards are shuffled and a committee is selected by drawing a card at random. Find P(commitee with 2 women and 1 man) and P(committee without a husband and wife).

Solution a) In a committee, order does not count, so the problem is that of selecting 3 people from 8 in all possible ways. From formula (2), the total number is

$$\binom{8}{3} = \frac{8!}{3!5!} = \frac{8 \times 7 \times 6}{1 \times 2 \times 3} = 56.$$

b) The 2 women can be selected in $\binom{4}{2}$ or 6 ways, and after they have been selected in any one of these ways, the 1 man can be selected in $\binom{4}{1}$ or 4 ways. Hence, by the multiplication principle of Section 2–1, the number of ways of selecting 2 women and 1 man is

$$\binom{4}{2} \times \binom{4}{1} = 6 \times 4 = 24.$$

c) If a husband and wife cannot both serve on the committee, then 3 couples must be represented on the committee. Three couples can be selected from 4 in $\binom{4}{3}$ ways. After the 3 couples have been selected, two choices can be made from the first couple (husband or wife), two from the second couple, and two from the third couple. By the multiplication principle, the total number of committees is

$$\binom{4}{3} \times 2 \times 2 \times 2 = 32.$$

Alternatively, there are 4 ways to select a couple and 6 ways to select the remaining member, or, in all, $6 \times 4 = 24$ ways to select a committee with a couple. Subtracting 24 from the total ways, 56, gives 32 committees without a couple. Frequently, counting unwanted cases and taking complements is easier than a direct count.

d) From (a) we see that there are 56 cards. Since each of the 56 possible drawings is equally likely, by part (b) we get:

$$P(\text{committee with 2 women and 1 man}) = 24/56 = 3/7.$$

Using part (c), we get:

$$P(\text{committee without husband and wife}) = 32/56 = 4/7.$$

Example 4 In how many ways can a selection of one or more books be made from 5 identical algebra books and 4 identical geometry books?

Solution Let us first deal with the algebra books. We can select 1 or 2 or 3 or 4 or 5 or none of them. Hence, the algebra books can be dealt with in 6 ways. After dealing with them in any one of these 6 ways, we can similarly deal with the geometry books in 5 ways. By the multiplication principle, we can make a selection from both kinds of books in 6×5, or 30, ways. These 30 ways include the case in which we take no algebra book and no geometry book. If we must take at least one book, then the number of selections is $30 - 1 = 29$.

2-12 Theorem: Pascal's Rule

$$\boxed{\binom{n+1}{r} = \binom{n}{r-1} + \binom{n}{r},} \qquad \text{for} \qquad 1 \le r \le n.$$

Proof The formula may be proved by substituting factorial symbols and simplifying. Another method of proof depends on the meanings of the symbols, as follows.

The number of selections of r objects that can be made from a given set of $n + 1$ objects, without restrictions, is $\binom{n+1}{r}$. Consider some specified object in the given set. If this specified object is included in the selection, the remaining $r - 1$ objects can be selected from the remaining n objects in $\binom{n}{r-1}$ ways. If the specified object is not included, the r objects must be selected from the remaining n objects, and this can be done in $\binom{n}{r}$ ways. The total number of selections is obtained by adding the number in which the specified object occurs

Table 2–3 Pascal's triangle for $\binom{n}{r}$, $0 \leq r \leq n \leq 10$

$$\binom{n+1}{r} = \binom{n}{r-1} + \binom{n}{r}, \qquad 1 \leq r \leq n$$

n \ r	0	1	2	3	4	5	6	7	8	9	10
0	1										
1	1	1									
2	1	2	1								
3	1	3	3	1							
4	1	4	6	4	1						
5	1	5	10	10 +	5	1					
6	1	6	15	20	15	6	1				
7	1	7	21	35	35	21	7	1			
8	1	8	28	56	70	56	28	8	1		
9	1	9	36	84	126	126	84	36	9	1	
10	1	10	45	120	210	252	210	120	45	10	1

to the number in which the specified object does not occur, since no other cases are possible. Therefore, the total number is

$$\binom{n}{r-1} + \binom{n}{r} = \binom{n+1}{r}. \quad \square$$

Note The foregoing proof is an application of the additional principle, not of the multiplication principle. The problem presents us with two operations, either of which is admissible separately, but not both simultaneously. It is a question of *either* this operation *or* that operation. Such operations are *mutually exclusive;* they cannot both occur together. Each operation forms the basis of a separate problem, and the final result is obtained by addition, not by multiplication.

Pascal's Rule gives a simple way of building a table of values of $\binom{n}{r}$, known as *Pascal's triangle.* Table 2–3 shows the part of Pascal's triangle for values of n from 0 through 10. The rows of the table correspond to values of n; the columns, to values of r. The first and last entries in each row are 1 because $\binom{n}{0} = \binom{n}{n} = 1$. The entry other than the first or last in each row is the sum of the entry immediately above it and the entry to the left of that one, by Pascal's Rule. Thus, for example, the little box illustrates

$$\binom{6}{4} = \binom{5}{3} + \binom{5}{4} \qquad \text{or} \qquad 15 = 10 + 5.$$

EXERCISES FOR SECTION 2–3

1. Evaluate the following: $\binom{9}{3}$, $\binom{9}{6}$, $\binom{m}{3}$, $\binom{k}{1}$, $\binom{5}{5}$.
2. Show that $\binom{n}{0} = 1$, and interpret it in terms of selections.

3. Solve the following equations for n:

   a) $\binom{n}{2} = 45;$ b) $\dfrac{{}_nP_4}{\binom{n-1}{3}} = 60;$ c) $\binom{n}{8} = \binom{n}{12}.$

4. In how many ways can a committee of 5 be chosen from 8 people?
5. A contractor needs 4 carpenters and 10 apply for the jobs. In how many ways can he pick out 4?
6. In how many ways can a selection of fruit be made from 7 plums, 4 lemons, and 9 oranges? (Assume that the 7 plums are indistinguishable; likewise for the lemons and for the oranges.)
7. How many selections of 1 or more letters can be made from 2 A's, 5 B's, and 9 C's.
8. Ten points are taken on the circumference of a circle. How many chords can be drawn by joining them in all possible ways? With these 10 points as vertices, how many triangles can be drawn? How many hexagons?
9. In how many ways can a selection of 4 musical records be made from 9? If a certain record must be chosen, in how many ways can the selection be made? In how many ways can it be made if a certain record must be left?
10. A company of 20 men is to be divided into 3 sections so that there are 3 men in the first, 5 in the second, and 12 in the third. In how many ways can this be done? (Don't multiply out.)
11. A sports car club plans to purchase 3 different cars selected from 8 different makes in the same price–performance group. Each club member is to vote for one 3-car selection. To facilitate matters, the ballot committee prepares a list of sixty 3-car selections. The committee then wonders whether the list has omissions or duplications. Give the committee some good advice.
12. Write a symbol for the number of combinations of 20 objects taken 4 at a time, and for the number of combinations of 100 objects taken 98 at a time. Compute the numerical value of each symbol, and find which is the greater.
13. A pack of playing cards contains 52 different cards. If a hand is made up of 5 cards, use the factorial notation to express the number of possible hands. (Disregard order in the hands.)
14. In how many ways can 2 booksellers divide between them 300 copies of one book, 200 copies of another, and 100 copies of a third, if neither bookseller is to get all the books? (Don't multiply out.)
15. A bridge deck of cards is made up of 13 spades, 13 hearts, 13 diamonds, and 13 clubs. How many different hands can be formed if each hand contains 5 spades, 4 hearts, 2 diamonds, and 2 clubs? (Don't multiply out.)
16. Six candidates contest an election for two similar offices. If a voter may mark his ballot either for one or for two candidates, in how many ways can he cast his vote?
17. How many 5-letter words, each consisting of 3 consonants and 2 vowels, can be formed from the letters of the word *equations*?
18. Twenty persons are to travel in a double-decker bus that can carry 12 passengers inside and 8 outside. If 4 of the persons will not travel inside, and 5 will not

travel outside, in how many ways can the passengers be seated (a) if the arrangement of the passengers inside, or outside, is not considered, and (b) if the arrangement inside and outside is considered?

19. In how many ways can 4 persons be selected from 5 married couples (a) if the selection must consist of 2 women and 2 men, and (b) if a husband and wife cannot both be selected?

20. (a) Find the number of ways in which at least one musical record can be selected from 4 identical stereo records and 8 identical monophonic records. (b) What would be the answer to part (a) if the 12 records were all different?

21. Verify that the entries for $n = 4$ in Table 2–3 satisfy the conditions $\binom{4}{0} = \binom{4}{4} = 1$ and $\binom{4}{r} = \binom{3}{r} + \binom{3}{r-1}$ for $1 \leq r \leq 3$.

22. Write out the entries that would be in the rows for $n = 11$ and $n = 12$ in Pascal's triangle, Table 2–3.

EXERCISES ON PERMUTATIONS AND COMBINATIONS

PART A

1. In how many ways can a man choose 3 gifts from 10 different articles?

2. A railway has 50 stations. If the names of the point of departure and the destination are printed on each ticket, how many different kinds of single tickets must be printed? How many kinds are needed if each ticket may be used in either direction between two towns?

3. In how many ways can 15 different objects be divided among A, B, and C, if A must receive 2 objects, B must receive 3 objects, and C must receive 10 objects?

4. Given 20 points, no three of which are in a straight line, find the number of straight lines that can be drawn by joining pairs of these points.

5. Given 4 noncoplanar points in space, how many planes can be determined by selecting triples of these points?

6. A ring of 8 boys is to be enlarged by the addition of 5 girls. In how many ways can this be done if no two girls are to stand beside each other? (Note that order counts here because people are distinguishable.)

7. A town council is made up of a mayor and 6 aldermen. How many different committees of 4 can be formed (a) if the mayor is on each committee, and (b) if the mayor is on no committee?

8. How many 4-letter words can be made from the letters of the word *zephyr*? How many of these words will not contain the letter r? How many will contain r? How many will begin with z and end with r?

9. In how many ways can a coach choose a team of 5 from 10 boys (a) if 2 specified boys must be included, and (b) if there are no restrictions?

10. A man has 8 different pairs of gloves. In how many ways can he select a right-hand glove and a left-hand glove that do not match?

11. A 35-mm colored slide is mounted in a $2'' \times 2''$ square holder. How many wrong ways are there of inserting the mounted slide into a projector?

PART B

12. In how many integers between 1000 and 9999 inclusive does the digit 3 occur?
13. How many words, each of 2 vowels and 2 consonants, can be formed from the letters of the word *involute*?
14. How many quadrilaterals can be formed, each having as its vertices 4 of the vertices of a given regular polygon of 20 sides, if no 2 of the selected 4 are opposite vertices of the given polygon?
15. Prove $\binom{n}{r} + \binom{n}{r-1} = \binom{n+1}{r}$ (Pascal's Rule) by using factorials to replace the symbols on the left, and simplifying.
16. There are 10 chairs in a row. In how many ways can 2 persons be seated? In how many of these ways will the 2 persons be sitting in adjacent chairs? In how many will they have at least one chair between them?
17. How many diagonals has a 20-sided polygon? How many sides has a polygon with 35 diagonals?
18. Four jobs of one kind can be held by women only, 5 jobs of another kind by men only, and 3 jobs of a third kind by either men or women. In how many ways can these jobs be filled from 18 applicants of whom 8 are women and 10 are men?
19. A football team has 5 basic plays for use on the first 3 downs. How many different sequences of 3 plays are available (a) if no play may be used twice in a sequence, (b) if there are no restrictions?
20. The Dates-by-Computer Company uses a preference questionnaire with 5 questions. The 5 questions have, respectively, 2, 3, 3, 4, and 5 mutually exclusive answers that lead to a description of the desired date. What is the minimum number of individuals needed if at least 3 men and 3 women are to be available for each possible description?
21. Fifteen points lie in a plane in such a way that 5 of the points are on one straight line and, apart from these, no 3 points are collinear. Find the total number of straight lines that can be obtained by joining pairs of the 15 points.
22. How many 6-digit numbers can be formed from the digits 1, 2, 3, 4, 5, 6, 7, 8, 9 if each number has 3 odd and 3 even digits and no two digits are alike?
23. In how many ways can one assign to 2 soldiers different 3-digit numbers? In how many ways can this be done if the 3-digit numbers are composed of even digits only (zero being considered even)?
24. In a set of 10 examinations, 2 are in mathematics. In how many different orders can the examinations be given if those in mathematics are not consecutive?
25. A stamp collector has 8 different Canadian stamps and 10 different United States stamps. Find the number of ways in which he can select 3 Canadian stamps and 3 United States stamps and arrange them in 6 numbered spaces in his album.
26. A symphony is recorded on 4 discs, both sides of each disc being used. In how many ways can the 8 sides be played on a phonograph so that some part of the symphony is played out of its correct order?
27. A railway coach has 10 seats facing backward and 10 facing forward. In how many ways can 8 passengers be seated, if 2 refuse to ride facing forward and 3 refuse to ride facing backward?

28. From a company of 20 soldiers, a squad of 3 men is chosen each night. For how many consecutive nights could a squad go on duty without two of the squads being identical? In how many of these squads would a given soldier serve?
29. Find the number of ways in which 8 persons can be assigned to 2 different rooms, if each room must have at least 3 persons in it.
30. Three Lexington mothers organize a car pool to take their children to a private school in Boston in the morning and bring them back in the afternoon. The pool operates 9 times each week, Monday through Friday, with no trip Friday afternoon. Each mother drives 3 times a week, and gets a different 3-trip assignment each week. Will they exhaust all possible trip assignments in the 13 years of schooling before college? (Assume 36 weeks of school per year.)

2-4 PERMUTATIONS OF THINGS THAT ARE NOT ALL DIFFERENT

In Sections 2-1 through 2-3, we considered arrangements of sets of objects that were different from each other. How will the number of possible permutations be affected if some objects in the given set are alike? . . . A little thought will doubtless convince you that if some objects in a set cannot be distinguished from others, the number of permutations is decreased. For example, how many different numbers can be formed by rearranging the digits 33332 among themselves? There are only 5 possibilities:

33332, 33323, 33233, 32333, 23333.

If the given 5 digits had been all different, we should have obtained 5!, or 120, arrangements.

Although the presence of indistinguishable objects appears to decrease the number of arrangements, the listing procedure once again has serious limitations. If you try to list all arrangements of the digits 3333332222, the novelty soon wears off because there are 210 possibilities. In Example 1, we examine a special case in order to discover a new, general method for counting permutations of objects that are not all different.

Example 1 In how many ways can the letters of the word *assess* be arranged, all at a time?

Solution The problem would be easy if the four s's were different from one another, for we know that there are 6! permutations of 6 *different* letters taken all together. We shall relate this familiar problem (letters all different) to our new problem (letters not all different) by making the four s's *temporarily* different, as described in the following.

Let the unknown total number of permutations of the letters of the word *assess* be x. Now consider any one of these permutations; for example,

$$s \quad s \quad s \quad s \quad a \quad e.$$

In this arrangement, if we replace the four s's by

$$s_1, \ s_2, \ s_3, \ s_4,$$

the original arrangement gives rise to $4!$ arrangements by permuting the four s's with subscripts (now different) without disturbing the other letters. In the same way, each of the original x permutations gives rise to $4!$ permutations. Thus the total number of permutations is $x(4!)$. Since the 6 letters

$$s_1, \ s_2, \ s_3, \ s_4, \ a, \ e$$

are now all different, $x(4!)$ is the number of permutations of 6 different letters, taken all together. Therefore,

$$x(4!) = 6!$$

or

$$x = \frac{6!}{4!}.$$

Recall that a similar type of reasoning was used to evaluate $\binom{n}{r}$. We can at once generalize this reasoning to show that the number of permutations of a set of n objects, taken all together, where r of the objects are alike and the rest are different, is $n!/r!$. Repeated applications of this principle yield the following theorem.

2–13 Theorem: Permutations of objects that are not all different

Given a set of n objects having n_1 elements alike of one kind, and n_2 elements alike of another kind, and n_3 elements alike of a third kind, and so on for k kinds of objects; then the number of permutations of the n objects, taken all together, is

$$\boxed{\frac{n!}{n_1!\, n_2! \cdots n_k!},} \tag{1}$$

where

$$n_1 + n_2 + \cdots + n_k = n.$$

2–14 Corollary: Permutations for two kinds of objects

If a set of n objects consists of r elements of one kind and $n - r$ elements of another, then the number of permutations of the n objects, taken all together, is

$$\boxed{\frac{n!}{r!(n-r)!} = \binom{n}{r} = \binom{n}{n-r}.} \tag{2}$$

Proof a) The proof follows at once from Theorem 2–13, when we set $n_1 = r$ and $n_2 = n - r$. Alternatively, this corollary can be proved as follows.

b) Suppose there are r A's and $n - r$ B's to be arranged in order. We think of n blank spaces to be filled, r with A's, and the rest with B's. The number of ways of selecting the r spaces for the A's is $\binom{n}{r}$ and, after this has been done, the A's can be arranged in the spaces in just one way. Next, the B's can be arranged in the remaining $n - r$ spaces in just $\binom{n-r}{n-r} \times 1$, or 1, way. Hence, the total number of arrangements is $\binom{n}{r}$. □

Remark The number of *permutations* of n objects, r alike of one kind and $n - r$ alike of another kind, is equal to the number of *combinations* of n different objects (the n blank spaces), taken r at a time. The foregoing proof shows why this particular equivalence between permutations and combinations occurs.

Example 2 How many arrangements can be made of the letters of the word *Mississippi*, taken all together?

Solution We have 11 letters in all, with one m, four i's, four s's, and two p's: $\{m\ iiii\ ssss\ pp\}$. Thus, $n_1 = 1$, $n_2 = 4$, $n_3 = 4$, and $n_4 = 2$. By Theorem 2–13, the total number of permutations of the 11 letters, taken all together, is

$$\frac{11!}{1!4!4!2!} = 34{,}650.$$

Example 3 a) A museum director wishes to arrange in a row 7 paintings, 3 by Cézanne and 4 by Picasso. The Picasso paintings are dated, and the director wishes them to appear in chronological order from left to right. In how many ways can the 7 paintings be arranged?

b) If the 7 paintings are hung at random what is the probability that the director's wishes will be met?

Solution a) Since the order of the Picasso paintings relative to one another cannot be changed, it follows that these paintings cannot be permuted among themselves, and so for the purposes of this problem they may be considered as identical. Hence, the problem is that of finding the number of permutations of 7 objects, taken all together, where 4 of the objects are identical. By Theorem 2–13, the number is

$$\frac{7!}{4!(1!)^3} = 210.$$

b) There are 7!, or 5040, possible arrangements in which the 7 paintings can be hung. Of these, 210 meet the director's wishes. Therefore, the probability that the paintings are in the required order is $210/5040 = 1/24$.

Example 4 *Paths in a lattice.* The streets of one section of Square City form a rectangular grid, or *lattice*, 6 blocks long and 4 blocks wide, as shown in Fig. 2–4. A pedestrian walks from O to A, always walking either north or east. (a) How many different paths are available? (b) If all paths are equally likely, what is the probability that he walks through B on his way to A?

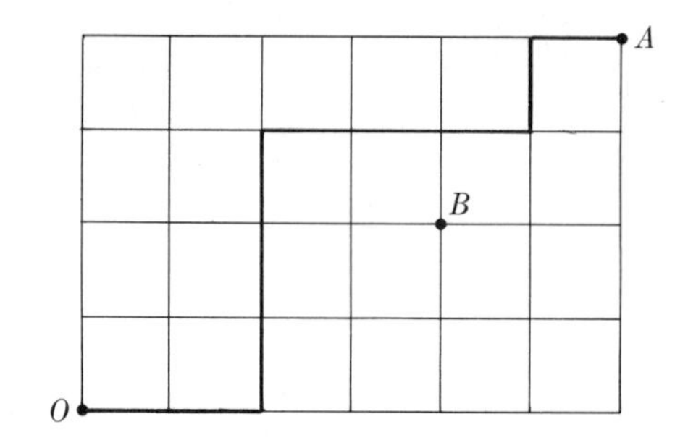

Fig. 2-4 A lattice path ($EENNNEEENE$).

Solution a) Each admissible path involves a walk of 6 blocks east and a walk of 4 blocks north. It follows that there is a 1–1 correspondence between the possible paths and the possible arrangements of the 10 letters $EEEEEENNNN$, taken all together, where E denotes a walk of 1 block east and N denotes a walk of 1 block north.

Hence, by Corollary 2–14, the number of possible paths is

$$\binom{10}{6} = 210.$$

b) Similarly, the number of paths from O to B is $\binom{6}{4}$, and the number of paths from B to A is $\binom{4}{2}$. Hence, by the multiplication principle, the number of paths from O to A by way of B is $\binom{6}{4} \cdot \binom{4}{2}$.

$$P(\text{path goes through } B) = \frac{\binom{6}{4}\binom{4}{2}}{\binom{10}{6}} = \frac{(15)(6)}{210} = \frac{3}{7}.$$

Comment The method exemplified in Example 4 is readily generalized to show that a walk from one corner of a lattice to another corner m blocks east and n blocks north can be taken in $\binom{m+n}{m}$ ways.

Example 5 *Stars and bars.* In how many ways can 10 identical objects be placed in 4 different boxes (a) if at least one object must go into each box, (b) if one or two or three of the boxes may be left empty?

Solution a) Imagine the 10 identical objects denoted by stars and arranged in a line, thus:

$$* \; * \; * \; * \; * \; * \; * \; * \; * \; *$$

We can divide the objects into 4 groups, at least one for each box, by inserting dividing bars among them. For example:

$$* \; * \; | \; * \; | \; * \; * \; * \; * \; | \; * \; * \; *$$

Among the 10 objects there are 9 spaces, 3 of which must be selected as positions for the dividing bars. Hence, we can mark off 4 groups, each of which contains at least one object, in $\binom{9}{3}$, or 84, ways.

b) If one or more boxes may be empty, the required number of ways equals the number of ways of placing 10 + 4, or 14, identical objects in 4 different boxes so that each box gets at least one object. Here is the reason: When the 14 identical objects are placed in the 4 boxes with no box empty, we can remove one object from each box, which leaves us with a placing of 10 objects in the 4 boxes, with the possibility that one or two or three boxes may be left empty. Moreover, all possible placings of the required type can be obtained in this way.

Because of the foregoing fact and the result of part (a), the required number of placings is $\binom{13}{3}$, or 286.

Homogeneous products Products such as x^2y^3z, xyz^4, and $x^2y^2z^2$ are said to be homogeneous because they are of the same degree, namely 6. How many homogeneous products of degree 6 can be formed with the variables x, y, and z?

The method of Example 5(b) can be used to answer such questions. In fact, the result of part (b) is the number of homogeneous products of degree 10 that can be formed with 4 variables. For example, consider

$$x^a y^b z^c w^d. \tag{3}$$

If, for any placing, a, b, c, and d are respectively the numbers of objects in the first, second, third, and fourth boxes, then

$$a + b + c + d = 10 \tag{4}$$

and expression (3) becomes a product of degree 10. Moreover, the total number of placings, $\binom{13}{3}$, is the total number of admissible sets of values of a, b, c, and d for which Eq. (4) holds. Since each set of values of a, b, c, and d for which Eq. (4) holds gives a product of degree 10, the total number of homogeneous products of degree 10 obtainable with the 4 variables x, y, z and w is $\binom{13}{3}$.

By generalizing the reasoning in part (b) of Example 5, we can show that the number of homogeneous products of degree n that can be obtained from k different variables is $\binom{n+k-1}{k-1}$. Details of the proof are left as an exercise. As a check on the formula, let us find the number of homogeneous products of degree 3 obtainable from the 2 variables x and y. Here $n = 3$ and $k = 2$, and the required number is $\binom{3+2-1}{2-1}$, or 4. Since the products are

$$x^3, \quad x^2y, \quad xy^2, \quad y^3,$$

the formula checks.

Example 6 Given $n + r$ letters, of which n are A's and r are B's, how many different sequences can be formed from the A's and B's, if each sequence must contain all n A's?

Solution There are $r + 1$ mutually exclusive cases because we may have n A's and no B's, or n A's and one B, or n A's and two B's, and so on. The $r + 1$

Table 2-4

Mutually exclusive cases	Number of sequences
n A's, no B's	$\binom{n}{0}$
n A's, one B	$\binom{n+1}{1}$
n A's, two B's	$\binom{n+2}{2}$
n A's, three B's	$\binom{n+3}{3}$
$\vdots$	$\vdots$
n A's, r B's	$\binom{n+r}{r}$

alternative cases are listed in Table 2-4, together with the number of sequences to which each case gives rise. For each case, the number of sequences is calculated by formula (2) of Corollary 2-14.

Since the cases are mutually exclusive, we get the total number of sequences by applying the addition principle. The number is

$$\binom{n}{0} + \binom{n+1}{1} + \binom{n+2}{2} + \cdots + \binom{n+r}{r}. \quad \square$$

Remark This sum equals $\binom{n+r+1}{r}$, as can be shown by successive applications of Pascal's Rule (Theorem 2-12, Section 2-3). We have

$$\binom{n}{0} + \binom{n+1}{1} = \binom{n+1}{0} + \binom{n+1}{1} = \binom{n+2}{1};$$

$$\binom{n+2}{1} + \binom{n+2}{2} = \binom{n+3}{2};$$

$$\binom{n+3}{2} + \binom{n+3}{3} = \binom{n+4}{3};$$

and so on. Finally, the rth step gives

$$\binom{n+r}{r-1} + \binom{n+r}{r} = \binom{n+r+1}{r}. \quad \square$$

EXERCISES FOR SECTION 2-4

1. Find the number of arrangements of the letters of the word *committee*, using all the letters in each arrangement.
2. How many different numbers can be obtained by arranging the digits 2233344455, all together, in all possible ways?
3. How many permutations can be made using the letters of the word *institution*, taken all at a time? How many of these begin with t and end with s?

4. In how many ways can 13 different cards be arranged in a row, if a certain 10 of them must always be in a specified order relative to each other?
5. Find the number of ways in which 6 plus signs and 4 minus signs can be arranged in a row.
6. Find the number of ways in which nine 3's and six 5's can be placed in a row so that no two 5's come together.
7. How many different numbers can be obtained by arranging the digits 123456789, all at a time, if the even digits must always remain in ascending order and the odd digits likewise?
8. Find the number of arrangements of the letters of the word *engineering*, taken all together. In how many of these are three *e*'s together? In how many are exactly two *e*'s together?
9. (Continuation of Exercise 8.) The word *engineering* is spelled out on a line of wooden blocks, one letter to each block. The blocks are thoroughly mixed and then placed at random in a row. Find (a) P(three *e*'s together), (b) P(exactly two *e*'s together).
10. A class consists of 12 girls and 10 boys. In how many ways can the class form a line, if the girls always remain in ascending order of height, and the boys likewise?
11. 5 girls and 3 boys line up at random to buy tickets for a basketball game. What are the chances that the girls are lined up in ascending order of height, and the boys likewise?
12. In Square City how many ways can one take a walk for 9 blocks, if he always walks 5 blocks west and 4 blocks north?
13. How many numbers greater than 3,000,000 can be formed from the digits 1, 1, 1, 2, 2, 3, 3?
14. In how many ways can 5 red balls, 4 black balls, and 4 white balls be placed in a row so that the balls at the ends of the row are of the same color?
15. Three identical sedans, 2 identical sports cars, and 1 station wagon are parked at random in a row. Find P(identical cars at ends of row).
16. (Refer to Example 4.) A pedestrian walks from A to O, always walking south or west. (a) How many different paths are available? (b) If all paths are equally likely, what is the probability that he walks through corner B on his way to O?

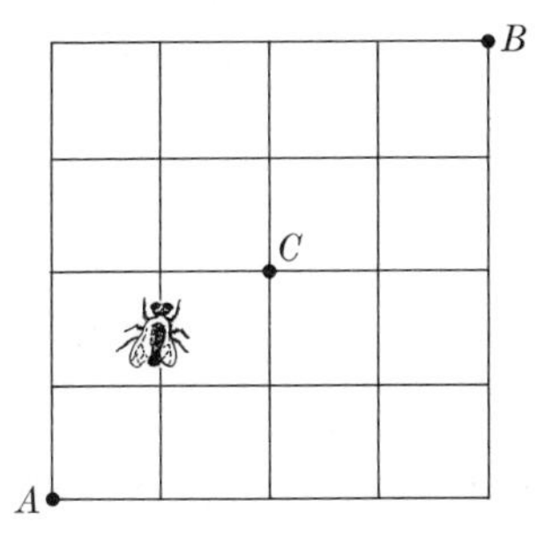

17. On the square wire lattice shown above, a fly goes from A to B, always walking to the right or upwards. If all paths are equally likely, find P(fly passes through C).

18. In how many ways can 10 similar flags be arranged on a row of 6 flagpoles, if at least 1 flag must appear on each pole?

19. In how many ways can 20 one-dollar bank notes be divided among 3 people (a) if each person gets at least \$1, (b) if one or two of the persons may get nothing.

20. How many homogeneous products of degree 15 can be formed with 5 variables?

21. The streets of a city consist of m streets running north and south and n streets running east and west. In how many ways can a man walk from the northeast to the southwest corner of the city, if he always walks the least possible number of blocks?

22. In how many ways can 9 identical cars be assigned to 3 dealers (a) if each dealer must get at least 1 car, (b) if one or two dealers may get no cars?

23. In how many ways can 5 different cars be assigned to 3 dealers, if each dealer must get at least 1 car?

24. Suppose that m places are to be filled from n candidates, where $n \geq m$. (a) How many ways are there of selecting m candidates from among the n available? (b) In how many of the different ways of part (a) is a particular candidate A included in the selection? (c) Assuming all selections of m from among the n are equally likely, what is the probability that candidate A is included? Discuss your result. Does it seem to be reasonable?

25. Use the method ot stars and bars to show that the number of homogeneous products of degree n that can be obtained from k variables is $\binom{n+k-1}{k-1}$.

Exercises 26, 27, and 28 involve the ideas of Example 6, page 63.

26. A ballot box contains $n + 1$ votes for candidate N and r votes for candidate R. Ballots are drawn from the box, one at a time, and the drawing stops as soon as all of N's ballots are removed. How many different orders of drawing are there? [*Note:* The last ballot drawn must show a vote for candidate N.]

27. The streets of Newville are laid out in a grid, running east–west and north–south. A man starts at the center of town and walks a block east or a block north, continuing in this manner at each intersection. He stops as soon as he gets $n + 1$ blocks north of his starting point. Find the number of ways he can get $n + 1$ blocks north before he goes $r + 1$ blocks east.

28. If the man of Exercise 27 walks until he reaches either $n + 1$ blocks north or $r + 1$ blocks east, and if all paths are equally likely, what is the probability that he gets $n + 1$ blocks north first?

REVIEW EXERCISES

1. A rat runs a branching maze, so constructed that he first must choose one of a pair of doors, beyond each of these he must choose one of 3 doors, and beyond each of these he must choose one of 4 doors. After passing through a door he cannot return. How many paths are there from start to finish?

2. A metallurgist, studying alloys, wants to study the effect of 3 different temperatures, 6 different heating times, and 4 different amounts of a copper compound. One experiment has one level for each variable. How many different experiments

must he perform if every triple of temperature level, heating time, and amount of copper is to be represented?

3. A dial safe has 100 positions on its dial and 3 settings are required for a combination. However, no setting can be fewer than 10 positions from the immediately preceding setting. How many combinations are there?

4. A key maker has 12 types of blanks. Each blank has 5 different positions where metal can be removed and there are 3 cutting depths at each position except the first, which has only 2. How many possible keys are there?

5. A soil chemist has 6 different treatments to study, and he can apply 3 different treatments, simultaneously, in a single experiment. How many experiments must he do to exhaust all triples of treatments?

6. If the soil chemist in the preceding exercise cannot have triples in which treatments A and B appear simultaneously, how many experiments are there?

7. A computing machine is used for the study of problem solving. It has 10 different steps it can use, and it does not use one it has previously used in the same attempt to solve a problem. The problems it solves require 4 different steps taken in the correct order. What is the largest number of attempts the computer may have to make before it solves a given problem?

8. An experimenter studying problem solving has designed a problem whose correct solution requires 6 steps taken in order. There are two steps of type A, two of type B, and two of type C. He has made a list of the possible orders,

$$AABBCC, \qquad AABCBC, \qquad \ldots, \qquad CCBBAA,$$

and he has 88 such orders in his list. Has he found them all?

9. In the World Series, the American League team, A, and the National League team, N, play until one team wins 4 games. If the sequence of winners is designated by letters ($NAAAA$ means National League won the first game and lost the next 4), how many different sequences of winners are possible?

2-5 THE BINOMIAL THEOREM

Expansions of positive, integral powers of the binomial $(a + x)$, such as

$$(a + x)^2 = a^2 + 2ax + x^2,$$
$$(a + x)^3 = a^3 + 3a^2x + 3ax^2 + x^3,$$

are of frequent use in algebra. Moreover, expansions of this kind are important for our future studies in this book, and are related to results already obtained in this chapter. We want to find a law or formula that readily produces such expansions.

Of course, we can always obtain them by ordinary multiplication. But the process soon becomes laborious. After we have shown by multiplication that

$$(a + x)^4 = a^4 + 4a^3x + 6a^2x^2 + 4ax^3 + x^4,$$

we may well begin to wonder if there isn't some better method of getting the result. If we study the expansions for $(a + x)^2$, $(a + x)^3$, and $(a + x)^4$, we soon note that part of the solution is easily guessed. The terms in the expansions, *apart from their coefficients*, can be written by inspection. For example, the expansion of $(a + x)^5$ has $5 + 1$, or 6, terms; without their coefficients, the terms are

$$a^5, \quad a^4x, \quad a^3x^2, \quad a^2x^3, \quad ax^4, \quad x^5.$$

Note that each of these terms is the product of 5 factors, where each factor is a or x. We say that such terms are of *degree* 5 in a and x. (In general, the *degree* of a term in a and x equals the number of factors a or x that the term contains.)

Proceeding along these lines, we see that the expansion of $(a + x)^n$ has $n + 1$ terms; without their coefficients, the terms are

$$a^n, \quad a^{n-1}x, \quad a^{n-2}x^2, \quad \ldots, \quad a^{n-r}x^r, \quad \ldots, \quad x^n,$$

where each term is of degree n in a and x. But there remains the question: How do we find the coefficients of these terms? Why, in the expansion of $(a + x)^4$, do we have $4a^3x$ and $6a^2x^2$? To answer such questions, let us examine the multiplication process.

Products of distinct binomials The use of subscripts may help to illuminate what is going on when we multiply binomials. Consider the following products:

$$(a_1 + x_1)(a_2 + x_2) = a_1a_2 + a_1x_2 + x_1a_2 + x_1x_2;$$

$$(a_1 + x_1)(a_2 + x_2)(a_3 + x_3)$$
$$= a_1a_2a_3 + \boxed{a_1a_2x_3} + \boxed{a_1x_2a_3} + \boxed{x_1a_2a_3}$$
$$+ a_1x_2x_3 + x_1a_2x_3 + x_1x_2a_3 + x_1x_2x_3.$$

The foregoing expansions illustrate three principles of multiplication:

1) the *degree*, in a and x, of each term in a product equals the number of factors multiplied;
2) the terms are obtained by *selecting exactly one* letter from each of the factors (note the subscripts on the right-hand sides);
3) the expansion consists of the sum of such terms *obtained in all possible ways.*

In these products of distinct binomials, if we drop subscripts and collect like terms, we obtain the expansions of $(a + x)^2$ and $(a + x)^3$, respectively. Note that principles (2) and (3) are the keys to the discovery of the coefficients. If, for example, we can select two a's and one x in 3 different ways, as shown by the boxes, then the term a^2x will occur 3 times and so have the coefficient 3 in the expansion.

Let us apply the three principles above to obtain the expansion of $(a + x)^4$. Since

$$(a + x)^4 = (a + x)(a + x)(a + x)(a + x),$$

the terms of the expansion are of degree 4 in a and x. The possibilities are listed once more for reference:

$$a^4, \qquad a^3x, \qquad a^2x^2, \qquad ax^3, \qquad x^4.$$

Each term is obtained by selecting exactly one letter from each of the 4 factors. To get the first term, a^4, we select no x's and four a's from the 4 factors. Because this can be done in $\binom{4}{0}$ or 1 way, a^4 occurs only once in the expansion, and its coefficient is 1. To get the second term, a^3x, we select an x from one of the factors and three a's from the remaining three factors. This can be done in $\binom{4}{1}$ or 4 ways. Thus the term a^3x occurs 4 times and has the coefficient 4. Similarly, the term a^2x^2 is obtained in $\binom{4}{2}$ or 6 ways; the term ax^3 in $\binom{4}{3}$ or 4 ways; and the term x^4 in $\binom{4}{4}$ or 1 way. These last three terms therefore have the coefficients 6, 4, and 1, respectively. The complete expansion is the *sum* of all these terms:

$$\begin{aligned}(a + x)^4 &= \binom{4}{0} a^4 + \binom{4}{1} a^3x + \binom{4}{2} a^2x^2 + \binom{4}{3} ax^3 + \binom{4}{4} x^4 \\ &= a^4 + 4a^3x + 6a^2x^2 + 4ax^3 + x^4.\end{aligned}$$

We now proceed to generalize the foregoing reasoning.

2–15 Theorem: The binomial theorem

If n is a positive integer, then

$$\begin{aligned}(a + x)^n = \binom{n}{0} a^n + \binom{n}{1} a^{n-1}x + \binom{n}{2} a^{n-2}x^2 + \cdots \\ + \binom{n}{r} a^{n-r}x^r + \cdots + \binom{n}{n} x^n.\end{aligned} \tag{1}$$

Proof Each term in the expansion of $(a + x)^n$ is of degree n in the variables a and x. Thus, if we ignore the coefficients, each term has the general pattern

$$a^{n-r}x^r, \qquad \text{when} \qquad r = 0, 1, 2, 3, \ldots, n.$$

The term $a^{n-r}x^r$ is obtained by selecting x from r of the factors and a from the remaining $n - r$ factors. This selection can be made in $\binom{n}{r}$ ways. (Cf. Theorem 2–10, Section 2–3.) Hence, the term $a^{n-r}x^r$ occurs $\binom{n}{r}$ times and its coefficient is $\binom{n}{r}$. Therefore, the complete general term is

$$\binom{n}{r} a^{n-r}x^r,$$

and the expansion is that shown in (1). In summation notation (see Appendix 1), this expansion may be written

$$(a+x)^n = \sum_{r=0}^{n} \binom{n}{r} a^{n-r}x^r. \quad \square$$

The coefficients $\binom{n}{r}$ are often called *binomial coefficients.*

Binomial formula with expanded coefficients Since

$$\binom{n}{0} = 1, \quad \binom{n}{1} = n,$$

$$\binom{n}{2} = \frac{n(n-1)}{2!},$$

$$\binom{n}{3} = \frac{n(n-1)(n-2)}{3!}, \qquad \text{etc.,}$$

the binomial expansion (1) may also be written as

$$\boxed{(a+x)^n = a^n + \frac{n}{1!}a^{n-1}x + \frac{n(n-1)}{2!}a^{n-2}x^2 + \cdots + x^n.} \tag{2}$$

Example 1 Expand $(2+x)^4$.

Solution Use formula (2), to obtain

$$\begin{aligned}(2+x)^4 &= 2^4 + \frac{4}{1}(2^3)x + \frac{4\cdot 3}{1\cdot 2}(2^2)x^2 + \frac{4\cdot 3\cdot 2}{1\cdot 2\cdot 3}(2)x^3 + \frac{4\cdot 3\cdot 2\cdot 1}{1\cdot 2\cdot 3\cdot 4}x^4\\ &= 16 + 32x + 24x^2 + 8x^3 + x^4.\end{aligned}$$

Example 2 Expand $(1-2x^2)^5$.

Solution From formula (2), we obtain

$$\begin{aligned}(1-2x^2)^5 &= (1+(-2x^2))^5\\ &= 1 + \frac{5}{1}(-2x^2) + \frac{5\cdot 4}{1\cdot 2}(-2x^2)^2 + \frac{5\cdot 4\cdot 3}{1\cdot 2\cdot 3}(-2x^2)^3\\ &\quad + \frac{5\cdot 4\cdot 3\cdot 2}{1\cdot 2\cdot 3\cdot 4}(-2x^2)^4 + (-2x^2)^5\\ &= 1 - 10x^2 + 40x^4 - 80x^6 + 80x^8 - 32x^{10}.\end{aligned}$$

Note The signs of the terms in this expansion alternate because of the negative sign of the second term of the binomial $(1-2x^2)$. For the same reason, the signs in the formal expansion of $(a-x)^n$ alternate.

Example 3 Prove that $\binom{n}{0} + \binom{n}{1} + \binom{n}{2} + \cdots + \binom{n}{n} = 2^n$.

Solution Since formula (1) is valid for all values of a and x, we may set $a = x = 1$. This gives

$$(1+1)^n = \binom{n}{0} + \binom{n}{1} + \binom{n}{2} + \cdots + \binom{n}{n},$$

which proves the proposition. □

Example 4 If nx is near zero, prove that

$$(1+x)^n \approx 1 + nx.$$

Solution The proof follows at once from formula (2) if we set $a = 1$ and ignore terms with x^2, x^3, and higher powers. These neglected terms are

$$\frac{n(n-1)}{2!}x^2, \qquad \frac{n(n-1)(n-2)}{3!}x^3, \qquad \text{and so on.}$$

Their absolute values are respectively less than the absolute values of

$$\frac{n^2x^2}{2!}, \qquad \frac{n^3x^3}{3!}, \qquad \text{and so on,}$$

because

$$n(n-1) < n^2, \qquad n(n-1)(n-2) < n^3, \qquad \text{and so on.}$$

If nx is near zero, these higher powers of nx are small compared with the first power. Therefore, we have

$$\boxed{(1+x)^n \approx 1 + nx.} \quad \square \tag{3}$$

In Exercise 42 of this section, you are asked to show that when $|nx| < 1$ the approximation (3) is in error by an amount that is less than $(nx)^2$. This estimate of the error shows that the nearer nx is to zero, the better the approximation. For example, consider

$$(1 + 0.003)^{20}.$$

Since $x = 0.003$ and $n = 20$, it follows that

$$(nx)^2 = (0.06)^2 = 0.0036.$$

Consequently, formula (3) gives a two-decimal approximation for 1.003^{20}:

$$1 + 20(0.003) = 1.06.$$

For nx near zero, the formula

$$(1+x)^n \approx 1 + nx$$

holds for nonintegral values of n as well as for integral values. For example,

$$(1 + x)^{1/2} \approx 1 + \tfrac{1}{2}x, \qquad \text{if } x \text{ is near zero.} \tag{4}$$

Example 5 Find $\sqrt{4.02}$, approximately.

Solution Transforming the radical and using (4), we get

$$\sqrt{4.02} = 2\sqrt{1.005} = 2(1 + 0.005)^{1/2} \approx 2(1 + 0.0025) = 2.005.$$

The negative binomial If we attempt to use the binomial formula (1) to expand $(1 - x)^{-n}$, where n is a positive integer, we get terms involving symbols such as $\binom{-n}{r}$. These symbols have no meaning for us, since they have not been defined. However, if we substitute in binomial formula (2), each term has a meaning and we get

$$(1 - x)^{-n} = 1 + \frac{-n}{1!}(-x) + \frac{-n(-n-1)}{2!}(-x)^2 + \frac{-n(-n-1)(-n-2)}{3!}(-x)^3 + \cdots,$$

or

$$(1 - x)^{-n} = 1 + \frac{n}{1!}x + \frac{n(n+1)}{2!}x^2 + \frac{n(n+1)(n+2)}{3!}x^3 + \cdots. \tag{5}$$

If $x \neq 0$, the series on the right-hand side of formula (5) has infinitely many terms, and is called an infinite series.

Two questions arise:

1. Does this infinite series have a sum?
2. If the infinite series has a sum, does this sum equal $(1 - x)^{-n}$?

We cannot deal with this problem in general at this time. We shall merely exhibit a particular case that may serve to make the facts plausible.

If $n = 1$, substitution in formula (5) gives

$$(1 - x)^{-1} = 1 + x + x^2 + x^3 + \cdots.$$

The right side is a geometric progression and, if $|x| < 1$, we have

$$1 + x + x^2 + x^3 + \cdots = \frac{1}{1 - x} = (1 - x)^{-1}.$$

Hence, in this particular case, formula (5) is valid if $|x| < 1$.

You are asked to accept without proof the following fact: If $|x| < 1$, then (5) is a valid formula for all real values of n. Formula (5) is called the *negative binomial.*

Example 6 If $|x| < 1$, find an expansion for $(1 - x)^{-2}$.

Solution Since $|x| < 1$, we use formula (5) and get

$$(1 - x)^{-2} = 1 + 2x + 3x^2 + 4x^3 + \cdots.$$

EXERCISES FOR SECTION 2–5

1. In the expansion of $(p + q)^7$, what is the degree of each term in p and q? Write, in ascending powers of q, the terms of this expansion without their coefficients. What is the general form of these partial terms?
2. How many terms are there in the expansion of $(m + n)^{100}$? Write the first three terms of this expansion, without their coefficients, in ascending powers of n.
3. In the expansion in ascending powers of x of $(1 + x)^{1000}$, what is (a) the 200th term? (b) the coefficient of the 375th term? (c) the coefficient of x^{625}?
4. In the expansion of $(a + x)^{100}$ in ascending powers of x, write, in unsimplified form, the 50th term. What is the 20th term? What is the term that contains x^{60}?
5. Expand $(q + p)^4$. If $p = \frac{1}{2}$ and $q = \frac{1}{2}$, find the values of the terms.

Use the binomial formula to expand the following:

6. a) $(1 + b)^5$; b) 1.01^5
7. a) $(1 - b)^5$; b) 0.98^5
8. $(1 + p)^7$
9. $(1 - 3a)^4$
10. $(1 - x^2)^5$
11. $(2 + 4m)^4$
12. $(x + y)^6$
13. $(3 - 6c)^5$
14. $(\frac{1}{2}a + 1)^4$
15. $(p + q)^7$
16. $(a + ax)^6$
17. $(1 - x^3)^5$
18. $(x^2 - x^3)^6$
19. $(x - y)^5$

Expand each of the following to 3 terms:

20. $(p + q)^{50}$
21. $(x - y)^{100}$
22. $(1 - a^2)^{40}$
23. $(2x - 3y)^8$

Use the binomial formula to find approximations for the following:

24. 1.002^{10}
25. 0.997^{20}
26. 1.004^5
27. 0.9998^{10}
28. $\sqrt{9.09}$
29. $\sqrt[3]{8.024}$
30. $\sqrt{26}$
31. $\sqrt[3]{999}$
32. $600^{3/4}$
33. $1/\sqrt[5]{3120}$

Find approximations for the following, given that $|x|$ is small:

34. $\sqrt{1 - 2x}$
35. $\sqrt[3]{1 + 6x}$
36. $\sqrt{4 + 8x}$
37. $\sqrt[3]{8 - 24x^2}$

Use the negative binomial to expand each of the following to 4 terms:

38. $(1 - x)^{-3}$
39. $(1 - x)^{-5}$
40. $(1 - q)^{-r}$
41. If n is even, use formula (1) with $a = 1$ and $x = -1$ to prove that

$$\binom{n}{0} + \binom{n}{2} + \binom{n}{4} + \cdots + \binom{n}{n} = \binom{n}{1} + \binom{n}{3} + \binom{n}{5} + \cdots + \binom{n}{n-1}.$$

42. If $|nx| < 1$, prove that the approximation formula,
$$(1 + x)^n \approx 1 + nx,$$
is in error by an amount less than $(nx)^2$. [*Hint:* $1/r! \leq 1/2^{r-1}$, $r \geq 1$.]
43. How many different terms are there in the expansion of $(a + b + c)^n$ for $n = 1$? for $n = 2$? for $n = 3$? for general n?

2–6 THE MANY USES OF $\binom{n}{r}$

In this chapter, we have encountered the number $\binom{n}{r}$ in several situations. Such multiple usage is not uncommon in mathematics and serves to underline the unity of the subject. The same mathematical model can often be used to solve a variety of problems. We may find that several problems that appear to be unrelated are actually different interpretations of a single mathematical model. When this happens, a number associated with the mathematical model provides an answer to each of the various problems that correspond to these interpretations. To illustrate this point, we shall consider four problems along with their solutions and generalizations.

1. A problem about arrangements The Piltdown Players plan a 5-night play festival. The play *Excelsior* (E) is to be offered on 2 nights, and the play *Now or Never* (N) is to be offered on 3 nights. In how many ways can the 5-night program be arranged?

Solution The problem is that of arranging the five letters
$$EENNN$$
in all possible ways. By Corollary 2–14, the number of possible arrangements is $\binom{5}{2}$.

In general: If n objects consist of r objects of one kind and $n - r$ objects of another kind, then the n objects can be arranged in a row in $\binom{n}{r}$ ways.

2. A problem about selections Five candidates run for the same office and 2 are to be elected. If there are no ties, in how many ways can the election turn out?

Solution It's a question of selecting, without regard to order, 2 persons from 5. By Theorem 2–10, the number of selections is $\binom{5}{2}$.

It's no accident that the answer is the same as that for Problem 1. Answers to both problems are provided by the same model. To show this, assign numbers to the 5 candidates, and let E stand for "elected" and N stand for "not elected". Then, for example, we can represent the election of candidates numbered 2 and 5 thus:

$$\begin{matrix} 1 & 2 & 3 & 4 & 5 \\ N & E & N & N & E \end{matrix}$$

Every arrangement of the 5 letters *EENNN* corresponds to an election result, and conversely. Hence, there are $\binom{5}{2}$ election results.

In general: r objects can be selected, without regard to order, from a set of n different objects in $\binom{n}{r}$ ways.

3. A problem about paths across a lattice If a pedestrian walks from O to A, always walking either north or east, how many different paths can be taken?

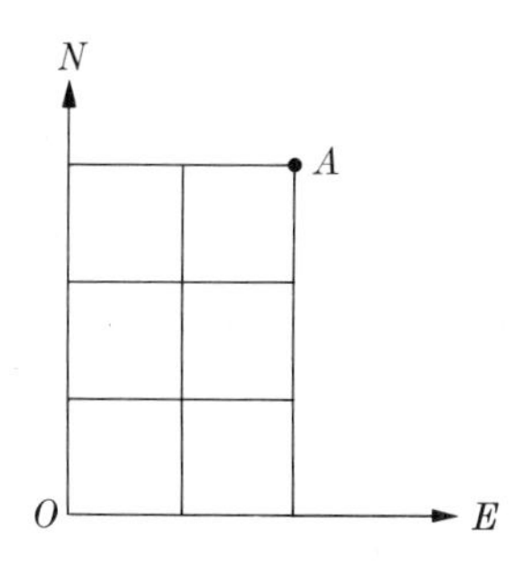

Solution As we have seen in Example 4 of Section 2–4, the number of possible paths is the number of arrangements of the letters *EENNN*, where E denotes "a walk of 1 block east" and N denotes "a walk of 1 block north". Again, the required number is $\binom{5}{2}$.

In general: If a rectangular lattice is r blocks long in an east–west direction and $n - r$ blocks long in a north–south direction, then the number of n-block paths from the southwest corner to the northeast corner of the lattice is $\binom{n}{r}$.

4. A problem about binomial coefficients What is the coefficient of a^3x^2 in the binomial expansion of $(a + x)^5$?

Solution From Theorem 2–15, the answer is once again $\binom{5}{2}$. To see why this answer is the same as that for the first three problems, recall that terms in the expansion of $(a + x)^5$ are constructed by choosing exactly 1 letter from each of the 5 factors $(a + x)$. Let E denote "x elected" and N denote "x not elected (that is, a elected)". Display the five factors of $(a + x)^5$ above an arrangement of *EENNN*, thus:

$$\begin{matrix} (a+x) & (a+x) & (a+x) & (a+x) & (a+x) \\ N & E & E & N & N \end{matrix}$$

This arrangement corresponds to the selection *axxaa*, and therefore to a term a^3x^2.

Similarly, each arrangement of the letters *EENNN* corresponds to a term a^3x^2, and conversely. Hence, the number of terms a^3x^2 is $\binom{5}{2}$, which is therefore the coefficient of a^3x^2 in the binomial expansion of $(a + x)^5$.

In general: The coefficient of $a^{n-r}x^r$ in the binomial expansion of $(a + x)^n$ is $\binom{n}{r}$.

The following scheme summarizes the foregoing facts:

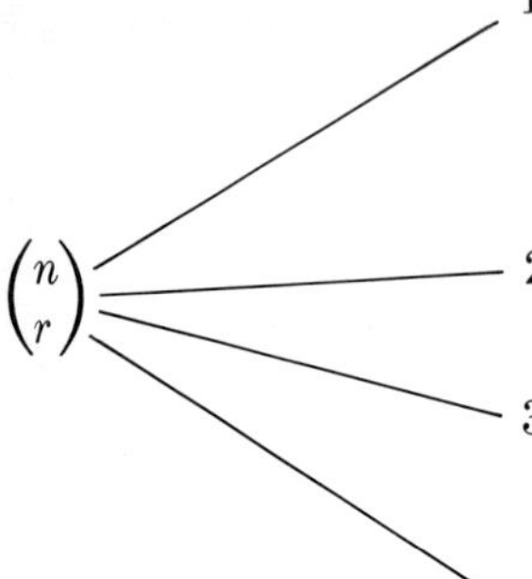

1) is the number of *permutations* of n objects, taken all together, where r of the objects are alike of one kind and $n - r$ of the objects are alike of another kind.

2) is the number of *combinations* of n different objects, taken r at a time.

3) is the number of n-block paths from one corner to the opposite corner of a rectangular lattice whose dimensions are r blocks by $n - r$ blocks.

4) is the coefficient of $a^{n-r}x^r$ in the binomial expansion of $(a + x)^n$.

EXERCISES FOR SECTION 2–6

1. In the example of the Piltdown Players, suppose that the program is to end with *Now or Never*. How many different 5-night programs of this type can be arranged, using *Excelsior* twice and *Now or Never* three times? [*Ans:* 6]
2. If 4 N's and 3 E's are arranged in a line at random, what is the probability that the 3 E's will be included in the first 5 positions (reading from left to right)? [*Ans:* 2/7]

3 Probability: equally likely outcomes

In this chapter, we shall continue the intuitive approach to probability that we began in Chapter 1. Our purpose is fourfold:

1) to gain an intuitive feeling for probabilities and some notion of how to work with them;
2) to make plausible the assumptions that we shall make later in the more formal mathematical treatment;
3) to become familiar with concepts and notations to be used later; and
4) to exploit the new methods of counting studied in Chapter 2.

We assume that equal probabilities are assigned to points in some finite sample space S. If A is a set from S, then $P(A)$ is the sum of the probabilities assigned to those points that are in A. For the empty set, ϕ, we always take $P(\phi) = 0$.

3–1 EVENTS AND SETS

As already indicated in Sections 1–6 and 1–8, when used as a technical word *an "event" is a subset of a sample space S of an experiment.* We have seen that subsets, or "events", may be described either verbally or by algebraic equations and inequalities. Such descriptions define subsets of S corresponding to the "events" under consideration.

A note on "or" and "and" In everyday English, expressions of the form "A or B" use the word "or" in two different ways:

1) in the *exclusive* sense, which connotes "A or B but not both" (for example, a coin falls "heads or tails");
2) in the *inclusive* sense, which connotes "A or B or both" (for example, consider the statement: "I may visit France or Italy this summer.").

Ordinarily, the context is a sufficient guide to the intended meaning. However, when we use the expression "A or B" in referring to *events,* the meaning is never in doubt because we *always* use the inclusive "or"; in other words, "A or B" means "A or B or both".

The foregoing usage agrees with the definition of $A \cup B$, that is, the *union* of sets A and B. For

$A \cup B$ is the set of all points belonging to A or to B or to both.

Thus, in this book, "the probability of A or B" always means

$$P(A \cup B).$$

The idea of simultaneous membership in two sets is connoted in our use of "and" when we talk about events. Thus if events A and B are subsets of a sample space, then "A and B" is their *intersection;* that is, the event "A and B" contains those sample points that belong to *both* A and B. For

the intersection of A and B, $A \cap B$, is the set of all elements belonging to both A and B.

When we use the verbal description "the probability of A and B", we mean

$$P(A \cap B).$$

We shall illustrate these ideas with further examples based on Table 1–11.

Example 1 (Refer to Section 1–8.) In the two-dice experiment, what is the probability that

$$r \leq 3 \quad \text{or} \quad c \leq 2?$$

Solution For the outcome $r \leq 3$, the red die must show either 1 or 2 or 3. The corresponding set A consists of the 18 points in the first three rows of Table 1–11. For $c \leq 2$, the clear die must show either 1 or 2, and the corresponding set B consists of the 12 points in the first two columns. The points in the union of A and B correspond to the event $r \leq 3$ or $c \leq 2$. To find the number of points in $A \cup B$, we must not add the number in A to the number in B, because there are 6 points that are in both sets and we must not count these twice. The correct count of points in $A \cup B$ is

$$18 + 12 - 6 = 24. \tag{1}$$

Therefore the probability of $r \leq 3$ or $c \leq 2$ is $\frac{24}{36}$, or $\frac{2}{3}$.

We notice in the above calculation that 18 is the number of points in A, 12 is the number in B, and 6 is the number in their intersection, $A \cap B$. Dividing all terms of Eq. (1) by 36, we get

$$\frac{18}{36} + \frac{12}{36} - \frac{6}{36} = \frac{24}{36}. \tag{2}$$

Thus, in this example, we may say that

$$P(A) + P(B) - P(A \cap B) = P(A \cup B); \tag{3}$$

for

$$P(A) = \tfrac{18}{36}, \qquad P(B) = \tfrac{12}{36},$$
$$P(A \cap B) = \tfrac{6}{36}, \qquad P(A \cup B) = \tfrac{24}{36}.$$

In the next section, we show that Eq. (3) is true in general.

EXERCISES FOR SECTION 3–1

Exercises 1 through 7 refer to the sample space for the two-dice experiment of Table 1–11.

1. What is the probability that $r > 2$ or $c > 3$?
2. What is the probability that $r > 2$ and $3c > 3$?
3. What is the probability that $r < 2$ or $c < 4$?
4. What is the probability that $r < 2$ and $c < 4$?
5. What is the probability that $r + c = 5$ or $r + c = 7$?
6. What is the probability that $r + c = 5$ and $r + c = 7$?
7. If A is the event "r is greater than 4" and B is the event "c is greater than 2", prove that
$$P(A) + P(B) - P(A \cap B) = P(A \cup B).$$

Exercises 8 through 14 refer to the sample space for Example 2, Section 1–8 (Table 1–13).

8. What is the probability that treatment a or c is chosen?
9. What is the probability that treatments a and c are chosen?
10. What is the probability that a and b *or* b and c are chosen?
11. What is the probability that a and b *and* b and c are chosen?
12. What is the probability that e is chosen or that b and c are chosen?
13. What is the probability that e is chosen and that also b and c are chosen?
14. Let A be the event "a is chosen" and B be the event "b is chosen". Show that
$$P(A) + P(B) - P(A \cap B) = P(A \cup B).$$

3–2 MUTUALLY EXCLUSIVE EVENTS

If two events cannot happen at the same time, they are said to be *mutually exclusive*. The computation of probabilities is especially simple when an event consists of other mutually exclusive events.

Example 1 In the two-dice example of Section 1–8, what is the probability that the sum $r + c$ is 7 or 10?

Solution (Refer to Table 1–11.) There are 6 sample points with $r + c = 7$, and 3 with $r + c = 10$. Since the corresponding sets do not overlap, there are 9 points with sum 7 or 10. Hence, the probability is $\frac{9}{36}$, or $\frac{1}{4}$.

In set language, if A is the set of points with $r + c = 7$, and B is the set with $r + c = 10$, then, *for this example*, we have

$$P(A) + P(B) = P(A \cup B). \tag{1}$$

The equality follows because

$$P(A) = \tfrac{6}{36}, \qquad P(B) = \tfrac{3}{36}, \qquad \text{and} \qquad P(A \cup B) = \tfrac{9}{36}.$$

Equation (1) is like Eq. (3) of the previous section, with $P(A \cap B) = 0$.

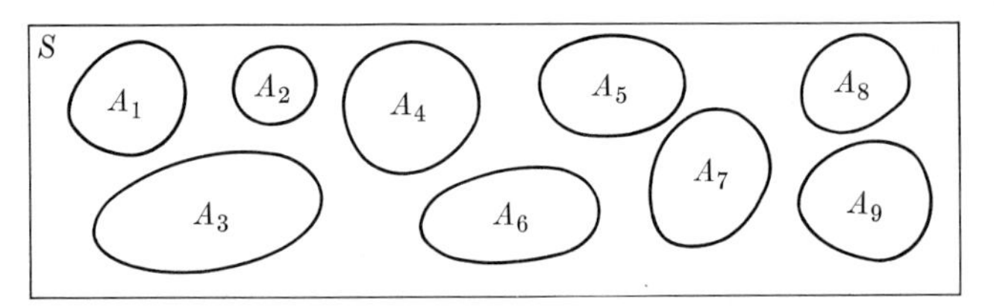

Fig. 3–1 Graph of mutually exclusive or disjoint subsets.

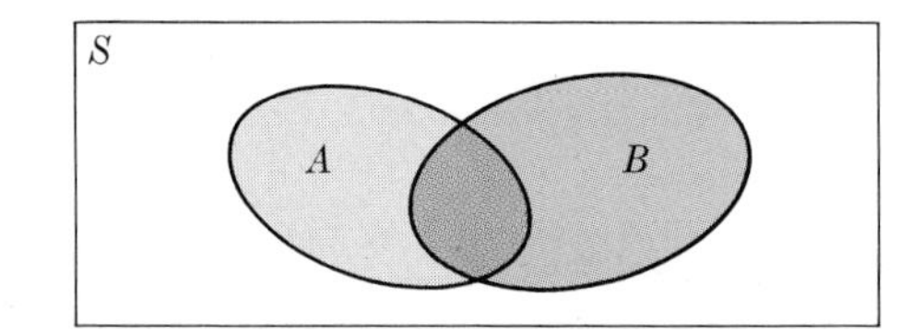

Fig. 3–2 Events in S.

3–1 Definition: Mutually exclusive events

If two events have no points in common, they are called mutually exclusive, or disjoint. And n events are mutually exclusive if no two of them have any points in common (Fig. 3–1).

We note a consequence of this definition: *the intersection of two or more mutually exclusive events is the empty set.*

3–2 Theorem: Probability of $A \cup B$

If A and B are events in a finite sample space S (Fig. 3–2), then

$$\boxed{P(A \cup B) = P(A) + P(B) - P(A \cap B).} \tag{2}$$

Proof The probability of $A \cup B$ is the sum of the probabilities of the points in $A \cup B$. Now $P(A) + P(B)$ is the sum of the probabilities of points in A plus the sum of the probabilities of points in B. Therefore $P(A) + P(B)$ includes the probabilities of points in the intersection $A \cap B$ twice. If we subtract this probability $P(A \cap B)$ once, we shall have the sum of the probabilities of all points in $A \cup B$, each taken just once. Hence

$$P(A \cup B) = P(A) + P(B) - P(A \cap B). \square \tag{3}$$

3-3 Corollary

If A and B are disjoint, then

$$\boxed{P(A \cup B) = P(A) + P(B).} \tag{4}$$

Equation (4) follows at once from Eq. (3) since, if A and B are disjoint, $A \cap B = \phi$, the empty set, and

$$P(A \cap B) = P(\phi) = 0.$$

3-4 Corollary: Several disjoint events

Let $A_1, A_2, \ldots, A_m$ be mutually exclusive events. Then

$$\boxed{P(A_1 \cup A_2 \cup \cdots \cup A_m) = P(A_1) + P(A_2) + \cdots + P(A_m).} \tag{5}$$

In words, the probability of A_1 *or* A_2 *or* . . . *or* A_m is the *sum* of their probabilities, provided the events are mutually exclusive.

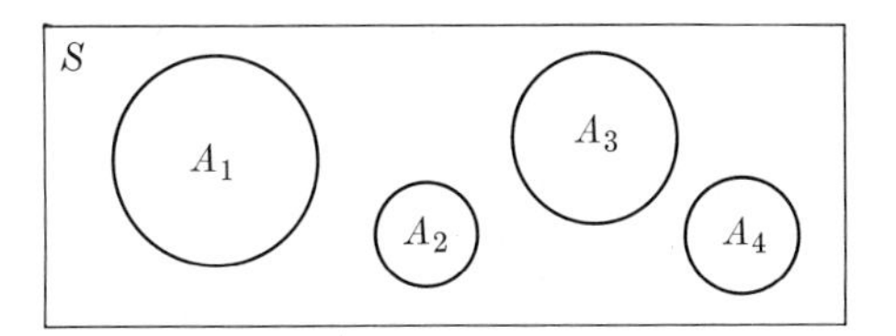

Fig. 3-3 Disjoint sets.

Proof (Cf. Fig. 3-3.) The probability of the union of A_1, A_2, and so on, is the sum of the probabilities of its points. The sum

$$P(A_1) + P(A_2) + \cdots + P(A_m)$$

is the sum of the probabilities of the points in A_1, plus the sum for A_2, and so on. Since the sets do not overlap, this sum includes the probabilities of the points in the union, once and only once for each point. $\square$

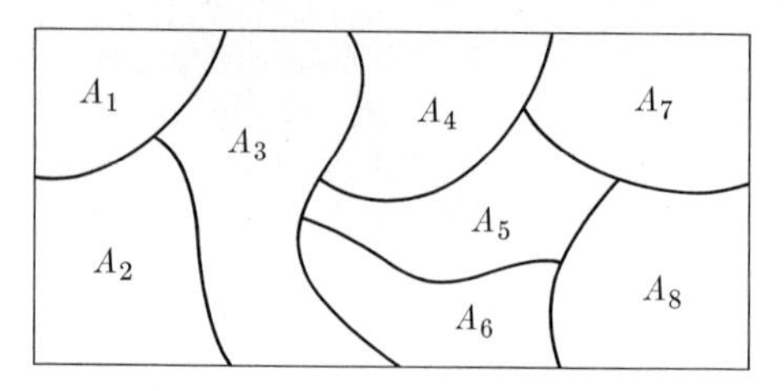

Fig. 3–4 A partition of S.

3–5 Definition: Partition

If the events $A_1, A_2, \ldots, A_m$ are mutually exclusive and exhaustive (i.e., their union contains all the sample points of S), then we say that the m events form a partition of the sample space S into m subsets.

For example, Fig. 3–4 illustrates a partition of S into 8 subsets:

$$S = A_1 \cup A_2 \cup A_3 \cup A_4 \cup A_5 \cup A_6 \cup A_7 \cup A_8.$$

3–6 Theorem: Probabilities under a partition

If $A_1, A_2, \ldots, A_m$ form a partition of a finite sample space S, then

$$\boxed{P(A_1) + P(A_2) + \cdots + P(A_m) = 1.}$$

Proof From Corollary 3–4, we have

$$\begin{aligned} P(A_1) + P(A_2) + \cdots + P(A_m) &= P(A_1 \cup A_2 \cup \cdots \cup A_m) \\ &= P(S) \\ &= 1. \ \square \end{aligned}$$

Example 2 A basketball coach has available three complimentary tickets to a professional basketball game. He decides to give the tickets to three players chosen at random from the five players of his first string team: Adams (a), Black (b), Carter (c), Davis (d), and Evans (e). What is the probability that both Adams and Black are chosen, or both Carter and Evans are chosen, or Black, Carter, and Davis are chosen?

Solution A sample space consists of the $\binom{5}{3}$, or 10, possible selections of the 5 players, taken 3 at a time:

$$S = \{abc, abd, abe, acd, ace, ade, bcd, bce, bde, cde\}.$$

We now select subsets of S that correspond to the three events in which we are interested. These subsets and their verbal descriptions are tabulated as follows:

Verbal description	Event
Adams and Black are chosen	$A_1 = \{abc, abd, abe\}$
Carter and Evans are chosen	$A_2 = \{ace, bce, cde\}$
Black, Carter, and Davis are chosen	$A_3 = \{bcd\}$

Since the three players are chosen at random, we assign to each sample point of S probability $\frac{1}{10}$. Then, because the events A_1, A_2, and A_3 are mutually exclusive, we have

$$\begin{aligned} P(A_1 \cup A_2 \cup A_3) &= P(A_1) + P(A_2) + P(A_3) \\ &= 0.3 + 0.3 + 0.1 = 0.7. \end{aligned}$$

Example 3 In the two-dice example of Section 1–8, what is the probability of *not* getting a double?

Solution There are 6 points in Table 1–11 that correspond to the event "throwing a double". Denote this set by A. Then

$$P(A) = \tfrac{6}{36} = \tfrac{1}{6}.$$

The desired probability of not getting a double is

$$P(\overline{A}) = 1 - P(A) = \tfrac{5}{6}.$$

The event "getting a double" and the event "not getting a double" are mutually exclusive. They are also said to be "complementary". The two events, getting a double and not getting a double, together exhaust all possible outcomes.

3–7 Definition: Complementary events

An event A and the event $\overline{A}$, consisting of all points of the same sample space not in A, are called *complementary events.*

Thus any event A and its complementary event $\overline{A}$ are mutually exclusive, and their union is the whole sample space. In other words, events A and $\overline{A}$ form a partition of S into two subsets.

3–8 Theorem: Complementary events

If A and $\overline{A}$ are complementary events, then

$$\boxed{P(\overline{A}) = 1 - P(A).} \tag{6}$$

Proof Since A and $\overline{A}$ are disjoint, formula (4) gives

$$P(A \cup \overline{A}) = P(A) + P(\overline{A}).$$

Since $A \cup \overline{A}$ is the entire sample space S, it follows that

$$P(A \cup \overline{A}) = P(S) = 1.$$

Therefore

$$P(\overline{A}) = 1 - P(A). \square$$

This formula was obtained in Section 1–6 as a consequence of the definition of probability of an event for sample spaces with equally likely outcomes. The present proof is also valid for more general sample spaces that will be studied in Chapter 4.

EXERCISES FOR SECTION 3–2

1. A die is rolled. Let E be the event "die shows 4", and F be the event "die shows even number". Are events E and F mutually exclusive?
2. A die is rolled. Let E be the event "die shows even number", and F be the event "die shows odd number". Are events E and F complementary? Are they mutually exclusive?
3. What is the probability of throwing a one or a two or a three with a single fair die?
4. If the probability that A wins a game is 0.6, what is the probability that A loses? Are the two events "A wins" and "A loses" mutually exclusive? Are they complementary? (What about a tie?)
5. Two coins are tossed. E is the event "getting two heads", and F is the event "getting two tails". Are events E and F mutually exclusive? Are they complementary? Evaluate $P(E \cup F)$.

   Exercises 6 through 16 refer to the two-dice experiment of Example 1 in Section 1–8. Find the probability that:

6. The sum of the spots is not 11.
7. The two dice show only 3 or 4 or both.
8. Neither 3 nor 4 appears.
9. Each die shows 3 or more spots.
10. At least one die shows fewer than 3 spots.
11. Both dice show fewer than 3 spots.
12. Only one die shows fewer than 3 spots.
13. $r + c$ is even or $r + c$ is odd.
14. $r + c = 4$ or $r + c = 11$.
15. $r \leq 2 + c$.
16. $r \neq c$.
17. Refer to Example 2 of this section. Let E be the event "Davis and Evans are chosen", F be the event "Black and Carter are chosen", and G be the event "Carter, Davis, and Evans are chosen". Find $P(E \cup F \cup G)$.

18. If the probability of Jackson's winning a race is $\frac{1}{3}$ and the probability of Taylor's winning is $\frac{1}{5}$, what is the probability that either Jackson or Taylor will win if they are in the same race?
19. Three coins are tossed. Find the probability of getting (a) no heads, (b) at least one head.
20. The integers 1, 2, 3, . . . , 20 are written on slips of paper which are placed in a bowl and thoroughly mixed. A slip is drawn from the bowl at random. What is the probability that the number on the slip is either prime or divisible by 3?

3–3 INDEPENDENT EVENTS

The present discussion introduces the notion, and leads to the definition, of *independent events*. When we say, in everyday language, that two events "have nothing to do with each other", we are describing what, in technical language, are called "independent events". We begin with an illustrative example.

Example 1 In the two-dice experiment of Section 1–8, what is the probability that $r \leq 3$ *and* $c \geq 5$?

Solution The event that concerns us requires that two conditions be satisfied simultaneously. If A is the set of points with $r \leq 3$ and B is the set with $c \geq 5$, then we want to know the number of points that these sets have in common—in short, their intersection, $A \cap B$. This intersection is the 3×2 array of points in the first three rows and the last two columns of Table 1–11. Hence we have

$$P(A \cap B) = \tfrac{6}{36} = \tfrac{1}{6}.$$

By counting, we find that

$$P(A) = \tfrac{18}{36} = \tfrac{1}{2} \qquad \text{and} \qquad P(B) = \tfrac{12}{36} = \tfrac{1}{3},$$

since A has 18 points and B has 12. Using these probabilities and the answer for $P(A \cap B)$, we verify that, *for this example*,

$$P(A \cap B) = P(A) \cdot P(B). \tag{1}$$

This multiplication formula, (1), agrees with the results obtained by an intuitive approach to the problem. Consider a very long series of throws of the two dice. We expect to find $r \leq 3$ in about half of these throws. Let us restrict our attention to this half of the throws. Of these throws, how many have $c \geq 5$? Since what happens on the red die does not affect the clear die, it seems reasonable that about $\frac{1}{3}$ of the throws with $r \leq 3$ will also have $c \geq 5$. Thus the fraction of throws with *both* $r \leq 3$ *and* $c \geq 5$ is about $\frac{1}{3}$ of $\frac{1}{2}$, or $\frac{1}{6}$.

Is formula (1) true in general? The answer is "no", as we shall see in Example 4. When formula (1) holds, the events A and B are called *independent events;* otherwise, they are called dependent events. Our intuition suggests that the fall of the red die is independent of the fall of the clear die; for it seems evident that the fall of the red die has nothing to do with the fall of the clear die, and our everyday usage of the word "independent" implies just that. Moreover, it turns out that when two everyday events "have nothing to do with each other", the probability that both events occur is obtained by the multiplication of their separate probabilities, as we have just seen in Example 1.

For technical purposes, however, we need a definition that frees us from the vagueness of the expression "have nothing to do with each other". Such a technical definition, suggested by the results of problems similar to Example 1, is now given.

3-9 Definition: Independent events

Events A and B are independent if and only if

$$\boxed{P(A \cap B) = P(A) \cdot P(B).} \tag{1}$$

The foregoing definition provides us with a clean-cut meaning for "independent events"; if two events A and B do not satisfy Eq. (1), the events are dependent.

3-10 Theorem

If A and B are independent events with nonzero probabilities, then sets A and B have a common sample point.

Proof Let ϕ represent the empty set. Either $A \cap B = \phi$ or $A \cap B \neq \phi$. If $A \cap B = \phi$, then $P(A \cap B) = 0$, and from (1) it follows that $P(A) = 0$ or $P(B) = 0$. Since this contradicts the hypothesis of the theorem, it follows that $A \cap B \neq \phi$. □

Example 2 *Independent events: coins.* Two coins are tossed. Show that the event "head on first coin" and the event "coins fall alike" are independent.

Solution A sample space for the experiment is

$$S = \{HH, HT, TH, TT\}.$$

Let event A be "head on first coin" and event B be "coins fall alike". Since the four outcomes in S are equally likely, we assign to each the probability $\frac{1}{4}$. Therefore we have

$$\begin{aligned} A &= \{HH, HT\}, & P(A) &= \tfrac{2}{4} = \tfrac{1}{2}, \\ B &= \{HH, TT\}, & P(B) &= \tfrac{2}{4} = \tfrac{1}{2}, \\ A \cap B &= \{HH\}, & P(A \cap B) &= \tfrac{1}{4}. \end{aligned}$$

Hence it follows that

$$P(A \cap B) = P(A) \cdot P(B),$$

and events A and B are independent, by Definition 3–9.

In the following examples, we exhibit first a case in which the definition of independence is satisfied, and second a case in which events do not satisfy the definition of independence.

Example 3 *Independent events: dice.* In the two-dice experiment of Section 1–8, what is the probability that the red die shows even and the clear die shows odd?

Solution Let us count points in the sample space S (see Table 1–11). There are 18 points (3 rows) with r even and 18 points (3 columns) with c odd. These two 18-point sets have 9 points in common—the 9 points where the three rows intersect the three columns. Hence there are 9 points with r even *and* c odd, and we have

$$P(r \text{ even } \textit{and } c \text{ odd}) = \tfrac{9}{36} = \tfrac{1}{4}.$$

Since

$$P(r \text{ even}) = \tfrac{18}{36} = \tfrac{1}{2}$$

and

$$P(c \text{ odd}) = \tfrac{18}{36} = \tfrac{1}{2},$$

the events "r even" and "c odd" are independent, by Definition 3–9.

Example 4 *Dependent events: dice.* In the two-dice experiment of Section 1–8, what is the probability that the sum on the two dice is 11 ($r + c = 11$) and, at the same time, $r \neq 5$? Are these two events independent?

Solution There are two points in Table 1–11 with $r + c = 11$: (5, 6) and (6, 5). If we denote this set of two points by E, then

$$P(E) = \tfrac{2}{36} = \tfrac{1}{18}.$$

Let F be the set defined by $r \neq 5$. Then F has 30 points and

$$P(F) = \tfrac{30}{36} = \tfrac{5}{6}.$$

Since the simultaneous event E *and* F has only the single point (6, 5), we have

$$P(E \cap F) = \tfrac{1}{36}.$$

Since $\frac{1}{36}$ is not equal to the product of $\frac{1}{18}$ and $\frac{5}{6}$, E and F are dependent events, by Definition 3–9.

Remark When three or more events are independent, the probability of their simultaneous occurrence is the product of their probabilities. Thus, for example, if E, F, and G are independent, then

$$P(E \cap F \cap G) = P(E) \cdot P(F) \cdot P(G). \tag{2}$$

For *complete* independence of the three events, we require something more than Eq. (2). This is discussed more fully in Chapter 4, Definition 4-7.

WARNING!

> There is a danger of confusing *mutually exclusive events* with *independent events.* A source of this confusion is the common expression "have nothing to do with each other". This expression provides a useful description of independence when applied to everyday events. But when applied mistakenly to *sets,* it suggests nonoverlapping; and nonoverlapping sets are mutually exclusive and are *not* independent. Indeed, in dealing with independent events A and B in a sample space, we know that the sets A and B must have a point in common if both A and B have nonzero probabilities. (Cf. Theorem 3-10.)

EXERCISES FOR SECTION 3-3

1. In the two-dice experiment of Example 1, Section 1-8, show that the event "$r > 4$" and the event "$c < 3$" are independent.
2. Three coins are tossed. Show that the event "heads on the first coin" and the event "tails on the last two" are independent. Show that the event "exactly two coins heads" and the event "three coins heads" are dependent.
3. If a coin is thrown four times, what is the probability that it will fall heads on the first throw, tails on the next two throws, and heads on the fourth throw?
4. A pair of dice is tossed twice. What is the probability that, on the second toss, each die shows spots different from those it showed on the first toss? Assume independence of the outcomes of the two tosses.
5. A die is tossed three times. What is the probability that the first toss will show odd, the second toss even, and the third toss a six? Assume independence of the outcomes of the three tosses.
6. In a certain school, examination results showed that 10% of the students failed mathematics, 12% failed English, and 2% failed both mathematics and English. A student is selected at random from the school roll. Are the event "student failed mathematics" and the event "student failed English" independent?
7. If E is any event in sample space S, show that E and S are independent. Are E and ϕ independent?
8. A bag contains 5 black marbles, 4 red marbles, and 3 white marbles. Three marbles are drawn in succession, each marble being replaced before the next one is drawn. What is the probability that the first marble is black, the second red, and the third white?

9. On September 25, 1967, The *New York Times* reported the following: "The odds against Alabama or Michigan State giving up 37 points to any team in the country would have been 100 to 1 Saturday morning. The odds against both of them doing so simultaneously would have been 1000 to 1". Correct this statement.

3–4 CONDITIONAL PROBABILITY

Often we deal with probabilities for part rather than all of a sample space. The probability that a person randomly selected from a population has blue eyes differs from the probability of blue eyes for a person randomly selected from the blondes in this population. For a set of students about to take a mathematics course, the probability that a randomly selected one will get an honor grade is lower than the probability for those who made honor grades in their last two mathematics courses. The chance of a serious fire in the next year in a warehouse selected at random from those in a large city differs from that in the subpopulation consisting only of fireproofed warehouses. Each of these examples focuses attention on the probability of an event in a subset of the original sample space, and emphasizes that the probability in the subset may differ from that in the whole space. The subpopulations are defined by extra conditions beyond those for the whole population, and probabilities associated with events in these subpopulations are called *conditional probabilities.*

To introduce the idea of conditional probability, we discuss the following example, based on the two-dice experiment of Section 1–8. Refer to Table 1–11 on page 27.

Example 1 Given that $r + c < 4$, find the probability that $r = 1$.

Discussion First, we need some idea of what such a probability means. Among all throws of two dice, some produce a sum $r + c$ that is less than 4, and others do not. We ignore all that do not and obtain a *reduced sample space,* S', consisting of three points:

$$\{(1, 1),\ (1, 2),\ (2, 1)\}.$$

Since these three outcomes were equally likely in the original sample space S, we assign them equal probabilities in the reduced sample space S'. Since they are the only points in S', we assign to each of them probability $\frac{1}{3}$. The event defined by $r = 1$ consists of the two points

$$(1, 1) \qquad \text{and} \qquad (1, 2).$$

Therefore, the probability of $r = 1$ in the reduced sample space S' is $\frac{2}{3}$. We call this result the *conditional probability* that $r = 1$, given that $r + c < 4$.

To study conditional probability further, consider, in the original sample space S, the sets that correspond to $r + c < 4$ and to $r = 1$. For convenience, these are tabulated in Table 3–1.

Table 3-1

Condition	Event
$r + c < 4$	$B = \{(1, 1),\ (1, 2),\ (2, 1)\}$
$r = 1$	$A = \{(1, 1),\ (1, 2),\ (1, 3),\ (1, 4),\ (1, 5),\ (1, 6)\}$

Since we want to know the chances of A, *given* B, we naturally are interested in the event $A \cap B$ corresponding to the set of points that are simultaneously in A and B. We have

$$A \cap B = \{(1, 1),\ (1, 2)\}.$$

What are the probabilities of events A, B, and $A \cap B$ in the *original* sample space S? They are

$$P(A) = \tfrac{6}{36}, \qquad P(B) = \tfrac{3}{36},$$
$$P(A \cap B) = \tfrac{2}{36}.$$

(Note that we purposely do not reduce these fractions to lowest terms. Sometimes such reduction obscures the pattern that we hope to discover.)

The usual notation for "the probability of A, given B" is

$$P(A|B).$$

The vertical bar is read "given".

In the foregoing example, our first solution led us to the result

$$P(A|B) = \tfrac{2}{3}.$$

We now observe that, *in this example*, we also have

$$P(A \cap B) = P(B) \cdot P(A|B), \tag{1}$$

since

$$\tfrac{2}{36} = \tfrac{3}{36} \cdot \tfrac{2}{3}.$$

Formula (1) suggests an alternative way of getting $P(A|B)$; for Eq. (1) implies

$$P(A|B) = \frac{P(A \cap B)}{P(B)}.$$

It is also interesting to consider the result obtained by interchanging A and B on both sides of Eq. (1). The event $A \cap B$ is the same as the event $B \cap A$. Thus, for this example, we wonder if it is also true that

$$P(A \cap B) = P(B \cap A) = P(A) \cdot P(B|A). \tag{2}$$

(See Exercise 2 at the end of this section.)

Example 2 Two a's and two b's are arranged in order. All arrangements are equally likely. Given that the last letter, in order, is b, find the probability that the two a's are together.

Solution A sample space of possible orders of the four letters is as follows:

$$S = \{aabb, abab, abba, baab, baba, bbaa\}.$$

Consider a reduced sample space, B, whose elements have b as the last letter:

$$B = \{aabb, abab, baab\}.$$

Since we want to know the probability that b is the last letter *and* that the two a's are together, we look for the points in B that contain aa. The required points are the intersection of B and A, where A is the set of all points that include aa:

$$A = \{aabb, baab, bbaa\}.$$

Therefore

$$A \cap B = \{aabb, baab\}.$$

If we treat all points of B as equally likely and as constituting a reduced sample space, we assign probability $\frac{1}{3}$ to each of them. Since two points of B contain aa, it follows that

$$P(A|B) = \tfrac{2}{3}.$$

Is Eq. (1) also satisfied in this example? In the original sample space S, we have

$$P(B) = \tfrac{3}{6}, \qquad P(A \cap B) = \tfrac{2}{6}.$$

Hence, *in this example,*

$$P(A \cap B) = P(B) \cdot P(A|B),$$

since

$$\tfrac{2}{6} = \tfrac{3}{6} \cdot \tfrac{2}{3}.$$

In Exercise 2 at the end of this section, you are asked to verify that Eq. (2) also holds for this example.

Remark Example 2 can be solved by making direct use of the ideas of Chapter 2; for if b is in the last position, the remaining three letters a, a, b can be arranged in $3!/2!$ or 3 ways. Of these three arrangements, only 2 ($aabb$ and $baab$) have the two a's together. Since the three possible arrangements with b in the last place are equally likely, we have

$$P(a\text{'s together}|b \text{ in last place}) = \tfrac{2}{3}.$$

The foregoing examples, and others like them, lead us to adopt the following definitions.

3-11 Definitions: Conditional probability and reduced sample space

The conditional probability of A, given B, is denoted by $P(A|B)$, and is defined by the equation

$$P(A|B) = \frac{P(A \cap B)}{P(B)}, \quad \text{if} \quad P(B) \neq 0. \tag{3}$$

The *reduced sample space* is B, the given event.

All probabilities are referred to some sample space, and $P(A)$ is an abbreviation for $P(A|S)$, where S is the whole sample space. But the S is ordinarily dropped as understood. When some subset of S, such as B, is known to contain all the outcomes of the experiment, then we need to be explicit and write $P(A|B)$. In particular, $P(B|B) = 1$.

Remark 1 The probabilities in the fraction on the right side of Eq. (3) are probabilities of the events in the *original* sample space S. Of course, we get the same result if we first convert to the *reduced* sample space, which is B. For we then increase the total probability in B to 1; and, if we compute the probabilities on the right side of Eq. (3) in sample space B, the denominator is 1 while the numerator is the probability of $A \cap B$ in B.

Remark 2 The restriction $P(B) \neq 0$ in Eq. (3) means that the *given* event B must not have probability zero. In finite sample spaces of equally likely outcomes, there must be a nonzero probability for B before it is useful to talk about the probability of A, given B.

In problems dealing with infinite sample spaces, events of probability zero can occur, and conditional probabilities, given such events, can be sensibly interpreted.

Remark 3 Equations (3) and (1) are essentially the same. We get (1) by multiplying both sides of (3) by $P(B)$. Conversely, assuming $P(B) \neq 0$, we get (3) by dividing both sides of (1) by $P(B)$.

Remark 4 Even if we know $P(A)$ and $P(B)$, there isn't any formula for computing $P(A \cap B)$ from them unless A and B are independent. We must treat A, B, and $A \cap B$ as three individual sets in S. It may not be a trivial matter to construct $A \cap B$ from the separate sets A and B unless the sets are small and we can list all the elements.

WARNINGS!

1) $P(A|B)$ is rarely the same as $P(A \cap B)$. Indeed, the conditional probability of A, given B, may be entirely different from $P(A)$ or from $P(A \cap B)$.
2) $A|B$ is *not* a symbol for a set.

For example, in connection with warning (1), consider sample space S of the two-dice experiment. (Cf. Table 1–11.) Let A denote the event "red die even", and let B denote the event "clear die 2". By counting sample points in S, we see that $P(A \cap B) = \frac{3}{36} = \frac{1}{12}$ and $P(A|B) = \frac{3}{6} = \frac{1}{2}$.

Example 3 In the two-dice experiment (cf. Table 1–11), if $r + c = 11$, what is the probability that the clear die shows 4?

Solution The probability is zero; it is impossible to have $c = 4$ if $r + c = 11$.

Example 4 In the two-dice experiment of Section 1–8, given that the red die shows 4, what is the probability that the clear die shows a number greater than 4?

Solution Let B be the event described by $r = 4$, and A the event described by $c > 4$. Then B contains 6 sample points (cf. Table 1–11), A contains 12, and their intersection $A \cap B$ contains 2 points, (4, 5) and (4, 6).

Therefore we have

$$P(B) = \tfrac{6}{36}, \qquad P(A \cap B) = \tfrac{2}{36},$$

whence Eq. (3) gives

$$P(A|B) = \frac{2/36}{6/36} = \frac{1}{3}.$$

Note that we obtain the result more directly by counting: 2 equally likely cases ($c = 5$ or 6) out of 6 equally likely cases in B yield a probability of $\frac{2}{6}$, or $\frac{1}{3}$.

Since A has 12 sample points, we also have

$$P(A) = \tfrac{12}{36} = \tfrac{1}{3}.$$

Thus, *in this example*, the conditional probability of A, given B, is the same as the probability of A. In other words, the information that the red die shows 4 does not change the probability that the clear die will show 5 or 6. The latter probability is $\frac{1}{3}$, regardless of the outcome on the red die.

The foregoing example illustrates a general theorem.

3–12 Theorem: Conditional probability of independent events

If A and B are independent events having nonzero probabilities, then

$$\boxed{P(A|B) = P(A) \qquad \text{and} \qquad P(B|A) = P(B).} \tag{4}$$

Proof Since A and B are independent, and since $A \cap B = B \cap A$, we have from Eq. (2), Section 3–3,

$$P(A \cap B) = P(B \cap A) = P(A) \cdot P(B).$$

Since neither $P(A)$ nor $P(B)$ is zero, we may use them as divisors. Equation (3) gives

$$P(A|B) = \frac{P(A \cap B)}{P(B)} = \frac{P(A) \cdot P(B)}{P(B)} = P(A).$$

The proof that $P(B|A) = P(B)$ is left as an exercise.

Remark Some authors prefer to define the independence of A and B with the condition $P(A|B) = P(A)$, and then prove as a theorem that

$$P(A \cap B) = P(A) \cdot P(B).$$

One advantage of this procedure is that the condition $P(A|B) = P(A)$ has a more intuitive connection with the idea of independence used in the everyday world, as, for example, when we say that the amount of precipitation is independent of the day of the week.

EXERCISES FOR SECTION 3–4

1. In the two-dice example of Table 1–11, given that $r + c \geq 10$, find the probability that $r = 5$. Given that $r + c = 8$, find the probability that $c \geq 4$.
2. For Example 1, show that

$$P(A \cap B) = P(B \cap A) = P(A) \cdot P(B|A).$$

3. Five-digit numbers are formed by permuting the digits 44433. All arrangements are equally likely. Given that a number is even, what is the probability that the two 3's are together?
4. Two dice are tossed. If the first die shows 5, what is the probability that the second die shows even?

Exercises 5 through 16 are based on the following data. Six men (Jones, Smith, Thoms, Dixon, Harris, and Peters) form a club. They decide to select from their number a committee of three. The selection process is to be by lot, so that all twenty possible committees are equally probable.

5. Verify that the number of committees of 3 that can be selected from 6 men is 20.
6. Describe a process of selecting such a committee by lot.
7. Set up a sample space of 20 points to represent the 20 possible committees.
8. What is the probability that Smith is on the committee? That Smith is not on the committee?
9. What is the probability that Smith is on the committee and Thoms is not?
10. What is the probability that neither Smith nor Thoms is on the committee?
11. Given that Smith is on the committee, what is the probability that Thoms is also on it?
12. What is the probability that Thoms, Dixon, and Harris are not all on the committee?

13. Given that Thoms and Dixon are on the committee, what is the probability that Harris is not?
14. What is the probability that Jones or Peters or both are on the committee?
15. Suppose that Jones and Smith are cousins, Thoms and Dixon are cousins, and Harris and Peters are cousins. What is the probability that the committee has two cousins on it?
16. Instead, suppose that Jones, Smith, and Peters are the only cousins. What is the probability that the committee has no two or more cousins on it?
17. Let p be the probability that an event will happen in one trial. Show that the probability of its happening in each of n independent trials is p^n.
18. If 2 persons are chosen from 10 and all choices are equally likely, what is the probability that two specified persons will both be chosen? That they will not both be chosen? That neither of them will be chosen?
19. Of 100,000 persons living at age 20, statistics show that 47,773 will be alive at 70. What is the probability that a person aged 20 will live to be 70? That he will die before he is 70?
20. The probability that A will die within the next 20 years is 0.025, and that B will die within the next 20 years is 0.030. What is the probability that both A and B will die within the next 20 years? That A will die and B will not die? That neither A nor B will die?
21. Seven persons form a line at random. What is the probability that two specified persons are next to each other? That these persons are not next to each other?
22. A basketball player has a probability of $\frac{1}{2}$ of scoring on a free throw. How many free throws would he have to take in order to make his probability of scoring one or more times at least 0.99?
23. If p is the probability that an event will happen in one trial, show that the probability that it will happen at least once in n independent trials is $1 - (1 - p)^n$.
24. A bag contains 3 white marbles and 4 black ones. In succession, three persons each draw a marble, without replacing it in the bag. The first person who draws a white marble wins. What are the respective chances of the person drawing first, the person drawing second, and the person drawing third? (They continue until someone wins.)
25. If you stop 3 people at random on the street, what is the probability that all were born on Friday? That two were born on Friday and the other on Tuesday? That none were born on Monday?
26. Two numbers are selected at random from 1, 2, 3, . . . , 10. What is the probability that the sum of the two numbers is even?
27. A buyer will accept a lot of 10 radios if a sample of 2, picked at random, contains no defectives. What is the probability that he will accept a lot of 10 if it contains 4 defectives?
28. A committee of 3 is chosen from a group of 20 people. What is the probability that a specified member of the group will be on the committee? That this specified member will not be on the committee?

29. A committee of 4 is chosen at random from 5 married couples. What is the probability that the committee will not include a husband and wife?
30. Seven-digit numbers are formed by permuting the digits 1, 2, 3, 4, 5, 6, 7. If all permutations are equally likely, what is the probability that, in a permutation selected at random, the odd digits will occur in ascending order?

3-5 SAMPLE SPACES WITH MANY ELEMENTS

When the number of elements in a sample space is very large, it is inconvenient to make a list. However, even without a list, the methods of counting developed in Chapter 2 may enable us to calculate probabilities for sample spaces with equally likely outcomes. The following examples illustrate the methods.

Example 1 *The first ace.* An ordinary bridge deck of 52 cards is thoroughly shuffled. The cards are then dealt face up, one at a time, until an ace appears. What is the probability that the first ace appears (a) at the fifth card? (b) at the kth card? (c) at the kth card or sooner?

Solution a) There are several possible sample spaces for this experiment. We choose one as follows. Once the cards are shuffled and in position in the deck, the only feature of each card that concerns us is whether it is an ace (A) or a non-ace (N). There are 48 non-aces and 4 aces, so we consider all possible arrangements of 48 N's and 4 A's in 52 numbered positions. There are

$$\frac{52!}{48!\,4!} = \binom{52}{4}$$

permutations of 48 N's and 4 A's, and each of these is a point in our sample space S. We assume that all points are equally likely and assign probability $1/\binom{52}{4}$ to each point.

Consider now an event E in sample space S, where E is described as "the first ace appears at the fifth card". If the first ace is in fifth place, then the first five symbols of every sample point in E are

$$N\ N\ N\ N\ A,$$

in that order. The number of points in E is, therefore, the number of ways of arranging the remaining 44 N's and 3 A's in the remaining 47 places. This number is

$$\frac{47!}{44!\,3!} = \binom{47}{3}.$$

Hence

$$P(E) = \frac{\binom{47}{3}}{\binom{52}{4}} = \frac{16{,}215}{270{,}725} \approx 0.060.$$

b) Similarly, if the first ace appears in the kth place in the row, then the remaining 3 aces and $48 - (k - 1)$ or $49 - k$ non-aces can be arranged in the last $52 - k$ positions in $\binom{52-k}{3}$ ways. Hence,

$$P\text{ (first ace in }k\text{th place)} = \frac{\binom{52-k}{3}}{\binom{52}{4}}, \qquad k = 1, 2, \ldots, 49.$$

c) Denote by F the event "first ace at kth card or sooner". Then the complementary event $\overline{F}$ is the event "4 aces after kth card". The first k symbols of every sample point in $\overline{F}$ are all N's. Therefore the number of sample points in $\overline{F}$ is the number of ways of arranging 4 A's and $48 - k$ N's in the remaining $52 - k$ places. This number is

$$\frac{(52-k)!}{(48-k)!\,4!} = \binom{52-k}{4}.$$

Hence

$$P(\overline{F}) = \frac{\binom{52-k}{4}}{\binom{52}{4}},$$

and

$$P(F) = 1 - P(\overline{F}) = 1 - \frac{\binom{52-k}{4}}{\binom{52}{4}}, \qquad k = 1, 2, \ldots, 48.$$

If $k = 9$, we obtain

$$P(F) \approx 1 - 0.46 = 0.54.$$

Thus there is a better than even chance that the first ace appears at or before the ninth card.

Remark The foregoing example affords an instance in which the so-called "maturation of chance" operates. In the light of this example, the student may wish to reread the discussion in Section 1–9.

Example 2 *The birthday problem.* There are k people in a room. What is the probability that at least two of these people have the same birthday, that is, have their birthdays on the same day and month of the year? What is the smallest value of k such that the probability is $\frac{1}{2}$ or better that at least two of the people have the same birthday? (Write down your guess.)

Solution We shall neglect February 29 and deal with a 365-day year. There are 365 possibilities for each person's birthday, and hence 365^k possibilities for the birthdays of k people. Thus our sample space S has 365^k points, each of which is an ordered k-tuple

$$(x_1, x_2, x_3, \ldots, x_k),$$

where x_1 represents the birthday of a first person, x_2 represents the birthday of a second person, ..., and x_k represents the birthday of the kth person. We assume that all of the 365^k possible outcomes are equally likely, and assign to each sample point probability $1/365^k$.

Table 3–2

Number of people in room	5	10	20	23	30	40	60
Probability that at least two birthdays are the same	0.027	0.117	0.411	0.507	0.706	0.891	0.994

Consider now an event E in sample space S, where E is described thus: "no two of the k people have the same birthday". Under this restriction, the birthday of a first person has 365 possible values, that of a second person 364 possible values, that of a third person 363 possible values, . . . , and that of the kth person $365 - (k - 1)$, or $365 - k + 1$, possible values. Therefore, by the multiplication principle, the number of possible sets of k birthdays with no two birthdays alike is

$$365 \cdot 364 \cdot 363 \cdots (365 - k + 1),$$

and this number is the number of sample points in E.

It follows that

$$P(E) = \frac{365 \cdot 364 \cdot 363 \cdots (365 - k + 1)}{365^k}.$$

Finally,

$$P(\text{at least 2 birthdays are the same}) = 1 - P(E).$$

Probabilities for specific values of k yield some rather startling information. Some results are given in Table 3–2. With as few as 23 people in the room, there is a better than even chance that two people have identical birthdays!

Remark Although in examples such as the foregoing no list is made, it is a good idea to think carefully about the nature of the individual points in the sample space, their number, and the subset that corresponds to a particular event whose probability is desired.

At times, it is advisable to split an event A into simpler subsets that are mutually exclusive, such as

$$A_1, A_2, A_3, \ldots, A_k.$$

In such a case, we have

$$A = A_1 \cup A_2 \cup A_3 \cup \cdots \cup A_k,$$

and (since these subsets are disjoint)

$$P(A) = P(A_1) + P(A_2) + \cdots + P(A_k). \tag{1}$$

As Example 3 illustrates, it sometimes happens that the subsets have equal probabilities; if so, Eq. (1) become

$$P(A) = kP(A_1).$$

Example 3 A small boy is playing with a set of 10 colored cubes and 3 empty boxes. If he puts the 10 cubes into the 3 boxes at random, what is the probability that he puts 3 cubes in one box, 3 in another box, and 4 in the third box?

Solution Imagine that the boxes have been lettered a, b, c to enable us to tell them apart. To help construct a sample space for the experiment, let us watch the boy perform the experiment and write down, in order, the letters of the boxes as he puts in the cubes, one after another. The result is a string of 10 letters; for example,

$$bbcaaaccba. \tag{2}$$

The particular sequence (2) corresponds to first cube in box b, second cube in box b, third cube in box c, fourth, fifth, and sixth cubes in box a, and so on. Thus the points of our sample space S [of which sequence (2) is one example] consist of all possible sequences of 10 letters, where each letter in the sequence may be a, or b, or c. From the multiplication principle of Chapter 2, we find that the number of points in this sample space is

$$n = 3 \times 3 \times 3 \times \cdots \times 3 = 3^{10}.$$

We assign probability $1/n$ to each sample point, since the boy puts the cubes into the boxes at random.

We next consider the event A described by saying that 3 cubes go into one box, 3 into another, and 4 into the third. Let us split this event into three mutually exclusive and exhaustive subsets, as follows:

A_1 is the event "3 cubes in box a, 3 cubes in box b, and 4 cubes in box c";
A_2 is the event "3 cubes in box a, 4 cubes in box b, and 3 cubes in box c";
A_3 is the event "4 cubes in box a, 3 cubes in box b, and 3 cubes in box c".

Since no two of these events can occur simultaneously, they are disjoint. Hence

$$P(A) = P(A_1) + P(A_2) + P(A_3).$$

Now let us focus attention on one of these subsets, say A_1. The points of S that are in A_1 have 3 a's, 3 b's, and 4 c's arranged in some order. By Theorem 2–13, Section 2–4, the total number of possible different arrangements of 3 a's, 3 b's, and 4 c's is

$$\frac{10!}{3!3!4!}. \tag{3}$$

Therefore the probability of A_1 is this number times $1/n$:

$$P(A_1) = \frac{10!}{3!3!4!} \times \frac{1}{3^{10}}.$$

Finally, it is clear that formula (3) also gives the number of points in A_2 or in A_3. Therefore the three events A_1, A_2, and A_3 have equal probabilities, and

$$P(A) = 3P(A_1) = 3 \times \frac{10!}{3!3!4!} \times \frac{1}{3^{10}} = \frac{1400}{3^8} \approx 0.213.$$

Example 4 *Sampling problem.* A school staff consists of 30 teachers: 20 women and 10 men. A random sample of 5 teachers is drawn for the discussion of school problems. What is the probability that the sample (a) is composed entirely of women, (b) has exactly 2 men?

Solution a) The number of possible equally likely samples is

$$\binom{30}{5}, \qquad \text{or} \qquad 142{,}506.$$

A sample composed entirely of women can be selected in

$$\binom{20}{5}, \qquad \text{or} \qquad 15{,}504, \quad \text{ways.}$$

Thus 15,504 points of S correspond to the event "sample composed entirely of women". Hence

$$P(5 \text{ women}) = \frac{\binom{20}{5}}{\binom{30}{5}} = \frac{15{,}504}{142{,}506} \approx 0.109.$$

b) A sample composed of exactly 2 men and 3 women can be selected in

$$\binom{10}{2} \times \binom{20}{3}, \qquad \text{or} \qquad 51{,}300, \quad \text{ways.}$$

Therefore

$$P(2 \text{ men and } 3 \text{ women}) = \frac{\binom{10}{2}\binom{20}{3}}{\binom{30}{5}} = \frac{51{,}300}{142{,}506} \approx 0.360.$$

Note The following generalization of the foregoing example is important. Suppose that we have a group of n objects, m A's and w $\overline{A}$'s $(m + w = n)$. From the n objects, we choose a sample of r. What is the probability that the sample contains exactly x A's? The data are collected in Table 3–3.

Table 3–3

	A	$\overline{A}$	Totals
In sample	x	$r - x$	r
Not in sample	$m - x$	$w - r + x$	$n - r$
Totals	m	w	n

We have $\binom{n}{r}$ possible samples. Of these, $\binom{m}{x}\binom{w}{r-x}$ have exactly x A's. Therefore

$$P(x\ A\text{'s}) = \frac{\binom{m}{x}\binom{w}{r-x}}{\binom{n}{r}}.$$

The formula just calculated tells how the probability is distributed among the possible 2×2 tables represented by Table 3–3. Each value of x gives a different table. The distribution of probabilities for such a set of 2×2 tables has a name: the *hypergeometric* distribution.

EXERCISES FOR SECTION 3–5

1. Refer to Example 1 of this section. What is the probability that the first ace appears at the 4th card? At the 47th card?
2. In the first-ace problem (Example 1, Section 3–5), what is the probability that the first ace appears at the 5th card or sooner? At the 49th card or sooner?
3. Refer to the birthday problem (Example 2, Section 3–5). Forty people are in a room. What is the probability that at least two of them have the same birthday? Ten people are accosted at random on the street and their birthdays noted. What is the probability that at least two of them have the same birthday? Estimate the probability that at least two members of the United States Senate have identical birthdays. (See Table 3–2.)
4. There are k people in a room. What is the probability that at least two of them have the same birthmonth? (Assume that all birthmonths are equally likely.)
5. Repeat Exercise 4 for the case in which $k = 5$.
6. Eight commuters drive their cars to the city each day and park at one of three parking lots. If the lots are selected at random, what is the probability that, on a given day, there will be 5 of these 8 cars in one parking lot, 2 in another, and 1 in the third?
7. From a lot of 20 radios a sample of 3 is randomly selected for inspection. If there are 6 defective radios in the lot, what is the probability that the sample (a) is composed entirely of defectives, (b) is composed entirely of nondefectives, (c) is composed of one defective and two nondefectives?
8. A class is made up of 35 students, 20 girls and 15 boys. It is decided to distribute 4 complimentary tickets by lot to 4 members of the class. What is the probability that (a) the tickets go to 4 girls, (b) the tickets go to 2 boys and 2 girls?
9. Show that

$$\binom{52-k}{4} + \sum_{j=1}^{k} \binom{52-j}{3} = \binom{52}{4}, \qquad k = 1, 2, \ldots, 48.$$

[*Hint:* Use Pascal's Rule to combine $\binom{52-k}{4}$ and $\binom{52-k}{3}$, etc.]

Remark. For a probabilistic interpretation of the formula in Exercise 9, divide both sides of the equation by $\binom{52}{4}$ and then refer to the formulas in Example 1, Section 3–5. You get

$$\frac{\binom{52-k}{4}}{\binom{52}{4}} + \sum_{j=1}^{k} \frac{\binom{52-j}{3}}{\binom{52}{4}} = 1.$$

The first term is P(4 aces after kth card), and the summation is P(first ace in 1st or 2nd or 3rd or . . . or kth place). Since the events in parentheses are mutually exclusive and exhaustive, the sum of their probabilities is 1.

10. There were 33 different presidents of the United States from 1789 to 1960. Before looking up their birthdays, what are the odds in favor of your finding that some pair of them had the same birthday? Now look in the *World Almanac*, or elsewhere, and determine the facts.
11. Visitors to historical sites often read inscriptions on tombstones. If a visitor selects a random sample of 30 tombstones, what is the probability of finding two dates of death that are the same month and day? Compare the dates of death of deceased United States presidents. Are any two of them the same month and day?

3-6 RANDOM DRAWINGS

In Chapter 1, we saw that certain physical considerations—fair tossing of symmetrical objects such as coins or dice, or thorough shuffling of identical cards with blindfold drawing—make it reasonable to assign equal probabilities to the points of the sample space. The fair tossing, the thorough shuffling, and the blindfold drawing are physical processes that we use in trying to achieve what is called "randomness"*; that is, in trying to give all outcomes equal chances or, mathematically, to give all points of the sample space equal probabilities.

In addition to the simple experiments presented above, there are more serious experiments where it is desirable to assign some chosen set of probabilities over the sample space of outcomes, and to make the outcomes occur in accordance with these assigned probabilities. In some of these experiments, such as the famous drawing of draft numbers in the fall of 1940, only one trial of the experiment may be made. In others, the experiment may be repeated many times, as in the simulation of the random-walk problem in Chapter 1.

In this section and in the next two, we give examples of such experiments. We discuss ways of achieving desired probabilities, whether equal or unequal, by physical processes, and traps to be avoided in attempting to produce probabilities by these processes. In Section 3-8, we present some physically sound processes for approximating required probability assignments.

Example 1 *Door prize.* At a school dance, a door prize is offered to the couple holding the winning ticket. The sample space consists of the k ticket numbers. The dance committee wants to give all couples an equal chance at the door prize, and therefore wishes each ticket to have the probability $1/k$ of being drawn. The tickets are placed in a bowl and mixed. Then a blindfolded person reaches into the bowl and draws out the winning ticket.

Criticism For the purpose at hand, the procedure seems adequate. If k is large, say 100 or more, the physical stirring is probably rather ineffective,

---

* The expressions "random", "at random", and "randomness" are not exclusively used for situations where equal probabilities are desired, but in everyday language this is usually what is meant. We use these expressions with the everyday meaning unless the text specifically states otherwise.

because slips of paper in a bowl are very difficult to mix thoroughly. One might wonder whether early, middle, and late arriving couples had equal chances, but an investigation of such a question by many repeated trials at successive dances is inappropriate. Each ticket may not have probability $1/k$ of being drawn, but we can only guess about the direction of bias, if one exists.

Example 2 *Selective Service numbers.* During World War II, it became necessary to choose an order in which to draft men into military service. Each man in a Selective Service District was given a number from 1 to 9000. (The significance of the number 9000 is that it was larger than the number of men in any one Selective Service District.)

Each number was placed in an opaque capsule, and the capsules were put into a bowl and stirred. The capsules were then drawn, one at a time, from the bowl until the supply of capsules was exhausted. A sample space of this experiment is the set of 9000! permutations of the numbers. No doubt the intention was to make each permutation have probability 1/9000!. After the stirring, high officials drew the numbers from the bowl, and the early numbers were announced by radio as they were drawn.

There is some question as to the effectiveness of the stirring. The resulting sequence of numbers had surprising properties, and some scientific papers have been written to prove that the drawing was not random. But from our previous discussion of the physical conditions under which we assign equal probabilities to sample points, we recognize that the real issue is whether or not the stirring was thorough. Only thorough stirring gives us confidence in the assignment of equal probabilities.

Here is the frequency distribution of the first 50 numbers drawn:

Numbers between	Frequency
1 — 1000	5
1001 — 2000	0
2001 — 3000	3
3001 — 4000	1
4001 — 5000	7
5001 — 6000	8
6001 — 7000	11
7001 — 8000	7
8001 — 9000	8
Total	50

Note that there are rather few numbers between 1 and 4000. We expect about $\frac{4}{9} \times 50$, or about 22, as opposed to the 9 observed. The actual finding is consistent with the notion that the capsules were in layers and not thoroughly stirred. It is also remarkable that among the first 50 drawn, all 5 numbers below 2000 were between 100 and 199. Of course, every set of 50 numbers drawn from 9000 would be remarkable in some way. It is the correspondence between the special remarkableness of these numbers and the special kind of outcome that we expect from inadequate stirring that raises doubts about the assignment of equal probabilities to the sample space. Better evidence would be a first-hand knowledge of the original mixing process. The moral, as every cook knows, is that thorough mixing is not as easy as it sounds.

Example 3 *Medical experiment.* A doctor proposes a new treatment for a certain disease. It is desired to compare the new treatment with the old. Of 20 patients available for the study, half will be given the new treatment and half the old. The 20 patients are grouped into 10 pairs, each pair consisting of two patients who have the disease in a similar state of advancement. The doctor plans to give one patient in each pair the new treatment, and the other patient the old. (This helps guard against the possibility that the half chosen for the new drug will be mainly severe cases or mainly light cases.)

So far so good. But how does the doctor pick from a pair the patient who is to receive the new treatment? One might think that the doctor's choice is immaterial, but he knows that the matching of pairs, though carefully done, is not perfect, and that he may, from his knowledge of the patients, subconsciously choose for the new treatment the patient who has the better chance of recovery from the disease. This would systematically bias the test in favor of the new drug. How can the doctor defend the experiment from this kind of bias?

The sample space consists of the $2^{10} = 1024$ possible ways of choosing one patient from each of the 10 pairs. Since there is no medical reason for preferring some of these choices to others, the doctor wants each to have probability $\frac{1}{1024}$. What physical process can he use to make his choices? One way is to list the 1024 choices on slips of paper and, after thorough mixing, draw one of the slips. Alternatively, he could arbitrarily assign to one patient of each pair the letter H and to the other, T, and for each pair toss a coin. If the coin falls heads, patient H gets the new drug; if it falls tails, patient T gets the new drug.

Empirically checking or solving a probability problem Sometimes, after working out a complicated problem in applied probability theory, the worker has some uneasiness in his mind about the accuracy of his solution. If the problem does not provide suitable special cases to give an adequate mathematical check, he may turn to random sampling as a check.

We illustrate with an easy problem. Two distinct numbers are drawn in order from the integers 1, 2, 3, . . . , 10, all ordered pairs being equally likely. What is the probability that the larger number of the ordered pair exceeds 5? After you have worked out your answer, check it by actually drawing two cards from 10 properly numbered cards, and then repeating this experiment a large number of times. Compare the empirical results with your theoretical answer.

Some probability problems are so complicated and mathematically intractable that to obtain numerical answers to a single problem, thousands of repetitions of an experiment are executed on high-speed computers. The numerical answer obtained from averaging, or otherwise analyzing, the many results is the one used for practical work. This technique is called the *Monte Carlo* method.

Random drawings for solving nonprobabilistic problems The Monte Carlo method is not reserved for problems in probability. An applied mathematician often finds it convenient to transform a nonprobabilistic problem into a probabilistic one. He then uses experimental methods not unlike those we have described. In other words, he uses empirical probabilistic methods to solve nonprobabilistic problems.

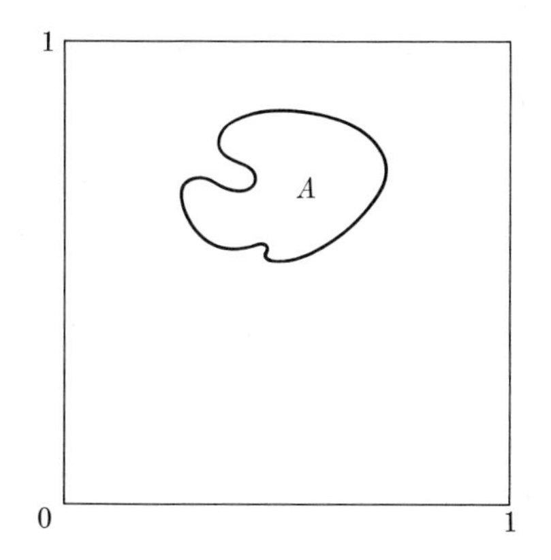

Fig. 3–5 Irregularly shaped region A, whose area is to be determined by a probabilistic process.

The following example illustrates this type of thinking. Suppose that we wish to approximate the area contained in the irregularly shaped region A of Fig. 3–5. Assume that it is possible to drop a point "at random" into the unit square. By "at random" we mean that every rectangular region of area p in the square has probability p of having the point fall in it. Thus the region of unknown area A has probability A that the point falls inside it.

When a point is dropped into the square, it either falls into the special region A, or it does not. Envisage dropping hundreds of points at random into the unit square. The fraction of points that fall inside A is a good estimate of its area. A rough and ready method of performing a suitable experiment is as follows. Draw the figure so that the unit square has sides of 2 inches, and then stand 8 or 10 feet away and throw darts at it. Only count throws where the dart hits the square. The area A is estimated by the ratio of the number of hits in A to the number of hits in the square.

If the original area A cannot be contained in a unit square, we must adjust the units in both dimensions so that it can. Then after getting the ratio of hits in the region to those in the unit square, we would have to multiply that ratio by the square of the ratio of the old units to the new. For example, if we had divided the original linear units by 100 for each dimension to get the region into the unit square, we would multiply the ratio of hits obtained from the experiment by 10,000 to get the final estimate of the area in the original units.

3–7 RANDOM NUMBERS

After one has tossed coins, drawn cards, thrown dice, and so on, for a large number of times, he begins to wish for faster and better methods of performing mathematically equivalent experiments. If we wish to draw 500 sets of three cards from a pack, the shuffling is slow and tiresome, and fatigue leads to poor shuffling and lack of randomness. Cards, marbles, and slips of paper are all very well if there are only a few to handle. If there are hundreds or, as often happens, thousands, the task gets out of hand. As a result of such considerations—slowness of handling physical objects and lack of randomness—random numbers were invented to provide a basis for mathematical experiments to simulate physical ones.

What are random numbers? Random numbers are formed from ordinary digits successively generated by a random process. The series of digits may be of almost any length desired. Published tables of random numbers have up to 1,000,000 digits.

Construction of a table of random digits Most tables of random digits are constructed by setting up a sample space consisting of the ten digits 0, 1, 2, 3, 4, 5, 6, 7, 8, 9. Some physical process is devised that gives good positive evidence that each of these digits has probability $\frac{1}{10}$ of occurring on each trial and that the separate trials are independent. Then the process is set in motion, and thousands of digits are generated and written down in the order in which they occur. Table 3–4 is a short table of random digits generated by such a process; Table I–A at the back of the book is a larger sample. One way to generate such digits is to roll a die and toss a coin, but ignore the two ordered pairs where the 6 appears on the die. Label

$$(H, 1),\ (H, 2),\ (H, 3),\ (H, 4),\ (H, 5),\ (T, 1),\ (T, 2),\ (T, 3),\ (T, 4),\ (T, 5)$$

with the digits

$$0,\ 1,\ 2,\ 3,\ 4,\ 5,\ 6,\ 7,\ 8,\ 9, \quad \text{respectively.}$$

If the coin and die are true, the probabilities are $\frac{1}{10}$ for each digit.

Persons wishing to make random drawings use these tables as described in Section 3–8. Thus they use the random process behind the table, instead of a process of their own devising. The tables are speedy to use, and are based on a better physical process than one we are likely to construct in a few minutes for ourselves.

It must be emphasized again that a set of digits is not in and of itself random or not random. Thus, if one writes the digits 825 and the digits 999, there is

Table 3–4 Brief table of random numbers*

Rows	Columns 1–5	6–10
1	22719	92549
2	17618	88357
3	25267	35973
4	88594	69428
5	60482	33679
6	30753	19458
7	60551	24788
8	35612	09972
9	43713	18448
10	73998	97374

no reason to say that the first set is random and the second not. We call a set of numbers random if the following conditions hold:

1) there are known probabilities in the sample space for those numbers; and
2) there is a physical process that generated the successive numbers with good assurance that each element of the sample space had the probability assigned to it.

The words "random digits" are an abbreviation for "randomly generated digits", where each digit has probability $\frac{1}{10}$.

3–8 USE OF TABLES OF RANDOM DIGITS

Random-digit tables have a great variety of uses. We shall offer a few examples by way of illustration.

To begin drawing random numbers from a random-digit table is sometimes an awkward matter. However, if you own your own table, the difficulty is easily overcome. Merely start at the beginning of the table and continue systematically until you have used as many digits as your problem requires. Then check off the used digits, and start the next problem with the next digit.

Example 1 In an earlier example, we wished to draw pairs of distinct numbers with equal probabilities from the 10 digits 1, 2, . . . , 10. In a random-digit table, it is often convenient to consider the digit "0" as "10"; we shall do so in this example.

---

* Reprinted by permission of the publisher, The Free Press of Glencoe, Illinois, from *A Million Random Digits with 100,000 Normal Deviates*, copyright 1955 by The RAND Corporation.

Let us start by sampling at the top left of Table 3–4 with the five digits 22719. Since the first digit is 2, in our first sample of two numbers the first number is 2. Reading across, we see that the next digit is 2. We ignore it, since the numbers in each of our ordered pairs must be distinct. The next digit is 7. Therefore, our first pair of numbers is (2, 7). Continuing, we note that the next digit in the table is 1, so we record 1 for the first number of our second ordered pair. The next digit in the table is 9, so our second pair is (1, 9).

Our line of five digits is now exhausted, so we proceed to the second row, consisting of the digits 17618. The first of these digits is 1 and the second is 7, so our next pair is (1, 7). We continue in this manner until we get as many pairs as we want.

Example 2 *Medical problem.* The doctor designing the medical experiment with 20 patients in pairs (Example 3, Section 3–6) might proceed as follows. He first lists his 10 pairs of patients in order in two columns:

Jones	Smith
Johnson	Williams
Hoffman	Wood
Ross	Farlow
Zanetti	Wilson

⋮

Suppose that he has used Table 3–4 through the first two sets of five digits. Then he begins with column 1, row 3, and proceeds to choose for the new treatment the patient in the first column if the digit is 0, 1, 2, 3, 4; otherwise, he chooses the patient in the second column. The random digits in the table are 25267. Therefore, in the first five pairs, Jones, Williams, Hoffman, Farlow, and Wilson are to be given the new treatment.

Example 3 *Coverage of blood donors.* Suppose that in a human population the 4 blood types A, B, C, and D represent 40%, 30%, 20%, and 10% of the people, respectively. It is proposed that every hospital have 4 donors, randomly selected, to give blood. Under this plan, what fraction of the population, on the average, would not find its type of blood available?

Solution Assign the digits 1, 2, 3, and 4 to A, the digits 5, 6, and 7 to B, the digits 8 and 9 to C, and the digit 0 to D. If we draw the first 4 digits from Table 3–4, we get 2271, which means that types A and B are represented, but not C or D. Hence 30% of the population is not covered in this sample. Ten such samples gave the following percentages:

$$30\%,\ 30\%,\ 30\%,\ 50\%,\ 0\%,\ 20\%,\ 60\%,\ 30\%,\ 30\%,\ 10\%.$$

The average is 29%, which means that 29% of the population would find the blood bank broke. Of course, to get a more accurate estimate of the percentage, we would have to take many more samples.

Example 4 *The problem of isolates.* Each of 10 individuals randomly chooses 2 of the others as "friends". A person whom no one chooses is called an "isolate". Estimate (a) the probability that no one is an isolate, (b) the mean or average number of isolates.

Solution Number the individuals 1, 2, 3, 4, . . . , 9, 0. Then allow each individual to draw 2 numbers, using the random numbers of Table 3–4 in the following manner. Individual 1 takes the first two numbers, other than his own, on line 1: he gets 2 and 7. Individual 2 uses line 2 similarly and gets 1 and 7; and so on. Table 3–5 illustrates one run of the experiment in which individual 9 was the only isolate.

Table 3–5 One run of the isolate experiment

Chooser	Individual chosen									
	1	2	3	4	5	6	7	8	9	0
1		x					x			
2	x						x			
3		x			x					
4					x			x		
5						x				x
6			x							x
7						x				x
8			x		x					
9			x	x						
0			x				x			
	x	x	x	x	x	x	x	x	0	x

Here is a summary of 10 runs of the experiment obtained using Table I at the back of the book:

Number of isolates:	0	1	2	3	Total
Frequency:	3	6	1	0	10

These data lead to the following estimates:

$$P(\text{no isolates}) = \tfrac{3}{10},$$

$$\text{average number of isolates} = \frac{6(1) + 1(2)}{10} = 0.8.$$

Further runs could nail these values down more precisely, and theory given later shows that the mean is exactly $10(\frac{7}{9})^9$, or 1.04, approximately.

Note Solving this problem by the methods of Chapter 2 is a challenge that you may wish to take up.

Example 5 *Not using all digits.* Suppose that we wish a distribution over a sample space of points A, B, and C with probabilities $\frac{1}{6}$, $\frac{2}{6}$, and $\frac{3}{6}$, respectively.

A convenient technique is to make the following correspondence:

Digit in table	Point in sample space
0	A
1	B
2	B
3	C
4	C
5	C
6, 7, 8, 9 (ignore)	No point

Thus we ignore the digits 6, 7, 8, 9 when they occur in the table. The only random digits of interest are 0, 1, 2, 3, 4, 5. They have total probability 1 and are equally likely, so each has probability $\frac{1}{6}$.

Example 6 *Obtaining finer probabilities.* Suppose that we require samples from a sample space of four points A, B, C, D with probabilities 0.11, 0.25, 0.34, 0.30, respectively. Instead of looking at the sample space of random digits with 10 equally likely points, we could consider two successive digits in the table as one of the 100 equally likely two-digit numbers 00, 01, 02, 03, . . . , 10, 11, . . . , 20, . . . , 99. Each of these has probability $\frac{1}{100}$ of occurring. (Why?)

We then set up a correspondence as follows:

Two-digit random number	Point in sample space
00–10	A
11–35	B
36–69	C
70–99	D

With two-digit numbers, most workers find it easier to read down a column than across a row. For example, if we start in row 4, columns 1 and 2, we read the random digits 88, 60, 30, 60, . . . , so our sample points are D, C, B, C, ...

Example 7 *Drawing a sample from a list.* Suppose that, for survey purposes, we wish to draw a sample of 200 students from the 800 students of a school. One way is to assign to each student one of the three-digit numbers 001, 002, . . . , 800, then enter the random-digit table and examine successive three-digit numbers. If, in Table 3–4, we start with columns 6, 7, 8 and row 1 and read down, the first three numbers obtained are 925, 883, and 359.

Each three-digit number is either the number of a student in the list, or it is not. If not, ignore the number, and proceed to the next. If the three-digit number belongs to a student in the list, then we check that student's name for the sample unless he has previously been checked, in which case we ignore that number and proceed to the next. The process is continued until 200 students have been checked. These students constitute the random sample from the population of 800 students.

EXERCISES FOR SECTIONS 3-6, 3-7, AND 3-8

Note. In the following exercises, use the random-digit table at the back of the book.

1. Describe a physical process for randomly choosing 2 persons from a group of 10. Set up a sample space and assign probabilities to it.
2. Write a description of a random drawing problem of your own, set up a sample space, and assign probabilities to it. Then describe a physical process for carrying out the drawings.
3. Compute the probability for the problem described near the bottom of page 104 and execute the experiment.
4. On a rectangular coordinate system, draw a square with vertices $(0, 0)$, $(0, 1)$, $(1, 1)$, and $(1, 0)$. With center at $(0, 0)$ and radius 1 unit, draw a quarter circle within the square. How can you use the Monte Carlo method, a table of random digits, and the foregoing figure to estimate the value of π?
5. Take a random sample of 30 pages of this book. Record whether or not each page of the sample has a figure or a table on it. Estimate the fraction of pages in the book that have figures or tables.
6. In Example 7 of Section 3-8, what is the sample space, and what is the probability attached to each sample point?
7. For phoning in connection with a TV program, it is desired to draw three names at random from a large telephone book (excluding the yellow pages). How would you draw the three names?
8. Suggest a method of using the random-digit table to obtain Selective Service numbers. (See Example 2, Section 3-6.)
9. Suppose, in the blood donor problem of Example 3, Section 3-8, that there were 8 donors, randomly selected, available in each hospital. Use 10 samples to estimate the fraction of the population that, on the average, would not be covered.
10. How would you modify Example 5 of Section 3-8 if the sample points were A, B, C, and D, with probabilities 0.2, 0.3, 0.4, and 0.1, respectively?
11. Modify Example 6, Section 3-8, to accommodate five sample points A, B, C, D, and E with probabilities 0.23, 0.32, 0.35, 0.06, and 0.04, respectively.
12. (Refer to Example 4 of Section 3-8.) Do 10 runs of the experiment on isolates. Use your data to estimate (a) P(no one is an isolate) and (b) the mean number of isolates.

3-9 CONCLUSION

By now we should realize that a mathematical theory does not always work out perfectly when applied to real-life situations, and that its value depends upon finding the conditions, if any, under which the theory is a close approximation to real life. Thus, in thinking about probabilities associated with the faces of a die, our mathematical die is a perfect homogeneous cube. Each face has probability $\frac{1}{6}$ of appearing. A brand new physical die bought from a reliable manufacturer is a close approximation to our theoretical cube. We expect the true probabilities for the physical die to be extremely close to, but not exactly equal to, $\frac{1}{6}$. A worn die might have probabilities rather far from $\frac{1}{6}$.

However, if we do not know the probabilities, all is not lost. It is one function of probability theory to state what the frequencies of various outcomes are when the initial probabilities are known. But it is also the function of statistics to make inferences about the values of the true probabilities on the basis of experimental results when the true probabilities are unknown.

All that is lost, as we know less and less about a die, are the values of the probabilities associated with its faces; *we do not lose the mathematical theory or the laws.* Later, we shall develop a more complete theory of probability for unknown probabilities and unequally likely events. From such theories, we can develop statistical methods for important problems.

It is fortunate that these methods work for initially unknown probabilities, because in most scientific and engineering work the probabilities are not known, but must be estimated from observations. When we come to real-life situations, we rarely assume that the ideal probabilities, obtained from counting possibilities, represent the physical situation. For example, we assume that a production process has some true, but unknown, probability of turning out a defective light bulb. We take observations and use them to estimate the unknown probability.

In practical work, idealized probabilities such as those obtained from counting are often treated as hypotheses that are available for a test. One might have the idealized notion that as many males are born as females, that is, that the probability of a male birth is $\frac{1}{2}$. After looking at the records for the United States in the years 1935–1952, one would soon be convinced that, consistently, more boys than girls are born. In 1950, there were 1,823,555 boys and 1,730,594 girls* born. We might then estimate that the probability of a male birth is about 1,823,555/3,554,149, or about 0.513. And we would abandon the notion that the true probability is 0.5, except as a rough approximation.

In the rest of this book, we shall consider both equally likely outcomes and unequally likely outcomes; and we shall thereby open the way to some new and challenging developments of the theory of probability and statistics.

---

* *The World Almanac—1956*, New York World Telegram, 1956, p. 302.

4 General theory of probability for discrete sample spaces

4–1 UNEQUALLY LIKELY OUTCOMES: THUMBTACK EXPERIMENTS

In Chapter 3 we discovered some general results in sample spaces with equally likely outcomes. For example, we found that

$$P(A \cup B) = P(A) + P(B) - P(A \cap B) \tag{1}$$

for events A and B.

In this chapter, we adopt a set of axioms and definitions that can be applied even when outcomes in a sample space are not equally likely. The axioms are reasonable and sufficient for proving general results like Eq. (1). But before stating these axioms, we consider a simple experiment to illustrate why some axioms, or assumptions, are needed. The experiment has these properties:

1) there are exactly two outcomes;
2) each outcome has a definite probability whose value is between zero and one;
3) the outcomes are not equally likely;
4) there is no obvious way to assign probabilities to the two outcomes.

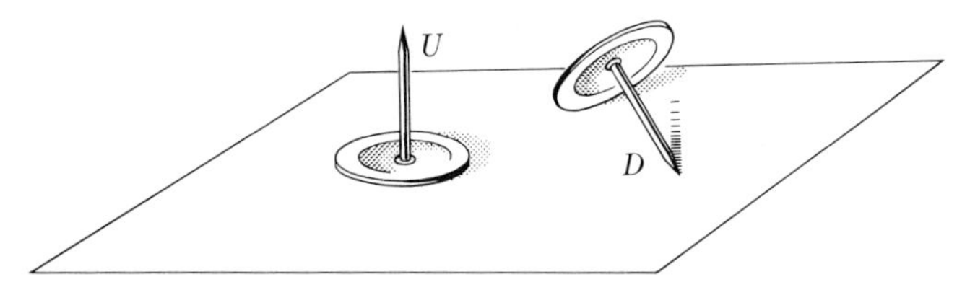

Fig. 4–1 Thumbtack.

Thumbtack experiment Imagine that an ordinary thumbtack is tossed or dropped onto a hard surface, where it bounces before coming to rest. When the thumbtack comes to rest, it points up (U) or down (D), as in Fig. 4–1.

These are the two possible outcomes of the experiment, just as head (H) and tail (T) are the two possible ways a coin can land. Each time the thumbtack is tossed, it seems reasonable to suppose that the outcomes U and D have fixed probabilities $P(U) = p$ and $P(D) = 1 - p$. But we cannot say, just by looking at the thumbtack, exactly what number between 0 and 1 is equal to p. In particular, there is no reason to believe that $p = \frac{1}{2}$, because the two cases U and D need not be equally likely.

How might we get some idea of the value of $P(U)$? Let us specify conditions and toss a thumbtack 50 times, say, then record the data and calculate the proportion of times the thumbtack falls U. This proportion is not $P(U)$: it is an *estimate* of $P(U)$. We cannot hope to get the probability exactly from such experiments. Even for apparently symmetrical coins, the proportion that actually falls heads in a sequence of 50 tosses may not be $\frac{1}{2}$. But if we specify that the coin be a thin flat disc, not out of shape like a bent bottle cap, and that it be given a vigorous toss into the air, with a spinning motion that turns it over and over many times before it lands, then it seems reasonable that the two cases, head and tail, are equally likely: $P(H) = P(T) = \frac{1}{2}$. If thumbtack tossing ever becomes a popular indoor sport, some physicist or statistician will no doubt develop a theory that predicts quite well, for a few given dimensions, materials, and tossing specifications, the probability that the tack falls U. The fact that we can arrive at reasonable theoretical probabilities for cards, dice, and coins, and cannot easily do so for a thumbtack does not lessen for us the reality of the probability $P(U)$.

Suppose our thumbtacks fall U about 40 times out of 100. We would estimate $P(U)$ to be 0.4. Now if 0.4 were the true probability, we could apply to this number the ideas we worked out earlier for coins and dice. Instead, let us suppose there is a true but unknown value for $P(U)$, say p.

Example 1 *Two tosses.* If we toss the thumbtack twice and its probability of falling U on a single toss is p, what is the probability that it falls U both times?

Solution Our sample space of ordered pairs is represented thus:

		Second toss	
		U	D
First toss	U	(U, U)	(U, D)
	D	(D, U)	(D, D)

Let event A be U on the first trial, and event B be U on the second trial. The point (U, U), whose probability we want, comprises the event $A \cap B$. Hence the probability that the thumbtack falls U both times is $P(A \cap B)$.

We assume that the tosses are independent. From our experience with sample spaces of equally likely outcomes, we know that in such spaces the probability of the intersection of independent events is the product of the individual probabilities. So we might assume this to be true more generally and assign probability

$$P(A) \cdot P(B) = p^2$$

to the point (U, U). Another line of reasoning that leads to the same result is the following. Consider a long sequence of pairs of tosses of a thumbtack. In this sequence, the proportion where the first toss of a pair results in U is approximately p. And approximately the proportion p of *these* fall U on the second toss as well. Hence we expect the long-run proportion that fall (U, U) to be about p^2. Thus both lines of reasoning suggest that we assign probability p^2 to the outcome (U, U); and we do so:

$$P(\{(U, U)\}) = p^2. \tag{2}$$

Notation The parentheses and braces in Eq. (2) are used in the following ways: the inside parentheses, in (U, U), are used in the same way we use parentheses to designate a point, say $(3, 4)$, in coordinate geometry. Thus (U, U) is a sample point. Next, the braces, in $\{(U, U)\}$, indicate a set whose only element is the point (U, U). Finally, the outer parentheses are used as they are, for example, in $P(E)$, denoting probability of a set E. However, the weird collection of braces and parentheses in Eq. (2) is almost too frightening to live with, so we shall adopt the logically less accurate, but typographically more pleasing, notation $P(U, U)$, and write simply

$$P(U, U) = p^2.$$

Similarly, we assign to the outcome (U, D) the probability

$$P(U, D) = pq,$$

where

$$q = 1 - p$$

is the probability that a thumbtack lands "point down". This method of assigning probabilities makes the total probability for all points in the sample space one because

$$p^2 + pq + qp + q^2 = (p + q)^2 = 1.$$

If, for example, we assume that 0.4 is the true probability $p = P(U)$, then

$$P(U, U) = (0.4)(0.4) = 0.16.$$

Similar considerations would give us

$$\begin{aligned} P(U, D) &= P(U) \cdot P(D) \\ &= (0.4)(0.6) = 0.24 \end{aligned}$$

and

$$P(D, D) = (0.6)(0.6) = 0.36.$$

EXERCISES FOR SECTION 4-1

1. In the thumbtack example, find the probabilities of the four possible outcomes on two tosses, assuming that $P(U) = 0.3$.
2. Use the results of Exercise 1 to find the probability that (a) at least one toss falls U; (b) the second toss falls D; (c) the second toss falls D, given that the first toss falls U [compare your answer with the answer to part (b)]; (d) both tosses fall alike.
3. How would you assign the probability that two tosses of a thumbtack fall U, given that both fall alike?
4. Suppose a thumbtack, with $P(U) = p$ and $P(D) = q = 1 - p$, is independently tossed three times in succession. List a sample space for the possible outcomes of this experiment. Assign probabilities to its points.
5. (Continuation.) In Exercise 4, take $P(U) = p = 0.4$, and find the probability that the thumbtack fell U twice and D once in the three tosses.
6. (Continuation.) Do the probabilities you assigned to the sample points in Exercise 4 add up to 1, as they should?
7. A thumbtack with $P(U) = p = 0.2$ is tossed four times. What probabilities would you assign to the following outcomes?

   a) $UUUD$ b) $UUDU$ c) $UDUU$ d) $DUUU$ e) $UUDD$
   f) $UDUD$ g) $DUUD$ h) $UDDU$ i) $DUDU$ j) $DDUU$
   k) U three times and D once l) U twice and D twice

8. Suppose that the length of the shaft of the thumbtack in Fig. 4-1 varies from 0 to some large positive value L. What would you guess $P(U)$ to be when the length of the shaft is 0? When it is L? Discuss.

4-2 SAMPLE SPACE AND PROBABILITY

In this section, we develop the axioms of probability in relation to the familiar notion of a *sample space* of an experiment. A sample space, we recall, is a set of elements such that any performance of the experiment produces a result that corresponds to exactly one element in the set. We restrict our attention to *finite* sample spaces (i.e., those with a finite number of sample points), or to *countably infinite* sample spaces (i.e., those whose sample points can be put into 1-1 correspondence with the positive integers). We call such sample spaces *discrete.* We begin by dealing with discrete sample spaces that are finite. Later, in Section 4-10, we shall informally extend our results to discrete sample spaces that are countably infinite. In a discrete sample space, every set of sample points is called an *event.* An *elementary* event contains exactly one sample point.*

* Hence different elementary events are always *mutually exclusive.* All the elementary events together form a *partition* of the sample space.

If a performance of the experiment produces a result that corresponds to a point in the subset E, we say that the *event E occurs*. The empty set is also an event, but it never occurs, since no sample points are in it.

The next example illustrates events in a sample space of 4 sample points.

Example 1 *Bond issue for new school.* A survey is made in connection with the planning for a new high school. Each of 100 voters is asked two questions:

1. Do you favor a bond issue to finance the building of the school?
2. Do you own property in the school district?

Discussion Each voter in the survey belongs to one of the following four categories:

e_1: favors issue and owns property,

e_2: favors issue and does not own property,

e_3: opposes issue and owns property,

e_4: opposes issue and does not own property.

The experiment of surveying 100 voters and classifying them is the same as 100 performances of the simpler experiment of asking just one voter and classifying him. The set

$$S = \{e_1, e_2, e_3, e_4\}$$

is an appropriate sample space for this single-voter experiment, since each performance must result in exactly one of these four possibilities. This sample space S also provides a scheme for tallying the results of the 100-voter experiment.

The nonempty subsets of S are

$$\begin{array}{llll} \{e_1\}, & \{e_1, e_2\}, & \{e_2, e_4\}, & \{e_1, e_3, e_4\}, \\ \{e_2\}, & \{e_1, e_3\}, & \{e_3, e_4\}, & \{e_2, e_3, e_4\}, \\ \{e_3\}, & \{e_1, e_4\}, & \{e_1, e_2, e_3\}, & \{e_1, e_2, e_3, e_4\}. \\ \{e_4\}, & \{e_2, e_3\}, & \{e_1, e_2, e_4\}, & \end{array}$$

Each of these subsets is an *event*. Technically, the empty set is also an event, though trivial. The subsets

$$E_1 = \{e_1\}, \qquad E_2 = \{e_2\}, \qquad E_3 = \{e_3\}, \qquad \text{and} \qquad E_4 = \{e_4\}$$

contain just one sample point apiece; they are the *elementary events*. Every event, other than the empty set, is the union of one or more distinct elementary events. These events can also be described verbally; for example, "favors the bond issue" describes the event $E_1 \cup E_2 = \{e_1, e_2\}$; and "owns property or opposes the issue" describes $\{e_1, e_3, e_4\}$. The event $\{e_1, e_2, e_3, e_4\}$ is the entire sample space S; it may be described by "person in the survey".

Later we shall want to see how we might attach probabilities to the sets in this sample space and continue the example. But we delay this in order to introduce the general idea of assigning probabilities to more general sample spaces.

Note In set theory, a logical distinction is made between a set E_1, that contains a single point e_1, and the point itself. We have made the distinction above by writing $E_1 = \{e_1\}$, to indicate that E_1 is the *set* whose only element is the *point* e_1. This permits us to write the probability of E_1 as $P(E_1)$ rather than as $P(\{e_1\})$. However, we shall not always make such distinctions, and may write this probability simply as $P(e_1)$, without the inner braces. This usage is an abbreviation.

Probability Given a sample space S, we need to assign probabilities to its events. We assume that the sample space has a finite number n of sample points:

$$S = \{e_1, e_2, \ldots, e_n\}.$$

To each event in S we assign a number, called its *probability*. We now adopt the following axioms, or postulates, about these probabilities.

AXIOMS FOR PROBABILITY IN FINITE SAMPLE SPACES

Axiom I *Positiveness.* The probability assigned to each event is positive or zero.

Axiom II *Certainty.* The probability of the entire sample space is 1.

Axiom III *Unions.* If A and B are mutually exclusive events, then

$$P(A \cup B) = P(A) + P(B).$$

We call the first of these the *positiveness* postulate because probabilities are never negative; they are either positive or zero. For most purposes, events with zero probability, in a finite sample space, can be deleted.

The second postulate is called the *certainty* postulate because it says, in effect, that the probability of an event that is bound to occur is 1. The entire sample space is just such a certain event because it contains all possible outcomes of the experiment.

The third postulate, concerning the probability of the *union* of two mutually exclusive events, permits us to focus attention on the *elementary* events when we are assigning probabilities. For, as the next theorem shows, as soon as we know the probabilities of the elementary events, the probabilities of all other events are uniquely determined by Axiom III.

4–1 Theorem

Let A be an event in a finite sample space S. If A is the empty set, then $P(A) = 0$. If A is nonempty then, $P(A)$ is the sum of the probabilities of the elementary events whose union is A.

Proof First, suppose $A = \phi$, the empty set. In Axiom III, take $A = \phi$ and $B = S$, the entire sample space. Then A and B are mutually exclusive, because

ϕ is empty, so

$$P(\phi \cup S) = P(\phi) + P(S). \tag{1a}$$

Also, since S is the entire sample space, the union of ϕ and S is S:

$$\phi \cup S = S.$$

Hence

$$P(\phi \cup S) = P(S). \tag{1b}$$

Subtracting Eq. (1b) from (1a), we have

$$0 = P(\phi),$$

and therefore $P(A) = 0$ if $A = \phi$.

Next, suppose A is nonempty and is the union of m distinct elementary events $E_1, E_2, \ldots, E_m$, where $E_i = \{e_i\}$, $i = 1, 2, \ldots, m$. For the purpose of the present proof, we assume that the sample points have been labeled in such a way that the m points in A are the first m points of the sample space. This simplifies the notation without affecting the validity of the proof.

If $m = 1$, then $A = E_1$ and $P(A) = P(E_1)$. If $m = 2$, then $A = E_1 \cup E_2$ is the union of two mutually exclusive events, because E_1 and E_2 are distinct elementary events. Axiom III gives the result,

$$P(A) = P(E_1) + P(E_2). \tag{2}$$

If $m = 3$, then $A = (E_1 \cup E_2) \cup E_3$ and, again by Axiom III,

$$P(A) = P(E_1 \cup E_2) + P(E_3).$$

Application of Eq. (2) leads to

$$P(A) = P(E_1) + P(E_2) + P(E_3).$$

The extension to values of m greater than 3 is readily made by mathematical induction. We assume the theorem is true for $m - 1$ elementary events and write A as the union of E_m and $E_1 \cup E_2 \cup \cdots \cup E_{m-1}$. When we apply Axiom III, we get

$$\begin{aligned} P(A) &= P(E_1 \cup E_2 \cup \cdots \cup E_{m-1}) + P(E_m) \\ &= P(E_1) + P(E_2) + \cdots + P(E_{m-1}) + P(E_m), \end{aligned}$$

the desired extension. □

In some applications, we feel that the n sample points are equally likely to occur, and then we assign to each elementary event the probability $1/n$. But in many applications, the elementary events have unequal probabilities. In the school bond example, we would assign to each category a probability equal to the proportion of voters in the school district who are in that category, if the proportions were known. Thus, if it were known that 40% favor the issue and own property, 20% favor the issue and do not own property, 30% oppose the issue and own property, and 10% oppose the issue and do not own property, we would assign probabilities to the elementary events as follows:

$$P(E_1) = 0.4, \qquad P(E_2) = 0.2, \qquad P(E_3) = 0.3, \qquad P(E_4) = 0.1.$$

We also arrange this information in the form of a two-by-two array giving the sample space and associated probabilities shown in Table 4-1. The purpose of this arrangement is to focus attention on the two attributes or characteristics that the survey is designed to study: namely, the state of property ownership, for one, and the attitude toward the bond issue, for the other. Each person in the survey either does or does not own property in the district, and either does or does not favor the bond issue. When we provide for a "yes" or "no" answer to each of the two survey questions, we get the four categories described earlier. Such a two-by-two table is often used to study a possible relationship between two characteristics or attributes.

Example 2 If the probabilities of the elementary events are those given in Table 4-1, what is the probability that a voter selected at random (a) is in favor of the bond issue? (b) is in favor of the bond issue or owns property in the district? (c) opposes the bond issue or does not own property in the district?

Solution Every event in the sample space can be expressed in terms of the events O and F and their complements $\overline{O}$ and $\overline{F}$, where

$$F = \{e_1, e_2\} = E_1 \cup E_2 \text{ corresponds to "favors bond issue"}$$

and

$$O = \{e_1, e_3\} = E_1 \cup E_3 \text{ corresponds to "owns property in the district".}$$

The events in question have these probabilities:

$$\begin{aligned} P(\text{person favors bond issue}) &= P(F) = P(E_1) + P(E_2) \\ &= 0.4 + 0.2 = 0.6, \end{aligned}$$

$$\begin{aligned} P(\text{favors bonds or owns property}) &= P(F \cup O) \\ &= P(E_1) + P(E_2) + P(E_3) \\ &= 0.4 + 0.2 + 0.3 = 0.9, \end{aligned}$$

$$\begin{aligned} P(\text{opposes bonds or does not own property}) &= P(\overline{F} \cup \overline{O}) \\ &= P(E_2) + P(E_3) + P(E_4) \\ &= 0.2 + 0.3 + 0.1 = 0.6. \end{aligned}$$

Theorems In Section 3-2, we proved the following theorems. (Note that their proofs, as given earlier, did not assume that the elementary events in S were equally likely. Those proofs are valid in any finite sample space, so we need not repeat them here.)

Table 4–1

	Owns property, O	Does not own property, $\overline{O}$
Favors bond issue, F	$E_1, p_1 = 0.4$	$E_2, p_2 = 0.2$
Opposes bond issue, $\overline{F}$	$E_3, p_3 = 0.3$	$E_4, p_4 = 0.1$

4–2 Theorem: *A* or *B* (or both)

$$P(A \cup B) = P(A) + P(B) - P(A \cap B). \tag{3}$$

4–3 Theorem: Mutually exclusive events

If $A_1, A_2, \ldots, A_m$ are mutually exclusive, then

$$P(A_1 \cup A_2 \cup \cdots \cup A_m) = P(A_1) + P(A_2) + \cdots + P(A_m). \tag{4}$$

4–4 Theorem: Complementary events

$$P(\overline{A}) = 1 - P(A). \tag{5}$$

In the bond example, we used elementary events to find the probability that a voter is in favor of the bond issue or owns property in the district:

$$P(F \cup O) = P(E_1) + P(E_2) + P(E_3) = 0.9.$$

The only sample points not in $F \cup O$ are those that belong neither to F nor to O; they therefore belong to $\overline{F} \cap \overline{O}$. Hence the complement of $F \cup O$ is $\overline{F} \cap \overline{O}$ and, by Theorem 4–4,

$$\begin{aligned} P(F \cup O) &= 1 - P(\overline{F} \cap \overline{O}) \\ &= 1 - 0.1 = 0.9. \end{aligned}$$

A third method uses Eq. (3):

$$\begin{aligned} P(F \cup O) &= P(F) + P(O) - P(F \cap O) \\ &= 0.6 + 0.7 - 0.4 = 0.9. \end{aligned}$$

Note that we would get the absurd result 1.3 if we forgot to subtract $P(F \cap O)$ from $P(F) + P(O)$, because the events F and O are *not* mutually exclusive.

EXERCISES FOR SECTION 4–2

1. Let A and B be events in a sample space S, such that

$$P(A) = 0.4, \qquad P(B) = 0.3, \qquad P(A \cap B) = 0.2.$$

Find the probabilities of:

a) $A \cup B$ b) $\overline{A}$ c) $\overline{B}$ d) $\overline{A} \cap B$ e) $A \cup \overline{B}$ f) $\overline{A} \cup \overline{B}$

2. In the two-dice example of Chapter 1, Table 1–11, the sample space is

$$S = \{(r, c) : r \text{ and } c \text{ are integers from 1 through 6}\}.$$

Let A be the event described by $r \leq 3$ and B the event described by $c \geq 4$. Find the probabilities of:

a) A b) B c) $A \cap B$ d) $A \cup B$
e) $\overline{A}$ f) $\overline{B}$ g) $\overline{A} \cup \overline{B}$ h) $\overline{A} \cap \overline{B}$

3. In Exercise 2 above, describe each of the following events in different mathematical symbols or in words:

a) $A \cap B$ b) $A \cup B$ c) $\overline{A} \cup \overline{B}$ d) $\overline{A} \cap \overline{B}$

4. *Color blindness.* Assume that 5% of males and 1% of females are color-blind. Assume, furthermore, that 50% of the population is male and 50% female. A person is to be selected at random from this population, and that person's sex and state of vision (color-blind or not) are to be recorded. List a sample space for the experiment of sampling one person. Assign probabilities to the elements of the sample space. What is the probability that (a) the person is male and color-blind? (b) the person is female and color-blind? (c) the person is color-blind? [Genetic theory suggests that if p is the proportion of color-blind males, p^2 is the proportion of color-blind females, so $\frac{1}{4}$% rather than 1% may be a more realistic figure for females in this example.]

5. Three events A, B, and C are mutually exclusive and their union is the sample space. If $P(A) = 3P(B)/2$ and $P(B) = 2P(C)$, find the probabilities of the three events.

6. A sample space is composed of n mutually exclusive events, of which $n - 1$ have identical probabilities and the remaining one has probability as large as $r + 1$ of the others. Find the probabilities of the two kinds of events.

7. The probabilities of the mutually exclusive events A and B are related as $P(B) = [P(A)]^2$, and $A \cup B = S$, the sample space. Find $P(A)$, (a) exactly, and (b) to two decimals.

8. The event C is twice as likely as A, and B is as likely as A and C together. The events are mutually exclusive and together they exhaust the sample space. Find their probabilities.

9. If one letter is chosen at random from the word *boot* and one letter from the word *toot*, what is the probability that the two letters are the same? (First, set up a sample space and assign probabilities to its elements.)

10. Two letters are drawn at random, without replacement, from the word *memento*. What is the probability that some arrangement of the two letters spells *me*?

11. Terry is batting in a ball game with no one on base. Assume the possible outcomes of this experiment, and the associated probabilities, to be as follows:

$P(\text{striking out}) = 0.35$, $P(\text{base on balls}) = 0.21$,
$P(\text{flying out}) = 0.17$, $P(\text{grounding out}) = 0.10$,
$P(\text{getting extra-base hit}) = 0.04$, $P(\text{being hit by the pitcher}) = 0.01$,
$P(\text{getting a single}) = 0.12$.

Find the following probabilities:

a) P(getting at least to first base safely),
b) P(having to hurry toward first base),
c) P(getting a hit),
d) P(getting put out).

12. In the school bond example (Table 4–1), if all the voters who do not own property in the district join the property owners who are opposed to the bond issue in voting against it, and everyone votes, how will the vote on the bond issue turn out?
13. Prove the theorem: If A is a subset of B and $P(B) = 0$, then $P(A) = 0$.
14. The statement "A implies B" means that every occurrence of A is also an occurrence of B. Explain why this is the same as saying that A is a subset of B in the sample space. Prove that if A implies B, then $P(A) \leq P(B)$.

4–3 INDEPENDENT EVENTS

The definitions of independence and dependence in Section 3–3 also apply in sample spaces where the outcomes are not equally likely.

Example 1 In the bond-issue example, show that F and O are dependent.

Solution Recalling that

$$F = \{e_1, e_2\}, \qquad O = \{e_1, e_3\}, \qquad F \cap O = \{e_1\},$$

we have

$$P(F) = 0.6, \qquad P(O) = 0.7, \qquad P(F \cap O) = 0.4$$

and

$$P(F \cap O) \neq P(F) \cdot P(O),$$

since

$$0.4 \neq 0.6 \times 0.7 = 0.42.$$

Example 2 A thumbtack with probability $P(U) = 0.4$ is tossed twice. If E is the event "first toss lands up" and F is the event "second toss lands up", show that the following pairs of events are independent:

a) E and F b) E and $\overline{F}$ c) $\overline{E}$ and F d) $\overline{E}$ and $\overline{F}$

Solution The possible outcomes and associated probabilities were discussed in Section 4–1. We reorganize the data here for reference.

		Outcome of second toss		
		F (U)	$\overline{F}$ (D)	Row sums
Outcome of first toss	E (U)	0.16	0.24	0.40
	$\overline{E}$ (D)	0.24	0.36	0.60
	Column sums	0.40	0.60	1.00

If E is the event "first toss lands up" and F the event "second toss lands up", then

$$P(E) = P(F) = 0.4$$

and

$$P(E \cap F) = P(U, U) = 0.16 = P(E) \cdot P(F),$$

so the events E and F are independent. It is also easy to verify that E and $\overline{F}$ are independent, as are $\overline{E}$ and F, and $\overline{E}$ and $\overline{F}$:

$$P(E \cap \overline{F}) = P(U, D) = 0.24 = P(E) \cdot P(\overline{F}),$$

$$P(\overline{E} \cap F) = P(D, U) = 0.24 = P(\overline{E}) \cdot P(F),$$

$$P(\overline{E} \cap \overline{F}) = P(D, D) = 0.36 = P(\overline{E}) \cdot P(\overline{F}).$$

We now restate the formal definition of independence, and prove a theorem suggested by the last example.

4-5 Definition: Independent events

Two events E and F are *independent* if and only if

$$\boxed{P(E \cap F) = P(E) \cdot P(F).} \tag{1}$$

4-6 Theorem: Independent events

Let E and F be independent events in a sample space S. Then E and $\overline{F}$ are independent, as are $\overline{E}$ and F, and $\overline{E}$ and $\overline{F}$.

Table 4-2 Independent events

	F	$\overline{F}$	Row sums
E	$P(E) \cdot P(F)$	$P(E) \cdot P(\overline{F})$	$P(E)$
$\overline{E}$	$P(\overline{E}) \cdot P(F)$	$P(\overline{E}) \cdot P(\overline{F})$	$P(\overline{E})$
Column sums	$P(F)$	$P(\overline{F})$	1

Proof Consider the two-way array in Table 4-2. We shall show that the entries in this table correctly give the probabilities of the corresponding compound events.

From the assumption that E and F are independent, Eq. (1) says that

$$P(E \cap F) = P(E) \cdot P(F),$$

and the entry in the upper left corner, corresponding to $P(E \cap F)$, is correct. Next, the row sums and column sums must be $P(E)$, $P(\overline{E})$, $P(F)$, and $P(\overline{F})$, as shown in Table 4-3.

Table 4–3

	F	$\overline{F}$	
E	$P(E) \cdot P(F)$		$P(E)$
$\overline{E}$			$P(\overline{E})$
	$P(F)$	$P(\overline{F})$	1

From the row sum $P(E)$, we see that

$$P(E \cap \overline{F}) = P(E) - P(E) \cdot P(F)$$
$$= P(E) \cdot [1 - P(F)] = P(E) \cdot P(\overline{F}), \tag{2}$$

which shows that E and $\overline{F}$ are independent.

Similarly, $\overline{E}$ and F are independent and

$$P(\overline{E} \cap F) = P(\overline{E}) \cdot P(F), \tag{3}$$

but we leave the proof as an exercise. Likewise $\overline{E}$ and $\overline{F}$ are independent:

$$P(\overline{E} \cap \overline{F}) = P(\overline{E}) - P(\overline{E} \cap F) = P(\overline{E}) - P(\overline{E}) \cdot P(F)$$
$$= P(\overline{E}) \cdot [1 - P(F)] = P(\overline{E}) \cdot P(\overline{F}). \tag{4}$$

Therefore, if E and F are independent, the probabilities of the compound events are those shown in the cells of Table 4–2. □

Note that the probability entered in any of the four main cells of Table 4–2 is just the product of the corresponding row and column probabilities. This property of probabilities of independent events is very easy to check when the probabilities are set up in a two-by-two table of this kind. If one entry can be filled in by multiplication of its row sum and its column sum, so can all the others.

Example 3 Are the events E and F independent if the probabilities are as shown in the following table?

	F	$\overline{F}$	
E	0.04	0.06	0.10
$\overline{E}$	0.08	0.82	0.90
	0.12	0.88	1.00

The answer is "no", because

$$0.04 \neq (0.12) \cdot (0.10) = 0.012.$$

We also observe that *every* entry is different from the product of its row sum and its column sum.

Remark In the bond-issue example, we have discussed ownership status (person owns or does not own property) and attitude toward the bond issue ("for" or "against"). If the events "owns property" and "for bond issue" were independent, then Theorem 4–6 would imply independence between such other pairs of events as "owns property" and "against bond issue", and so on. It would then be convenient to speak of independence of "ownership status" and "attitude on bond issue". So, in general, when independence works for one cell of a two-by-two table, we say that the characteristic, or label, associated with the rows is independent of that associated with the columns. And indeed, even in a larger table of m rows and n columns, if the probability in every cell is the product of its row total and column total, we continue to say that the row label is independent of the column label.

Independence of three or more events If we study three or more events we may represent them by $E_1, E_2, \ldots, E_m$. (In this discussion, the E_i's are not elementary events.) It is natural to say that these m events are *independent* provided the probability of their intersection is equal to the *product* of their probabilities:

$$P(E_1 \cap E_2 \cap \cdots \cap E_m) = P(E_1) \cdot P(E_2) \cdots P(E_m). \tag{5}$$

But if $m \geq 3$, Eq. (5) alone is not sufficient to guarantee the truth of the equations that we get by replacing some of these events by their complements on both sides of Eq. (5), as is the case when $m = 2$. To achieve this desired goal, we need to require *complete independence.*

4–7 Definition: Complete independence

The m events are said to be *completely independent* if and only if every *combination* of these events, taken any number at a time, is independent.

When $m = 3$, *complete independence* of E_1, E_2, E_3 means that the following equations are satisfied:

$$\begin{aligned} P(E_1 \cap E_2 \cap E_3) &= P(E_1) \cdot P(E_2) \cdot P(E_3), \\ P(E_1 \cap E_2) &= P(E_1) \cdot P(E_2), \\ P(E_1 \cap E_3) &= P(E_1) \cdot P(E_3), \\ P(E_2 \cap E_3) &= P(E_2) \cdot P(E_3). \end{aligned} \tag{6}$$

And if equations (6) are satisfied, so is any equation we get by replacing an event by its complement on both sides of one of the original equations. For instance,

$$P(E_1 \cap \overline{E}_2 \cap E_3) = P(E_1) \cdot P(\overline{E}_2) \cdot P(E_3). \tag{7}$$

Or we may replace any two, or three, events by their complements on both sides of the equation and get a true result.

Remark It might be supposed that 3 events are independent if every pair of them is independent. However, such pairwise independence does *not* imply independence of the three events, as the following example shows.

Example 4 *Three pairwise independent events that are not independent.*

Discussion Two coins are tossed. If E_1 is the event "head on first coin", E_2 the event "head on second coin", and E_3 the event "the coins match; both are heads or both tails", then

$$P(E_1) = P(E_2) = P(E_3) = \tfrac{1}{2}$$

and

$$P(E_1 \cap E_2) = P(E_1 \cap E_3) = P(E_2 \cap E_3) = \tfrac{1}{4}.$$

Hence the events are independent in pairs. But

$$P(E_1 \cap E_2 \cap E_3) = \tfrac{1}{4} \neq P(E_1) \cdot P(E_2) \cdot P(E_3),$$

so they are *not* independent when taken all together.

Example 5 Independently, a coin is tossed, a card is drawn from a deck, and a die is thrown. What is the probability that we observe a head on the coin, an ace from the deck, and a five on the die?

Solution

$$P(\text{head}) = \tfrac{1}{2}, \qquad P(\text{ace}) = \tfrac{1}{13}, \qquad P(5 \text{ on die}) = \tfrac{1}{6}.$$
$$P(\text{head and ace and } 5) = \tfrac{1}{2} \times \tfrac{1}{13} \times \tfrac{1}{6} = \tfrac{1}{156}.$$

Example 6 *Flawless shoes.* In a shoe factory, uppers, soles, and heels are manufactured separately and randomly assembled into single shoes. Five percent of the uppers, four percent of the soles, and one percent of the heels have flaws. What percent of the pairs of shoes are flawless in these three parts?

Solution Let U, S, and H stand for unflawed upper, sole, and heel, respectively, and $\overline{U}$, $\overline{S}$, and $\overline{H}$ stand for the flawed parts. For a single shoe,

$$P(U) = 1 - 0.05 = 0.95, \qquad P(S) = 1 - 0.04 = 0.96,$$
$$P(H) = 1 - 0.01 = 0.99,$$
$$P(U \cap S \cap H) = 0.95 \times 0.96 \times 0.99 \approx 0.903.$$

This is the probability that one shoe is unflawed. Assuming that pairs are also randomly assembled, we would have

$$\begin{aligned} P(\text{both shoes unflawed}) &= P(\text{left and right unflawed}) \\ &= P(\text{left unflawed}) \cdot P(\text{right unflawed}) \\ &\approx 0.903 \times 0.903 \\ &\approx 0.815. \end{aligned}$$

Example 7 *Light bulbs.* Light bulbs are produced by a sequence of machine operations. When the machine is in good working order, it produces one defective bulb per thousand. The outcomes for successive bulbs are independent. What is the probability that the next two bulbs produced are nondefective?

Solution Let E be "first bulb nondefective", F be "second bulb nondefective".

$$P(E) = P(F) = 1 - 0.001 = 0.999,$$
$$P(E \cap F) = 0.999 \times 0.999 = (1 - 0.001)^2 \approx 1 - 2(0.001) = 0.998.$$

After this section, when we speak of *independence*, we shall mean *complete independence.*

EXERCISES FOR SECTION 4–3

1. If two events E and F are mutually exclusive and have probabilities different from zero, prove that they are dependent.
2. Give examples of events E and F like those described in Exercise 1, based upon the two-dice example, Table 1–11.
3. Prove that if E and F are independent, then $\overline{E}$ and F are also independent.
4. Prove that if
$$P(E \cap F) \neq P(E) \cdot P(F),$$
then
$$P(\overline{E} \cap F) \neq P(\overline{E}) \cdot P(F),$$
$$P(E \cap \overline{F}) \neq P(E) \cdot P(\overline{F}),$$
$$P(\overline{E} \cap \overline{F}) \neq P(\overline{E}) \cdot P(\overline{F}).$$
5. Three ordinary dice are thrown. Assuming the outcomes on the dice are completely independent, find the probability that the sum of the numbers on the top faces is five.
6. A certain automatic machine makes bolts and fills boxes with them. If 1 box in 100 has at least one defective bolt in it and the outcomes are independent, what is the probability that each of the next 3 boxes has one or more defective bolts? That all have no defective bolts?
7. The probability that a man is hospitalized during the next month is 0.01. If we consider three men who are strangers to each other, what is the probability that during the next month exactly one of them goes to the hospital?
8. There are three traffic lights spaced several miles apart on a highway between towns A and B. The cycles of the three lights are one minute each. The three lights show green 30, 40, and 50 seconds, respectively. Assuming that a car strictly observes traffic-light regulations, what is the probability that the car makes the trip from A to B without being stopped by any of these three traffic lights? That the car will be stopped by exactly one light? By exactly two lights? By all three? (Assume that this is the only car on the road from A to B.)

9. Two ordinary dice are independently thrown and the outcomes on the top faces are observed. Show that the events

E_1: first die shows an even number,

E_2: second die shows an odd number,

E_3: sum of the results is odd

are pairwise independent, but not completely independent.

10. Let $S = \{e_1, e_2, e_3, e_4, e_5, e_6\}$ be the sample space of an experiment. Suppose the probabilities of the elementary events are

$$p_1 = \tfrac{1}{8}, \quad p_2 = \tfrac{5}{16}, \quad p_3 = \tfrac{1}{16}, \quad p_4 = \tfrac{3}{8}, \quad p_5 = p_6 = \tfrac{1}{16},$$

where $p_i = P(\{e_i\})$. Let $E = \{e_1, e_4\}$, $F = \{e_1, e_2, e_5\}$, $G = \{e_1, e_2, e_3\}$. Show that E, F, and G are independent, but not completely independent.

11. In matches between two teams, teams A, B, and C score points in games, independently of whom they play, according to the following probability table:

	Points					
Team	0	1	2	3	4	5
A		0.5			0.5	
B	0.2			0.8		
C			0.8			0.2

The team with the most points wins. Show that $P(A$ beats $B)$, $P(B$ beats $C)$, and $P(C$ beats $A)$ are all greater than $\frac{1}{2}$. That is, A usually beats B, B usually beats C, and C usually beats A. Thus the relation "usually beats" need not be transitive.

ADDITIONAL EXERCISES FOR SECTION 4-3

WORLD SERIES EXERCISES

In a World Series, Teams A and B play until one team has won 4 games. Let p be the probability that team A wins any individual game played with B. Then $q = 1 - p$ is the probability that B wins. Use this information to answer the questions in Exercises 1 through 9.

1. What is the probability that A wins the first 4 games? That B wins the first 4 games? That the series ends at 4 games? [*Ans:* p^4, q^4, $p^4 + q^4$]
2. What is the probability that A wins the series in the 5th game? That the series ends at 5 games? [*Ans:* $4p^4q$, $4pq(p^3 + q^3)$]
3. What is the probability that A wins the series in the 6th game? That the series ends at 6 games? [*Ans:* $10p^4q^2$, $10p^2q^2(p^2 + q^2)$]
4. What is the probability that A wins the series in the 7th game? That the series ends at 7 games? [*Ans:* $20p^4q^3$, $20p^3q^3$]

5. Using the results of Exercises 1 through 4, construct a sample space for the experiment of playing a World Series and assign probabilities to the sample points. What is the probability that team A wins the series? That team B wins? (Express the answer for B's winning in two different ways.) [*Ans:* $P(A \text{ wins}) = p^4(1 + 4q + 10q^2 + 20q^3)$]
6. In Exercise 5, suppose $p = \frac{2}{3}$, $q = \frac{1}{3}$, so that team A is "twice as good" as team B.* Is A's chance of winning the series also twice the probability that B wins? If not, what are the odds in favor of A's winning the series? [*Ans:* $P(A \text{ wins}) = \frac{1808}{2187}$, $P(B \text{ wins}) = \frac{379}{2187}$; $\approx 4.77{:}1$]
7. If, in Exercise 5, $p = q = \frac{1}{2}$, what is the probability that the series ends in 4 games? 5? 6? 7? [*Ans:* $\frac{2}{16}, \frac{4}{16}, \frac{5}{16}, \frac{5}{16}$]
8. In Exercise 5, assume that $p = \frac{2}{3}$. What is the probability that the series ends in 4 games? 5? 6? 7? [*Ans:* $\frac{153}{729}, \frac{216}{729}, \frac{200}{729}, \frac{160}{729}$]
9. In Exercise 8, with $p = \frac{2}{3}$, which is more likely, that the series is over before the 6th game, or that it is not over then? What are the relative odds?

4-4 EXAMPLES OF BINOMIAL EXPERIMENTS

Some experiments are composed of repetitions of independent trials, each with *two* possible outcomes. When the probabilities of the two outcomes are unchanged from one trial to another, we name the trials *Bernoulli trials*, in honor of James Bernoulli who wrote in Latin, about the year 1700, an important work on probability titled *Ars Conjectandi*. A *binomial experiment* consists of a fixed number, n, of Bernoulli trials. We introduce the binomial distribution not only because it is a mathematical model for an enormous variety of real-life phenomena, but also because it has important properties that recur in many other probability models. We begin with a few examples of binomial experiments.

Marksmanship example A trained marksman shooting five rounds at a target, all under practically the same conditions, may hit the bull's-eye from 0 to 5 times. In repeated sets of five shots his numbers of bull's-eyes vary. What can we say of the probabilities of the different possible numbers of bull's-eyes?

Inheritance in mice In litters of eight mice from similar parents, the number of mice with straight instead of wavy hair is an integer from 0 to 8. What probabilities should be attached to these possible outcomes?

Aces (ones) with three dice When three dice are tossed repeatedly, what is the probability that the number of aces is 0 (or 1, or 2, or 3)?

General binomial problem More generally, suppose that an experiment consists of a number of independent trials fixed in advance, that each trial results

---

* Historical results suggest that these figures approximate the relative strengths of the teams in actual play.

in either a "success" or a "non-success" ("failure"), and that the probability of success remains constant from trial to trial. In the examples above, the occurrence of a bull's-eye, a straight-haired mouse, or an ace could be called a "success". In general, any outcome we choose may be labeled "success".

The major question in this section and the next is: What is the probability of exactly x successes in n Bernoulli trials?

In Chapter 3 we answered questions like those in the examples, usually by counting points in a sample space. Fortunately, a general formula of wide applicability solves all problems of this kind. Before deriving this formula, we explain what we mean by "problems of this kind".

Experiments are often composed of several identical trials, and sometimes experiments themselves are repeated. In the marksmanship example, a trial consists of "one round shot at a target" with outcome either one bull's-eye (success) or none (failure). Further, an experiment might consist of five rounds, and several sets of five rounds might be regarded as a super-experiment composed of several repetitions of the five-round experiment. If three dice are tossed, a trial is one toss of one die and the experiment is composed of three trials. Or, what amounts to the same thing, if one die is tossed three times, each toss is a trial, and the three tosses form the experiment. Mathematically, we shall not distinguish the experiment of three dice tossed once from that of one die tossed three times. These examples are illustrative of the use of the words "trial" and "experiment", but they are quite flexible words and it is well not to restrict them too narrowly.

Example 1 *Student football managers.* Ten students act as managers for a high-school football team, and of these managers a proportion p are licensed drivers. Each Friday one manager is chosen by lot to stay late and load the equipment on a truck. On three Fridays the coach has needed a driver. Considering only these Fridays, what is the probability that the coach had drivers all 3 times? Exactly 2 times? 1 time? 0 time?

Discussion Note that there are 3 trials of interest. Each trial consists of choosing a student manager at random. The 2 possible outcomes on each trial are "driver" or "nondriver". Since the choice is by lot each week, the outcomes of different trials are independent. The managers stay the same, so that $p = P(\text{driver})$ is the same for all weeks. These are Bernoulli trials. We now generalize these ideas for general binomial experiments.

For an experiment to qualify as a *binomial experiment,* it must have four properties:

1) there must be a fixed number of trials,

2) each trial must result in a "success" or a "failure",

3) all trials must have identical probabilities of success (Bernoulli trials),

4) the trials must be independent of each other.

Below we use our earlier examples to describe and illustrate these four properties. We also give, for each property, an example where the property is absent. The language and notation introduced are standard.

1. There must be a fixed number n of repeated trials For the marksman, we study sets of five shots ($n = 5$); for the mice, we restrict attention to litters of eight ($n = 8$); and for the aces, we toss three dice ($n = 3$).

Experiment without a fixed number of trials Toss a die until an ace appears. Here the number of trials is a variable, not a fixed number. These are still Bernoulli trials even though the experiment is not a binomial experiment.

2. Binomial trials Each of the n trials is either a success or a failure. "Success" and "failure" are just convenient labels for the two categories of outcomes when we talk about binomial trials in general. These words are more expressive than labels like "A" and "not-A". It is natural from the marksman's viewpoint to call a bull's-eye a success, but in the mice example it is arbitrary which category corresponds to straight hair in a mouse. The word "binomial" means "of two names" or "of two terms", and both usages apply in our work: the first to the names of the two outcomes of a binomial trial, and the second to the terms p and $(1 - p)$ that represent the probabilities of "success" and "failure". Sometimes when there are many outcomes for a single trial, we group these outcomes into two classes, as in the example of the die, where we have arbitrarily constructed the classes "ace" and "not-ace".

Experiment without the two-class property We classify mice as "straight-haired" or "wavy-haired", but a hairless mouse appears. We can escape from such a difficulty by ruling out the animal as not constituting a trial, but such a solution is not always satisfactory.

3. All trials have identical probabilities of success Each die has probability $p = \frac{1}{6}$ of producing an ace; the marksman has some probability p, perhaps 0.1, of making a bull's-eye. Note that we need not know the value of p, for the experiment to be binomial.

Experiment where p is not constant During a round of target practice the sun comes from behind a cloud and dazzles the marksman, lowering his chance of a bull's-eye.

4. The trials are independent Strictly speaking, this means that the probability for each possible outcome of the experiment can be computed by multiplying together the probabilities of the possible outcomes of the single Bernoulli trials. Thus in the three-dice example $P(\text{ace}) = p = \frac{1}{6}$, $P(\text{not-ace}) = 1 - p = \frac{5}{6}$, and the independence assumption implies that the probability that the three dice fall ace, not-ace, ace in that order is $(\frac{1}{6})(\frac{5}{6})(\frac{1}{6})$. Experimentally, we expect independence when the trials have nothing to do with one another.

Table 4–4 Couples voting for pictures A and B

		Girls' votes		
		A	B	
Boys' votes	A	0.45	0.15	0.6
	B	0.15	0.25	0.4
	Total	0.6	0.4	1

Examples where independence fails A family of five plans to go together either to the beach or to the mountains, and a coin is tossed to decide. We want to know the number of people going to the mountains. When this experiment is viewed as composed of five binomial trials, one for each member of the family, the outcomes of the trials are obviously not independent. Indeed, the experiment is better viewed as consisting of one Bernoulli trial for the entire family. The following is a less extreme example of dependence. Consider couples visiting an art museum. Each person votes for one of a pair of pictures to receive a popular prize. Voting for one picture may be called "success", for the other "failure". An experiment consists of the voting of one couple, or two trials. In repetitions of the experiment from couple to couple, the votes of the two persons in a couple probably agree more often than independence would imply, because couples who visit the museum together are more likely to have similar tastes than are a random pair of people drawn from the entire population of visitors. Table 4–4 illustrates the point. The table shows that 0.6 of the boys and 0.6 of the girls vote for picture A. Therefore, under independent voting, 0.6×0.6 or 0.36 of the couples would cast two votes for picture A, and 0.4×0.4 or 0.16 would cast two votes for picture B. Thus in independent voting, $0.36 + 0.16$ or 0.52 of the couples would agree. But Table 4–4 shows that $0.45 + 0.25$ or 0.70 agree, too many for independent voting.

Each performance of an n-trial binomial experiment results in some whole number from 0 through n as the value of the variable X, where

$$X = \text{total number of successes in } n \text{ binomial trials.}$$

Thus, in the marksmanship example, we are interested in the number of bull's-eyes, not which shots were bull's-eyes. A binomial experiment can produce scores other than the number of successes. For example, the marksman gets 5 shots, but suppose we take his score to be the number of shots *before* his first bull's-eye, that is, 0, 1, 2, 3, 4 (or 5, if he gets no bull's-eye). Then we do not score the number of bull's-eyes, and his score is not the number of successes.

The constancy of p and the independence are the conditions most likely to give trouble in practice. Obviously, very slight changes in p do not change the probabilities much, and a slight lack of independence may not make an ap-

Table 4–5

Sample point	X, number of U's	Probability
DDD	0	q^3
DDU	1	pq^2
DUD	1	pq^2
UDD	1	pq^2
DUU	2	p^2q
UDU	2	p^2q
UUD	2	p^2q
UUU	3	p^3

preciable difference. On the other hand, even when the binomial model does not describe well the physical phenomenon being studied, the binomial model may still be used as a baseline for comparative purposes; that is, we may discuss the phenomenon in terms of its departures from the binomial model.

Summary A *binomial experiment* consists of n (≥ 1) independent binomial trials, all with the same probability p ($0 \leq p \leq 1$) of yielding a success. The outcome of the experiment is X successes. The variable X takes the values $x = 0, 1, \ldots, n$ with probabilties $P(X = x)$ or, more briefly, $P(x)$.

We shall find a formula for the probability of exactly x successes for given values of p and n. When each number of successes x is paired with its probability of occurrence $P(x)$, the set of pairs $(x, P(X))$, $x = 0, 1, \ldots, n$, is a function called a *binomial distribution*. The choice of p and n determines the binomial distribution uniquely, and different choices always produce different distributions (except when $p = 0$; then the number of successes is always 0). The set of all binomial distributions is called *the family of binomial distributions*, but in general discussions this expression is often shortened to "the binomial distribution", or even "the binomial" when the context is clear.

We turn now to another simple example that illustrates the features of a binomial experiment and its associated binomial distribution.

Example 2 *Binomial experiment with three thumbtacks.* When a thumbtack is tossed it can land point up (U), or point down (D). Suppose three thumbtacks, 1 red, 1 white, 1 blue, but otherwise alike, are tossed. We want the probabilities for the variable X, the number of tacks landing U. The sample space of possible outcomes for the three thumbtacks contains 8 sample points, with associated values of X, and probabilities as shown in Table 4–5. The first, second, and third letters of any sample point indicate in order the outcomes for the red, white, and blue tacks. Thus UDD means that the red tack landed point up, and the other two landed point down.

Let the probabilities of U and D be p and q, respectively, where $p + q = 1$. Then we have

$$P(U) = p, \qquad P(D) = q. \tag{1}$$

We assume that outcomes on the three tacks are independent. Hence the probability assigned to any one of the sample points is obtained by multiplying three probabilities. For instance,

$$P(UDU) = P(U) \cdot P(D) \cdot P(U) = pqp = p^2q.$$

We let $b(x; 3, p)$ denote the probability of getting x U's when there are three tacks each with probability p of landing U. [For $b(x; 3, p)$ read "b of x when $n = 3$ and probability of success is p".] We can summarize the preceding results by writing the binomial distribution we have just obtained in the form of a *probability table:*

Probability, $b(x; 3, p)$	q^3	$3pq^2$	$3p^2q$	p^3
Value, x	0	1	2	3

(2)

Since the coefficients 1, 3, 3, 1 in the top line of table (2) are binomial coefficients, we can write a formula for $b(x; 3, p)$, as follows:

$$b(x; 3, p) = \binom{3}{x} p^x q^{3-x}, \qquad x = 0, 1, 2, 3, \tag{3}$$

where $\binom{3}{x} = 3!/x!(3 - x)!$ is the number of permutations of x U's and $(3 - x)$ D's (Corollary 2–14).

Although we derived Eq. (3) for the thumbtack problem, the formula is more general. The set of four probabilities arising from formula (3), together with the associated values of x, form the binomial distribution for 3 independent trials, each having p as the probability of success. The probability table (2) also displays this binomial distribution.

Note that the right-hand side of formula (3) is the product of two factors:

1) the binomial coefficient $\binom{3}{x}$, which counts the number of different arrangements of exactly x successes and $3 - x$ failures in 3 trials;
2) The factor $p^x q^{3-x}$, which gives the probability for any one of the different ways of getting x successes and $3 - x$ failures.

When we derive the generalization of formula (3) in the next section, we exploit the fact that binomial probabilities are always products of two such parts.

Adding all of the probabilities given in formula (3) for $x = 0, 1, 2, 3$, we get

$$q^3 + 3q^2p + 3qp^2 + p^3.$$

This sum is the binomial expansion of $(q + p)^3$, since

$$(q + p)^3 = q^3 + 3q^2p + 3qp^2 + p^3. \tag{4}$$

This is another reason for calling the set of probabilities obtained from formula (3) a "binomial" distribution. And because $q + p = 1$ and $1^3 = 1$, Eq. (4) shows that the sum of the four probabilities is 1.

Another way to represent the probabilities in a binomial distribution is to make a graph. Thus, if we erect the ordinates q^3, $3q^2p$, $3qp^2$, p^3 at $x = 0, 1, 2, 3$, respectively, we obtain the *probability graph.*

Because the four probabilities q^3, $3q^2p$, $3qp^2$, p^3 add up to 1, the ordinates also add up to 1. In Fig. 4–2, $p = 0.3$, $q = 0.7$.

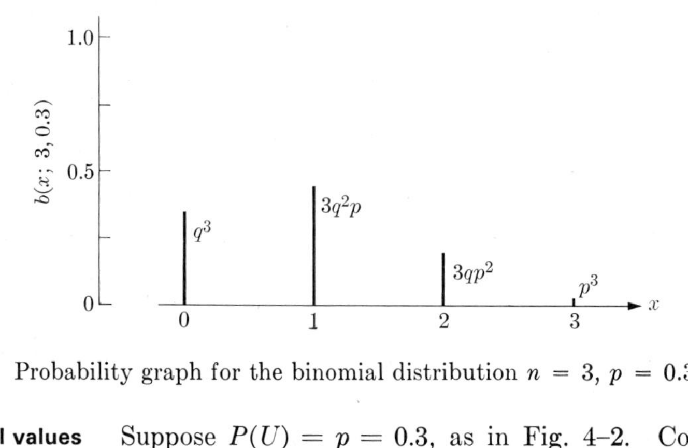

Fig. 4–2 Probability graph for the binomial distribution $n = 3$, $p = 0.3$, $q = 0.7$.

Numerical values Suppose $P(U) = p = 0.3$, as in Fig. 4–2. Compute the probabilities of getting 0, 1, 2, 3 U's when the three tacks are tossed.

We use formula (3), with $p = 0.3$, $q = 1 - p = 0.7$, and substitute in succession $x = 0, 1, 2, 3$. From these calculations we get the following probability table for this binomial distribution:

$b(x; 3, 0.3)$	0.343	0.441	0.189	0.027
x	0	1	2	3

(5)

In this same example, we can calculate still other probabilities. For instance, the probability of getting *at least* 2 U's is

$$\begin{aligned} P(X \geq 2) &= 3qp^2 + p^3 \\ &= b(2; 3, 0.3) + b(3; 3, 0.3) = 0.189 + 0.027 = 0.216. \end{aligned}$$

The probability of getting not more than one U is

$$\begin{aligned} P(X \leq 1) &= q^3 + 3pq^2 \\ &= b(0; 3, 0.3) + b(1; 3, 0.3) = 0.343 + 0.441 = 0.784. \end{aligned}$$

Remark In examples like that of the marksman hitting the bull's-eye or that of the thumbtacks, the value of p cannot be readily guessed. However, a good estimate of p in such problems can be made if we record the results of several hundred shots, or tosses, and take as the estimate the ratio of the number of bull's-eyes, or U's, to the total number of trials.

EXERCISES FOR SECTION 4-4

1. Verify that the probability table (5) for the thumbtack example is correct, and check by addition that the sum of the probabilities is 1.
2. Verify that formula (3) is correct by substituting $x = 0, 1, 2, 3$ in succession, and thus check the entries in probability table (2).
3. A bag contains 1 red and 2 white balls, identical except for color. If 3 balls are randomly drawn one at a time with replacement after each draw, find the exact binomial distribution of the number of white balls in the sample of 3.
4. Three candidates run for different offices in different states. Each has 1 chance in 3 of being elected in his state. What is the chance that at least one of them is elected?
5. If $n = 3$ in a binomial experiment, what *values* can p have if $P(0)$ is to equal $P(1)$? Determine p if $P(0) = P(3)$.
6. Three dice were thrown 648 times and the number of times a "5 or 6" appeared was tabulated as follows:

Number of "5 or 6"'s	Observed frequency
0	179
1	298
2	141
3	30
Total	648

   Obtain the theoretical probability for each outcome for perfect dice, multiply by 648, and compare the resulting theoretical frequencies with the observed ones.
7. Toss 3 coins 24 times and compare the observed numbers of heads with their theoretical frequencies. (Be sure each coin is flipped separately.)
8. In a binomial experiment with $n = 3$ show that $P(X = 1 \text{ or } 2) = 3pq$.
9. (For students without calculus.) Make a graph of $b(2; 3, p)$ as p varies from 0 to 1 and estimate the value of p that maximizes it.
   (For students with some calculus.) Find the value of p that makes $b(2; 3, p)$ a maximum and evaluate the maximum. [*Hint:* Replace q by $1 - p$ before differentiating.]
10. A thumbtack that can fall point up, U, or point down, D, with $P(U) = p$ and $P(D) = q = 1 - p$, is independently tossed 4 times. List a sample space for

the possible outcomes of this experiment. Assign probabilities to its points. Show that the sum of these probabilities is 1. Find the probability distribution of the number that fall point up.

11. *Puzzle.* In seeking the probability of either 3 heads or 3 tails in a single throw of 3 coins, it has been reasoned that of 3 coins at least 2 must show like faces, and the probability that the third coin is the same as the other 2 is $\frac{1}{2}$, the desired probability. What is the correct probability? Try to find the flaw in the reasoning.

4-5 EXTENSION OF THE BINOMIAL EXPERIMENT TO n TRIALS

In the thumbtack example, if n thumbtacks are tossed, there are two different outcomes for each toss, U or D. Therefore, by the multiplication rule, there are $2 \times 2 \times \cdots \times 2$ (n factors), or 2^n, different outcomes for the experiment (Section 2-1). Hence the sample space S for this experiment has 2^n distinct points. Each point determines a value of the variable X, where X is the number of U's in the sample point. What is the probability of x U's, where x is any one of the numbers $0, 1, 2, \ldots, n$? That is, what is $b(x; n, p)$?

In Section 4-4 we found that the binomial probabilities for 3 binomial trials had two parts: a coefficient $\binom{3}{x}$ and a factor $p^x q^{3-x}$. We proceed to find the two parts of $b(x; n, p)$ in general, using the language of trials, successes, and failures instead of tosses, U's and D's.

First, in how many ways can we get exactly x successes in n trials? From the theory of permutations of two kinds of objects (Section 2-4) the number of ways is $\binom{n}{x}$; that is, the sample space S has exactly $\binom{n}{x}$ points representing outcomes with x successes and $n - x$ failures.

Second, what is the probability of x successes and $n - x$ failures in a given order? If we assume that the outcomes for the n trials are independent, then the probability of x successes and $n - x$ failures *in any given order* is the product of x p's and $n - x$ q's. We had the identical pattern in the set of probabilities in the thumbtack example of Section 4-4 [calculations following Eq. (1) of that section]. The reason we get the same probability for each arrangement of the x successes and $n - x$ failures is that independence implies multiplication, and multiplication is commutative. Thus the desired probability for any given order of x successes and $n - x$ failures is

$$p^x q^{n-x}. \tag{1}$$

This is true when x is any one of the numbers $0, 1, 2, \ldots, n$.

Among the 2^n points in S, there are $\binom{n}{x}$ points with x successes, each point having the probability $p^x q^{n-x}$ assigned to it. Therefore

$$b(x; n, p) = \binom{n}{x} p^x q^{n-x}, \qquad x = 0, 1, 2, \ldots, n. \tag{2}$$

Note that the probability $b(x; 3, p)$ [Section 4-4, Eq. (3)] is a special case of formula (2) with $n = 3$.

We observe that $\binom{n}{x} p^x q^{n-x}$ is the $(x+1)$st term in the binomial expansion of $(q+p)^n$, because the binomial expansion (Section 2-5) can be displayed as follows:

$$(q+p)^n = q^n + \binom{n}{1} pq^{n-1} + \cdots + \binom{n}{x} p^x q^{n-x} + \cdots + p^n. \tag{3}$$

Since $q + p = 1$, $(q+p)^n = 1$. This result is reassuring, because it shows that our derivation has accounted for all the probability in the sample space. The set of ordered pairs.

$$\left(x, \binom{n}{x} p^x q^{n-x}\right), \qquad x = 0, 1, \ldots, n,$$

is the general *binomial distribution.* We have proved the following general theorem about binomial experiments:

4-8 Theorem: Binomial distribution

If an experiment consists of n independent binomial trials, each with probability p of success and probability q $(=1-p)$ of failure, then the probability that the experiment results in exactly x successes and $n-x$ failures is

$$\boxed{b(x; n, p) = \binom{n}{x} p^x q^{n-x}, \qquad x = 0, 1, 2, \ldots, n.} \tag{4}$$

Example 1 *Five coin tosses.* In tossing a coin, the probability of a head is assumed to be $\frac{1}{2}$. If the coin is tossed 5 times, what is the probability (a) of exactly two heads? (b) of more than one head?

Solution Let X be the number of heads on the 5 tosses.

a) By Eq. (2),

$$P(X = 2) = b\left(2; 5, \frac{1}{2}\right) = \binom{5}{2}\left(\frac{1}{2}\right)^2\left(\frac{1}{2}\right)^3 = 10 \cdot \frac{1}{4} \cdot \frac{1}{8} = \frac{10}{32} = \frac{5}{16}.$$

b) $P(X > 1)$ is found most easily by using complementary events. The various mutually exclusive events are 0, 1, 2, 3, 4, or 5 heads. Therefore

$$\begin{aligned} P(X > 1) &= 1 - P(X \le 1) \\ &= 1 - b(0; 5, \tfrac{1}{2}) - b(1; 5, \tfrac{1}{2}) \\ &= 1 - (\tfrac{1}{2})^5 - 5(\tfrac{1}{2})^5 = \tfrac{26}{32} = \tfrac{13}{16}. \end{aligned}$$

Example 2 *Batter's problem.* Suppose the probability that a batter gets a hit is $\frac{1}{4}$. At first glance, some people interpret this figure to mean that the batter is sure to get a hit if he bats four times. What is the probability?

Solution In 4 times at bat, the probability of *at least one hit* is

$$\begin{aligned} P(X \geq 1) &= 1 - b(0; 4, \tfrac{1}{4}) \\ &= 1 - (\tfrac{1}{4})^0(\tfrac{3}{4})^4 \\ &= \tfrac{175}{256}. \end{aligned}$$

The answer is about 0.68, which is far from a certainty. The confusion arises from the fact that the mean number of hits is one, a confusion between a mean and a probability.

Example 3 *Two- and four-engine planes.* Suppose that, in flight, airplane engines fail with probability q, independently from engine to engine, and that a plane makes a successful flight if at least half of its engines run. For what values of q is a two-engine plane to be preferred to a four-engine one? The probability an engine does not fail is $p = 1 - q$.

Solution We begin by computing the probabilities of successful flights for the two types of planes. Let X be the number of engines that do not fail.

Two-engine plane	Four-engine plane
P (successful flight) $=$	P (successful flight) $=$
$P(X \geq 1) = 1 - P(0)$	$P(X \geq 2) = 1 - P(0) - P(1)$
$= 1 - b(0; 2, p)$	$= 1 - b(0; 4, p) - b(1; 4, p)$
$= 1 - q^2$	$= 1 - q^4 - 4pq^3$
	$= 1 - q^4 - 4(1 - q)q^3$
	$= 1 - 4q^3 + 3q^4$

1. *Graphical approach.* In Fig. 4–3, a graph is given of the probabilities of successful flights for the two kinds of planes as a function of q, the probability that a single engine fails. The crossing point of the two curves cannot be read precisely (but it is near $q = \frac{1}{3}$), and so the following algebraic approach may be preferred.

2. *Algebraic approach.* The inequality that implies that the probability of successful flight for the two-engine plane is greater than or equal to the corresponding probability for the four-engine plane is

$$1 - q^2 \geq 1 - 4q^3 + 3q^4.$$

Subtracting $1 - q^2$ from both sides of this inequality, we have the following equivalent relation:

$$0 \geq q^2 - 4q^3 + 3q^4.$$

Factoring q^2 from the expression on the right yields

$$0 \geq q^2(1 - 4q + 3q^2).$$

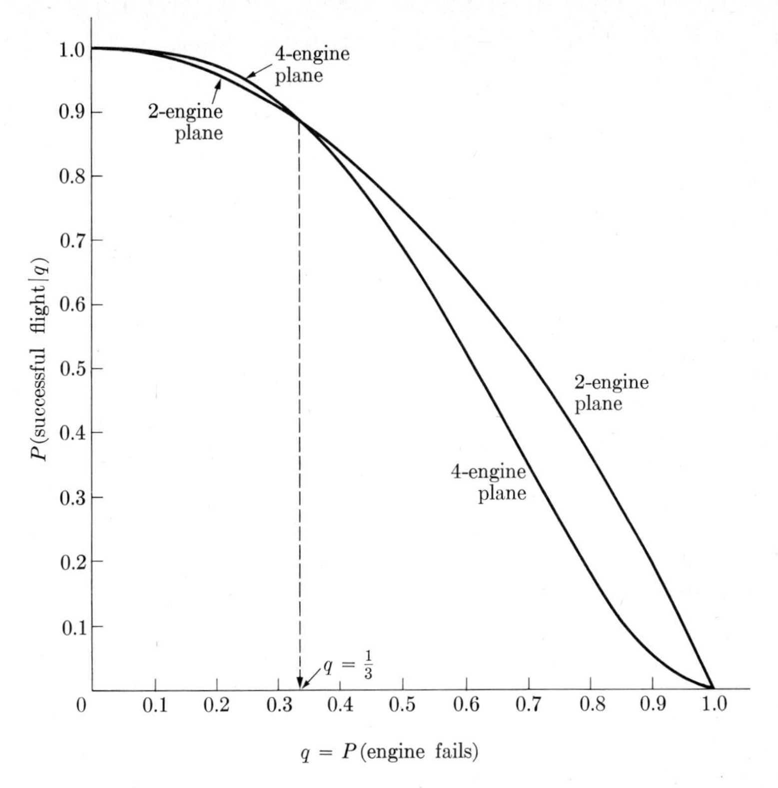

Fig. 4–3 Probabilities of successful flights plotted against q, the probability of failure for a single engine.

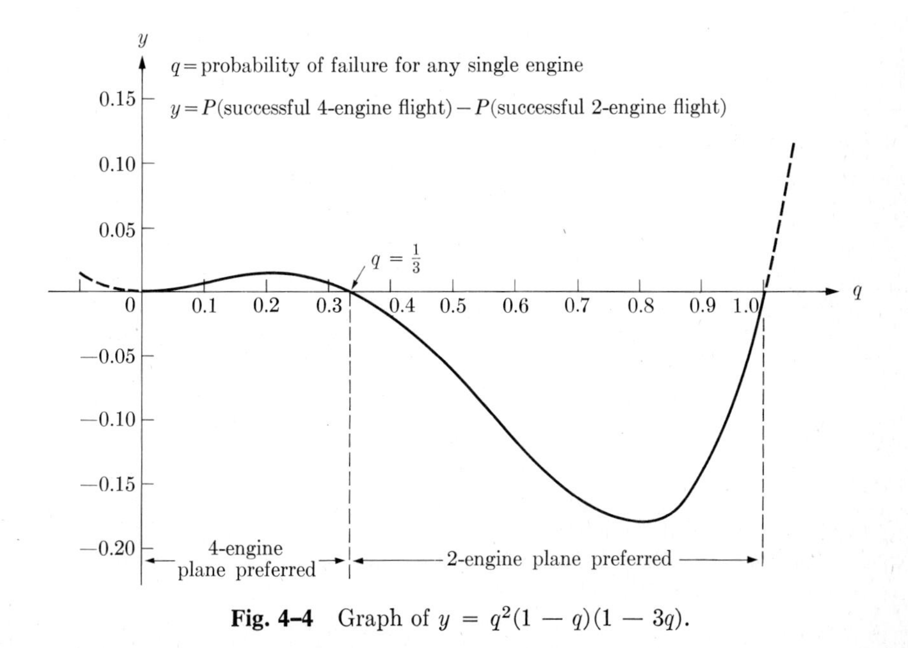

Fig. 4–4 Graph of $y = q^2(1 - q)(1 - 3q)$.

Finally, we factor the right-hand side further and obtain

$$0 \geq q^2(1 - q)(1 - 3q). \tag{5}$$

If $q = 0$, $q = 1$, or $q = \frac{1}{3}$, the right-hand member is zero and the two kinds of planes have equal chances of successful flights.

Figure 4–4 graphs the right-hand side of inequality (5) against q. The graph also shows that equal chances of successful flights for the two types of plane occur when $q = 0$, $q = 1$, or $q = \frac{1}{3}$. The graph further shows that if $\frac{1}{3} < q < 1$, then the right side of inequality (5) is less than zero because the curve falls below the q-axis. Similar reasoning shows that for $0 < q < \frac{1}{3}$ the curve is above the q-axis and the four-engine plane is to be preferred. The foregoing facts can also be obtained by studying the signs of the factors of the right-hand member of inequality (5) for various values of q. Needless to say, the practical situation is that q, the probability that any one engine fails, is very much less than $\frac{1}{3}$.

EXERCISES FOR SECTION 4–5

1. A baseball player's batting average is .300. What is the probability that he gets exactly 2 hits in 4 times at bat? Are there considerations that make you doubt that the number of hits in 4 times at bat is binomially distributed? If so, what are they?
2. A thumbtack falls point up 40% of the time. Compute the probabilities for each possible number of times it falls point up in 5 tosses. Display the results in a probability table and in a graph.
3. Why do you think the number of baseball games won by the home team in a double-header may fail to be binomially distributed?
4. If n coins are independently tossed, show that the probability that all or all but one will fall with the same face up is $(n + 1)/2^{n-1}$, if $n > 2$. What is the correct answer when $n = 2$?
5. Suppose an amateur rifleman has a probability of 0.05 of hitting a bull's-eye on a single shot. What is the probability that in 20 shots he never hits the bull's-eye? (Use logarithms.) Also set up, but do not evaluate, the probability calculation for hitting it 4 or more times in 20 shots. (The setup of this latter problem is straightforward, but the numerical computations are excessively long. In the next section we show how such problems can be solved easily by the use of tables.)
6. Compare the performance of a one-engine plane with that of a two-engine plane, using the same assumptions as those of Example 3.
7. In a binomial experiment consisting of 3 trials, the probability of exactly 2 successes is 12 times as great as that of 3 successes. Find p.
8. One-third of the male freshmen entering a college are at least 6 feet tall. If roommates are assigned randomly for freshmen, 4 to a room, what is the probability that at least 3 of the 4 in one room are under 6 feet? (Ignore the fact that the actual sampling is without replacement.)

9. Two-thirds of the secretaries in a large stenographic pool are licensed motor-vehicle operators. If 4 secretaries are drawn at random to go on a trip, what is the probability that at least 2 are licensed drivers?
10. A quiz has 6 multiple-choice questions, each with 3 alternatives. Sheer guesswork yields what probability of 5 or more right?
11. A risky operation used for patients with no other hope of survival has a survival rate of 80%. What is the probability that exactly 80% of the next 5 patients operated upon survive?
12. (For students with some calculus.) For given values of x and n, consider $b(x; n, p)$ as a function of p and show that it is maximized when $p = x/n$. [*Hint:* First solve the problem assuming that x is neither 0 nor n, then handle those two cases separately.] *Remark.* This is one reason for using the observed number of successes divided by the total number of trials as an estimate of p. That estimate, x/n, $x = 0, 1, \ldots, n$, is called the *maximum likelihood estimate* of p because of the maximizing property you are to prove.
13. Consider two binomial experiments each with $p = \frac{1}{2}$, one of size $n = 2m$ trials, the other of size $n = 2m - 1$ trials, where m is a positive integer. Show that $P(m)$ is the same for both experiments.
14. For what values of q is a one-engine plane to be preferred to a three-engine plane? Use the assumptions of Example 3.
15. For what values of q is a two-engine plane to be preferred to a three-engine plane? Use the assumptions of Example 3.
16. Suppose three- and five-engine planes fly if more than half their engines work. If q is the probability of failure for a single engine and engines perform independently, find the values of q for which the three-engine plane is to be preferred.
17. Two independent binomial experiments, one of n and the other of m trials, both have probability p of success on each trial. Show that the probability of a total of exactly x successes in the two experiments combined is

$$\binom{n+m}{x} p^x (1-p)^{n+m-x},$$

and interpret this result.

4-6 BINOMIAL PROBABILITY TABLES

It is a dreary task to compute the probability of every outcome for a large set of binomial trials. Extensive tables are available for the binomial distribution, and we present a small one (Table IV) for your use.

This table, at the back of the book, is in two parts. Part A gives $b(x; n, p)$, the probability of observing exactly x successes in a binomial experiment composed of n trials. Values are given for all x for $n = 2$ to 25 and for $p = .01$, .05, .10, .20, .30, .40, .50, .60, .70, .80, .90, .95, .99.

Part B of the table gives, for the same binomial distributions, the probability of observing r *or more* successes. Thus this part of the table gives the "cumulative" probability from r through n, rather than the probability of a single number of successes. Many applications require sums rather than single probabilities. Symbolically, this part of the table gives

$$P(X \geq r) = b(r; n, p) + b(r+1; n, p) + \cdots + b(n; n, p)$$
$$= \sum_{x=r}^{n} b(x; n, p).$$

Each 3-digit entry in the table should be read with a decimal preceding it. The symbol 1− means a probability larger than 0.9995, but less than 1. The symbol 0+ means a probability less than 0.0005, but greater than 0.

Let us check a value in the table. To find the probability of 4 or more successes in 5 trials when $p = 0.8$, we compute

$$b(4; 5, 0.8) + b(5; 5, 0.8)$$

and get

$$\binom{5}{4}(.8)^4(.2) + \binom{5}{5}(.8)^5(.2)^0 = .40960 + .32768 = .73728.$$

We read in Table IV–A that $b(4; 5, 0.8) \approx 0.410$ and that $b(5; 5, 0.8) \approx 0.328$, and in the cumulative Table IV–B for $n = 5$, $r = 4$, $p = 0.8$, we read 0.737. All three tabled probabilities agree to three decimal places with our calculated values.

Example 1 For $n = 10$, $p = 0.4$, find the probability of 3 or more successes.

Solution Reading directly from Table IV–B, we find that the probability to three decimal places is 0.833.

Example 2 *Interpolation in the tables.* With $n = 25$, find the value of p that makes $P(X \geq 8) = 0.4$.

Solution A tabular array assists with such a problem.

Values of p	$P(X \geq 8)$, $n = 25$
$p = 0.20$	0.109
$p = ?$	0.400
$p = 0.30$	0.488

By ordinary interpolation, we have

$$\frac{p - 0.2}{0.3 - 0.2} \approx \frac{0.400 - 0.109}{0.488 - 0.109},$$

whence

$$p \approx 0.2 + \frac{0.400 - 0.109}{0.488 - 0.109}(0.3 - 0.2) \approx 0.28.$$

This result agrees, to two decimal places, with the value obtained from a bigger table than ours.

Example 3 If $n = 15$, $p = 0.05$, find the probability of 2 or fewer successes.

Solution 1

$$P(X \leq 2) = 1 - P(X \geq 3)$$
$$\approx 1 - 0.036 = 0.964.$$

Solution 2 We could, instead, focus on the number of failures. Two or fewer successes is equivalent to 13 or more failures. We would then enter the table with p appropriate to failures, 0.95. Then we read directly

$$P(X \geq 13) \approx 0.964.$$

Other tables A brief list of more extensive tables of the binomial distribution follows. The notation $n = 1[1]10[5]100$ means that n goes from 1 to 10 in steps of 1 and from 10 to 100 in steps of 5.

1. Harvard Computation Laboratory, *Tables of the Cumulative Binomial Probability Distribution*, Harvard University Press (1955). Cumulatives only for $n = 1[1]50[2]100[10]200[20]500[50]1000$, $p = 0.00[0.01]0.50$ and $\frac{1}{3}, \frac{1}{6}, \frac{1}{8}, \frac{3}{8}, \frac{1}{12}, \frac{5}{12}, \frac{1}{16}, \frac{3}{16}, \frac{5}{16}, \frac{7}{16}$.
2. National Bureau of Standards, *Tables of the Binomial Probability Distribution*, Applied Mathematics Series 6 (1950). Gives both cumulative and single terms for $n = 1[1]49$, $p = 0.00[0.01]0.50$.
3. Harry G. Romig, *50–100 Binomial Tables*, New York: John Wiley & Sons, Inc. (1953). Gives both cumulative and single terms for $n = 50[5]100$, $p = 0.00[0.01]0.50$.
4. Ordnance Corps, *Tables of the Cumulative Binomial Probabilities*, Ordnance Corps Pamphlet ORDP 20–1, U. S. Government Printing Office (September, 1952). Gives cumulative only for $n = 1[1]150$, $p = 0.00[0.01]0.50$.

EXERCISES FOR SECTION 4–6

In Exercises 1 through 9, let X denote the number of successes in n binomial trials, with probability p of success on each trial.

1. For $n = 15$ and $p = 0.6$, find (a) $P(X \geq 7)$, (b) $P(X = 7)$.
2. For $n = 25$ and $p = 0.8$, find (a) $P(X > 19)$, (b) $P(X = 19)$.
3. For $n = 20$ and $p = 0.3$, find (a) $P(X \geq 6)$, (b) $P(X = 6)$.
4. For $n = 25$ and $p = 0.65$, find (a) $P(X \geq 11)$, (b) P(11 or more failures), (c) $P(X = 11)$.
5. With $n = 22$, find the value of p that makes $P(X \geq 8) = 0.4$.
6. With $n = 20$, find the value of p that makes $P(X \geq 7) = 0.5$.
7. With $n = 15$, find the value of p that makes $P(X \geq 10) = 0.8$.

8. Given that $n = 12$ and $p = 0.8$, find (a) $P(X = 8)$, (b) $P(X \leq 8)$, (c) $P(X \geq 8)$.
9. For $n = 6$ and $p = 0.2$, find the value of $P(X = 2)$.
10. In shooting a rifle, the probability that John hits the target is 0.95, and the probability that he gets a bull's-eye is 0.20. He shoots 25 times. What is the probability that he hits the target more than 20 times? That he gets exactly 5 bull's-eyes? That he gets 5 or more bull's-eyes?
11. (Continuation.) Suppose John shoots only 22 times. What is the probability of exactly 10 bull's-eyes? Fewer than 10? More than 10? Check that the three results add to 1.
12. A die is tossed 12 times. What is the probability of more than 4 aces?
13. If the probability of seven or more successes in 25 trials in a binomial experiment is 0.5, what is the probability of success on each trial? (Give answer to two decimal places.)
14. Twenty-five coins are poured from a sack onto a table. What is the probability that the number of heads is between 8 and 17, inclusive?
15. If 40% of the voters in a large town favor candidate A, what is the probability that in a random sample of 25 voters, the majority in the sample will favor him?
16. A census of a United States town of 25,000 showed that 75% of the families owned refrigerators. Twenty families were randomly selected for intensive sociological and economic investigation. Approximately what is the probability that 10 or fewer of these families have refrigerators? (A binomial calculation here is approximate because we ignore the fact that the sampling is done without, instead of with, replacement. Since the sample size is small compared with the population size, the approximation is a good one.)
17. What is the probability of exactly 8 successes in a binomial experiment of 11 trials if the probability of success on each trial is 0.8?
18. Use Table IV to work Exercise 5 of Section 4-5.
19. For $n = 25$, use binomial tables to find the two values of p that satisfy $P(X = 8) = 0.075$.
20. Five balls were drawn, one at a time, with replacement, from a bag containing an equal number of black and white balls. The number of black balls was then tabulated for 819 sets of consecutive drawings to give the following observed frequency distribution:

Number of black balls	Observed frequency
0	30
1	125
2	277
3	224
4	136
5	27
Total	819

Obtain the theoretical frequencies from the binomial and compare them with the observed values.

21. Assume that you serve on the school committee for your community, and that you know that the population of fourth-grade schoolchildren has 95% right-handed and 5% left-handed children. You observe that the fourth-grade classroom has 20 tablet armchairs, all with the tablet on the right arm. Assume that the 20 students assigned to this room are a random sample of the fourth-grade school population.

    a) What is the probability of one or more left-handed students in a class of 20?
    b) Suppose that you influence the school committee to exchange one of the chairs for a left-armed one. What is the probability that the chairs just come out even with the students: 1 left-armed chair, 1 left-handed student, 19 right-armed chairs, 19 right-handed students?
    c) How much have you improved the probability that everyone's handedness is provided for?
    d) The chairs are permanently installed. You could arrange that there be 21 chairs, one left-armed, for 20 students. Now what is the probability that everyone's handedness is provided for?

22. (a) Compare $b(1; 4, .30)$ with $b(3; 4, .70)$. (b) Compare $b(6; 18, .40)$ with $b(12; 18, .60)$.

23. (Continuation.) Prove $b(r; n, p) = b(n - r; n, q)$ if $q = 1 - p$. This shows that tables like Table IV–A need not tabulate for values of p in excess of 0.5.

24. (a) Use Table IV–B, $n = 23$, to compare $P(X \geq 11)$ for $p = 0.6$ with $1 - P(X \geq 13)$ for $p = 0.4$. (b) Use Table IV–B, $n = 5$, to compare $P(X \geq 2)$ for $p = 0.2$ with $1 - P(X \geq 4)$ for $p = 0.8$.

25. (Continuation.) Prove

$$\sum_{x=r}^{n} b(x; n, p) = 1 - \sum_{x=n-r+1}^{n} b(x; n, q).$$

    This shows that cumulative binomial tables need not have values of p in excess of 0.5. However, such values are a convenience.

26. Suppose 5 cards are drawn from an ordinary bridge deck, with replacement and reshuffling after each card is drawn. Find the probability of getting 0, 1, 2, 3, 4, or 5 red cards in the sample of 5 cards.

27. *The accident fallacy.* Let us assume that each factory worker has a probability $p = 0.01$ of having an industrial accident independently in each 2-week period, $n = 25$ periods a year. Find the percentage of people who have no accidents and then find the value of x in the statement, "x percent of the people are causing 100% of the accidents". What percentage will have 2 accidents in the year? 3 accidents? Shall we label those with 3 accidents accident-prone? Why?

4–7 CONDITIONAL PROBABILITY

In Section 3–4, conditional probabilities were studied for sample spaces whose sample points have equal probabilities. In this section we extend the notion of conditional probability to more general sample spaces. In the next two sections of this chapter, we study two classes of applications of conditional probability:

a) its use in assigning probabilities in a sample space,

b) its use in modifying our "degree of belief" in various alternative hypotheses as a result of experimental evidence.

Example 1 An irregular tetrahedron is tossed into the air. The four faces, numbered 1, 2, 3, 4, have corresponding probabilities 0.1, 0.2, 0.3, 0.4 of being on the bottom when the tetrahedron comes to rest. Given that either face 1 or face 2 is down, what is the probability that it is face 1?

Solution Since we are *given* that either face 1 or face 2 is down, we can ignore the other two possibilities and consider a reduced sample space consisting solely of the outcomes "face 1 down" and "face 2 down". The probability of face 2 being down is twice that of face 1. Hence, in a large number of performances of the experiment resulting in one of these two faces being on the bottom, we expect face 1 to be down about $\frac{1}{3}$ of the time and face 2 to be down about $\frac{2}{3}$ of the time. Therefore

$$P\text{ (face 1}|\text{face 1 or face 2)} = \frac{1}{3} = \frac{0.1}{0.1 + 0.2}.$$

The result has the form

$$P(A|A \text{ or } B) = \frac{P(A)}{P(A) + P(B)},$$

where A and B are mutually exclusive events "face 1 down" and "face 2 down".

Example 2 *Color blindness.* Assume that 5% of males and 1% of females are color-blind, and that males and females each form 50% of the population. A researcher studying color blindness selects a color-blind person at random. What is the probability that the person so selected is (a) male, (b) female?

Solution The given data provide us with the probabilities shown in Table 4–6. For instance, 5% of 50% of the population, or 2.5%, is both male and color-blind, so 47.5% is male and not color-blind. Similarly, 1% of 50%, or $\frac{1}{2}$ of 1%, is female and color-blind, and 49.5% is female and not color-blind. A sample of 1000 persons having exactly these percentages would contain 25 color-blind males and 5 color-blind females—a total of 30 color-blind persons. Since males make up $\frac{25}{30}$ of this group and females $\frac{5}{30}$, it seems reasonable to say that the probability of selecting a male is $\frac{25}{30}$, and the probability of selecting a female is $\frac{5}{30}$.

Table 4–6 Color blindness

		Color-blind, C	Normal color vision, N	Row sums
Male	M	0.025	0.475	0.500
Female	F	0.005	0.495	0.500
Column sums		0.030	0.970	1.000

We write the conditional probability of the event "person selected is male", given the event "person is color-blind", as

$$P\,(\text{male}|\text{color-blind}) = P(M|C) = \tfrac{25}{30} = \tfrac{5}{6}.$$

Note that $\frac{25}{30}$ is also the same as

$$\frac{0.025}{0.030} = \frac{P(M \cap C)}{P(C)},$$

and we have, for this example,

$$P(M|C) = \frac{P(M \cap C)}{P(C)}. \tag{1}$$

Similarly,

$$P(F|C) = \frac{P(F \cap C)}{P(C)} = \frac{0.005}{0.030} = \frac{1}{6}.$$

Since the axioms do not treat conditional probability, we require a definition.

4-9 Definition: Conditional probability

The conditional probability of an event A, given B, is denoted by $P(A|B)$ and is defined by

$$\boxed{P(A|B) = \frac{P(A \cap B)}{P(B)},} \tag{2}$$

where A, B, and $A \cap B$ are events in a sample space S, and $P(B) \neq 0$.

Remark If we multiply both sides of Eq. (2) by $P(B)$, we get

$$\boxed{P(A \cap B) = P(B) \cdot P(A|B).} \tag{3}$$

Order is not important in $A \cap B$, because

$$A \cap B = B \cap A.$$

Hence we also have

$$P(B \cap A) = P(B) \cdot P(A|B), \tag{4}$$

and by interchanging A and B in Eq. (4), we get

$$\boxed{P(A \cap B) = P(A) \cdot P(B|A).} \tag{5}$$

Equation (5) is used in Section 4-8 to assign probabilities. It can also be extended to three or more events. For example, the probability of the joint occurrence of three events A, B, and C is

$$P(A \cap B \cap C) = P(A) \cdot P(B|A) \cdot P(C|A \cap B). \tag{6}$$

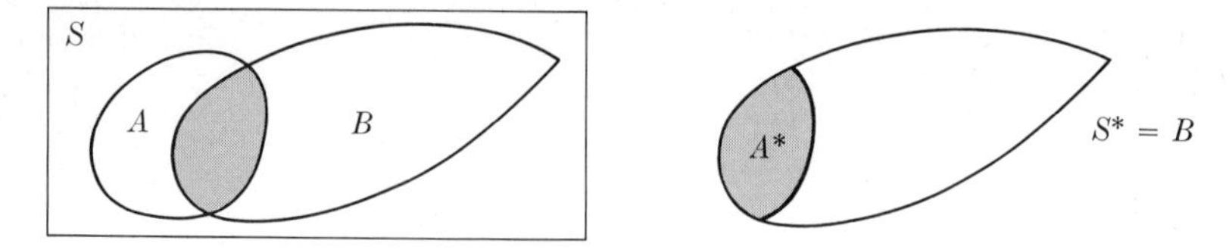

Fig. 4–5 New sample space, $S^* = B$.

Another way of looking at Eq. (2) may be helpful. Consider a sample space S, and the events A, B, and their intersection $A \cap B$ (shown shaded in Fig. 4–5). If we are given B, we ignore all other possible outcomes in S, and think of B as constituting a new, reduced sample space S^*. (See Fig. 4–5.) If we were to assign to points of S^* the same probabilities they had in S, these would add up to only $P(B)$. We wish that the total probability in the new sample space S^*, which is just B, were 1. We achieve this desired goal by enlarging all probabilities p_i of points in B by multiplying each probability by the constant factor $1/P(B)$. When we assign these new probabilities to points of B,

$$p_i^* = \frac{p_i}{P(B)}, \tag{7}$$

and sum both sides of Eq. (7) over all values of i corresponding to sample points in $B = S^*$, we find that the total probability in S^* is

$$\sum p_i^* = \frac{\sum p_i}{P(B)} = \frac{P(B)}{P(B)} = 1,$$

as desired.

Finally, to get the probability of any event A, given B, we add the p^* probabilities of the points of A that are in the reduced sample space $S^* = B$. These are the points in the intersection $A \cap B$; we denote their probabilities in S by $p_1, p_2, \ldots, p_m$. Then, summing both sides of Eq. (7) for values of i from 1 through m, we get

$$P(A|B) = \sum_{i=1}^{m} p_i^* = \sum_{i=1}^{m} \frac{p_i}{P(B)} = \frac{P(A \cap B)}{P(B)}.$$

Note that the conditional probability of A, given B, is proportional to the probability of $A \cap B$, the proportionality factor being $k = 1/P(B)$, just as it is in Eq. (7) for individual points.

Example 3 A coin is tossed until a head appears, or until it has been tossed three times. Given that the head does not occur on the first toss, what is the probability that the coin is tossed three times?

Solution A sample space is given by

$$S = \{H,\ TH,\ TTH,\ TTT\},$$

with associated probabilities

$$P(H) = \tfrac{1}{2}, \qquad P(TH) = \tfrac{1}{4}, \qquad P(TTH) = P(TTT) = \tfrac{1}{8}.$$

These add to 1. Let B be the given event, "no head on first toss". Then

$$B = \{TH,\ TTH,\ TTT\}$$

and

$$P(B) = \tfrac{1}{4} + \tfrac{1}{8} + \tfrac{1}{8} = \tfrac{1}{2}.$$

Next, let A be the event "coin is tossed three times". Then

$$A = \{TTH,\ TTT\}, \qquad P(A) = \tfrac{1}{4},$$

and

$$A \cap B = A, \qquad P(A \cap B) = \tfrac{1}{4}.$$

Hence

$$P(A|B) = \frac{P(A \cap B)}{P(B)} = \frac{1/4}{1/2} = \frac{1}{2}.$$

Example 4 The integers from 1 through n are assigned probabilities proportional to their sizes. (a) Find the probabilities. (b) Find the conditional probability of 1, given that 1 or n occurs.

Solution a) The total probability must be 1, and the probability of any integer i from 1 through n is proportional to i:

$$P(i) = k \times i, \qquad i = 1, 2, \ldots, n.$$

Then

$$\sum_{i=1}^{n} ki = k[1 + 2 + 3 + \cdots + n] = 1.$$

But

$$1 + 2 + 3 + \cdots + n = \frac{n(n+1)}{2}.$$

Hence

$$k = \frac{2}{n(n+1)}$$

and

$$P(i) = \frac{2i}{n(n+1)}.$$

b)

$$P(1|1 \text{ or } n) = \frac{P(1 \cap 1 \text{ or } n)}{P(1 \text{ or } n)} = \frac{P(1)}{P(1) + P(n)}$$

$$= \frac{k \times 1}{k \times 1 + k \times n} = \frac{1}{1+n}.$$

Note that we did not need to know the value of the proportionality factor k to solve part (b).

EXERCISES FOR SECTION 4-7

1. If A and B are *mutually exclusive* and $P(B)$ is not zero, what can you say about $P(A|B)$? Interpret your result.
2. If A always occurs when B does, then every sample point in B is also in A; B is a subset of A. What can you say about $P(A|B)$ in such circumstances? Interpret your result.
3. If A and B are independent and $P(B) \neq 0$, what can you say about $P(A|B)$? Does this seem reasonable?
4. In Example 1, what is the probability that face 3 is down, given that face 4 is not down?
5. In Example 2, a person is selected at random from the population of people with normal color vision. What is the probability that the selected person is (a) male? (b) female?
6. In Example 3, given that the coin was tossed at most 2 times, what is the probability that it was tossed exactly twice?
7. A coin is tossed until a head first appears, or until it has been tossed 4 times. Given that a head did not appear on either of the first 2 tosses, find the probability that (a) the coin was tossed 4 times, and (b) it was tossed just 3 times.
8. By Eq. (3), if $P(B) \neq 0$, then $P(A \cap B) = P(B) \cdot P(A|B)$. If $P(B) = 0$, $P(A|B)$ is undefined. But $P(A \cap B) = P(B) \cdot P(A|B)$ is still true in some sense. Why?
9. In a sophomore class of 180 students, all of whom took both English and History, 15 failed History, 10 failed English, and 5 failed both. Find the probability that a student chosen at random from this class failed History and passed English. Find the probability that he failed English and passed History.
10. Twenty boys went on a picnic. Five got sunburned, 8 got bitten by mosquitoes, and 10 got home without mishap. What is the probability that a sunburned boy was ignored by the mosquitoes? What is the probability that a bitten boy was also burned?
11. An insurance company finds that about one check in a thousand is drawn on insufficient funds, and that such checks are invariably postdated. The company also finds that about one check in one hundred drawn on sufficient funds is postdated. If a postdated check is received, what is the conditional probability that it comes from a customer having insufficient funds?
12. In the baseball problem, Exercise 11 of Section 4-2, find the theoretical batting average.
13. Suppose 2 bad light bulbs get mixed up with 10 good ones, and that you start testing the bulbs, one by one, until you have found both defectives. What is the probability that you will find the last defective on the 7th testing?
14. (a) If, in the two-dice experiment of Table 1-11, it is known that at least one die has fewer than 3 spots showing, what is the probability that the other die has 3 or more spots? (b) If we are given that $r \leq c + 2$, what is the probability that $r + c = 10$?

15. If a family having 4 children is known to have at least 1 boy, what is the probability that it has exactly 2 boys? (Assume that boys and girls have an equal chance of being born.) What additional unstated assumptions are you making?
16. The integers from 1 through $2n$ are assigned probabilities proportional to their logarithms. (a) Find the probabilities. (b) Show that the conditional probability of the integer 2, given that an even integer occurs, is

$$\frac{\log 2}{n \log 2 + \log (n!)}.$$

17. Prove: If A and B are mutually exclusive and $P(A \cup B)$ is not zero, then

$$P(A|A \cup B) = \frac{P(A)}{P(A) + P(B)}.$$

Which examples in the text illustrate applications of this result?

4-8 USING THE PRODUCT RULE TO ASSIGN PROBABILITIES IN A SAMPLE SPACE

We now illustrate how the rule

$$P(A \cap B) = P(B \cap A) = P(A) \cdot P(B|A) \tag{1}$$

is used in assigning probabilities when A and B are not assumed to be independent.

Example 1 Jimmy likes to go shopping with his mother because he can sometimes get her to buy him a toy. The probability that she takes him along on her shopping trip this afternoon is 0.4; and if she does, the probability that he gets a toy is 0.8. What is the probability that she takes him shopping and buys him a toy?

Solution

$$\begin{aligned} P(\text{shopping and toy}) &= P(\text{shopping}) \cdot P(\text{toy}|\text{shopping}) \\ &= 0.4 \times 0.8 = 0.32. \end{aligned}$$

Example 2 A magazine advertiser estimates that the probability that his ad will be read by a subscriber is 0.4, and that if it is read, the probability that the reader will buy his product is 0.01. Using these estimates, find the probability that a subscriber will read the ad and buy the product.

Solution

$$\begin{aligned} P(\text{read ad and buy product}) &= P(\text{read ad}) \cdot P(\text{buy product}|\text{read ad}) \\ &= 0.4 \times 0.01 = 0.004. \end{aligned}$$

Table 4–7(a) Drawings without replacement

		Second ball B	Second ball R	Row totals
First ball	B	$\frac{2}{21}$	$\frac{5}{21}$	$\frac{1}{3}$
	R	$\frac{5}{21}$	$\frac{9}{21}$	$\frac{2}{3}$
Column totals		$\frac{1}{3}$	$\frac{2}{3}$	1

Example 3(a) *Drawing without replacement.* An urn contains 5 black balls and 10 red balls. Two balls are drawn at random, one after the other, without replacement. Set up a sample space for the possible outcomes of the experiment, with appropriate probabilities.

Solution As a sample space for the experiment, we take

$$S = \{(B, B),\quad (B, R),\quad (R, B),\quad (R, R)\},$$

where, for example, (B, R) means "first ball black, second ball red". Since balls are drawn at random, all balls in the urn at any drawing are equally likely to be drawn. For both balls to be black, the first one drawn must be black ($p_1 = \frac{5}{15}$), and the second one drawn must also be black ($p_2 = \frac{4}{14}$). Therefore

$$\begin{aligned} P(B, B) &= P(\text{1st ball black}) \cdot P(\text{2nd ball black}|\text{1st ball black}) \\ &= \tfrac{5}{15} \times \tfrac{4}{14} = \tfrac{2}{21}. \end{aligned}$$

Likewise,

$$\begin{aligned} P(B, R) &= P(\text{1st ball black}) \cdot P(\text{2nd ball red}|\text{1st ball black}) \\ &= \tfrac{5}{15} \times \tfrac{10}{14} = \tfrac{5}{21}, \end{aligned}$$

$$\begin{aligned} P(R, B) &= P(\text{1st ball red}) \cdot P(\text{2nd ball black}|\text{1st ball red}) \\ &= \tfrac{10}{15} \times \tfrac{5}{14} = \tfrac{5}{21}, \end{aligned}$$

$$\begin{aligned} P(R, R) &= P(\text{1st ball red}) \cdot P(\text{2nd ball red}|\text{1st ball red}) \\ &= \tfrac{10}{15} \times \tfrac{9}{14} = \tfrac{9}{21}. \end{aligned}$$

The results are summarized in Table 4–7(a).

Note that $\frac{1}{3} \times \frac{1}{3} \neq \frac{2}{21}$; the outcome for the second ball is not independent of the outcome for the first ball.

Example 3(b) *Drawing with replacement.* If the sampling in Example 3(a) is done with replacement (we put the first ball back before drawing the second

Table 4–7(b) Drawings with replacement

Second ball

		B	R	Row totals
First ball	B	$\frac{1}{9}$	$\frac{2}{9}$	$\frac{1}{3}$
	R	$\frac{2}{9}$	$\frac{4}{9}$	$\frac{2}{3}$
Column totals		$\frac{1}{3}$	$\frac{2}{3}$	1

ball), then the probability of a black ball on the second drawing is independent of the outcome on the first:

$$P(\text{2nd ball black}) = P(\text{1st ball black}) = \tfrac{5}{15} = \tfrac{1}{3}.$$

The probabilities of sample points in this new experiment are shown in Table 4–7(b). Each cell entry is the product of the corresponding row total and column total.

Example 4 *Two urns.* An ordinary die is thrown once. If a 1 or 6 appears, a ball is then drawn from urn I; otherwise, a ball is drawn from urn II. Urn I contains 2 red balls, 1 white ball, and 2 blue balls. Urn II contains 3 white balls, 1 blue ball, and no red balls. Set up a sample space for the possible outcomes of the experiment and find the probability (a) that a white ball is drawn, and (b) that urn I was used, given that a white ball was drawn.

Solution The experimental conditions imply the following probabilities:

$$\begin{aligned} P(\text{I}) &= \tfrac{1}{3}, & P(R|\text{I}) &= \tfrac{2}{5}, & P(W|\text{I}) &= \tfrac{1}{5}, & P(B|\text{I}) &= \tfrac{2}{5}, \\ P(\text{II}) &= \tfrac{2}{3}, & P(R|\text{II}) &= 0, & P(W|\text{II}) &= \tfrac{3}{4}, & P(B|\text{II}) &= \tfrac{1}{4}. \end{aligned}$$

Using these data, we construct a sample space showing the urn used and the color of ball drawn. Table 4–8 shows probabilities of the possible outcomes

Table 4–8 Two urns

Color of ball

		R	W	B	Row sums
Urn	I	$\frac{2}{15}$	$\frac{1}{15}$	$\frac{2}{15}$	$\frac{1}{3}$
	II	0	$\frac{2}{4}$	$\frac{2}{12}$	$\frac{2}{3}$
Column sums		$\frac{2}{15}$	$\frac{17}{30}$	$\frac{18}{60}$	1

Table 4–9 Average results for 100,000 tests for a rare disease

Test outcome	Have disease	Do not have disease	Totals
Positive	98	999	1,097
Negative	2	98,901	98,903
Totals	100	99,900	100,000

of the experiment. Hence, $P(W) = \frac{17}{30}$, and

$$P(\text{I}|W) = \frac{P(\text{I} \cap W)}{P(W)} = \frac{1/15}{17/30} = \frac{2}{17}.$$

The chance of urn I, given that a white ball was drawn, is no longer 1 in 3; now it is only 2 in 17.

Example 5 *Test for a rare disease.* Some medical tests turn out either "positive" or "negative". "Positive" indicates that the person tested has the disease in question; "negative" indicates that he does not have it. Suppose that such a test for a rare disease sometimes makes mistakes: 1 in 100 of those free of the disease have positive test results, and 2 in 100 of those having the disease have negative test results. The rest are correctly identified. One person in 1000 has the disease. Find the probability that a person with a positive test has the disease.

Solution The test sounds reliable, for it makes only 1 or 2 errors per 100 persons tested. In order to find the probabilities, we record in Table 4–9 the average results for 100,000 cases. Of an original 100,000, only 100 have the disease. On the average, 2 of these give negative tests, and of those free of the disease, 999 give positive tests. If the test is positive, then

$$P(\text{disease}|\text{positive test}) = \frac{98}{1097} \approx \frac{1}{11}.$$

This result may well surprise you, because it does not seem reasonable that a test that rarely makes a mistake could be wrong in its diagnosis 10 times out of 11. This surprise stems from a misinterpretation of the meaning of the test. Far from being worthless, the evidence of a positive test increases $P(\text{disease})$ from 1/1000 to about 1/11. Roughly, the original probability has been multiplied by 90. Viewed in this way, the change has been marked. Similarly, before the test $P(\text{disease}) = 100/100{,}000$, but a negative test reduces $P(\text{disease})$ to 2/98,903. Roughly, the probability has been divided by 50. The effect of the information from this test is to multiply or divide the probability of having the disease by substantial factors.

EXERCISES FOR SECTION 4–8

1. In the two-urn example, Table 4–8, verify the probabilities of the following events:

   a) $\text{II} \cap W$ b) $\text{I} \cap B$ c) $\text{I} \cap R$ d) $\text{II} \cap B$

2. In the two-urn example, Table 4–8, find the following conditional probabilities:

   a) $P(\text{II}|W)$ b) $P(\text{I}|R)$ c) $P(\text{II}|R)$ d) $P(\text{I}|B)$ e) $P(\text{II}|B)$

3. Suppose that the urn in Example 3, Table 4–7(a), contains b black balls and r red balls. Set up a sample space and assign probabilities to points in it, assuming that two balls are drawn, one after the other, without replacement. Using these probabilities, compute:

   a) $P(\text{2nd ball red}|\text{1st ball black})$, b) $P(\text{2nd ball red})$,
   c) $P(\text{1st ball red}|\text{2nd ball red})$.

4. Repeat Exercise 3, assuming that the sample is drawn with replacement.

5. (This exercise should be worked before going on to Section 4–9.) In a certain factory, machine A produces 30% of the output, machine B produces 25%, and machine C produces the rest. One percent of the output of machine A is defective, as is 1.2% of B's output, and 2% of C's. In a day's run, the three machines produce 10,000 items. What is the probability that one item drawn at random from these 10,000 is defective? If it is defective, what is the probability that it was produced by A? by B? by C?

6. Suppose $P(E) = 0.3$, $P(F) = 0.2$, and $P(E \cup F) = 0.4$. Make a two-by-two table showing probabilities of $E \cap F$, $\overline{E} \cap F$, $E \cap \overline{F}$, $\overline{E} \cap \overline{F}$. What are the following probabilities equal to?

   a) $P(E \cap F)$ b) $P(E|F)$ c) $P(F|E)$
   d) $P(\overline{E}|\overline{F})$ e) $P(E \cup \overline{F})$ f) $P(\overline{E} \cup \overline{F})$

7. In rolling a die repeatedly, what is the probability that a 1 appears for the first time on the 4th roll?

8. In dealing cards from a bridge deck, what is the probability that the first spade occurs at the 5th card?

9. In Example 1, suppose that the probability that Jimmy's mother will buy him a toy when she does not take him with her is 0.3. If the other probabilities are unchanged, what is the probability that she gets him a toy when she goes shopping?

10. In Example 2, make the further assumption that the probability that a nonsubscriber will read the ad is 0.003, and that if a nonsubscriber reads the ad, the probability that he will buy the product is 0.008. What is the probability that a randomly chosen person will read the ad and buy the product? Assume there is one chance in 20 that a person is a subscriber.

11. From twelve tickets numbered from 1 through 12, two tickets are drawn, one after the other, without replacement. What is the probability that (a) both numbers are even? (b) both numbers are odd? (c) the first number is even and the second number is odd? (d) one number is even and the other number is odd?

12. In preparation for an examination, a student has been given two sets of questions to study, with 5 questions in each set. At the time of the examination, he knows the answers to all of the questions in the first set, and to 4 of the 5 questions in the second set. If the examination consists of 3 questions, 2 chosen at random from one set, and 1 chosen at random from the other set, and the examiner tosses a coin to decide which set to take the two problems from, what is the probability that the student will be able to answer all of the questions? That he can answer only 2 of them?
13. Suppose, in Exercise 12, that there are 10 questions in each set and, at the time of the examination, the student knows the answers to 9 questions in the first set and 8 questions in the second set. If the other conditions of Exercise 12 are unchanged, what is the probability that the student can answer all 3 questions on the examination? That he can answer none of the questions?
14. In Example 5, the test for rare diseases, suppose that the error rates remain the same but that half the people tested have the disease and half do not have it. Among those tested, what is the probability that a person whose test is positive has the disease? That a person whose test is negative is disease-free? Should these two probabilities add to 1? Do the results accord more with your intuition than those of Example 5? If so, why?
15. Rework Exercise 14, assuming that 1 person in 10 of those tested has the disease.

4–9 BAYES'S THEOREM

At the start of the test in the rare-disease example at the end of Section 4–8, the probabilities of testing a diseased or a nondiseased person were 1/1000 and 999/1000, respectively. These probabilities measure the chances that a diseased person, or a nondiseased person, *will be* tested, and are often called *a priori*, or *prior*, probabilities. They are the probabilities *prior* to any information that the test may yield.

Suppose now that we know the conditions of the test setup, and that we know, in addition, only the outcome of a patient's test. If a prize is offered for correctly guessing whether or not the patient has the disease, how should we guess? We need the conditional probabilities of having the disease given the outcome of the test. These are called *a posteriori*, or *posterior*, probabilities because they are probabilities *after* the result of the "experiment" is known.

From Table 4–9, we have:

$$P(\text{disease-free}|\text{positive test}) = \frac{999}{1097} \approx 0.9107 \approx \frac{10}{11},$$

$$P(\text{disease-free}|\text{negative test}) = \frac{98{,}901}{98{,}903} \approx 0.9999798 \approx 1 - \frac{2}{100{,}000}.$$

Hence, whether the test is negative or positive, we have a much better chance of guessing correctly if we choose "disease-free". But this is the guess that any intelligent person would make without the test result! If a patient's test is positive, why might a physician decide to continue testing this patient with other kinds of tests? One reason is that it may be worth more to society and to the patient to detect the presence of a disease than to state that the disease is absent.

Use the data of Exercise 14, Section 4–8, where the disease is not so rare among those tested, to show that the posterior probability of disease, given a positive test, and of disease-free, given a negative test, both exceed $\frac{1}{2}$.

Example 1 *Two urns.* In the two-urn example, Example 4 of Section 4–8, find the posterior probabilities of urns I and II, given that the first ball drawn is replaced and, after mixing, a second ball is drawn from the same urn as the first ball, and that both balls are white.

Solution We forgo listing the sample space S consisting of sample points like $(I; R, R)$, $(I; R, W)$, and so on to $(II; B, B)$, where $(I; R, W)$ means "urn I is used, the first ball drawn is red, and the second ball drawn is white", and so on. But we have such a sample space in mind, and we assign probabilities to its points in accord with the laws of conditional probability. For example, since the first ball is replaced before the second is drawn,

$$\begin{aligned} P(\mathrm{I}; R, W) &= P(\mathrm{I}) \cdot P(R|\mathrm{I}) \cdot P(W|\mathrm{I}) \\ &= \tfrac{1}{3} \cdot \tfrac{2}{5} \cdot \tfrac{1}{5} = \tfrac{2}{75}. \end{aligned}$$

The purpose of this example is to introduce the notation used in the general Bayes's Theorem, so we do not leap at once to the numerical solution.

Let E be the event "two white balls are drawn":

$$E = \{(\mathrm{I}; W, W), (\mathrm{II}; W, W)\}.$$

Also, let H_1 be the event "urn I was used", and H_2 the event "urn II was used". Note that H_1 and H_2 are mutually exclusive and that their union is S. We want the conditional probabilities

$$P(H_1|E) \qquad \text{and} \qquad P(H_2|E).$$

The formula for conditional probability tells us that

$$P(H_1|E) = \frac{P(H_1 \cap E)}{P(E)}, \tag{1}$$

and we can write a similar equation with H_2 in place of H_1. It is easy to compute $P(H_1 \cap E)$ and $P(H_2 \cap E)$; their values are

$$P(H_1 \cap E) = P(\mathrm{I}; W, W) = \frac{1}{3} \cdot \frac{1}{5} \cdot \frac{1}{5} = \frac{1}{75}, \tag{2a}$$

$$P(H_2 \cap E) = P(\mathrm{II}; W, W) = \frac{2}{3} \cdot \frac{3}{4} \cdot \frac{3}{4} = \frac{3}{8}. \tag{2b}$$

Moreover,

$$P(E) = P(H_1 \cap E) + P(H_2 \cap E) = \frac{233}{600}, \tag{3}$$

since E must occur either with H_1 or with H_2, and it cannot occur simultaneously with both. If we now substitute from Eqs. (2) and (3) into Eq. (1) we get

$$P(H_1|E) = \frac{P(H_1 \cap E)}{P(H_1 \cap E) + P(H_2 \cap E)} = \frac{1/75}{233/600} = \frac{8}{233} \approx 0.03,$$

$$P(H_2|E) = \frac{P(H_2 \cap E)}{P(H_1 \cap E) + P(H_2 \cap E)} = \frac{3/8}{233/600} = \frac{225}{233} \approx 0.97.$$

Note how the evidence provided by the outcome "both balls white" is reflected in the high posterior probability of urn II, where white balls predominate.

Bayes's Theorem, which generalizes the results of examples like the foregoing, can be used in scientific work in the following way. Suppose there are several mutually exclusive and exhaustive hypotheses $H_1, H_2, \ldots, H_n$ to account for a phenomenon that is subject to test by experiment. Before a particular experiment begins, it may be very hard to assign probabilities, i.e., prior probabilities, to these hypotheses. An experimenter might assign probabilities that are in some way proportional to the "intensity of belief" he has in the various hypotheses.* (Another investigator might assign quite different probabilities.) An experiment is performed, with the aim of discovering evidence to modify these prior probabilities. Such evidence may even assign such low *posterior* probabilities to some of the hypotheses as to eliminate them from further consideration, just as the drawing of a red ball in Example 4 of Section 4–8 eliminates urn II.

Each new experiment can begin with *a priori* probabilities of the remaining hypotheses proportional to the *a posteriori* probabilities that resulted from the previous experiments. In this way, scientific evidence accumulates and modifies our beliefs, weakening our intensity of belief in some hypotheses, strengthening it in others. And the more evidence that accumulates, the less does it matter what the original *a priori* probabilities were, provided they were all tenable and that no possible hypothesis was assigned prior probability 1 or 0.

4–10 Bayes's Theorem

Let $H_1, H_2, \ldots, H_n$ be mutually exclusive events whose union is the sample space S of an experiment. Let E be an arbitrary event of S such that $P(E) \neq 0$. Then

$$\boxed{P(H_1|E) = \frac{P(H_1 \cap E)}{P(H_1 \cap E) + P(H_2 \cap E) + \cdots + P(H_n \cap E)},} \tag{4}$$

and similar results hold for H_2, H_3, and so on.

---

* See, for example, *Inference and Disputed Authorship: The Federalist*, by F. Mosteller and D. L. Wallace, Addison-Wesley Publishing Co., Inc., 1964. Chapter 3, pp. 46–91.

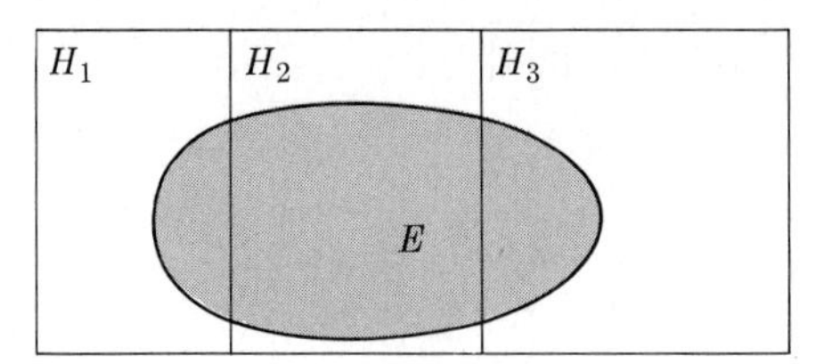

Fig. 4–6 Partitioning where $n = 3$: $S = H_1 \cup H_2 \cup H_3$, $E = (H_1 \cap E) \cup (H_2 \cap E) \cup (H_3 \cap E)$.

Proof The proof will be given for the case $n = 3$. Figure 4–6 and Tables 4–10 and 4–11 illustrate this case. The three hypotheses H_1, H_2, and H_3 are mutually exclusive and exhaustive; their union is S. The part of E that is in H_1 is $H_1 \cap E$, the part in H_2 is $H_2 \cap E$, and the part in H_3 is $H_3 \cap E$. The entire event E is the union of these three mutually exclusive events—similarly for the complementary event $\overline{E}$, which we include for completeness but which plays no part in the proof.

Table 4–11 gives the probabilities of the joint events in the cells of Table 4–10.

Since the H's are mutually exclusive and exhaustive, the first column sum is $P(E)$:

$$P(E) = P(H_1 \cap E) + P(H_2 \cap E) + P(H_3 \cap E). \tag{5}$$

Table 4–10 Partition of sample space

		Event		
		E	$\overline{E}$	Unions of rows
Hypothesis	H_1	$H_1 \cap E$	$H_1 \cap \overline{E}$	H_1
	H_2	$H_2 \cap E$	$H_2 \cap \overline{E}$	H_2
	H_3	$H_3 \cap E$	$H_3 \cap \overline{E}$	H_3
Unions of columns		E	$\overline{E}$	S

Table 4–11 Probabilities for Table 4–10

		Event		
		E	$\overline{E}$	Row sums
Hypothesis	H_1	$P(H_1 \cap E)$	$P(H_1 \cap \overline{E})$	$P(H_1)$
	H_2	$P(H_2 \cap E)$	$P(H_2 \cap \overline{E})$	$P(H_2)$
	H_3	$P(H_3 \cap E)$	$P(H_3 \cap \overline{E})$	$P(H_3)$
Column sums		$P(E)$	$P(\overline{E})$	1

By the law of conditional probability,

$$P(H_1|E) = \frac{P(H_1 \cap E)}{P(E)}$$
$$= \frac{P(H_1 \cap E)}{P(H_1 \cap E) + P(H_2 \cap E) + P(H_3 \cap E)}.$$

This completes the proof for $n = 3$. The proof for $n = 2$, or $n \geq 4$, follows the same pattern and leads to Eq. (4) in each case. □

Example 2 (See Exercise 5, Section 4–8.) In a factory, machine A produces 30% of the output, machine B produces 25%, and machine C produces the remaining 45%. One percent of the output of machine A is defective, as is 1.2% of B's output, and 2% of C's. In a day's run, the three machines produce 10,000 items. An item drawn at random from a day's output is defective. What is the probability that it was produced by A? by B? by C?

Solution You have already applied Bayes's Theorem if you solved this exercise in Section 4–8. The connection is made by taking E, H_1, H_2, and H_3 to be the following events:

E: defective item,

H_1: item produced by machine A,

H_2: item produced by machine B,

H_3: item produced by machine C.

Then $P(H_1|E)$ is the probability that the item was produced by A, *given* that it was defective. $P(H_1 \cap E)$ is the probability of the event "produced by A *and* defective", with similar meanings for $P(H_2 \cap E)$ and $P(H_3 \cap E)$. The data give the following probabilities for an item selected at random from the total day's production:

$$P(H_1) = 0.30, \qquad P(E|H_1) = 0.010,$$
$$P(H_2) = 0.25, \qquad P(E|H_2) = 0.012,$$
$$P(H_3) = 0.45, \qquad P(E|H_3) = 0.020.$$

From these data we may compute

$$P(H_1 \cap E) = P(H_1) \cdot P(E|H_1) = 0.003$$
$$P(H_2 \cap E) = P(H_2) \cdot P(E|H_2) = 0.003$$
$$P(H_3 \cap E) = P(H_3) \cdot P(E|H_3) = 0.009$$
$$\text{Total:} \quad P(E) = 0.015.$$

Before an item is drawn from the population and examined, the probabilities of its having been produced by machines A, B, and C are 0.30, 0.25, and 0.45, in that order. Bayes's Theorem is useful in telling us how these probabilities are modified when we have the additional information that the item drawn was defective. The new probabilities are

$$P(H_1|E) = \frac{P(H_1 \cap E)}{P(E)} = \frac{0.003}{0.015} = 0.20,$$

$$P(H_2|E) = \frac{P(H_2 \cap E)}{P(E)} = \frac{0.003}{0.015} = 0.20,$$

$$P(H_3|E) = \frac{P(H_3 \cap E)}{P(E)} = \frac{0.009}{0.015} = 0.60.$$

We summarize these results:

	Machine		
	A	B	C
A priori probability (before information that item is defective)	0.30	0.25	0.45
A posteriori probability (after information that item is defective)	0.20	0.20	0.60

This example illustrates one of the chief applications of Bayes's Theorem. We start with a set of prior probabilities associated with the possibilities H_1, H_2, and so on. Next we perform an experiment and observe that event E has occurred. Then we use this information to modify the set of prior probabilities, replacing

$$P(H_1) \quad \text{by} \quad P(H_1|E),$$
$$P(H_2) \quad \text{by} \quad P(H_2|E),$$

and so on, with the help of Eq. (4).

Remark 1 In order to compute $P(H_1 \cap E)$ we use the *a priori* probability $P(H_1)$ and the conditional probability of E, given H_1, because

$$P(H_1 \cap E) = P(H_1) \cdot P(E|H_1). \tag{6a}$$

Similarly for the other "hypotheses", H_2 and so on,

$$P(H_i \cap E) = P(H_i) \cdot P(E|H_i). \tag{6b}$$

If we use Eqs. (6a,b) to evaluate the numerator and denominator in Eq. (4), we also have

$$P(H_1|E) = \frac{P(H_1) \cdot P(E|H_1)}{P(H_1) \cdot P(E|H_1) + P(H_2) \cdot P(E|H_2) + \cdots + P(H_n) \cdot P(E|H_n)}. \tag{7}$$

The left side of Eq. (7) is the *a posteriori* probability of H_1, given E; on the right side appear the *a priori* probabilities of $H_1, H_2, \ldots, H_n$ together with the probabilities of E given H_1 of E given H_2, and so on.

Remark 2 In the rare-disease example, the prior odds are 999 to 1 in favor of the test patient's being disease-free. The posterior odds of his being disease-free, given that his test was negative, are 98,901 to 2. The result in Bayes's Theorem can always, as here, be expressed in terms of *odds*. The *prior* odds are proportional to the *prior* probabilities

$$P(H_1), \qquad P(H_2), \qquad \ldots, \qquad P(H_n).$$

The *posterior* odds are proportional to the numerators of Eq. (4) and the other equations like it, since they all have the same denominator. Hence the *posterior* odds are proportional to

$$P(H_1 \cap E), \qquad P(H_2 \cap E), \qquad \ldots, \qquad P(H_n \cap E).$$

EXERCISES FOR SECTION 4-9

1. Sixty percent of the members of a club are men. Eighty percent of the men and 75% of the women have activity tickets for all the club activities. A ticket is found and turned in to the club's lost-and-found department. What is the probability that it belongs to a woman? To a man?
2. Three girls, Alice, Betty, and Charlotte, wash the family dishes. Since Alice is the oldest, she does the job 40% of the time. Betty and Charlotte share the other 60% equally. The probability that at least one dish will be broken when Alice is washing them is 0.02; for Betty and Charlotte the probabilities are 0.03 and 0.02. The parents don't know who is washing the dishes, but one night they hear one break. What is the probability that Alice was washing? Betty? Charlotte?
3. An experiment consists of throwing a three-sided die and then, depending upon the outcome of the throw, selecting a ball from one of two urns. If the die falls "1 or 2", the ball is drawn from an urn containing 1 red ball and 4 black balls; if the die falls "3", the ball is drawn from an urn with 3 red and 2 black balls. You didn't see the die thrown, but you observed that a red ball was drawn. What is the probability that it came from the first urn? From the second?
4. Suppose, in the two-urn problem of Example 1, the first ball is replaced, and a second ball is drawn from the same urn as the first. If both balls are white, what is the probability that urn I was used? Urn II?
5. (Continuation.) In Exercise 4 above, suppose that 2 blue balls are drawn, without replacement. What are the posterior probabilities of the two urns?
6. (Continuation.) In Exercise 5, suppose that the first ball is replaced before the second is drawn. Find the posterior probabilities of the urns if both balls are blue.
7. A fair coin is tossed and if it falls "heads" we draw a ball from urn I; if "tails", from urn II. Urn I contains 3 red balls and 1 white ball. Urn II contains 1 red ball and 3 white balls. What are the prior and posterior probabilities of the two urns, assuming (a) that a red ball is drawn, (b) that a white ball is drawn?

8. Solve Exercise 7 under the modified assumptions that the prior probabilities are 0.1 for the first urn and 0.9 for the second.

9. In Exercise 8, suppose the experiment continues for n drawings, the ball being replaced and the contents of each urn thoroughly mixed before the next drawing. If all n balls drawn are red, what are the posterior probabilities of the urns? For what value, or values, of n are these posterior probabilities approximately equal? What happens to these probabilities if n is very large? What is your interpretation of this result?

10. Assume that 1 coin in 10,000,000 has two heads; the rest are legitimate. If a coin, chosen at random, is tossed 10 times and comes up "heads" every time, what is the probability that it is two-headed?

11. (Continuation.) In Exercise 10, suppose the coin falls "heads" n times in a row. How large must n be to make the odds approximately even that the coin is two-headed?

12. A commuter who works in Boston must either go through the Callahan tunnel or across the Mystic River bridge to get home. He varies his route, choosing the tunnel with probability $\frac{1}{3}$, the bridge with probability $\frac{2}{3}$. If he goes by tunnel, he gets home by 6 o'clock 75% of the time; if he goes by bridge, he gets home by 6 o'clock only 70% of the time, but he likes the scenery better that way. If he gets home after 6 o'clock, what is the probability that he used the bridge?

13. An automobile insurance company classifies drivers as class A (good risks), class B (medium risks), and class C (poor risks). They believe that class A risks constitute 30% of the drivers who apply to them for insurance, class B 50%, and class C 20%. The probability that a class A driver will have one or more accidents in any 12-month period is 0.01, for a class B driver the probability is 0.03, and for a class C driver it is 0.10. The company sells Mr. Jones an insurance policy and within 12 months he has an accident. What is the probability that he is a class A risk? Class B? Class C?

14. (Continuation.) If a policyholder, in Exercise 13, goes n years without an accident, and years are independent, what are the odds that he belongs to class A? Class B? Class C?

15. In a factory, machine A produces 40% of the output and machine B produces 60%. On the average, 9 items in 1000 produced by A are defective and 1 item in 250 produced by B is defective. An item drawn at random from a day's output is defective. What is the probability that it was produced by A? by B?

16. Friends of yours play two games about equally often. One game is played with one die, the other with two dice. The score in either game is the number of dots on the top face, or faces. You hear the score of a throw announced as 2. What is the chance they are playing the one-die game?

17. What is the answer to the question of Exercise 16 if the announced score is 6? If it is 7? If it is 1?

18. Under hypothesis H_1 a rare event E has the very small probability p of occurring, while under a second hypothesis, H_2, its probability is p^2. (a) If the two hypotheses are equally likely, and are the only ones, and E occurs, find $P(H_1|E)$. Interpret. (b) Suppose that, instead of E, $\overline{E}$ occurs. Find $P(H_1|\overline{E})$, compare it with $P(H_1)$, and comment on the value of one $\overline{E}$ observation.

19. Events $A_1, A_2, \ldots, A_n$ are mutually exclusive, exhaustive, and equally likely prior hypotheses. The conditional probability of E, given A_i, is

$$P(E|A_i) = \frac{i}{n}, \qquad i = 1, 2, \ldots, n.$$

If, in two independent trials, EE occurs, find $P(A_i|EE)$. Evaluate for $i = n$, $n = 10$. [You may use the formulas

$$\sum_{i=1}^{n} i = n(n+1)/2, \qquad \sum_{i=1}^{n} i^2 = n(n+1)(2n+1)/6.]$$

★4-10 DISCRETE SAMPLE SPACES THAT ARE COUNTABLY INFINITE

In Section 4-2, we introduced axioms for probability in finite sample spaces. Then we used these axioms to deal with events in such sample spaces.

In order to deal with events in countably infinite sample spaces, we need only modify Axiom III.

AXIOMS FOR PROBABILITY IN DISCRETE SAMPLE SPACES

Axiom I *Positiveness.* The probability assigned to each event is positive or zero.

Axiom II *Certainty.* The probability of the entire sample space is 1.

Axiom III′ *Unions.* If A and B are mutually exclusive events, then

$$P(A \cup B) = P(A) + P(B);$$

and, more generally, if $A_1, A_2, A_3, \ldots$ is any finite or countably infinite collection of mutually exclusive events, then

$$P(A_1 \cup A_2 \cup A_3 \cup \cdots) = P(A_1) + P(A_2) + P(A_3) + \cdots.$$

The foregoing set of axioms has an immediate consequence: A *finite* sample space may have sample points with equal or with unequal probabilities, as we have seen earlier in this chapter; but a sample space that has an infinite number of sample points with nonzero probabilities cannot have these probabilities equal. The reason is this: an infinite number of equal, positive numbers cannot sum to 1, and therefore cannot satisfy Axiom II.

In assigning probabilities to the points of a countably infinite sample space, we use the three axioms as illustrated in the following example.

Example 1 A coin is tossed until a head appears. (a) Construct a sample space S for this experiment. (b) Assign probabilities to the sample points in S. (c) Show that $P(S) = 1$.

Solution a) Since the first head may appear at the first toss, or the second toss, or the third toss, and so on, our sample space may be indicated as follows:

$$S = \{H, TH, TTH, TTTH, \ldots\}.$$

b) Assuming that $P(H) = \frac{1}{2}$ and that the tosses are independent, we assign probabilities to the sample points thus:

$$P(H) = \tfrac{1}{2}; \qquad P(TH) = \tfrac{1}{2} \times \tfrac{1}{2} = (\tfrac{1}{2})^2;$$
$$P(TTH) = \tfrac{1}{2} \times \tfrac{1}{2} \times \tfrac{1}{2} = (\tfrac{1}{2})^3; \qquad P(TTTH) = \tfrac{1}{2} \times \tfrac{1}{2} \times \tfrac{1}{2} \times \tfrac{1}{2} = (\tfrac{1}{2})^4;$$

and so on. In general, the nth sample point is

$$TTT \cdots (\text{to } n - 1\ T\text{'s})H$$

and

$$P(TTT \cdots (\text{to } n - 1\ T\text{'s})H) = (\tfrac{1}{2})^n.$$

This method assigns a nonzero probability to infinitely many sample points.

c) Since the sum of the probabilities assigned in (b) gives rise to an infinite geometric progression with $r < 1$, we expect the total probability in S to be

$$\frac{1}{2} + \frac{1}{4} + \frac{1}{8} + \cdots = \frac{1/2}{1 - 1/2} = 1.$$

Theorems for events in a countably infinite sample space The proofs of the theorems in Section 4–2, given for finite sample spaces, are valid with little modification for countably infinite sample spaces. For convenience, we list these theorems and note the modifications that are needed to extend the validity of their proofs to discrete sample spaces with an infinite number of sample points.

4–1 Theorem (extended): Probability of an event

Let A be an event in a discrete sample space S. If A is the empty set, then $P(A) = 0$. If A is nonempty, then $P(A)$ is the sum of the probabilities of the elementary events whose union is A.

The statement of this extended theorem differs from that of Theorem 4–1 in Section 4–2 only because the word "finite" has been replaced with "discrete". For the extended theorem, the first part of the proof in Section 4–2 (for $A = \phi$) remains valid. The second part of the proof (for $A \neq \phi$) is as follows:

If A is nonempty, denote the elementary events of sample space S by

$$E_1,\ E_2,\ E_3,\ \ldots .$$

Then the subset of these elementary events that comprise A can be relabeled as

$$F_1,\ F_2,\ F_3,\ \ldots,$$

where

$$A = F_1 \cup F_2 \cup F_3 \cup \cdots .$$

Since the F's are elementary events in S, they are therefore mutually exclusive and obey Axiom III′. Hence,

$$P(A) = P(F_1) + P(F_2) + P(F_3) + \cdots . \ \square$$

4–2 Theorem (extended): *A* or *B* (or both)

$$P(A \cup B) = P(A) + P(B) - P(A \cap B).$$

The proof given in Section 3–2 is valid for all finite sample spaces, and it can be readily extended to countably infinite sample spaces with the aid of Axiom III′ and the extended Theorem 4–1.

4–3 Theorem (extended): Mutually exclusive events

If $A_1, A_2, \ldots, A_n$ are mutually exclusive events, then

$$P(A_1 \cup A_2 \cup \cdots \cup A_n) = P(A_1) + P(A_2) + \cdots + P(A_n).$$

If, in this theorem, the number of mutually exclusive events becomes countably infinite, then the theorem is identical with Axiom III′:

$$P(A_1 \cup A_2 \cup A_3 \cup \cdots) = P(A_1) + P(A_2) + P(A_3) + \cdots.$$

4–4 Theorem (extended): Complementary events

$$P(\overline{A}) = 1 - P(A).$$

As in Theorem 4–2, with the aid of Axiom III′ we can extend the proof in Section 3–2 to countably infinite sample spaces.

EXERCISES FOR SECTION 4–10

For each of the experiments described in Exercises 1 through 4, (a) construct a sample space S based upon the number of trials required; (b) assign probabilities to the sample points of S; and (c) show that $P(S) = 1$.

1. A die is thrown until the top face shows 6; associate probabilities with the number of the last throw.
2. A card is drawn at random, with replacement, from a well-shuffled bridge deck until a heart is obtained. Associate probabilities with the number of the last draw.
3. Of 100 well-mixed lottery tickets, 10 are prizewinners. A ticket is drawn at random, with replacement, until a winner is obtained.
4. Johnson's probability of sinking a free throw in basketball is p. He tries free throws until he sinks one. (Assume that p is constant and that the throws are independent.)
5. Given that $P(x$ loaves of bread are sold$) = (\frac{1}{2})^{x+1}$, where $x = 0, 1, 2, 3, \ldots$. Show that (a) $P(S) = 1$; (b) P(even number of loaves are sold) $= 2/3$; (c) P(10 or more loaves are sold) $= 1/2^{10}$.

★6. A point is awarded to product A or to product B, whichever wins a trial. Trials continue until one product is two points ahead of the other, and the leading product is then considered the winner of the test. If trials are independent and p is the probability that product A wins a given trial, what is the probability that product A wins the whole test?

MISCELLANEOUS EXERCISES FOR CHAPTER 4

For Exercises 1 through 3, use the following data. A box contains 5 books. A boy randomly takes out one book and then replaces it. He does this 5 times.

1. What is the probability that he has had every book out of the box?
2. What is the probability that he has taken exactly 4 different books from the box?
3. What is the probability that the number of different books he takes out of the box is exactly 3? Exactly 2? Exactly 1?

In Exercises 4 through 7, 4 dice are thrown, and we want to know the probability that:

4. All four dice show the same number.
5. No two are alike.
6. Two are alike of one kind and two are alike of another kind.
7. Two are alike and the other two differ from these and from each other.

In Exercises 8 through 15, use the following information about a game played with two regular dice. A player throws two dice, and if he scores 7 or 11, he wins. If he scores 2, 3, or 12, he loses. But if he scores 4, 5, 6, 8, 9, or 10, he throws the dice again, and keeps on throwing until (a) he gets a 7, in which case he loses, or (b) he gets the score that he got on his first throw, in which case he wins. Find the probability that:

8. He loses on the first throw.
9. He wins on the first throw.
10. He scores 4 on the first throw, and goes on to win.
11. He scores 5 on the first throw, and goes on to win. (*Note.* Throws other than 5 and 7 can be ignored, once the 5 is thrown.)
12. He scores 8 on the first throw, and wins.
13. He scores 9 on the first throw, and wins.
14. He scores 10 on the first throw, and wins.
15. He wins.

In Exercises 16 through 19, assume that 4 cards are drawn, without replacement, from an ordinary bridge deck. Find the probability that:

16. All 4 suits are represented.
17. Exactly 3 suits are represented.
18. All cards are from the same suit.
19. Exactly 2 suits are represented.
20. Draw 4 cards from a shuffled pack, then put them back and repeat until 25 hands are drawn. Record the numbers of different suits represented in each hand. Compare the experimental relative frequencies with the theoretical results you got in Exercises 16 through 19.

In Exercises 21 through 26, assume that 5 cards are drawn, without replacement, from a bridge deck with the four 2's removed. Find the probability that the number of red cards in the hand is:

21. 5   22. 4   23. 3   24. 2   25. 1   26. 0

27. Use a bridge deck with the four 2's removed, and draw a hand of 5 cards, without replacement. Record the number of red cards. Replace the 5 cards that were drawn, shuffle the pack, and repeat the experiment until 25 hands have been drawn and the results recorded. Compare the observed relative frequencies with the theoretical probabilities of Exercises 21 through 26.

In Exercises 28 through 31, use the following information. Three students, A, B, C, have equal claims for an award. They decide that each will toss a coin, and that the man whose coin falls unlike the other two wins. (The "odd man" wins.) If all three coins fall alike, they toss again.

28. Describe a sample space for the result of the first toss of the three coins, and assign probabilities to its elements. What is the probability that A wins on the first toss? That B does? That C does? That there is no winner on the first toss?

29. Given that there is a winner on the first toss, what is the probability that it is A?

30. What is the probability that no winner is decided in the first 2 tosses? In the first n tosses?

31. Given that no winner is decided in the first n tosses, what is the probability that A wins on the next toss?

32. Refer to Example 5, Section 4–8. Given that a, where $0 < a < 1$, is the fraction of the population with the disease, and that the test is performed on 1000 people, find the set of values of a such that the conditional probability of having the disease, given a positive test, is greater than $\frac{1}{2}$.

5 Numbers determined by experiments: random variables

5–1 RANDOM VARIABLES AND THEIR PROBABILITY FUNCTIONS

This chapter introduces two important new concepts: *random variable* and *probability function.* The idea of a sample space is familiar, and we use examples based on this idea to show how random variables and their probability functions arise. The examples point the way to general definitions, and we then go on to study some properties of random variables.

Table 5–1 Three coins

Sample point	Number of heads	Probability
HHH	3	$\frac{1}{8}$
HHT	2	$\frac{1}{8}$
HTH	2	$\frac{1}{8}$
THH	2	$\frac{1}{8}$
HTT	1	$\frac{1}{8}$
THT	1	$\frac{1}{8}$
TTH	1	$\frac{1}{8}$
TTT	0	$\frac{1}{8}$

Example 1 Three coins are tossed. How many fall "heads"?

Discussion The answer is a number determined by the outcome of the experiment. The number may be 0, 1, 2, or 3. Although we cannot predict the outcome exactly, we can say what the possibilities and probabilities are. A sample space for the experiment is shown in the first column of Table 5–1. The second column shows the number of heads for each sample point, and the third column shows the probabilities of the sample points.

Table 5-2 Three coins: probability function for number of heads

Probability	$\frac{1}{8}$	$\frac{3}{8}$	$\frac{3}{8}$	$\frac{1}{8}$
Number of heads	0	1	2	3

The information about the possible numbers of heads, and their probabilities, is collected in Table 5-2. The probability of getting exactly 2 heads is found by adding the probabilities of HHT, HTH, THH, and similarly for other possibilities. Figure 5-1 shows the graph of the probability function whose values are given in Table 5-2. In the graph, vertical bars with lengths proportional to the probabilities are added to guide the eye. The actual graph consists of just the four points indicated by the dots at the tops of these bars.

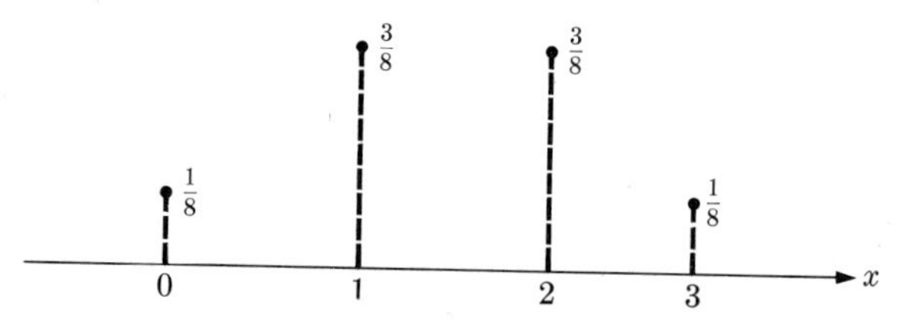

Fig. 5-1 Graph of probability function for number of coins that fall heads when three coins are tossed.

If we let the variable X represent the number of heads, then Table 5-2 shows the possible values that X can have, and the probability of each value. This set of ordered pairs, each of the form

(number of heads, probability of that number),

is the *probability function* of X. Since the value of X is a number determined by the outcome of an experiment, X is called a *random variable.*

It may seem a bit awkward at first, but we often wish to distinguish between a random variable X and one of its values. To help us make such a distinction we use the capital letter X for the random variable and the small letter x for one of its values. And we use $f(x)$ (read "f at x") for the probability that the random variable X takes on the value x:

$$f(x) = P(X = x).$$

Thus, in the three-coin experiment,

$$f(0) = P(X = 0) = \tfrac{1}{8}, \qquad f(2) = P(X = 2) = \tfrac{3}{8},$$
$$f(1) = P(X = 1) = \tfrac{3}{8}, \qquad f(3) = P(X = 3) = \tfrac{1}{8}.$$

Note In this example, the values of $f(x)$ are

$$1 \times \tfrac{1}{8}, \qquad 3 \times \tfrac{1}{8}, \qquad 3 \times \tfrac{1}{8}, \qquad 1 \times \tfrac{1}{8}.$$

The coefficients 1, 3, 3, 1 are the binomial coefficients $\binom{3}{x}$ for $x = 0, 1, 2, 3$. Consequently all values of $f(x)$ are given by the following formulas:

$$f(x) = P(X = x) = \binom{3}{x}\left(\frac{1}{2}\right)^3$$

$$= \frac{3!}{x!(3-x)!}\left(\frac{1}{2}\right)^3, \qquad x = 0, 1, 2, 3.$$

Example 2 *Sums.* A three-sided die is made from an engineer's ruler by painting the numbers 1, 2, and 3 on the three long faces. If such a die is thrown twice, what is the probability function of the random variable X, where X is the sum of the two face-down digits?

Table 5–3 Sample space for 2 throws of three-sided die

		Outcome of second throw		
		1	2	3
Outcome of first throw	1	(1, 1) 2	(1, 2) 3	(1, 3) 4
	2	(2, 1) 3	(2, 2) 4	(2, 3) 5
	3	(3, 1) 4	(3, 2) 5	(3, 3) 6

Solution All pairs of faces are equally likely. The sample space of outcomes can be conveniently listed in a square array, as in Table 5–3. In that table (2, 1), for example, indicates that 2 was the outcome of the first throw and that 1 was the outcome of the second throw. We enter the value of the sum, X, below each outcome pair.

Because each cell has probability $\frac{1}{9}$, the probability function of X, the sum of the numbers on the two bottom faces, is obtained by counting the number of ways each sum can happen and multiplying by $\frac{1}{9}$. The result is the probability function in the following table:

Probability, $f(x)$	$\frac{1}{9}$	$\frac{2}{9}$	$\frac{3}{9}$	$\frac{2}{9}$	$\frac{1}{9}$
Sum, x	2	3	4	5	6

Figure 5–2 shows the graph of the probability function for this example.

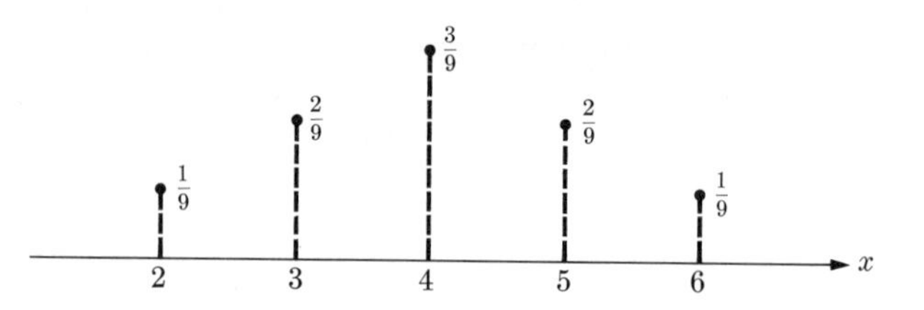

Fig. 5-2 Graph of probability function for the sum in the three-sided ruler example.

You may have noticed that in this example X is the sum of two other random variables, the outcome on the first throw and the outcome on the second throw, which might be labeled U and V, respectively. Thus $X = U + V$.

Example 3 We select one of the integers 1 through 10 at random and count its divisors, or factors. What is the probability that it has exactly 2 divisors? Exactly 1 divisor? Exactly 4 divisors? More than 4?

Solution We first explain the terminology which is conventional in the theory of numbers. We say that one integer is a *divisor*, or *factor*, of a second integer if the second is a whole number times the first. Thus the divisors of 6 are 1, 2, 3, and 6; and the divisors of 7 are 1 and 7. Here we are interested only in positive divisors.

Table 5-4 Numbers of divisors of integers 1 through 10

Integer	1	2	3	4	5	6	7	8	9	10
Number of divisors	1	2	2	3	2	4	2	4	3	4

The first row of Table 5-4 is a sample space for the experiment of selecting an integer from 1 through 10 at random. Let X be the number of divisors of the selected integer. The second row shows the value of the random variable for each sample point. Each integer has probability 0.1 of being drawn because the expression "at random" means that the integers 1 through 10 are equally likely.

Next, we combine cases according to the number of divisors and add their probabilities, 0.1 for each sample point, thus obtaining Table 5-5. We let x stand for any one of the possible numbers of divisors, and $f(x)$ for the probability that X takes the value x. Thus with the value $x = 2$, we associate the probability $f(2) = P(X = 2) = 0.4$.

Table 5-5 Numbers of divisors and their probabilities (for integer selected at random from 1 through 10)

Probability, $f(x)$	0.1	0.4	0.2	0.3
Number of divisors, x	1	2	3	4

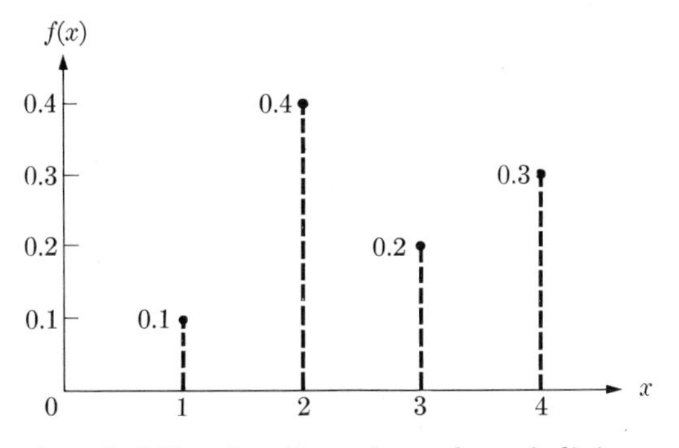

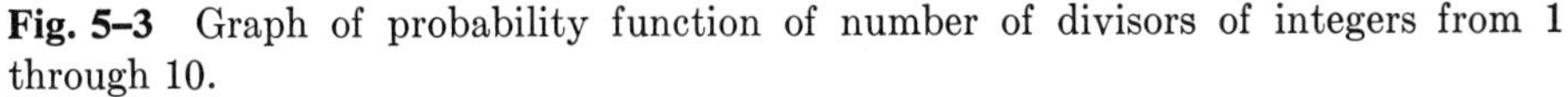
Fig. 5–3 Graph of probability function of number of divisors of integers from 1 through 10.

A graph of the points with coordinates x and $f(x)$ is shown in Fig. 5–3. This graph represents the *probability function* of the random variable X, where X is the "number of divisors of an integer from 1 through 10 chosen at random".

Using Table 5–5, we can easily answer the four original questions about the number of divisors:

$$\begin{aligned} P(X = 2) &= f(2) = 0.4, \\ P(X = 1) &= f(1) = 0.1, \\ P(X = 4) &= f(4) = 0.3, \end{aligned}$$

and, since no integer from 1 through 10 has more than 4 divisors,

$$P(X > 4) = 0.$$

With these examples to guide us, we now formulate the following general definitions.

5–1 Definitions

1) *Random variable.* A variable whose value is a number determined by the outcome of an experiment is called a *random variable.*

2) *Probability function.* Let X be a random variable with possible values $x_1, x_2, \ldots, x_t$ and associated probabilities $f(x_1), f(x_2), \ldots, f(x_t)$. Then the set f whose elements are the ordered pairs

$$(x_i, f(x_i)), \qquad i = 1, 2, \ldots, t,$$

is called the *probability function* of X.

Thus the number of divisors of an integer chosen at random from 1 through 10 is a random variable with $t = 4$ possible values:

$$x_1 = 1, \qquad x_2 = 2,$$
$$x_3 = 3, \qquad x_4 = 4.$$

The associated probabilities are those given in Table 5-5. The probability function for this example is the set of ordered pairs of numbers represented by dots in the graph of Fig. 5-3:

$$f = \{(1, 0.1),\ (2, 0.4),\ (3, 0.2),\ (4, 0.3)\}.$$

Note Although in this example the subscripts on x happen to be identical with the values (that is, $x_i = i$), this need not always be the case. For example, if our random variable had as its values the squares of the integers 0, 1, 2, 3, 4, we might define $x_1 = 0$, $x_2 = 1$, $x_3 = 4$, $x_4 = 9$, $x_5 = 16$. The subscript i and value of x_i would then be given by the formula $x_i = (i - 1)^2$ for $i = 1, 2, 3, 4, 5$. In general, no simple formula connects the value of x_i with the subscript i; but when available, a simple formula may well be helpful.

In this book, we rarely list probability functions as sets of ordered pairs. It is more convenient, and equally valid, to show the probability function by means of a formula for $f(x)$, or by means of a table like Table 5-5.

Probabilities are assigned to events (sets of sample points), and the assignment of a probability to each of the possible events is called the *probability distribution* over the sample space. In finite sample spaces with n elementary events, there are 2^n possible sets; hence if n is large, it is inconvenient to list the probabilities for 2^n sets. Instead, we usually give the *probability function*, which lists a probability for each of the n *elementary* events. Thus the probability function is one way of summarizing the probability distribution. In discussing probabilities generally, it is common to speak of the probability distribution interchangeably with the probability function.

Example 4 *Trials to first success.* Imagine that a sequence of binomial trials is continued until the first success occurs. If a trial results in a *success*, we record an S for that trial; if the result is a failure, we record an F. Thus a sample space for the experiment might well be the set

$$\{S, FS, FFS, FFFS, \ldots, \underbrace{F \ldots F}_{n-1\ F\text{'s}}S, \ldots\}.$$

If $P(\text{success}) = p$ and $P(\text{failure}) = q = 1 - p$, the probabilities associated with the sample points are

$$p,\ qp,\ q^2p,\ q^3p,\ \ldots,\ q^{n-1}p,\ \ldots,$$

because successive trials are independent and the probabilities are obtained by multiplication. Note that the *first* success occurs at trial n if and only if the first $n - 1$ trials are failures and the nth trial is a success. If we let X be the number of trials up to and including the first success, then the probability function for X is given by

$$P(X = x) = f(x) = q^{x-1}p, \qquad \text{for } x = 1, 2, 3, \ldots. \tag{1}$$

The *domain* of the probability function is the set of positive integers 1, 2, 3, ...

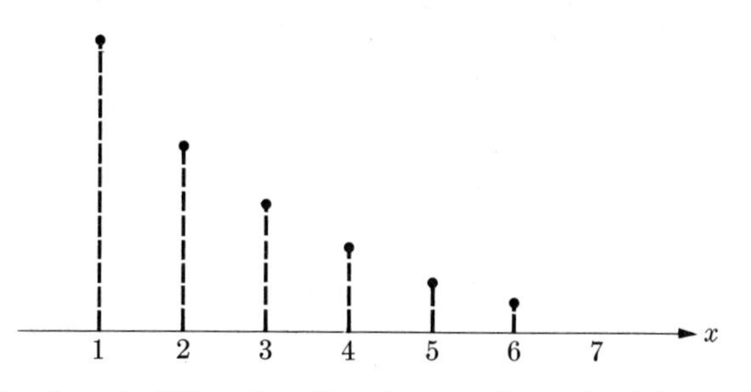

Fig. 5–4 Graph of probability function for number of trials to first success: the geometric distribution ($p = \frac{1}{3}$).

Because the probabilities form a geometric progression, this probability distribution is called the *geometric distribution*. Observe that the total probability is one, as it should be, since the sum is

$$\sum_{x=1}^{\infty} f(x) = p + qp + q^2p + q^3p + \cdots = \frac{p}{1-q} = \frac{p}{p} = 1,$$

provided $p = 1 - q \neq 0$. Hence, if p, the probability of success on each individual trial, is not zero, the probability that a success will occur *sometime* is one. Figure 5–4 shows part of a graph of a geometric distribution.

Comment A random variable is like any other variable except that we may know more about the random variable, namely the probability that it takes any one of its possible values.

Example 5 *Matching historical events and dates.* A student is to match three historical events (battle of Lexington and Concord, Columbus's discovery of America, battle of Hastings) with three dates (1775, 1492, 1066). If he guesses, with no knowledge of the correct answers, what is the probability function of the number of answers he gets right?

Solution A sample space S for the experiment of giving the student this test could be the following six permutations of the three dates:

e_1: 1066, 1492, 1775 e_4: 1492, 1775, 1066
e_2: 1066, 1775, 1492 e_5: 1775, 1066, 1492
e_3: 1492, 1066, 1775 e_6: 1775, 1492, 1066

If he answers strictly by guessing, then each permutation has probability $\frac{1}{6}$.

Next, we associate with each element of S the number of correct answers X that it provides. With no loss of generality, we may assume that the events are listed in the order (1) battle of Hastings, (2) Columbus's discovery of America, (3) battle of Lexington and Concord. If the student chooses e_1 as his answer, he gets all 3 right. If he chooses e_2, e_3, or e_6, he gets 1 right. If he chooses e_4 or e_5, he gets 0 right. Table 5–6 shows the sample space of permutations

Table 5–6 Matching dates with historical events

Permutation e_i	Probability of e_i	Number of correct answers, X
e_1	$\frac{1}{6}$	3
e_2	$\frac{1}{6}$	1
e_3	$\frac{1}{6}$	1
e_4	$\frac{1}{6}$	0
e_5	$\frac{1}{6}$	0
e_6	$\frac{1}{6}$	1

Table 5–7 Probability function

Probability, $f(x)$	$\frac{2}{6}$	$\frac{3}{6}$	$\frac{1}{6}$
Number of correct answers, x	0	1	3

$e_1, e_2, \ldots, e_6$ and the number of correct answers in each. Table 5–7 organizes the data in a form that shows the probability function of the random variable X (= "number of correct answers").

Remark A random variable may also be considered as a *function* that assigns a real number to each sample point. Table 5–6 illustrates this idea; the random variable X may be thought of as a function defined on the domain

$$\{e_1, e_2, e_3, e_4, e_5, e_6\},$$

with the values (see the third column of Table 5–6):

$$X(e_1) = 3, \qquad X(e_2) = 1, \qquad X(e_3) = 1,$$
$$X(e_4) = 0, \qquad X(e_5) = 0, \qquad X(e_6) = 1.$$

Considering X as a *function* from the sample space S to the real numbers, we note that this function maps the event $\{e_1\}$ onto the real number 3, the event $\{e_2, e_3, e_6\}$ onto the number 1, and the event $\{e_4, e_5\}$ onto the number 0. The first of these events has probability $\frac{1}{6}$, the second has probability $\frac{3}{6}$, and the last has probability $\frac{2}{6}$. As an alternative sample space, we could have listed the *possible* outcomes of the experiment in terms of the possible number of correct answers, thus: $\{0, 1, 2, 3\}$. Our analysis shows that the associated probabilities are $\frac{2}{6}, \frac{3}{6}, 0, \frac{1}{6}$, respectively. Because the probability is zero for just 2 correct answers, we delete that possibility and get the probability function given in Table 5–7.

For our present purposes, it is sufficient to consider a random variable as a variable whose value is a number determined by the outcome of an experiment.

Idea of a run If you toss a coin 9 times and it comes up

$$T\ H\ H\ H\ H\ H\ H\ H\ T,$$

in that order, you may wonder if something in the construction of the coin, or in the way it was tossed, caused so few *runs* (only 3 in this example). Any unbroken sequence of like letters is called a *run*, even though the sequence has only 1 letter (length 1), as at the beginning and end of the foregoing example. The middle run of H's has length 7. Some statistical tests for randomness are based on runs. For instance, in the next example a large number of runs might suggest that people visiting the soda fountain prefer not to sit side by side.

Example 6 *Runs of two kinds of elements.* A small soda fountain has 5 seats in a row, 3 of which are occupied. Assuming that all seating arrangements of 3 persons are equally likely, find the probability function of the number of runs of occupied seats (O) and empty seats (E). (Example: $\underline{O}\ \underline{E\ E}\ \underline{O\ O}$ has three runs, as indicated by the underlining.)

Solution The experiment is to have 3 people come into the soda fountain when all 5 seats are empty, and sit down. A sample space for this experiment is a list of all possible arrangements of three O's and two E's, corresponding to the three occupied seats and two empty seats. The number of sample points is

$$\binom{5}{3} = \frac{5!}{3!2!} = 10,$$

since that is the number of permutations of 5 things, of which three are O's and two are E's. If people choose seats at random, each sample point has probability $\frac{1}{10}$.

The *random variable* of interest to us in this experiment is the number of runs of O's and E's in the sample point that represents the seating arrangement. A list of sample points together with the number of runs in each is given below. We are concerned only with "occupied" or "empty", not with the different persons seated. Thus the seating arrangement designated $O\ O\ O\ E\ E$ means that seats numbered 1 through 3 are occupied, seats 4 and 5 are empty.

Seating	Runs	Seating	Runs
$O\,O\,O\,E\,E$	2	$O\,E\,E\,O\,O$	3
$O\,O\,E\,O\,E$	4	$E\,O\,O\,O\,E$	3
$O\,O\,E\,E\,O$	3	$E\,O\,O\,E\,O$	4
$O\,E\,O\,O\,E$	4	$E\,O\,E\,O\,O$	4
$O\,E\,O\,E\,O$	5	$E\,E\,O\,O\,O$	2

Counting in this list the number of ways to get each possible number of runs, we obtain the probability function of the random variable X, which denotes the number of runs (see Table 5–8).

Table 5–8 The probability function of the number of runs of five elements, three of one kind and two of another

Probability, $f(x)$	0.2	0.3	0.4	0.1
Number of runs, x	2	3	4	5

Idea of turning points The next example deals with a topic used in studying economic time series, such as daily stock market averages or weekly production of automobiles. A time series is a set of observations or measurements arranged in the order in which they were made. If there is no trend, these measurements should fluctuate about a mean value, some above and some below. If they continually increase or decrease or follow some cyclical pattern, it may be possible to predict the future behavior of the series.

If among three successive numerical measurements the middle one is the least or the greatest of the three, it is called a *turning point* of the sequence. Thus in the sequence

$$3,\ 5,\ 4,\ 7$$

the numbers 5 and 4 are turning points because 5 is the greatest of 3, 5, 4 and 4 is the least of 5, 4, 7. In random fluctuations there are more likely to be many turning points in the successive measurements than there would be if the measurements were in general increasing or decreasing.

Example 7 *Turning points.* If all permutations of four different measurements are equally likely, what is the probability function of the random variable X, where X is the number of turning points?

Solution For the purpose of counting the number of turning points, there is no loss of generality if we replace the measurements, in order of increasing magnitude, by the numbers 1, 2, 3, and 4. Then any sequence of four different measurements provides some permutation of the four numbers 1, 2, 3, 4. Thus 1 4 2 3 indicates that the smallest measurement is first, the largest second, and so on. And this permutation has two turning points: 4 and 2. We list the 4! permutations, together with the numbers of turning points, in Table 5–9.

Table 5–9 Permutations of 1, 2, 3, 4 and numbers of turning points

Permutation	Turning points	Permutation	Turning points	Permutation	Turning points	Permutation	Turning points
1 2 3 4	0	2 1 3 4	1	3 1 2 4	1	4 1 2 3	1
1 2 4 3	1	2 1 4 3	2	3 1 4 2	2	4 1 3 2	2
1 3 2 4	2	2 3 1 4	2	3 2 1 4	1	4 2 1 3	1
1 3 4 2	1	2 3 4 1	1	3 2 4 1	2	4 2 3 1	2
1 4 2 3	2	2 4 1 3	2	3 4 1 2	2	4 3 1 2	1
1 4 3 2	1	2 4 3 1	1	3 4 2 1	1	4 3 2 1	0

Table 5-10

Probability, $f(x)$	$\frac{2}{24}$	$\frac{12}{24}$	$\frac{10}{24}$
Number of turning points, x	0	1	2

Counting up the frequency of each number of turning points, we get the probability function shown in Table 5-10.

Thus we get no turning points only $\frac{1}{12}$ of the time, one turning point half the time, and two turning points nearly half the time.

Example 8 *Binomial probability function.* In Sections 4-4 and 4-5 we studied the binomial distribution. Here we recall the formula

$$b(x; n, p) = \binom{n}{x} p^x q^{n-x}, \qquad x = 0, 1, 2, \ldots, n, \tag{2}$$

which expresses the probability that the number of successes obtained in an experiment of n binomial trials will be x. The number of trials is fixed, n. The random variable X that we study is the number of successes. When specific values are assigned to the parameters n and p, the probability for x is

$$f(x) = P(X = x) = b(x; n, p).$$

★Example 9 *The negative binomial distribution.* (Number of trials until rth success.) Suppose that a sequence of Bernoulli trials with probability p of success on each trial is continued until exactly the rth success occurs. Let X be the number of trials required. What is the probability function for X?

Solution The rth success can occur at any trial from the rth on. Suppose that x is an integer greater than or equal to r, and suppose that the rth success occurs precisely at trial x. Then there must be $r - 1$ successes and $x - r$ failures in the first $x - 1$ trials. The total number of ways of permuting $r - 1$ S's and $x - r$ F's in a row of $x - 1$ symbols is $(x - 1)!/(r - 1)!(x - r)!$, or, alternatively, $\binom{x-1}{r-1}$. We don't need to write out the sample space in detail: we just need to visualize it as made up of sequences of S's and F's, with exactly r S's, one of which occurs last. Any sample point that has a total of x symbols in it has probability $q^{x-r}p^r$, because the trials are independent and there are $x - r$ failures and r successes. Multiplying the number of sample points for the event $\{X = x\}$ by the probability of each of its points, we get the probability

$$f(x) = P(X = x) = \binom{x-1}{r-1} q^{x-r}p^r, \qquad x = r, r + 1, r + 2, \ldots. \tag{3}$$

Remark Since $|q| < 1$, if $p \neq 0$, the negative binomial formula, Eq. (5) of Section 2-5, gives

$$p^r(1-q)^{-r}$$

$$= p^r\left[1 + rq + \frac{(r+1)r}{2!}q^2 + \frac{(r+2)(r+1)r}{3!}q^3 + \cdots\right]$$

$$= p^r\left[\binom{r-1}{r-1}q^{r-r} + \binom{\overline{r+1}-1}{r-1}q^{\overline{r+1}-r} + \binom{\overline{r+2}-1}{r-1}q^{\overline{r+2}-r} + \cdots\right]$$

$$= p^r\sum_{x=r}^{\infty}\binom{x-1}{r-1}q^{x-r}.$$

This result enables us to show that the sum of all the probabilities for the negative binomial distribution of Example 9 is one; for if $p \neq 0$, then $|q| < 1$ and the sum of all the probabilities for the negative binomial is

$$\sum_{x=r}^{\infty} P(X = x) = \sum_{x=r}^{\infty} p^r\binom{x-1}{r-1}q^{x-r} = p^r(1-q)^{-r} = p^r(p^{-r}) = 1.$$

In words: for any assigned positive integer r, the probability is one that a sequence of binomial trials will eventually produce r successes, provided the probability of success on each trial is not zero. Because the probability function for the number of trials to the rth success is related in the foregoing way to the expansion of $(1-q)^{-r}p^r$, the family of distributions corresponding to the parameters r and p is called the *negative binomial distribution.*

Summary The foregoing examples illustrate the steps in constructing the probability function of a random variable.

1. Construct for the given experiment the sample space of possible outcomes, along with their associated probabilities.
2. List the value of the random variable that corresponds to each sample point.
3. List the possible values $x_1, x_2, \ldots, x_t$ of the random variable, and list the associated probabilities $f(x_1), f(x_2), \ldots, f(x_t)$. (Compute the probability of x_i by adding together the probabilities of all sample points that correspond to x_i.)

   Then the set of ordered pairs

$$(x_i, f(x_i)), \qquad i = 1, 2, \ldots, t,$$

is the probability function of the random variable. The probability function is usually displayed either as a table, like Tables 5-2, 5-5, 5-7, 5-8, and 5-10, or as a formula for $f(x)$, like Eqs. (2) and (3).

Further comment on notation The expression $P(X = x_i)$ denotes the probability that the random variable X takes the value x_i. Usually we shall introduce the probability function and write $f(x_i)$ for the probability $P(X = x_i)$. Sometimes it is convenient to abbreviate $P(X = x_i)$ to $P(x_i)$ when no confusion would develop. If we need to talk about more than one random variable, we may introduce such other letters as Y or Z, having respective values y or z and probability functions g or h. Thus we might have $f(x_i) = P(X = x_i)$, $g(y_j) = P(Y = y_j)$, and $h(z_k) = P(Z = z_k)$.

EXERCISES FOR SECTION 5-1

1. An ordinary six-sided die is thrown once. Find the probability function of the number of dots appearing on the top face. Graph the probability function.
2. Throw an ordinary die 50 times and record the numbers of times it falls with 1, 2, . . . , 6 dots up. (Or use your random numbers, Table I-A, to simulate this experiment.) Divide these numbers by 50 to convert them to relative frequencies and plot the graph of $(x, r(x))$, where $x = 1, 2, \ldots, 6$ and $r(x)$ equals the observed relative frequency. Compare with the graph of the probability function in Exercise 1 above.
3. Suppose a number is selected at random from the integers 1 through 20. Let x be the number of its divisors. Construct the probability function of X and graph it. What is the probability that there will be 4 or more divisors?
4. A coin is tossed 3 times. Let X be the number of runs in the sequence of outcomes: first toss, second toss, third toss. Find the probability function of X and construct its graph. What values of X are most probable?
5. Do a "turning points" example for the case of 3 measurements. What number of turning points has the greatest probability? Least?
6. Two ordinary six-sided dice are thrown (see Table 1-11). Find the probability function for the total score on their top faces and graph it. Do you think the points on the graph lie on any simple curve, or curves? Discuss.
7. (Continuation.) A white die and a red die are thrown at the same time and the difference $R - W$ is observed, where R is the number on top of the red die and W is that on top of the white. Find the probability function of this difference and sketch its graph. What values of $R - W$ are most probable? Least probable? Compare this probability function and its graph with those obtained in Exercise 6 above. Comment.
8. The three-sided engineer's ruler of Example 2 is thrown 3 times. Find the probability function of the sum of the 3 face-down digits. Sketch and discuss its graph. [*Hint:* Let X be the sum on the first two throws, Y the score on the third throw, and consider pairs of values of X and Y as constituting the sample space; then form the sum $X + Y$. Use the probability function of X given in Example 2.]
9. Using the table of random numbers at the back of the book, record a sequence of 25 numbers. The results correspond to random sampling, with replacement, from the digits 0, 1, . . . , 9. However, interpret "0" as "10", and beside each number record the number of its divisors. Compute the relative frequencies of

the various numbers of divisors observed, and plot the graph of relative frequencies against number of divisors. Compare with the probability function of Example 3 and its graph.

10. Throw an ordinary die. Record, in order, the number on top, then the number on the side nearest you (the "front"), next the one on the bottom, and finally the one on the side farthest from you. Repeat this operation 10 times, each time recording a sequence of 4 numbers according to the method described. Next, compute the number of turning points in each sequence and plot their relative frequencies on a graph. Compare with the graph of the probability function of turning points in Example 7 and comment.

11. Graph the probability function of the number of runs in Example 6.

12. Simulate the seating arrangement experiment of Example 6 as follows. Shuffle 3 red cards and 2 black cards from a bridge deck, then deal them out one at a time. Record R for red and B for black. Regard reds as "occupied" and black as "empty". Repeat the operation 25 times, shuffling the cards well before each deal. Compute the number of runs obtained in each sequence of 5, and the relative frequencies of various numbers of runs obtained in the 25 sequences. Plot, and compare with the graph of the probability function of Example 6.

13. Suppose 4 coins are tossed. If X is the number of tails, find the probability function of X and graph it.

14. Repeat Exercise 13 letting X be the number of heads minus the number of tails.

15. From a lot of 10 television sets containing 4 defectives, a sample of 3 sets is drawn at random without replacement. Let X be the number of defectives in the sample. (a) Describe a sample space for this experiment. (b) How many points are there in your sample space? (c) Tabulate the probability function of X. (d) Graph this probability function.

16. In the example dealing with historical events and dates, suppose that the number of events is an arbitrary positive integer n $(n > 2)$. If a student tries to match the n events and n dates by guessing, what is the probability that he gets (a) all n answers correct? (b) exactly $n - 1$ answers correct? (c) exactly $n - 2$ answers correct?

17. Given the following probability function:

x	0	1	2	3	4	5	6	7
$f(x)$	0	c	$2c$	$2c$	$3c$	c^2	$2c^2$	$7c^2 + c$

(a) Find c. (b) Evaluate $P(X \geq 5)$ and $P(X < 3)$. (c) If $P(X \leq k) > \frac{1}{2}$, what is the minimum value of k?

18. An ordinary die is tossed until a 3 appears for the first time. (a) What is the probability function for the random variable X = number of tosses? (b) For what positive integer m is $P(X \leq m) \approx P(X > m) \approx \frac{1}{2}$?

19. For the geometric distribution of Example 4, show that $P(X > x) = q^x$. [*Suggestion:* recall that the sum of a geometric series $a + ar + ar^2 + \cdots$ is $a/(1 - r)$ if $|r| < 1$.]

20. Show that the individual terms of the geometric probability distribution $f(x) = q^{x-1}p$ are the same as the terms you get by dividing p by $1 - q$ using ordinary algebraic long division.

21. Multiply the series $1 + q + q^2 + q^3 + \cdots$ by itself and arrange the terms in increasing powers of q. Next, multiply each term by p^2 and compare your answer with the probability in Example 9 for the number of trials to the second success with $r = 2$. [This relates to the Remark following Example 9 for the special case $r = 2$.]

5–2 MATHEMATICAL EXPECTATION OF A RANDOM VARIABLE: POPULATION MEAN

In this section, we introduce the concept of the *mean value* of a random variable. It is closely related to the notion of the arithmetic mean, or average.

Example 1 A bowl contains 300 tags; 150 are numbered 1, 100 are numbered 2, and 50 are numbered 3. A tag is drawn at random from the bowl, its number X is recorded, the tag is returned to the bowl, and the tags are thoroughly mixed. This process is repeated 500 times. What is the arithmetic average of the values of the random variable X that are thus recorded?

Solution Let n_1 of the tags that were drawn have the number 1, n_2 have 2, and n_3 have 3. Then the arithmetic average $\bar{x}$ (read "x bar") is

$$\bar{x} = \frac{1 \times n_1 + 2 \times n_2 + 3 \times n_3}{n_1 + n_2 + n_3}. \tag{1}$$

The numerator of this expression is a "weighted sum" of 1's, 2's, and 3's, each "weighted" by a factor n_1, n_2, or n_3 that is equal to the number of times the given number is drawn. This average (1) can also be expressed in terms of the *proportions* n_1/n, n_2/n, n_3/n, with

$$n = 500 = n_1 + n_2 + n_3,$$

to yield

$$\bar{x} = 1 \times \frac{n_1}{n} + 2 \times \frac{n_2}{n} + 3 \times \frac{n_3}{n}. \tag{2}$$

The expression (2) exhibits the average as another weighted sum of 1's, 2's, and 3's; here the three numbers are weighted by their relative proportions. Of course, we don't know the exact values of these proportions (unless we actually perform the experiment), so we can't say in advance just what the average produced by this particular experiment is. But since the probability of drawing a 1 is $f(1) = \frac{1}{2}$, of drawing a 2 is $f(2) = \frac{1}{3}$, and of drawing a 3 is $f(3) = \frac{1}{6}$, we might suppose that the proportions n_1/n, n_2/n, and n_3/n are approximately equal to $f(1)$, $f(2)$, and $f(3)$, respectively. Thus, for a value of

n as large as 500, we might expect an average near

$$1 \times f(1) + 2 \times f(2) + 3 \times f(3) = 1 \times \tfrac{1}{2} + 2 \times \tfrac{1}{3} + 3 \times \tfrac{1}{6} = \tfrac{5}{3}.$$

The arithmetic mean of *all* the tags in the bowl is also $\frac{5}{3}$:

$$\frac{1 \times 150 + 2 \times 100 + 3 \times 50}{150 + 100 + 50} = \frac{500}{300} = \frac{5}{3}.$$

For a *small* sample of tags, we would not expect the sample average necessarily to be near this population mean, but for a large sample most people do expect it, and it usually is near.

Example 2 The number of divisors of an integer from 1 through 10, chosen at random, is a random variable X. What is its expected value?

Solution Table 5-5 provides the probability function for the number of divisors X. We use this probability function to compute the "population mean" in the way indicated in Example 1 above. The result is

$$1 \times 0.1 + 2 \times 0.4 + 3 \times 0.2 + 4 \times 0.3 = 2.7.$$

This is the *average* result that we might expect from a *large number* of performances of the experiment. Of course, no number has 2.7 divisors; moreover, on just *one* performance of the experiment, the most likely number of divisors is 2, since that has the highest probability.

The foregoing examples lead us to the following definitions.

5-2 Definition: Sample average

Let X be a random variable whose possible values are $x_1, x_2, \ldots, x_t$. Suppose that a sample of n observations produces n_1 values of X that are equal to x_1, n_2 that are equal to x_2, ..., n_t that are equal to x_t:

Frequency	n_1	n_2	$\cdots$	n_t	Total: n
Value of X	x_1	x_2	$\cdots$	x_t	

Then the *average* value of X for this *sample* is

$$\bar{x} = \frac{x_1 n_1 + x_2 n_2 + \cdots + x_t n_t}{n_1 + n_2 + \cdots + n_t} = \frac{\sum x_i n_i}{\sum n_i},$$

or

$$\boxed{\bar{x} = \frac{1}{n} \sum_{i=1}^{t} x_i n_i.} \tag{3}$$

(See Appendix 1 for a discussion of the summation symbol, $\sum$.)

The set of ordered pairs (x_i, n_i), $i = 1, 2, \ldots, t$, displayed as a table in Definition 5–2, is called the *frequency distribution* of the sample values, or the *observed frequency distribution.* In dealing with samples, the frequency distribution plays an important role, just as the probability function does in dealing with populations. The frequency distribution is one way to report the sample.

5–3 Definition: Mathematical expectation — population mean

Let X be a random variable with probability function as follows:

Probability, $f(x)$	$f(x_1)$	$f(x_2)$	$\cdots$	$f(x_t)$
Value of X, x	x_1	x_2	$\cdots$	x_t

The *mathematical expectation* of X, denoted by $E(X)$, is defined to be

$$E(X) = x_1 f(x_1) + x_2 f(x_2) + \cdots + x_t f(x_t),$$

or

$$\boxed{E(X) = \sum_{i=1}^{t} x_i f(x_i).} \tag{4}$$

$E(X)$ is also called the *mean* of X, or the *population mean*, or the mean of the distribution f.

Remark The mean is also abbreviated μ (read "mew" and spelled "mu"), the Greek letter for "m", the first letter of the word "mean". Sometimes several random variables $X, Y, \ldots$ are being studied together. We may use these letters as subscripts on μ to indicate the means. Thus we would write

$$\mu_X = E(X) \qquad \text{and} \qquad \mu_Y = E(Y).$$

When only one random variable is being considered, the subscript is usually omitted.

We may express the result of Eq. (4) in words:

To compute the mean of a random variable, multiply each possible value of the variable by its probability and add these products.

Equations (3) and (4) are not identical, but they are similar. In particular, the proportions n_i/n in Eq. (3) vary from one sample to another, and it is only a coincidence if n_i/n is equal to the probability $f(x_i)$. However, it is true that

$$\frac{n_i}{n} \approx f(x_i),$$

and therefore

$$\bar{x} \approx \mu.$$

These approximations are usually better when n is large, and, of course, become equalities if the sample coincides with the entire population.

Example 3 One die is thrown. What is the mathematical expectation of the number of dots on the top face?

Solution Let the random variable X denote the number of dots on the top face of the die. The possible values are 1, 2, . . . , 6, each with probability $\frac{1}{6}$. Hence, by Eq. (4),

$$\begin{aligned}\mu &= E(X)\\ &= 1 \times \tfrac{1}{6} + 2 \times \tfrac{1}{6} + 3 \times \tfrac{1}{6} + 4 \times \tfrac{1}{6} + 5 \times \tfrac{1}{6} + 6 \times \tfrac{1}{6}\\ &= \tfrac{21}{6} = 3.5.\end{aligned}$$

5–4 Theorem

Let X be the number of successes in n trials of a binomial experiment where the probability of success on each trial is p. Then the mean number of successes is

$$\mu_X = E(X) = np. \tag{5}$$

Proof If x is any of the integers 0, 1, . . . , n the probability that X takes the value x is

$$f(x) = b(x; n, p) = \binom{n}{x} p^x q^{n-x}, \qquad [q = 1 - p].$$

Therefore, by Definition 5–3,

$$E(X) = \sum_{x=0}^{n} xf(x) = 0f(0) + 1f(1) + 2f(2) + \cdots + nf(n). \tag{6}$$

What do we know that could help us? About all that we know is that $\sum_{x=0}^{n} b(x; n, p) = 1$, *for every* n and p. That is, we know how to sum binomials. In discrete problems it frequently happens that summations can be manipulated so as to change them to a related form that can be summed by reducing or increasing the value of one or more of the parameters. Let us try this trick on the terms of Eq. (6).

The zero term contributes nothing to the sum in Eq. (6), and for $x \geq 1$, the term $xf(x)$ can be expressed as:

$$xf(x) = x\,\frac{n!}{x!(n-x)!}\,p^x q^{n-x} = (np)\,\frac{(n-1)!}{(x-1)!(n-x)!}\,p^{x-1} q^{n-x}.$$

We can see a multiple of $b(x-1; n-1, p)$ shaping up. The binomial coefficient is already in hand, the exponent of p matches one of the denominators and that of q the other, but q's exponent does not emphasize that the size of the new experiment is $n-1$. We rewrite the equation as

$$xf(x) = (np)\binom{n-1}{x-1} p^{x-1} q^{(n-1)-(x-1)}, \qquad \text{for } 0 \leq x-1 \leq n-1. \tag{7}$$

If we sum from $x = 1$ to $x = n$, we note that $x - 1$ goes from 0 to $n - 1$ and the factor np is common to all the summands. Therefore, we get

$$\begin{aligned} E(X) = \sum_{x=0}^{n} xf(x) &= \sum_{x=1}^{n} (np) \binom{n-1}{x-1} p^{x-1} q^{(n-1)-(x-1)} \\ &= np \sum_{x-1=0}^{n-1} b(x-1; n-1, p) \\ &= np(q+p)^{n-1} = np. \ \square \end{aligned}$$

★5–5 Theorem: Mean of geometric distribution

In a sequence of Bernoulli trials, let p be the probability of success on each trial. Let X be the number of trials to and including the first success. Then, if $p \neq 0$, the mean or expected number of trials is

$$\mu_X = E(X) = \frac{1}{p}. \tag{8}$$

Proof For any positive integer x, the probability that the first success occurs on the xth trial is

$$P(X = x) = f(x) = q^{x-1}p,$$

where $q = 1 - p$ and $x = 1, 2, \ldots$. Therefore, if we assume the extension of Definition 5–3, the expected number of trials is

$$E(X) = \sum_{x=1}^{\infty} xf(x) = 1f(1) + 2f(2) + 3f(3) + \cdots.$$

By substitution, we get

$$E(X) = p + 2qp + 3q^2p + 4q^3p + \cdots. \tag{9a}$$

To sum the series on the right side of Eq. (9a), we use the same trick that works for geometric series; we multiply both sides by q and then subtract the new series from the old:

$$qE(X) = \qquad qp + 2q^2p + 3q^3p + \cdots, \tag{9b}$$

$$\begin{aligned} (1-q)E(X) &= p + qp + q^2p + q^3p + \cdots \\ &= \frac{p}{1-q} = \frac{p}{p} = 1, \quad \text{if } p \neq 0. \end{aligned}$$

Therefore, if $p \neq 0$, then $q \neq 1$, and

$$E(X) = \frac{1}{1-q} = \frac{1}{p}. \ \square \tag{10}$$

Remark In words, Eq. (10) says that the expected waiting time, that is, the mean number of trials, to the first success is $1/p$. If $p = 1$, the first success occurs at the first trial (with certainty). For a coin, the probability of heads on any toss is $\frac{1}{2}$, so the mean number of trials until the coin falls heads for the first time is $1/p = 2$. For a regular die, with $p = \frac{1}{6}$, the mean or expected number of tosses until any particular score first appears is $1/p = 6$. However, in any particular experiment, the actual number of trials until first success occurs can be very large. In Exercise 19 at the end of Section 5–1, you were asked to show that $P(X > x) = q^x$ for the geometric distribution. One proof is as follows:

$$\begin{aligned} P(X > x) &= f(x+1) + f(x+2) + f(x+3) + \cdots \\ &= q^x p + q^{x+1} p + q^{x+2} p + \cdots \\ &= q^x p(1 + q + q^2 + \cdots) = q^x p/(1-q) = q^x p/p = q^x. \end{aligned} \quad (11)$$

For the die with $p = \frac{1}{6}$ and $q = \frac{5}{6}$, the probability that *more* than six trials will be required to produce the first success is $(\frac{5}{6})^6 = (0.833\ldots)^6 \approx 0.334$. In other words, about $\frac{1}{3}$ of the experiments will require more than the mean number of trials.

Example 4 What is the mathematical expectation of the number of runs when 3 things of one kind and 2 things of another kind are arranged at random in a row?

Solution Here the random variable X is the number of runs whose probability function is given in Table 5–8. Using it, we compute

$$\mu = E(X) = 2 \times 0.2 + 3 \times 0.3 + 4 \times 0.4 + 5 \times 0.1 = 3.4.$$

Remark Note that in this example, again, the mathematical expectation of the number of runs is 3.4, not an integer, and not any value that the random variable could actually have. The same is true of the mathematical expectation of the number of dots on the top face of the die in Example 3 and of the number of divisors in Example 2. We mention this because the term "mathematical expectation" is often abbreviated "expectation". The examples show that this *"expectation" is not something we "expect" in the ordinary sense of the word, except that the long-run average over repeated experiments is likely to be close to it.* Again, the term "expected value" is sometimes used as a synonym for "mathematical expectation", but there should be no implication that this value is frequent, highly probable, or even possible. It is merely the weighted mean of the possible values, each weighted by its probability.

Example 5 According to an American experience mortality table, the probability that a 25-year-old man will survive one year is 0.992, and that he will die within a year is 0.008. An insurance company offers to sell such a man a \$1000 one-year term life insurance policy for a premium of \$10. What is the company's expected gain?

Solution The "gain", X, is a random variable that may take the value +\$10 (if the man lives) or −\$990 (if he dies). The probability function is as follows:

$f(x)$	0.992	0.008
x	+10	−990

and

$$\mu = E(X) = 10 \times 0.992 - 990 \times 0.008 = 2.$$

It is important that the expected gain (before administrative expenses and taxes) be positive in order to enable the insurance company to stay in business and to build up reserves to pay its beneficiaries and policyholders.

Example 6 *One-armed bandit.* A simplified slot machine has 2 dials. Each dial has 3 kinds of pictures on it, identified as "apples", "bells", and "cherries". The machine is rigged so that the 2 dials operate independently, and after they are spun, each comes to rest with 1 of the 3 pictures showing in a window on the front of the machine. The probabilities of the possible outcomes, for each dial, are

Outcome	Apples	Bells	Cherries
Probability	0.1	0.4	0.5

Each play costs five cents. A play consists of pulling a lever that spins the dials, resulting in one of the 9 possible combinations of 2 pictures, 1 on each dial. The machine pays off as follows:

for 2 apples, 50¢;
for 2 bells, 10¢;
for 2 cherries, 5¢;
for anything else it pays nothing.

Find the mathematical expectation of net profit (in money) to a person who plays once.

Solution The random variable X here equals the number of cents won. Table 5–11 shows a sample space of possible outcomes, their probabilities, and the corresponding profit in cents. The three entries in the upper left corner,

(a, a): 45
0.01

mean that the outcome "two apples" has an associated profit of 45 cents, and occurs with probability 0.01.

Table 5–11 One-armed bandit

		Second dial		
		Apples 0.1	Bells 0.4	Cherries 0.5
	Apples 0.1	(a, a): 45 0.01	(a, b): −5 0.04	(a, c): −5 0.05
First dial	Bells 0.4	(b, a): −5 0.04	(b, b): 5 0.16	(b, c): −5 0.20
	Cherries 0.5	(c, a): −5 0.05	(c, b): −5 0.20	(c, c): 0 0.25

The expected value of the random variable X, the profit on one play, is

$$\begin{aligned}\mu_X = E(X) &= 45 \times 0.01 + 5 \times 0.16 + 0 \times 0.25 - 5 \times 0.58 \\ &= .45 + .80 - 2.90 = -1.65 \text{ (cents).}\end{aligned}$$

In ten plays, the expected loss is 16.5 cents; in 100 plays, \$1.65.

EXERCISES FOR SECTION 5–2

1. a) Given the table:

Frequency	1	2	3
Value of x	3	4	5

Use formula (3) of Definition 5–2 to compute $\bar{x}$.

b) The ungrouped sample values in the foregoing table are

$$3,\ 4,\ 5,\ 5,\ 4,\ 5.$$

Find the arithmetic average of these six numbers.

c) In general, the ungrouped sample values are

$$z_1, z_2, z_3, \ldots, z_n.$$

Use the summation symbol $\sum$ to express the sample average $\bar{z}$.

2. In Example 2, Section 5–1, what is the expected sum of the 2 face-down digits in the experiment with the engineer's ruler?
3. In Example 5, Section 5–1, what is the expected number of correct answers?
4. In Example 6, Section 5–1, what is the expected number of runs?
5. From a bag of 7 marbles, 5 red and 2 blue, 3 marbles are drawn at random without replacement. Check that the expected number of blue marbles is $3 \times \frac{2}{7}$.
6. For Exercise 17, Section 5–1, compute the expected value of X.
7. For Exercise 15, Section 5–1, find the expected number of defective television sets.

8. Fifteen dice are thrown. What is the expected number of aces?

9. The probability of a thumbtack landing point up is 0.3. If 15 thumbtacks are tossed, what is the expected number that land with points up?

10. How many dice must be tossed if the expected number of aces is to be 5?

11. Two binomial experiments are performed: 13 cards are randomly drawn from a bridge deck, with replacement after each draw, and 12 dice are rolled. Find the expected total number of aces (ones) and deuces (twos) in the two experiments combined.

12. In a binomial experiment, if μ must be at least a distance $3\sqrt{npq}$ from both 0 and n, show that $n \geq 9$ times the larger of p/q and q/p.

13. In a binomial experiment with $n = 2$, $p = \frac{1}{200}$, find the probability of 1 or more successes, and compare the result with $\mu = np$. Make a similar comparison for $n = 3$, $p = \frac{1}{300}$; for $n = 3$, $p = 0.1$. Comment on these results.

14. (Continuation.) For a binomial experiment, show that if μ is near zero,

$$P(X \geq 1) \approx \mu.$$

[*Hint:* Recall the approximation for $(1 + x)^n$, Section 2–5, Example 4.]

15. (Continuation.) Use the results of Exercise 14 to find, approximately, $P(X \geq 1)$, in a binomial experiment with $n = 50$, $p = \frac{1}{2000}$.

16. Refer to Example 6, Section 5–2: one-armed bandit. Find the expected number of cents profit for one play on a two-independent-dial slot machine, given the following data for one dial:

Outcome	Apples	Bells	Cherries
Probability	0.1	0.3	0.6

Payoffs: for 2 apples, 25¢   for 2 cherries, 5¢
for 2 bells, 10¢   for anything else, zero

17. *Problem of points.* To decide who wins a \$4 prize, A and B play the following game. A coin is tossed. If the coin falls heads, A gets a point; if it falls tails, B gets a point. The first person to get 3 points wins. After 3 tosses, A has 2 points and B has 1. Make a sample space for the rest of the game. Let X be A's winnings. What is the expected value of A's winnings when he has 2 points and B has 1?

18. In a lottery, 100 tickets are sold at 25 cents each. There are 4 cash prizes, worth \$10, \$3, \$2, and \$1, respectively. What is the expected net gain for a purchaser of two tickets?

19. Four identical light bulbs are temporarily removed from their sockets and placed in a box. The bulbs are then taken at random from the box and put back in the sockets. What is the expected number of bulbs that will be replaced in their original sockets?

20. A baseball player has probability $p = \frac{1}{4}$ of getting a hit whenever he bats. (a) What is the expected number of hits for him in 4 times at bat? (b) What is the probability that he gets no hit in 4 times at bat? (c) Does your answer to part (b) contradict your answer to part (a)? Explain.

21. Find the expected value of the sum of the numbers of dots on the top faces of two ordinary cubical dice, on one throw.

22. *Roulette.* A roulette wheel has 38 equally spaced openings numbered 00, 0, 1, 2, 3, . . . , 35, 36. A gambler may bet \$1 on any number. The croupier spins the roulette wheel and drops a small ball onto it while it is spinning. If the ball comes to rest on the number the gambler has bet on, he receives \$35 in addition to his bet of \$1, but otherwise he loses his \$1. Find the mathematical expectation of his gain.

23. Find the expected number of turning points in a series of 4 different measurements. (See Table 5–9 or 5–10.)

24. The number of accidents that occur at a particular intersection between 4:30 and 6:30 p.m. on Fridays is 0, 1, 2, or 3, with corresponding probabilities 0.94, 0.03, 0.02, 0.01. Find the expected number of accidents (a) during one such Friday period, (b) during 100 such periods.

25. Player A pays B \$1, and 3 unbiased dice are rolled fairly. A receives \$2 from B if 1 ace appears, \$4 if 2 aces appear, and \$8 if 3 aces appear; otherwise he gets nothing. Is this a fair game? (That is, do A and B have the same expectation of gain?) If not, how much should A receive from B when 3 aces appear, to make the game fair?

26. In the World Series, suppose one team is stronger than the other and has probability $\frac{2}{3}$ of winning each game, independent of the outcomes of any other games. Under these assumptions, it is possible to show that the probabilities that the series ends in 4, 5, 6, or 7 games are about .21, .30, .27, or .22, respectively. Find the expected number of games in the series, under these assumptions.

27. The probability that a man aged 50 will live another year is 0.988. How large a premium should the insurance company charge him for a \$1000 term life insurance policy for one year (not including insurance company charges for administration, profit, etc.)?

28. In one play of the game called "chuck-a-luck", the player wins 15, 10, 5, or -5 cents (-5 means he loses 5 cents), with probabilities $\frac{1}{216}$, $\frac{15}{216}$, $\frac{75}{216}$, and $\frac{125}{216}$, respectively. (Cf. Wallis and Roberts, *Statistics, a New Approach*, The Free Press, 1956, p. 332.) Find the mathematical expectation of the player's gain (a) on one play of the game, (b) on 100 plays.

29. A sample of 4 balls is drawn without replacement from an urn containing 3 red and 5 white balls. If the sample contains 2 or more red balls, the player receives one dollar; otherwise, he loses fifty cents. What is the mathematical expectation of his gain? (First set up an appropriate sample space for the experiment.)

30. A fair coin is tossed until the first time a tail comes up or until three heads occur. Write out a sample space for this experiment and assign probabilities to its elements. Find the expected number of tosses in one performance of the experiment.

31. A farmer estimates that during the coming year his hens will produce 10,000 dozen eggs. He further estimates that, after taking into account his various costs and the seasonal price fluctuations, he may gain as much as 6 cents per dozen, or lose as much as 2 cents per dozen, and that the probabilities associated with these

possibilities are as follows:

Gain (in cents per dozen)	6	4	2	0	−2
Probability	0.20	0.50	0.20	0.06	0.04

What should he estimate to be his expected gain (a) in cents per dozen, (b) on the 10,000 dozen?

32. The possible values of a random variable X are the integers from n through $n + m$. If these possibilities are equally likely, find $E(X)$.

33. The random variable X has values 0 and n with probabilities $(n - 1)/n$ and $1/n$, in that order. Find $E(X)$. Describe the graph of the probability function of X (a) for $n = 5$, (b) for $n = 20$, (c) for $n = 1000$, (d) for $n \to$ "infinity". What is the limit of $E(X)$ as $n \to$ "infinity"?

34. The possible values of a random variable X are the integers $1, 2, 3, \ldots, n$ and $P(X = x) = cx$ for some constant c. (The probability function has a triangular shaped graph.) Show that $c = 2/n(n + 1)$. Find $E(X)$. Is $E(X) \approx \frac{2}{3}n$ when n is large? Discuss and interpret the result.

35. A regular six-sided die is thrown once and the number of dots on the top face is recorded. The die is then thrown again. If the number obtained on this throw is different from the recorded number, call the result a *success*, and stop. If the originally recorded number appears, call the result a *failure*, and continue throwing the die. What is the mean number of throws (following the first throw) up to and including the first success?

36. A regular die is tossed until a 6 first appears. What is the probability that 10 or more tosses are required?

37. A sample of size n is randomly drawn without replacement from a population of N objects, m defective and $N - m$ nondefective. Show that the expected number of defectives in the sample is nm/N.

5–3 MEAN OF A FUNCTION OF A RANDOM VARIABLE

Suppose that X is a random variable, a variable whose value is a number determined by the outcome of an experiment. If the value of X is increased by 5, the result is again a number determined by the outcome of that experiment: a number that is a value of the new random variable $X + 5$. Or, if the value of X is squared, the result is a value of the random variable X^2. In this section, we study random variables that are related to X: variables such as aX, $X + c$, $aX + c$, X^2, and $(X - c)^2$, where a and c are constants. Each of these random variables has a probability function, which we can get from the probability function of X, and each has a mean. In the next example, we show how these means are computed directly from the probability function of X without going through the intermediate step of finding the probability function of the related random variable.

Example 1 The random variable X has a probability function as follows:

Probability, $f(x)$	0.2	0.3	0.5
Value of X, x	-1	0	1

Compute the following means: $E(X)$, $E(2X)$, $E(X+1)$, $E(2X+1)$, $E(X^2)$, and $E[(X-0.3)^2]$.

Solution a) $E(X) = -1 \times 0.2 + 0 \times 0.3 + 1 \times 0.5 = 0.3$. Thus, the mean of X is 0.3.

b) The possible values of $2X$, and their probabilities, are as follows:

Probability	0.2	0.3	0.5
Value of $2X$	-2	0	2

Note that $P(2X = -2)$ is the same as $P(X = -1)$, and so on. If we multiply each possible value of $2X$ by its probability and add these products, we get the mean, or expected value, of $2X$:

$$\begin{aligned}\mu_{2X} = E(2X) &= -2 \times 0.2 + 0 \times 0.3 + 2 \times 0.5 \\ &= 0.6 = 2E(X).\end{aligned}$$

Doubling every number doubles the mean.

c)
$$\begin{aligned}\mu_{X+1} &= E(X+1) \\ &= (-1+1) \times 0.2 + (0+1) \times 0.3 + (1+1) \times 0.5 \\ &= 1.3 = E(X) + 1.\end{aligned}$$

Obviously, if we add 1 to every number, the mean is increased by 1.

d)
$$\begin{aligned}E(2X+1) &= (-2+1) \times 0.2 + (0+1) \times 0.3 + (2+1) \times 0.5 \\ &= 1.6 = 2E(X) + 1.\end{aligned}$$

Doubling every number and adding 1 doubles the mean and adds 1 to the result.

e) The random variable X^2 has only two possible values: 0 and 1. The associated probabilities are

$$\begin{aligned}P(X^2 = 0) &= P(X = 0) = 0.3, \\ P(X^2 = 1) &= P(X = -1) + P(X = 1) = 0.7.\end{aligned}$$

Therefore

$$E(X^2) = 0 \times 0.3 + 1 \times 0.7 = 0.7 \neq [E(X)]^2.$$

Note that the mean of the square is not the square of the mean.

f) $$E[(X-0.3)^2] = (-1.3)^2 \times 0.2 + (-0.3)^2 \times 0.3 + (0.7)^2 \times 0.5$$
$$= 0.61 = E(X^2) - [E(X)]^2.$$

There is something special about 0.3 in this example; it is the mean of X.

A common feature in all these examples, except (e), is this: we have computed the mean of a function of X by substituting in the formula for the function the possible values of X (in these cases -1, 0, 1), multiplying the results by the probabilities of these values of X (here 0.2, 0.3, 0.5), and adding the products. We formalize this procedure in the following theorem.

5-6 Theorem: Mean of a function

Let X be a random variable whose probability function is as follows:

Probability, $f(x)$	$f(x_1)$	$f(x_2)$	$\cdots$	$f(x_t)$
Value of X, x	x_1	x_2	$\cdots$	x_t

Let H be a function of X. Then the mean, or expected value, of the new random variable $H(X)$ is given by

$$E[H(X)] = H(x_1)f(x_1) + H(x_2)f(x_2) + \cdots + H(x_t)f(x_t), \tag{1}$$

or

$$E[H(X)] = \sum_{i=1}^{t} H(x_i)f(x_i). \tag{2}$$

Proof Suppose that $Y = H(X)$ takes the value y_1 for m distinct values of X, say for $X = x_1, x_2, \ldots, x_m$. [For example, both $X = -1$ and $X = +1$ yield the value $X^2 = +1$ in Example 1(e) above.] Then

$$y_1 = H(x_1) = H(x_2) = \cdots = H(x_m). \tag{3}$$

The corresponding contribution to the mean of Y is

$$y_1 \cdot P(Y = y_1).$$

But

$$P(Y = y_1) = f(x_1) + f(x_2) + \cdots + f(x_m),$$

so that

$$y_1 \cdot P(Y = y_1) = y_1 f(x_1) + y_1 f(x_2) + \cdots + y_1 f(x_m),$$

and when we take Eq. (3) into account, we see that

$$y_1 \cdot P(Y = y_1) = H(x_1)f(x_1) + H(x_2)f(x_2) + \cdots + H(x_m)f(x_m). \tag{4}$$

Similarly, if another set of values of X corresponds to y_2, a third set to y_3, and so on, we can group the terms on the right side of Eq. (1) into terms corresponding to

$$y_1 P(Y = y_1) + y_2 P(Y = y_2) + \cdots = E(Y).$$

Thus Eq. (1) allows us to use the probability function of X to get the same result that we would get by computing the mean of $Y = H(X)$ from the probability function of Y. □

★**Remark 1** The foregoing proof tacitly assumes that the sample space for X is finite. We shall assume, without proof, the fact that the theorem is also true for a countably infinite sample space, if we sum over all possible values of X. Similar simple changes would be necessary in Eqs. (3) and (4) if infinitely many values of X should happen to correspond to the same value of Y.

Remark 2 In Example 1, part (e) appeared as an exception to our method of computing the mean. But this exception is removed by Theorem 5–6. Applying that theorem, we now get, for Example 1(e), the following:

$$\begin{aligned} E(X^2) &= \sum_{i=1}^{3} x_i^2 f(x_i) \\ &= (-1)^2 \times 0.2 + (0)^2 \times 0.3 + (1)^2 \times 0.5 = 0.7, \end{aligned}$$

which is the result we obtained earlier.

Remark 3 Without Theorem 5–6 at our disposal, the mean of the new random variable $Y = H(X)$ would need to be computed from the probability function of Y by multiplying each possible value of Y by its probability and adding these products. In Example 1(b) above, we have illustrated this for $Y = 2X$. The possible values of Y are

$$\begin{aligned} y_1 &= 2x_1 = 2 \times (-1) = -2, \\ y_2 &= 2x_2 = 2 \times 0 = 0, \\ y_3 &= 2x_3 = 2 \times 1 = 2, \end{aligned}$$

and their probabilities are

$$\begin{aligned} P(Y = y_1) &= P(Y = -2) = P(X = -1) = f(x_1), \\ P(Y = y_2) &= P(Y = 0) = P(X = 0) = f(x_2), \\ P(Y = y_3) &= P(Y = 2) = P(X = 1) = f(x_3). \end{aligned}$$

Hence we find, for that example,

$$\begin{aligned} E(Y) &= y_1 P(Y = y_1) + y_2 P(Y = y_2) + y_3 P(Y = y_3) \\ &= y_1 f(x_1) + y_2 f(x_2) + y_3 f(x_3) \\ &= 2x_1 f(x_1) + 2x_2 f(x_2) + 2x_3 f(x_3), \end{aligned}$$

which corresponds to the result given by Eq. (1). Note that we do *not* get

$$2x_1 f(2x_1) + 2x_2 f(2x_2) + 2x_3 f(2x_3),$$

because the probability that $2X$ takes the value $2x_i$ is the same as the probability that X takes the value x_i, and this is $f(x_i)$, not $f(2x_i)$.

Example 1 has illustrated some further results that we now state as theorems, since they are true in general. We also provide algebraic proofs.

5–7 Theorem

Let X be a random variable with a finite mean. Then

$$\boxed{E(aX + b) = aE(X) + b,} \tag{5}$$

for any numerical constants a and b.

Proof Suppose the probability function of X is

$$\{(x_i, f(x_i)) : i = 1, 2, \ldots, t\}.$$

Then, by Theorem 5–6,

$$E(aX + b) = (ax_1 + b) f(x_1) + (ax_2 + b) f(x_2) + \cdots + (ax_t + b) f(x_t).$$

We expand the right side of this equation, factor out a and b, and get

$$\begin{aligned} E(aX + b) = {} & a[x_1 f(x_1) + x_2 f(x_2) + \cdots + x_t f(x_t)] \\ & + b[f(x_1) + f(x_2) + \cdots + f(x_t)]. \end{aligned}$$

The first of the bracketed expressions is $E(X) = \sum x_i f(x_i)$, and the second is 1, since $\sum f(x_i) = 1$. Therefore we have the desired result,

$$E(aX + b) = aE(X) + b. \ \square$$

★ **Remark 4** In our proof of Theorem 5–7, we assumed that X has only a finite set of possible values. If the sample space for X is countably infinite, then infinite series replace finite sums, and the proof depends on theory that we shall not go into. We shall assume, without proof, the fact that the theorem holds when X has a countably infinite sample space, provided that X has a finite mean.

5–8 Corollary

Let X be a random variable with mean $E(X) = \mu$. Then $E(X - \mu) = 0$.

Proof Take $a = 1$, $b = -\mu$ in Theorem 5–7:

$$E(X - \mu) = E(X) - \mu = \mu - \mu = 0. \ \square \tag{6}$$

Remark 5 The expected value of $X - c$ is often called the *first moment* of X, taken about c. The reason for this terminology is that $E(X - c)$ is, by Theorem 5-6,

$$(x_1 - c)f(x_1) + (x_2 - c)f(x_2) + \cdots + (x_t - c)f(x_t), \tag{7}$$

and this has the following physical interpretation. If we imagine a light but rigid bar with weights (in some system of units) equal to $f(x_1)$ at x_1, $f(x_2)$ at x_2, and so on, $f(x_t)$ at x_t, then formula (7) represents the sum of products of these weights each multiplied by the length of the "lever arm" from c to that weight. (See Fig. 5-5.)

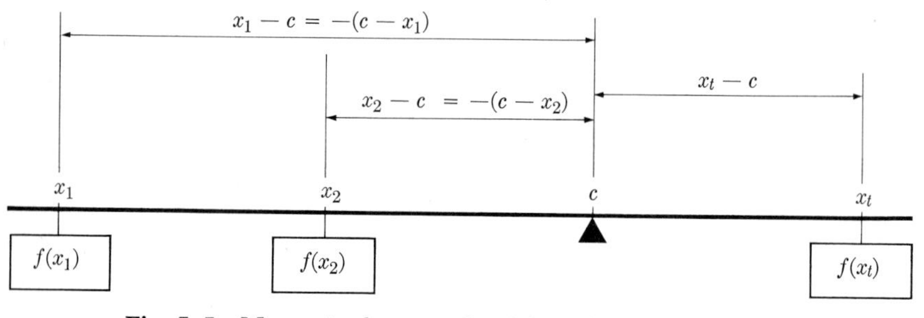

Fig. 5-5 Moment, about c, of weights $f(x_1), f(x_2), \ldots, f(x_t)$.

In physics, expression (7) is called the first moment of the system of weights about c. If c is at the center of mass of the system, the first moment is zero, and there is no tendency for the system to rotate about a support placed at that point. Thus children on a seesaw can balance by placing the support directly under the center of mass. Equation (6) tells us that if we place the fulcrum at the mean $\mu = E(X)$, the first moment about μ is zero. Conversely, if

$$E(X - c) = 0,$$

then $c = E(X)$; the mean is the only point about which the first moment is zero. This is one sense in which the mean is used to represent the "location", or the "middle", of the domain of the probability function. It isn't always at an equal distance from the ends of the domain—just as the point of balance of a seesaw isn't always halfway between the two children. If one child is much heavier than the other, the support must be nearer that child. Likewise, if lots of the probability is piled up near one end of the domain of a random variable, the mean is usually near that end.

★**Remark 6** For a countably infinite sample space, we speak of the existence of the mean of a random variable X only if $\sum_{i=1}^{\infty} x_i f(x_i)$ gives the same finite sum for every possible rearrangement of its terms. The totals of some infinite series explode as does the one in the St. Petersburg Paradox Exercise 10 given at the close of this section. When some x_i are positive and some negative, different arrangements of terms can even lead to different totals. These horrors perk up one's interest in the convergence of infinite series in general, but we cannot treat the matter here. We shall not prove, but it is cheering to know, that if we let a_i represent a term in an infinite series, and if $\sum_{i=1}^{\infty} |a_i|$ has a finite sum, then $\sum_{i=1}^{\infty} a_i$ also has a finite sum and rearranging the terms of the latter series does not change its sum.

However, for the special case of the geometric distribution, we have seen that the number of trials, X, to and including the first success can be 1, 2, 3, . . . ; and $x_i = i$ has the associated probability

$$P(X = x_i) = f(x_i) = q^{i-1}p.$$

Thus, in Theorem 5-5, the series that defines $E(X)$ leads to our old friend the infinite geometric progression, which converges (has a finite sum) if $p \neq 0$ and enables us to show that its sum, $E(X)$, is $1/p$.

For an ordinary six-sided die, if X is the number of trials to and including the first 5, then $p = \frac{1}{6}$ and the mean of X is 6. Consider an experiment in which we toss three dice simultaneously and define success as meaning that the sum of the scores is 5. The three dice can land in $6^3 = 216$ ways, and only 6 of these ways (namely the triples 113, 131, 311, 122, 212, 221) give the required sum 5. Hence, $P(\text{success}) = \frac{1}{36}$. By Theorem 5-5, the mean of the new distribution for X, the number of trials to and including the first sum 5, is $1/p = 36$.

For experiments in which p is very small, the mean $1/p$ is very large.

5-9 Theorem

Let X be a random variable and let G and H be functions of X such that $E[G(X)]$ and $E[H(X)]$ both exist. Then $G(X) + H(X)$ has an expected value, and

$$E[G(X) + H(X)] = E[G(X)] + E[H(X)]. \tag{8}$$

Proof We shall assume that X has a finite range of possible values $x_1, x_2, \ldots, x_t$ with associated probabilities $f(x_i)$, where $i = 1, 2, \ldots, t$. Applying Theorem 5-4 to the function $G + H$, we get

$$\begin{aligned} E[G(X) + H(X)] &= \sum_{i=1}^{t} [G(x_i) + H(x_i)]f(x_i) \\ &= \sum_{i=1}^{t} G(x_i)f(x_i) + \sum_{i=1}^{t} H(x_i)f(x_i) \\ &= E[G(X)] + E[H(X)]. \ \square \end{aligned}$$

Remark 7 If X has a countably infinite range of possible values, and if $E[G(X)]$ and $E[H(X)]$ exist, it can be proved that Theorem 5-9 still holds. We shall assume this fact without proof.

EXERCISES FOR SECTION 5-3

For Exercises 1 through 5, use the following data:

Probability, $f(x)$	0.2	0.1	0.3	0.3	0.1
Value of X, x	-2	-1	0	1	2

1. Compute $E(X)$.
2. Find the probability table for the function $3X - 1$ and then compute $E(3X - 1)$. Compare your answer with $3E(X) - 1$.
3. Find the probability table for the function $2X + 3$. Compute $E(2X + 3)$ and compare your answer with $2E(X) + 3$.
4. Find the probability function for X^2 and compute $E(X^2)$.
5. Find the probability function for $X^2 + 1$ and compute $E(X^2 + 1)$.

For Exercises 6 through 9, use the following data:

Probability, $f(x)$	0.2	0.3	0.2	0.2	0.1
Value of X, x	1	2	3	4	5

6. Find $E(X)$.
7. Compute, as easily as possible, (a) $E(3X - 7)$, (b) $E(X - 2.7)$, and (c) $E(10X)$.
8. Compute $E(X^2)$.
9. Compute $E[(X - 2.7)^2]$ and then show that your answer is equal to

$$E(X^2) - [E(X)]^2.$$

10. A fair coin is fairly tossed until a head appears. If the first head appears on an odd-numbered trial, the player receives 10 cents; otherwise, he pays 15 cents. What is the expected value of his winnings?
11. A carnival man sells 3 balls to be tossed into 9 boxes organized in a three-by-three array like a tic-tac-toe board. Each ball fills exactly one of the 9 boxes. The tosser wins a prize if the three balls are in a straight line. With random tossing, what is the probability of winning? At 25 cents for one play of the three balls, what is the expected cost of a prize? (Consider how many times he must play to win.)

★12. *Skiing accidents.* Five skiers per 1000 have reportable accidents on an average skiing day at a major slope. On the other hand, the average number of skiing days per accident for skiers is 440 (and not 250 = 1000/4). Suppose that there are three grades of skiers: "poor" having one accident per 100 days, "good" having one accident per 500 days, and "excellent" having one accident per 1000 days. What proportion of each kind of skier would account for the data?

ST. PETERSBURG PARADOX EXERCISES

Description of play. A player tosses a coin until it falls tails, or until he has tossed n times without a tail. Let X be the number of heads in one play of the game.

1. Find the probability function of X for $n = 2$.
2. Repeat Exercise 1 for $n = 3$.
3. Repeat Exercise 1 for $n = 4$.
4. Repeat Exercise 1 for a general value of n.

Description of payoff. In the game described above, the number of dollars the player receives is the random variable $Y = 2^X$.

5. For Exercise 1 find $E(Y)$.
6. For Exercise 2 find $E(Y)$.
7. For Exercise 3 find $E(Y)$.
8. For Exercise 4 find $E(Y)$. [*Hint:* Recall that $p + p^2 + \cdots + p^{n-1}$ is the sum of a geometric progression.]
9. Discuss the behavior of $E(Y)$ as n grows large.
10. Lift the restriction to n trials, and consider the expected payoff when the player tosses until he gets a tail. (This is the original St. Petersburg problem, and it is satisfactory to say that the expectation is infinite.)

Fair game. Recall that a game between two persons is said to be fair if the expected value to both persons is zero. Our player plays against a bank (or gambling house), tossing until he gets a tail, with payoff $Y = 2^X$ dollars, as before.

11. If the game is to be fair, and if the bank has unlimited resources, how much should the player pay the bank for one play of the game?
12. Suppose the bank has only 2^{20} dollars (\$1,048,476). What should the player pay the bank to make it a fair game?
13. Do Exercise 12 supposing that the bank has 4×10^{11} dollars. (This is about the size of the national debt of the United States of America in 1970.)

Note. Although it is amusing to see the modest payments required to play a fair game against a bank with astronomical resources, the result in Exercise 11 is somewhat shocking. The importance of that result is not its literal interpretation, i.e., that no one can pay an infinite amount. Rather, consideration of this and similar problems led people to realize that expected dollar value is not the only measure of worth, because a man will not invest a large amount of money in an enterprise with an even larger expected value if the probability that he gets his money back is tiny. Most of us would not care to risk \$10,000 for a 1/10,000 chance at a tax-free billion, even though the expected gain is \$90,000. Economists introduced the notion of utility to explain such behavior. Reference: H. Chernoff and L. E. Moses, *Elementary Decision Theory*. John Wiley and Sons, Inc., 1959, pp. 79–89.

6 Variability: measures of spread

6–1 VARIABILITY

We recall that the probability function of a random variable X tells us the possible values that X can have and the probabilities of those values. For many practical purposes, it is convenient to have a quick summary of the information that the probability function furnishes. The *mean*, or expected value $E(X)$, is one such summary; *it tells where the center of mass of the probability function is located.* Thus the mean number of dots on the top face of a die is 3.5, the mean number of divisors of an integer from 1 through 10 is 2.7, the mean number of heads when two coins are tossed is 1. *The mean is useful in giving us a quick picture of the long-run average result when an experiment is performed over and over.* But it tells us nothing about how outcomes spread out from one performance of the experiment to another. We shall now consider various alternative ways of measuring such variability, or spread, and then introduce the two most commonly used measures of spread, the *standard deviation* and the *variance*. (Either of these measures determines the other because the variance is the square of the standard deviation.)

Idea of spread, or variability To gain some experience with the idea of variability, we consider six random variables $X_A, X_B, \ldots, X_F$ whose probability graphs are shown in order $A, B, \ldots, F$ in Fig. 6–1. These probability functions are symmetrical about the value $x = 0$; their means are all equal to zero. We consider various ways of measuring their spreads about this common mean.

The first measure of variability, or spread, that suggests itself is the *range*, defined as follows. Consider those values of X that have probabilities greater than zero. Then the range of X is the largest of these values minus the smallest. In examples A, B, and C the range is 2; in examples D and E the range is 4; and in example F the range is 6. We might prefer a measure of variability that distinguishes among A, B, and C; particularly so since we wish to measure variability around the mean, and X_A has $\frac{6}{8}$ of its probability concentrated at the mean, while X_C has none.

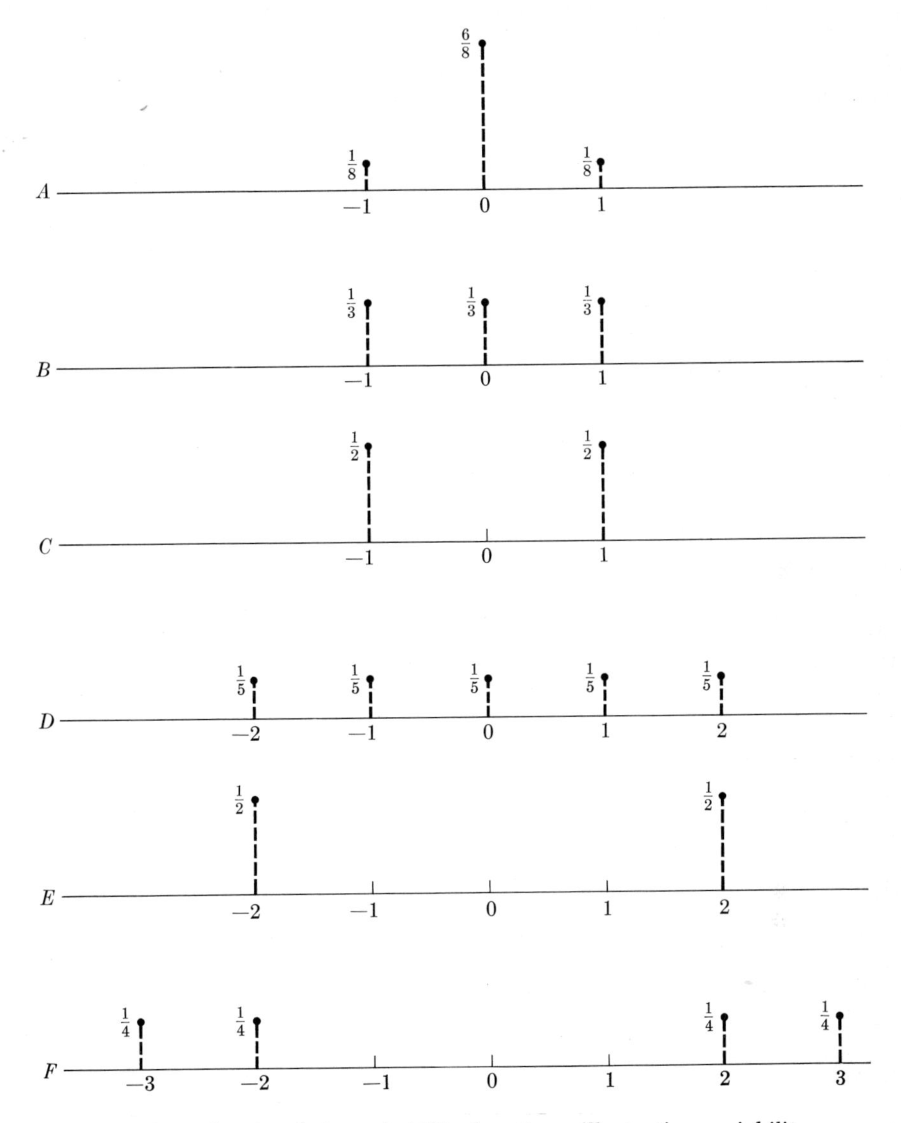

Fig. 6–1 Graphs of six probability functions, illustrating variability.

How would we compare the spread about the mean of the random variables X_A and X_B? First let us consider X_A, the outcome on an experiment A. In 1200 performances of the experiment, we expect 0 as the outcome about 900 times, $+1$ about 150 times, and -1 about 150 times. By contrast, experiment B would yield about 400 zeros, 400 plus ones, and 400 minus ones—results that seem to jump around more than the results of experiment A. Thus it seems reasonable that any measure of variability that is proposed should say that B is more variable than A.

The comparison between B and C is less obvious. However, since we use the *mean* as a measure of *location* of the probability distribution, we shall measure *variability about the mean.* Now B gives a result equal to the mean, 0, about $\frac{1}{3}$ of the time, and a result 1 unit away from the mean about $\frac{2}{3}$ of the time. By contrast, C always gives a result that is 1 unit away from the mean. Hence C seems more variable than B, when variability is measured about the mean.

Clearly, E seems more variable than C, and F more variable than E, but D and C are harder to compare.

Let us try to compare C and D. In C, the outcome is always either $+1$ or -1, and hence is 1 unit away from the mean, 0. In D, on the other hand, the outcome is at the mean about $\frac{1}{5}$ of the time, is 1 unit away from the mean about $\frac{2}{5}$ of the time and is 2 units away from the mean the remaining $\frac{2}{5}$ of the time. Hence the mathematical expectation of these "distances away from the mean" is

$$0 \times \tfrac{1}{5} + 1 \times \tfrac{2}{5} + 2 \times \tfrac{2}{5} = \tfrac{6}{5},$$

which is slightly greater than the corresponding value for C. Thus, by this line of reasoning, D is more variable than C.

The line of reasoning applied to D in the previous paragraph introduces the mathematical expectation of the absolute distance of X from its mean as a measure of variability. Thus, applying Eq. (1) of Theorem 5–6 to the function

$$H(X_D) = |X_D - \mu|,$$

we have

$$\begin{aligned} \mathrm{E}(|X_D - \mu|) &= |-2 - 0| \cdot \tfrac{1}{5} + |-1 - 0| \cdot \tfrac{1}{5} + |0 - 0| \cdot \tfrac{1}{5} \\ &\quad + |1 - 0| \cdot \tfrac{1}{5} + |2 - 0| \cdot \tfrac{1}{5} \\ &= 2(\tfrac{1}{5}) + 1(\tfrac{1}{5}) + 0(\tfrac{1}{5}) + 1(\tfrac{1}{5}) + 2(\tfrac{1}{5}) = \tfrac{6}{5}. \end{aligned}$$

The *mean absolute deviation* is defined as follows:

6–1 Definition: Mean absolute deviation

Let X be a random variable with mean μ, that is, $E(X) = \mu$. Then the *mean absolute deviation* of X, about μ, is the expected value of $|X - \mu|$:

$$\boxed{\text{mean absolute deviation of } X = E(|X - \mu|).} \tag{1}$$

In the third column of Table 6–1, we exhibit the mean absolute deviations for the examples shown in Fig. 6–1.

Although the mean absolute deviation gives a sensible measure of variability, it is not mathematically tractable. The absolute values are hard to combine algebraically, so the first thing that suggests itself is to remove them. But the ordinary mean deviation is zero, by Corollary 5–8.

The big advantage of the absolute values of the deviations is that they all count in the same direction; since none is negative, they can't cancel each other.

Table 6–1 Measures of variability for the examples of Fig. 6–1

Example	Probability function	Mean absolute deviation	Variance	Standard deviation
A	$f(x)$: $\frac{1}{8}$ $\frac{6}{8}$ $\frac{1}{8}$ x : -1 0 1	$\frac{1}{4}$	$\frac{1}{4}$	0.500
B	$f(x)$: $\frac{1}{3}$ $\frac{1}{3}$ $\frac{1}{3}$ x : -1 0 1	$\frac{2}{3}$		0.816
C	$f(x)$: $\frac{1}{2}$ $\frac{1}{2}$ x : -1 1	1		1.000
D	$f(x)$: $\frac{1}{5}$ $\frac{1}{5}$ $\frac{1}{5}$ $\frac{1}{5}$ $\frac{1}{5}$ x : -2 -1 0 1 2	$\frac{6}{5}$	2	1.414
E	$f(x)$: $\frac{1}{2}$ $\frac{1}{2}$ x : -2 2	2		2.000
F	$f(x)$: $\frac{1}{4}$ $\frac{1}{4}$ $\frac{1}{4}$ $\frac{1}{4}$ x : -3 -2 2 3	$\frac{5}{2}$		2.550

Another function that has this useful feature is the *squared deviation*, $(X - \mu)^2$. And this turns out to be much more tractable mathematically. As we become acquainted with properties of the *variance*, which uses the squares of deviations from the mean to measure variability, we shall see that there are two fundamental reasons for using it rather than some other measure:

1. *Additivity.* The variance of the sum of two independent random variables is the sum of their variances, and even when the two variables are dependent the variability of their sum has a simple formula.

2. *Central limit theorem.* The limiting behavior of a random variable that is the sum of a large number of independent random variables depends upon the *variances* of these random variables.

Of course, it isn't just the biggest squared deviation that counts, but rather the weighted mean of all the squared deviations, each weighted according to its probability. Statisticians call this *mean squared deviation*, $E[(X - \mu)^2]$, the "variance", and sometimes denote it by Var (X).

For example *D*, the computation of the variance goes as follows, since $\mu = 0$:

Probability, $f(x)$	$\frac{1}{5}$	$\frac{1}{5}$	$\frac{1}{5}$	$\frac{1}{5}$	$\frac{1}{5}$
Value of X, x	-2	-1	0	1	2
Value of $X - \mu$, $x - 0$	-2	-1	0	1	2
Value of $(X - \mu)^2$, x^2	4	1	0	1	4
Value of $x^2 f(x)$	$\frac{4}{5}$	$\frac{1}{5}$	0	$\frac{1}{5}$	$\frac{4}{5}$

$$\text{Var}(X) = \tfrac{4}{5} + \tfrac{1}{5} + 0 + \tfrac{1}{5} + \tfrac{4}{5} = 2$$

Hence, for D, the variance is 2. A similar computation for distribution C, which you are asked to perform in Exercise 2, shows that its variance is 1.

6–2 Definition: Variance

Let X be a random variable with mean $E(X) = \mu$. The variance of X, denoted by Var (X), is defined by

$$\operatorname{Var}(X) = E[(X - \mu)^2] = \sum_{i=1}^{t} (x_i - \mu)^2 f(x_i). \tag{2}$$

In words, the variance of X is the mean squared deviation of X from its mean.

One final adjustment is necessary to get from the variance of X to a measure of variability expressed in the original X-units. The units of Var (X) are squares of the units of X, so we recover the original units by taking the positive square root of the variance. The number so obtained is called the *standard deviation* of X. The standard deviation of X is denoted by σ_X (read "sigma sub-X"), or by the small Greek letter σ (read "sigma") without a subscript, if it is clear from the context what the random variable is.

6–3 Definition: Standard deviation

Let X be a random variable with mean μ. The *standard deviation* of X is the positive square root of the variance, and is given by

$$\sigma_X = \sqrt{\operatorname{Var}(X)} = \sqrt{E[(X - \mu)^2]}. \tag{3}$$

Remark Obviously, the variance of X is the square of the standard deviation:

$$\operatorname{Var}(X) = \sigma_X^2. \tag{4}$$

Example 1 The fifth column of Table 6–1 shows the standard deviations of the examples A through F of Fig. 6–1. Note that both the mean absolute deviation in column 3 and the standard deviation in column 5 assign measures of variability that increase as we read down the table.

Example 2 Let X represent the number of heads that appear when one coin is tossed and Y the number of heads when two coins are tossed. Compare the variances of the random variables X and Y. The probability functions are:

Probability, $f(x)$	$\frac{1}{2}$	$\frac{1}{2}$
Value of X, x	0	1

Probability, $f(y)$	$\frac{1}{4}$	$\frac{1}{2}$	$\frac{1}{4}$
Value of Y, y	0	1	2

Solution We first compute the means:

$$\mu_X = E(X) = 0 \times \tfrac{1}{2} + 1 \times \tfrac{1}{2} = \tfrac{1}{2},$$
$$\mu_Y = E(Y) = 0 \times \tfrac{1}{4} + 1 \times \tfrac{1}{2} + 2 \times \tfrac{1}{4} = 1.$$

Then the variances are

$$\begin{aligned}\text{Var}(X) &= \sigma_X^2 = E[(X - \mu_X)^2] \\ &= (0 - \tfrac{1}{2})^2 \times \tfrac{1}{2} + (1 - \tfrac{1}{2})^2 \times \tfrac{1}{2} = \tfrac{1}{8} + \tfrac{1}{8} = \tfrac{1}{4},\end{aligned}$$

$$\begin{aligned}\text{Var}(Y) &= \sigma_Y^2 = E[(Y - \mu_Y)^2] \\ &= (0 - 1)^2 \times \tfrac{1}{4} + (1 - 1)^2 \times \tfrac{1}{2} + (2 - 1)^2 \times \tfrac{1}{4} \\ &= \tfrac{1}{4} + \tfrac{1}{4} = \tfrac{1}{2} = 2\sigma_X^2.\end{aligned}$$

The variance of the number of heads for two coins is double the variance of the number of heads for one coin.

Example 3 A single six-sided die is tossed. Find the mean and variance of the number of dots on the top face.

Solution Let X represent the number of dots on the top face. The probability function of X is:

Probability, $f(x)$	$\frac{1}{6}$	$\frac{1}{6}$	$\frac{1}{6}$	$\frac{1}{6}$	$\frac{1}{6}$	$\frac{1}{6}$
Value, x	1	2	3	4	5	6

The mean, as we have found before, is

$$\begin{aligned}\mu_X &= E(X) = 1 \times \tfrac{1}{6} + 2 \times \tfrac{1}{6} + 3 \times \tfrac{1}{6} + 4 \times \tfrac{1}{6} + 5 \times \tfrac{1}{6} + 6 \times \tfrac{1}{6} \\ &= 21 \times \tfrac{1}{6} = \tfrac{7}{2}.\end{aligned}$$

The variance is

$$\begin{aligned}\sigma_X^2 &= E[(X - \mu_X)^2] \\ &= (1 - \tfrac{7}{2})^2 \times \tfrac{1}{6} + (2 - \tfrac{7}{2})^2 \times \tfrac{1}{6} + (3 - \tfrac{7}{2})^2 \times \tfrac{1}{6} \\ &\quad + (4 - \tfrac{7}{2})^2 \times \tfrac{1}{6} + (5 - \tfrac{7}{2})^2 \times \tfrac{1}{6} + (6 - \tfrac{7}{2})^2 \times \tfrac{1}{6} = \tfrac{35}{12}.\end{aligned}$$

We shall soon prove a general formula that is usually simpler for computing the variances. The formula is

$$\sigma_X^2 = E(X^2) - [E(X)]^2. \tag{5}$$

We check that Eq. (5) gives the correct result for the variance of the score on the die:

$$E(X^2) = 1^2 \times \tfrac{1}{6} + 2^2 \times \tfrac{1}{6} + 3^2 \times \tfrac{1}{6} + 4^2 \times \tfrac{1}{6} + 5^2 \times \tfrac{1}{6} + 6^2 \times \tfrac{1}{6} = \tfrac{91}{6},$$
$$[E(X)]^2 = (\tfrac{7}{2})^2 = \tfrac{49}{4},$$

so that

$$E(X^2) - [E(X)]^2 = \frac{91}{6} - \frac{49}{4} = \frac{182 - 147}{12} = \frac{35}{12}.$$

The result agrees with our previous calculation of the variance.

Equation (5) says that *the variance of X is the mean of the square of X minus the square of the mean of X.* We state this important result as a theorem, give a proof for any random variable that takes only three distinct values, and then indicate the proof in general.

6-4 Theorem: Variance

Let X be a random variable with mean $E(X) = \mu$ and variance

$$\text{Var}(X) = \sigma^2.$$

Then

$$\boxed{\sigma^2 = E(X^2) - [E(X)]^2 = E(X^2) - \mu^2.} \tag{6}$$

Proof Suppose the probability function of X is as follows:

Probability, $f(x)$	$f(x_1)$	$f(x_2)$	$\cdots$	$f(x_t)$
Value of X, x	x_1	x_2	$\cdots$	x_t

We temporarily assume that the number of values of X is $t = 3$. The proof for smaller or larger values of t is similar.

By definition,

$$\begin{aligned}\text{Var}(X) &= \sigma^2 = E[(X - \mu)^2] \\ &= (x_1 - \mu)^2 f(x_1) + (x_2 - \mu)^2 f(x_2) + (x_3 - \mu)^2 f(x_3).\end{aligned} \tag{7}$$

We expand the squares, and get

$$\begin{aligned}(x_1 - \mu)^2 f(x_1) &= x_1^2 f(x_1) - 2\mu x_1 f(x_1) + \mu^2 f(x_1), \\ (x_2 - \mu)^2 f(x_2) &= x_2^2 f(x_2) - 2\mu x_2 f(x_2) + \mu^2 f(x_2), \\ (x_3 - \mu)^2 f(x_3) &= x_3^2 f(x_3) - 2\mu x_3 f(x_3) + \mu^2 f(x_3).\end{aligned}$$

Summing both sides of these equations, and collecting terms on the right according to the powers of μ, we get

$$\begin{aligned}\sum(x_i - \mu)^2 f(x_i) &= [x_1^2 f(x_1) + x_2^2 f(x_2) + x_3^2 f(x_3)] \\ &\quad - 2\mu[x_1 f(x_1) + x_2 f(x_2) + x_3 f(x_3)] \\ &\quad + \mu^2[f(x_1) + f(x_2) + f(x_3)] \\ &= \sum x_i^2 f(x_i) - 2\mu \sum x_i f(x_i) + \mu^2 \sum f(x_i).\end{aligned} \tag{8}$$

By definition of mathematical expectation, we have

$$\sum x_i^2 f(x_i) = E(X^2), \tag{9a}$$

$$\sum x_i f(x_i) = E(X), \tag{9b}$$

and, since the sum of the probabilities is 1,

$$\sum f(x_i) = 1. \tag{9c}$$

If we introduce the right-hand terms from Eqs. (9a, b, c) into the right-hand side of the last line of Eq. (8), and recall that $E(X) = \mu$, we get

$$\begin{aligned}\sum (x_i - \mu)^2 f(x_i) &= E(X^2) - 2\mu E(X) + \mu^2 \\ &= E(X^2) - 2E(X) \cdot E(X) + [E(X)]^2 \\ &= E(X^2) - [E(X)]^2 = E(X^2) - \mu^2.\end{aligned} \tag{10}$$

If $t = 3$, the index i in the sums in Eqs. (8), (9), and (10) goes from 1 to 3; more generally, it goes from 1 to t. Since, by definition, the left-hand side of Eq. (10) is $E[(X - \mu)^2] = \sigma^2$, the proof of the theorem is thus completed. □

Example 4 Find the mean and variance of the number of divisors X in an integer from 1 through 10 chosen at random.

Solution The random variable X has the probability function

Probability, $f(x)$	0.1	0.4	0.2	0.3
Value, x	1	2	3	4

As we found earlier, the mean, or expected value, of X, is

$$\mu = E(X) = \sum x_i f(x_i) = 2.7.$$

It is not convenient to compute $E[(X - 2.7)^2]$ directly. But it is easy to apply Eq. (5):

$$\begin{aligned}\sigma_X^2 &= E(X^2) - \mu^2 \\ &= 1^2 \times 0.1 + 2^2 \times 0.4 + 3^2 \times 0.2 + 4^2 \times 0.3 - (2.7)^2 \\ &= 8.3 - 7.29 = 1.01.\end{aligned}$$

When computing variances for some discrete random variables, especially those whose possible values are integers (number of successes in a binomial experiment, or number of trials to first success, for example), it is useful to replace X^2 in Eq. (6) by $X(X - 1) + X$. The two expressions are equal for all values of X. If we apply the result of Theorem 5–9,

$$E[G(X) + H(X)] = E[G(X)] + E[H(X)],$$

with

$$G(X) = X(X-1), \qquad H(X) = X,$$

we get

$$E[X^2] = E[X(X-1)] + E[X].$$

When we combine this result with that expressed by Eq. (6), we have the following corollary.

6-5 Corollary: Alternative form for variance

Let X be a random variable with mean $E(X) = \mu$ and variance

$$\text{Var}\,(X) = \sigma^2.$$

Then

$$\sigma^2 = E[X(X-1)] + \mu - \mu^2. \tag{11}$$

To illustrate how Eq. (11) works, suppose that n is a positive integer, and let X be the number of successes in n binomial trials, where p is the probability of success on each individual trial. We know from Theorem 5-4 of Section 5-2 that the mean of the binomial is

$$E(X) = \mu = np.$$

To complete the calculation of the variance, we apply Eq. (2) of Theorem 5-6 with $H(X) = X(X-1)$ and get

$$E[H(X)] = \sum_{x=0}^{n} x(x-1)\binom{n}{x} p^x q^{n-x}. \tag{12}$$

The terms for $x = 0$ and $x = 1$ in the sum contribute zero because $x(x-1) = 0$ for both these values of x. Therefore, for $n \geq 2$, we can sum from $x = 2$ to $x = n$. [For $n = 1$ the right side of Eq. (11) reduces to

$$p - p^2 = p(1-p) = pq,$$

and we shall see later that this fits the general formula, which is $\sigma^2 = npq$.] Suppose, therefore, that $n \geq 2$ and $2 \leq x \leq n$. Then

$$x(x-1)\binom{n}{x} p^x q^{n-x} = \frac{n(n-1)(n-2)!}{(x-2)!(n-x)!} p^x q^{n-x}$$

$$= n(n-1)p^2 \binom{n-2}{x-2} p^{x-2} q^{(n-2)-(x-2)}. \tag{13a}$$

When we sum both sides of Eq. (13a) letting x run from 2 to n, $x - 2$ goes from 0 to $n - 2$ and the term $n(n-1)p^2$ is a common factor that does not depend upon the index of summation:

$$\sum_{x=2}^{n} x(x-1)\binom{n}{x} p^x q^{n-x} = n(n-1)p^2 \sum_{x=2}^{n} \binom{n-2}{x-2} p^{x-2} q^{(n-2)-(x-2)}$$

$$= n(n-1)p^2(q+p)^{n-2}$$

$$= (n^2 - n)p^2, \tag{13b}$$

because $q + p = 1$. We substitute this for $E[X(X - 1)]$ in Eq. (11) and replace μ by its value np to get

$$\begin{aligned}\sigma^2 &= (n^2 - n)p^2 + np - n^2p^2 \\ &= np - np^2 \\ &= np(1 - p) = npq.\end{aligned}$$

The result is stated formally in the following theorem.

6-6 Theorem: Mean and variance for the binomial

Let X be the number of successes in n binomial trials with probability p of success on each trial. Let $q = 1 - p$. Then the mean of X is np and its variance is npq:

$$\boxed{\mu = E(X) = np, \qquad \sigma^2 = \text{Var}(X) = npq.} \tag{14}$$

Example 5 In a binomial experiment, the probability p of success on each trial is unknown. (The results of the experiment are to be used to estimate p.) Let X be the number of successes in n trials. Show that the standard deviation σ_X is less than or equal to $\frac{1}{2}\sqrt{n}$.

Solution From Eq. (14) we know that

$$\sigma_X^2 = npq = np(1 - p)$$

and we also know that the probability p is between 0 and 1, inclusive:

$$0 \leq p \leq 1.$$

By simple algebra, adding and subtracting $\frac{1}{4}$ to complete the square, we get

$$p(1 - p) = \tfrac{1}{4} - (\tfrac{1}{4} - p + p^2) = \tfrac{1}{4} - (\tfrac{1}{2} - p)^2.$$

From this it is easy to see that the maximum value of $p(1 - p)$ for $0 \leq p \leq 1$ occurs when $p = \frac{1}{2}$, because a positive amount gets subtracted from $\frac{1}{4}$ for any $p \neq \frac{1}{2}$. Therefore

$$\sigma_X^2 = np(1 - p) \leq n(\tfrac{1}{4}) \qquad \text{and} \qquad \sigma_X \leq \tfrac{1}{2}\sqrt{n}.$$

$\frac{1}{4}$ 0 $\frac{1}{2}$ 1 p

□

Remark Theorem 6-6 shows and Example 5 illustrates that the standard deviation of a binomial random variable, considered as a function of n, grows more slowly than its mean. The reason is that the mean, $\mu = np$, grows like the first power of n, while the standard deviation, $\sigma = \sqrt{npq}$, grows like the square root of n.

Sometimes another way to cut down the labor involved in computing a variance is to shift the origin of the domain of values of the variable, or to change the scale. The addition of a constant to each value of a random variable shifts the mean by that same constant, but does not change the variance. But the multiplication of each value of the variable by a positive constant is equivalent to a change in units (for example, from tons to pounds, or from feet to miles). Such a change in units multiplies both the mean and the standard deviation by the same factor; however, it multiplies the variance by the square of that factor, since variance is measured in *squares* of the units of the variable.

The following theorem states how the variance and standard deviation are affected by such transformations of the random variable. A proof of the theorem is called for in Exercise 9.

6-7 Theorem

Let X be a random variable with variance σ^2. Let c be a number. Then

$$\boxed{\sigma_{cX}^2 = c^2\sigma_X^2, \qquad \sigma_{cX} = |c|\sigma_X} \tag{15}$$

and

$$\boxed{\sigma_{X+c}^2 = \sigma_X^2, \qquad \sigma_{X+c} = \sigma_X.} \tag{16}$$

Example 6 Let the probability function of X be as follows:

Probability, $f(x)$	0.3	0.2	0.5
Value, x	2025	2050	2075

Find σ_X^2.

Solution Subtract 2050 from each value of X, and divide the results by 25. The new random variable is

$$Y = \frac{X - 2050}{25}, \tag{17}$$

and its probability function is

Probability, $g(y)$	0.3	0.2	0.5
Value, y	-1	0	1

We compute the mean of Y and of Y^2:

$$\mu_Y = E(Y) = -1 \times 0.3 + 0 \times 0.2 + 1 \times 0.5 = 0.2,$$
$$E(Y^2) = (-1)^2 \times 0.3 + 0^2 \times 0.2 + 1^2 \times 0.5 = 0.8.$$

Therefore, the variance of Y is

$$\sigma_Y^2 = E(Y^2) - \mu_Y^2$$
$$= 0.8 - 0.04 = 0.76.$$

From Eq. (17) we see that

$$X = 25Y + 2050.$$

Hence

$$\begin{aligned} \sigma_X^2 &= \sigma_{(25Y+2050)}^2 \\ &= \sigma_{25Y}^2 && \text{[by Eq. (16)]} \\ &= 625\sigma_Y^2 && \text{[by Eq. (15)]} \\ &= 625 \times 0.76 \\ &= 475. \end{aligned}$$

The following corollary can be proved by the method used in the foregoing example.

6-8 Corollary

Let X be a random variable with variance σ_X^2. Let a and b be numbers. Then the variance of $aX + b$ is $a^2\sigma_X^2$:

$$\boxed{\operatorname{Var}(aX + b) = a^2 \operatorname{Var}(X),} \tag{18a}$$

or

$$\boxed{\sigma_{aX+b}^2 = a^2\sigma_X^2.} \tag{18b}$$

Standard deviate We frequently want to construct a new random variable Z whose distribution has the same shape as that of another random variable X, but has a specified mean and variance. When the mean is required to be zero and the variance unity, the new variable Z is called a *standardized random variable*, briefly a *standardized variable*, or a *standard deviate*. If X has mean μ_X and variance σ_X^2, then, by Corollary 6-8,

$$Z = \frac{X - \mu_X}{\sigma_X} = \left(\frac{1}{\sigma_X}\right) X - \frac{\mu_X}{\sigma_X}$$

has mean 0 and variance 1, and so Z is a standardized random variable or a standard deviate. When a variable is standardized to have other than mean 0 and variance 1, some mention must be made of the values of these quantities, because otherwise 0 and 1 are understood.

Obviously these standardized variables with mean 0 and variance 1 offer numerical and algebraic simplification for theoretical work. We often prove theorems first for standardized variables and then, by an addition and a multiplication, adjust the answers to get full generality.

★In advanced work, both in the physical sciences and in economics, we sometimes need to know the distribution of the number of times a random function crosses the axis. The following example illustrates the idea.

Example 7 *Random polynomial.* Certain cubic equations of the form

$$Ax^3 + Bx^2 + Cx + D = 0$$

have random coefficients A, B, C, and D, which independently take the values 1 or -1 with probability $\frac{1}{2}$. Find the distribution of the number of real roots, and compute the mean and the standard deviation of the distribution.

Solution Although we can make $2^4 = 16$ equally likely cubics, 8 of them can be obtained by multiplying each of the other 8 by -1. Each member of a pair will have the same number of roots. We can therefore restrict our discussion to the following 8 cubics:

	Real roots	Number of real roots
a) $+x^3 + x^2 + x + 1 = 0$	-1	1
b) $+x^3 + x^2 + x - 1 = 0$	≈ 0.54	1
c) $+x^3 + x^2 - x + 1 = 0$	≈ -1.84	1
d) $+x^3 - x^2 + x + 1 = 0$	≈ -0.54	1
e) $-x^3 + x^2 + x + 1 = 0$	≈ 1.84	1
f) $+x^3 + x^2 - x - 1 = 0$	$-1, -1, +1$	3
g) $+x^3 - x^2 + x - 1 = 0$	1	1
h) $-x^3 + x^2 + x - 1 = 0$	$1, 1, -1$	3

Obviously 1 and -1 are attractive numbers to try as roots for these equations. Equations (a), (f), (g), and (h) factor readily to give the indicated results. A theorem from the theory of equations says that the roots of (e) are the reciprocals of those of (b), and those of (d) are the reciprocals of those of (c). Numerical work gives the results shown for (b) to (e).

Since the equations are equally likely, the distribution of N, the number of real roots, is

N	1	3
$P(N)$	$\frac{3}{4}$	$\frac{1}{4}$

The mean number of real roots is $1\frac{1}{2}$ because

$$\mu_N = E(N) = 1 \times \tfrac{3}{4} + 3 \times \tfrac{1}{4} = \tfrac{3}{2}.$$

The expected value of N^2 is

$$E(N^2) = 1^2 \times \tfrac{3}{4} + 3^2 \times \tfrac{1}{4} = 3,$$

so the variance is

$$\sigma_N^2 = E(N^2) - \mu_N^2 = 3 - \tfrac{9}{4} = \tfrac{3}{4}.$$

Finally, the standard deviation is

$$\sigma_N = \tfrac{1}{2}\sqrt{3} \approx 0.866.$$

EXERCISES FOR SECTION 6–1

1. (a) Compute the variance and standard deviation for example B of Table 6–1. (b) Compute the mean absolute deviation and compare your answer with that given in the table.
2. (a) Compute the variance and standard deviation for example C of Table 6–1. (b) Compute the mean absolute deviation and compare your answer with that given in the table.
3. (a) Compute the variance and standard deviation for example E of Table 6–1, and compare them with the corresponding results for example C. Comment. (b) Compute the mean absolute deviation for example E of Table 6–1. Compare your answer with that given in the table. Also compare with example C, and comment.
4. The random variable X takes the values -1, 0, and 1 with probabilities 0.3, 0.2, and 0.5, respectively. Find (a) the mean, μ, (b) the mean absolute deviation of X about μ, (c) the variance σ^2, (d) the standard deviation σ.
5. In the medical experiment, Example 2, Section 1–8, Table 1–13 (selections of 3 from 5 treatments), let X be the number of times that treatment a appears in the listing of the 3 chosen. That is, $X = 0$ if a is not among those chosen, and $X = 1$ if a is chosen. Compute $E(X)$ and Var (X).
6. In Example 6, Section 5–1, for runs of 2 E's and 3 O's, let X be the number of runs in the sample point representing the outcome of the experiment. Compute $E(X)$ and σ_X^2.
7. In Example 6, Section 5–1, Table 5–10, on turning points for 4 different measurements, compute $E(X)$ and Var (X), where X represents the number of turning points in a sample point. Let $\sigma = \sqrt{\text{Var}(X)}$. What is the probability that $X \geq \mu + \sigma$? That $\mu - 2\sigma \leq X \leq \mu + 2\sigma$?
8. Using Eq. (5) of the text and the formulas

$$1 + 2 + 3 + \cdots + n = \frac{n(n+1)}{2},$$

$$1^2 + 2^2 + 3^2 + \cdots + n^2 = \frac{n(n+1)(2n+1)}{6},$$

show that the mean and variance of a random variable that takes the values 1, 2, 3, ..., n, each with probability $1/n$, are

$$\mu = \frac{n+1}{2}, \qquad \sigma^2 = \frac{n^2 - 1}{12}.$$

9. (a) Prove that $\text{Var}(cX) = c^2 \text{Var}(X)$. (b) Prove that $\text{Var}(X + c) = \text{Var}(X)$.
10. Use Theorem 6–7 to prove Corollary 6–8.

In each of the following exercises, 11 through 14, the probability function of a random variable is given. Find the mean, the variance, and the standard deviation.

11.

Probability, $f(x)$	0.1	0.2	0.3	0.4
Value, x	9998	9999	10,000	10,001

12.

Probability, $f(x)$	0.6	0.3	0.1
Value, x	0.0016	0.0032	0.0064

13.

Probability, $f(x)$	0.25	0.35	0.15	0.25
Value, x	−300	−200	−100	0

14.

Probability, $f(x)$	0.3	0.3	0.3	0.1
Value, x	2.75	3.00	3.25	4.00

For each of the following probability functions of X, calculate the mean, variance, and standard deviation:

15.

Probability, $f(x)$	0.4	0.2	0.4
Value of X, x	−1	0	1

16.

Probability, $f(x)$	0.1	0.3	0.4	0.2
Value of X, x	1	2	3	4

17.

Probability, $f(x)$	0.1	0.2	0.4	0.2	0.1
Value of X, x	−2	−4	6	4	2

18.

Probability, $f(x)$	0.1	0.4	0.5
Value of X, x	650	700	750

19. An engineer's ruler with triangular cross section has one of the numbers 1, 2, and 3 printed on each of its three faces. Imagine rolling the ruler on the floor, and let X be the number of the face on the bottom when the ruler comes to rest. Use the result given in Exercise 8 to find the mean, the variance, and the standard deviation of X.

20. Consider the experiment of Exercise 19 with two such rulers. Let Y be the sum of the numbers on the bottom faces of the rulers when they come to rest. Find the mean and the variance of Y.

21. A regular tetrahedron is a symmetrical solid with four faces. The faces are numbered 1, 2, 3, 4, and the tetrahedron is rolled on the floor. Let the random variable X be the number on the bottom face after the tetrahedron is rolled. Use the result of Exercise 8 to find the mean, variance, and standard deviation of X.

22. Consider rolling two tetrahedrons like the one described in Exercise 21. Let the random variable Y be the maximum face-down number when the two tetrahedrons come to rest. Find the mean, variance, and standard deviation of Y.

23. (Continuation.) In the experiment of Exercise 22, let the random variable Z be the minimum number on a bottom face when the tetrahedrons come to rest. Find the mean, variance, and standard deviation of Z.

24. If the variance of a random variable X is 0.76, what is the variance of the random variable $10X$? Of $2X$? Of $X/2$?

25. If the variance of the random variable Y is 15, what is the variance of $Y + 7$? $Y - 3$?

★26. The random variable X has for its probability function the *geometric distribution* $f(x) = q^{x-1}p$, for $x = 1, 2, 3, \ldots$, where $0 < p < 1$. Verify the calculations in the following outline and thereby determine the variance σ_X^2.

a) $E(X) = \sum_{x=1}^{\infty} xq^{x-1}p = p[1 + 2q + 3q^2 + 4q^3 + \cdots] = p/(1-q)^2 = 1/p.$

b) $E[X(X-1)] = \sum_{x=1}^{\infty} x(x-1)q^{x-1}p = p[1 \cdot 0 + 2 \cdot 1q + 3 \cdot 2q^2 + \cdots],$

so that

$$qE[X(X-1)] = p[1 \cdot 0q + 2 \cdot 1q^2 + 3 \cdot 2q^3 + \cdots]$$

and, by subtraction, $(1-q)E[X(X-1)]$ can be written as

$$2p[q + 2q^2 + 3q^3 + \cdots] = 2pq[1 + 2q + 3q^2 + \cdots] = 2pq/(1-q)^2 = 2q/p.$$

Therefore, $E[X(X-1)] = 2q/p^2$.

c) From the results in (a) and (b) above, and Corollary 6–5, show that the variance of the geometric distribution is $\sigma_X^2 = q/p^2$.

27. Using the results for the geometric distribution given in Exercise 26, determine which is larger, its mean or its standard deviation.

★28. The quadratic equation

$$Ax^2 + Bx + C = 0$$

has random coefficients A, B, C that take values 1 and -1 independently with probability $\frac{1}{2}$. Find the probability that the equation has 2 real roots. Find the mean and standard deviation of the number of real roots.

29. A binomial distribution has mean μ and standard deviation σ. Express the parameters n (number of trials) and p (probability of success on each trial) in terms of μ and σ. Explain why it must be true that $\sigma^2 \leq \mu$ and μ^2 is divisible by $\mu - \sigma^2$ if $\sigma^2 < \mu$. [*Ans:* $p = (\mu - \sigma^2)/\mu$; $n = \mu/p = \mu^2/(\mu - \sigma^2)$]

30. Given a binomial probability distribution with mean μ and standard deviation σ, how large should n be in order that $\mu - 3\sigma$ and $\mu + 3\sigma$ fall between 0 and n for all values of p such that $0.1 \leq p \leq 0.9$.

6–2 AVERAGE AND VARIANCE IN A SAMPLE

In Chapter 5 and the first section of this chapter, we have learned about random variables, probability functions, means, and variances. These ideas apply to *theoretical* outcomes of experiments. They help us to predict what is *likely to*

Table 6-2 Red cards in hands of five

Number of red cards	Number of hands
0	1
1	6
2	10
3	7
4	5
5	1
	Total 30

happen as the result of an experiment, provided we know the probability function, but rarely can they tell us exactly what will happen.

In this section, we study the results that actually did happen in some experiments. There are two main reasons for such a study:

1. A comparison of observed results with predicted theoretical results gives us a better understanding of the theory and of its reliability when used for making predictions.

2. In many experiments, we don't know the probability distribution of the random variable under study. In the school bond-issue example (Section 4-2), the proportion of people in the district who own property and favor the bond issue is unknown at the time the survey is planned. So we can't use that proportion to predict the outcome of the survey. In fact, we do just the opposite; we use the outcome of the survey to estimate the proportions of people in the four categories of interest. Or, we might wish to estimate the probability distribution of heights of American men of age 20. It would be costly in time and money to make a complete analysis of heights of all American men of age 20, so a sample is studied; and inferences about the average height and the variability of heights in the population are based on the average height and variability of heights in the sample.

Example 1 From an ordinary bridge deck of 52 cards, a hand of 5 cards is dealt without replacement. The number of red cards is tallied. The cards are reshuffled and the experiment is repeated 29 more times, giving a total of 30 hands. The results are shown in Table 6-2. What is the average number of red cards per hand? What is the standard deviation?

Solution The average number of red cards per hand is found as follows:

$$\text{average} = \frac{\text{total number of red cards in 30 hands}}{\text{total number of hands in 30 hands}}$$

$$= \frac{0 \times 1 + 1 \times 6 + 2 \times 10 + 3 \times 7 + 4 \times 5 + 5 \times 1}{30} = \frac{72}{30} = 2.4.$$

We denote this sample average by $\bar{x}$ (read "x bar"). Thus $\bar{x} = 2.4$.

Table 6–3 Calculation of variance for data of Table 6–2 ($\bar{x} = 2.4$)

Number of red cards, x_i	Number of hands, n_i	Deviation, $x_i - \bar{x}$	Squared deviation, $(x_i - \bar{x})^2$	Product, $(x_i - \bar{x})^2 n_i$
0	1	−2.4	5.76	5.76
1	6	−1.4	1.96	11.76
2	10	−0.4	0.16	1.60
3	7	+0.6	0.36	2.52
4	5	+1.6	2.56	12.80
5	1	+2.6	6.76	6.76
Totals	30			41.20

Next we compute the sample variance, i.e., the average squared deviation from $\bar{x}$, for this sample. In Table 6–2, the first column gives the possible values $x_i = 0, 1, 2, 3, 4, 5$ for the number of red cards per hand; the second column shows the frequency n_i with which the value x_i occurred. The squared deviations $(x_i - \bar{x})^2$ occur with these same frequencies, as shown in Table 6–3.

Multiplying each squared deviation by the number of times it occurs, and adding, we get 41.20, the *sum of the squared deviations* for all 30 hands. The *average* squared deviation is called the *sample variance*, and is denoted by s^2. Thus, for this example,

$$s^2 = \frac{41.20}{30} \approx 1.37.$$

The *sample standard deviation* s is the positive square root of the variance:

$$s \approx \sqrt{1.37} \approx 1.17.$$

Thus, for the data of Table 6–2, we have found

$$\text{average number of red cards per hand} = \bar{x} = 2.4,$$
$$\text{standard deviation of numbers of red cards per hand} = s \approx 1.17.$$

The sample average and standard deviation together provide a useful, quick summary of the frequency distribution in the sample. The average is a measure of *location:* it tells where the "center" of the sample is located. The standard deviation measures the dispersion, or spread, around the average. In the present example, $\bar{x} = 2.4$ is almost exactly halfway between the extreme values 0 and 5. And those extreme values, in turn, are at distances 2.4 and 2.6 from $\bar{x}$. If we measure these distances in standard deviation units, we find

$$2.4/1.17 \approx 2.05 \qquad \text{and} \qquad 2.6/1.17 \approx 2.22.$$

Thus all the values of x, in this example, are within 2.22 standard deviations of the sample average. It is usually true that all, or nearly all, of the observations in a sample lie within 3 standard deviations of the sample average.

We now make the following formal definitions:

6–9 Definitions: Sample variance and standard deviation

Given a set of n observations or measurements in which the value x_1 occurs n_1 times, x_2 occurs n_2 times, and so on, x_t occurs n_t times:

Frequency, n_i	n_1	n_2	$\cdots$	n_t	Total: n
Value, x_i	x_1	x_2	$\cdots$	x_t	

Let $\bar{x}$ be the average of the measurements:

$$\boxed{\bar{x} = \frac{1}{n} \sum x_i n_i.} \tag{1}$$

The *variance* s_x^2 is defined by

$$s_x^2 = \frac{(x_1 - \bar{x})^2 n_1 + (x_2 - \bar{x})^2 n_2 + \cdots + (x_t - \bar{x})^2 n_t}{n_1 + n_2 + \cdots + n_t},$$

or

$$\boxed{s_x^2 = \frac{1}{n} \sum (x_i - \bar{x})^2 n_i.} \tag{2}$$

The *standard deviation* s_x is the positive square root of the variance.

Computational formula The sample variance, Eq. (2), is the average of the squares of the deviations of the observations from their average: briefly, the *average squared deviation.* For computations it is often easier to use the following formula, which is analogous to formula (5) of Section 6–1:

$$\boxed{s_x^2 = \frac{1}{n} \sum x_i^2 n_i - \bar{x}^2} \tag{3a}$$

or

$$\boxed{s_x^2 = \text{Ave}\,(x^2) - \bar{x}^2.} \tag{3b}$$

In Eq. (3a), we have omitted the limits of summation. [As in Eqs. (1) and (2), i goes from 1 through t. See Appendix 1, Section A1–1, following Eq. (4), for a discussion of omission of limits of summation.] In Eq. (3b), we have used the notation Ave (x^2) to denote the average value of x^2:

$$\text{Ave}\,(x^2) = \frac{1}{n} \sum x_i^2 n_i.$$

A proof of Eq. (3a) is asked for in Exercise 1 at the end of this section.

Remark 1 The definitions of Eqs. (1) and (2) for the sample mean $\overline{x}$ and sample variance s_x^2 are used when the data have been grouped as in the frequency distribution exhibited in Definition 6–9. If the data are not grouped and $x_1, x_2, \ldots, x_n$ are the n observed values or measurements, then the sample average or sample mean is given by

$$\overline{x} = \frac{1}{n} \sum x_i \tag{4a}$$

and the sample variance is

$$s_x^2 = \frac{1}{n} \sum (x_i - \overline{x})^2 = \frac{1}{n} \sum x_i^2 - \overline{x}^2. \tag{4b}$$

We can think of Eqs. (4a) and (4b) as derived from Eqs. (1) and (2), respectively, if we note that $n_i = 1$ for each group and do not require that each group necessarily have a different value of x. Back in Definition 5–2 and here in Definition 6–9, we rather implied that the x_i's were distinct, but the formulas still work even if some x_i's have identical values. These formulas are used so frequently they should be committed to memory.

Remark 2 If samples of size n are drawn at random from a population with variance σ^2, the sample variance s^2 varies from sample to sample. Its long-run average can be shown to be $(n - 1)\sigma^2/n$. Some authors define the sample variance by dividing by $n - 1$ in Eq. (2) rather than by n. Then their sample variance across many samples averages to σ^2. However, $(n - 1)/n$ is close to 1 when n is large, so the two definitions are practically identical for large samples.

The numbers $\overline{x}$, s_x^2, and s_x are called *sample average*, *sample variance*, and *sample standard deviation*, respectively, to distinguish them from the corresponding features of the *population*. The *sample* values are computed from the observed measurements. Any set of measurements can be thought of as a "sample" from the "population" of all possible sets of measurements obtainable or imaginable under comparable experimental conditions. In the example of 30 hands of 5 cards each (Section 6–2, Example 1), the 30 hands are a "sample" drawn with replacement from the "population" of $\binom{52}{5}$ possible hands. For each hand the cards are dealt without replacement. A different sample would usually have a different average and a different variance. Also, the *sample* average and variance are usually different from the theoretical mean and variance of the *population*. The sample quantities are called *statistics* and the population quantities are called *parameters*. Thus $\overline{x}$ is a statistic used to estimate the parameter μ.

Example 2 Compare the sample mean and variance with the population mean and variance for the problem in Example 1.

Solution For Example 1, the sample characteristic is the number of red cards in a hand of 5 cards. This is a random variable X whose possible values are 0, 1, 2, 3, 4, 5. There are 52 cards in the deck, composed of 26 red cards and 26 black cards. A hand of 5 cards can be chosen in $\binom{52}{5}$ ways. A hand of 5 cards containing x red cards and $5 - x$ black cards can be selected in $\binom{26}{x}\binom{26}{5-x}$ different ways. Hence

$$P(X = x) = \frac{\binom{26}{x}\binom{26}{5-x}}{\binom{52}{5}}. \tag{5}$$

The values of the right-hand side of Eq. (5), for $x = 0, 1, \ldots, 5$, are shown in the following table (probabilities accurate to three decimal places):

Probability, $f(x)$	0.025	0.150	0.325	0.325	0.150	0.025
Value, x	0	1	2	3	4	5

The theoretical mean number of red cards in a hand of 5 is

$$\mu \approx 0(.025) + 1(.150) + 2(.325) + 3(.325) + 4(.150) + 5(.025) = 2.5.$$

This result can also be obtained at once by noting that the probability function is symmetric about $x = 2.5$.

We compute the variance from the formula

$$\sigma^2 = E(X^2) - \mu^2.$$

To compute $E(X^2)$, we square each possible value of X, multiply the result by the probability of that value, and add, to get

$$E(X^2) \approx 7.400.$$

Therefore, to three decimals, the theoretical (population) variance is

$$\sigma^2 \approx 7.400 - 6.250 = 1.150.$$

Recall that the sample average and standard deviation for the sample of 30 hands were

$$\bar{x} = 2.4 \quad \text{and} \quad s \approx 1.17,$$

while

$$\mu = 2.5 \quad \text{and} \quad \sigma \approx 1.07.$$

We see that the sample average and standard deviation serve as reasonable *estimates* of the theoretical mean and standard deviation of the population.

It is well at this point to summarize and compare some relevant characteristics of populations and samples.

Population	Sample
Possible value: $x_1, x_2, \ldots, x_t$	Observed value: $x_1, x_2, \ldots, x_t$
Probability: $f(x_1), f(x_2), \ldots, f(x_t)$	Relative frequency: $\frac{n_1}{n}, \frac{n_2}{n}, \ldots, \frac{n_t}{n}$
Mean: $\mu = \sum x_i f(x_i)$	Average: $\bar{x} = \frac{1}{n} \sum x_i n_i$
Variance: $\sigma^2 = \sum (x_i - \mu)^2 f(x_i)$	Variance: $s_x^2 = \frac{1}{n} \sum (x_i - \bar{x})^2 n_i$

Remark 3 When we group the data, as we have in the frequency tallies, the x_i's in the sample go from x_1 through x_t, just as in the population.

But often we don't group the observed data, but list them as

$$x_1, x_2, \ldots, x_n.$$

Then the x_i's in the sample are the values observed, usually in the order of their occurrence if there is an order. We then think of x_i as an observed value of a random variable X_i, for each i from 1 through n. Thus one full sample,

$$(x_1, x_2, \ldots, x_n),$$

produces values for all of the random variables $(X_1, X_2, \ldots, X_n)$. The sample average

$$\bar{x} = \frac{1}{n} \sum x_i$$

is an observed value of the random variable

$$\bar{X} = \frac{1}{n} \sum X_i.$$

Usually, in a sample, all of the n random variables X_i have the same probability function: that of the random variable in the population being sampled.

EXERCISES FOR SECTION 6-2

Compute $\bar{x}$, s^2, and s for each of the following sets of measurements:

1. 1, 1
2. 1, 2, 3
3. $-1, 0, 1$
4. $+2, -2$
5. $4, -5, -6$
6. .1, .3, .6
7. Five measurements are 1's, 3 measurements are 2's, and 1 measurement is a 3. Find $\bar{x}$, s^2, and s.
8. If half the measurements have value 1 and half have value 3, find the variance and standard deviation.

In each of the following exercises, 9 through 13, values are given for n (the number of observations in a sample), $\sum x_i$, and $\sum x_i^2$. Using these data, find the sample average, variance, and standard deviation. If you think that the given data are inconsistent, state your reason for thinking so.

	n	$\sum x_i$	$\sum x_i^2$
9.	10	35	140
10.	8	−56	408
11.	25	100	400
12.	12	30	65
13.	100	3	.90

In Exercises 14 through 17, we use the following notation: $x_1, x_2, \ldots, x_n$ and $y_1, y_2, \ldots, y_n$ are sets of measurements whose means are $\bar{x}$ and $\bar{y}$ and whose standard deviations are s_x and s_y, respectively; c and k are constants.

14. If $y_i = x_i + k$, show that $\bar{y} = \bar{x} + k$, $s_y = s_x$, and $s_y^2 = s_x^2$.
15. If $y_i = cx_i$, show that $\bar{y} = c\bar{x}$, $s_y = |c|s_x$, and $s_y^2 = c^2 s_x^2$.
16. If $y_i = cx_i + k$, show that $\bar{y} = c\bar{x} + k$, $s_y = |c|s_x$, and $s_y^2 = c^2 s_x^2$.
17. If $z_i = x_i + y_i$, then $\bar{z} = \bar{x} + \bar{y}$.
18. Without calculation, explain why the numbers 100, 101, 200 have the same variance as the numbers 1000, 1001, 1100.
19. Without calculation, explain why the standard deviation of the numbers 1, 2, 3, 4 is half the standard deviation of the numbers 2, 4, 6, 8.
20. In a certain neighborhood, 3 families have no car, 20 families have 1 car, 15 families have 2 cars, and 2 families have 3 cars. Find the mean and standard deviation of the number of cars per family.
21. The following frequency distribution was obtained in a breeding experiment with mice:

Number in litter:	1	2	3	4	5	6	7	8	9
Frequency:	7	11	16	17	26	31	11	1	1

Find the mean, variance, and standard deviation of the distribution.

22. The following frequency distribution gives the lengths of 800 ears of corn in inches, to the nearest half inch.

Length of ear:	4.0	4.5	5.0	5.5	6.0	6.5	7.0	7.5	8.0	8.5	9.0	9.5	10.0
Frequency:	1	1	8	33	70	110	176	172	124	61	32	10	2

a) Compute the mean and standard deviation of the distribution.
b) What percent of the measurements are within s of $\bar{x}$? Within $2s$? Within $3s$?

23. Ernest Thompson Seton gives, in *The Arctic Prairies*, the numbers of antelopes in 26 bands seen along the Canadian Pacific Railroad in Alberta, within a stretch of 70 miles, as follows:

8, 4, 7, 18, 3, 9, 14, 1, 6, 12, 2, 8, 10,
1, 3, 4, 6, 18, 4, 25, 4, 34, 6, 5, 16, 4.

Find the average number in a band, the standard deviation, and the percent of bands within s of $\bar{x}$ and the percent within $2s$ of $\bar{x}$.

24. Show that Eqs. (3a, b) in the text are valid. Compare with Eq. (5), Section 6–1.
25. In a sample of 100 random polynomials of degree 48, whose coefficients were each independently randomly chosen as 1 or -1, W. Fairley found the observed distribution of the number of real roots to be

Number of real roots:	0	2	4	6	Total
Frequency:	10	61	28	1	100

Estimate the mean and variance of the distribution of the number of real roots. Comment on the mean in view of the degree of the equations.

6–3 CHEBYSHEV'S THEOREM FOR A PROBABILITY DISTRIBUTION

Up to this point, we have discussed the mean, the variance, and the standard deviation for *probability distributions*, and the sample average, variance, and standard deviation for *observed sets of measurements*. We now wish to show how the standard deviation can be used to provide information about the way probability accumulates in intervals centered on the mean as their widths grow. We have an intuitive feeling that when the standard deviation is small the probability piles up near the mean, and when the standard deviation is large the probability spreads out more. With the aid of a remarkable theorem due to Chebyshev, which we study in this section, we shall be able to answer questions like the following:

What percent of the total probability lies in a given interval centered at the mean?

How wide an interval about the mean is needed to guarantee that, for example, three-quarters of the total probability of the random variable is included in that interval?

Before stating the theorem, however, we look at a simple example.

Example 1 Consider the random variable X having the following probability function:

Probability, $f(x)$	$\frac{27}{64}$	$\frac{27}{64}$	$\frac{9}{64}$	$\frac{1}{64}$
Value of X, x	0	1	2	3

Find the probability that is associated with values of X:

a) at or within 1 standard deviation from the mean,
b) at or within 2 standard deviations from the mean,
c) at or within 3 standard deviations from the mean.

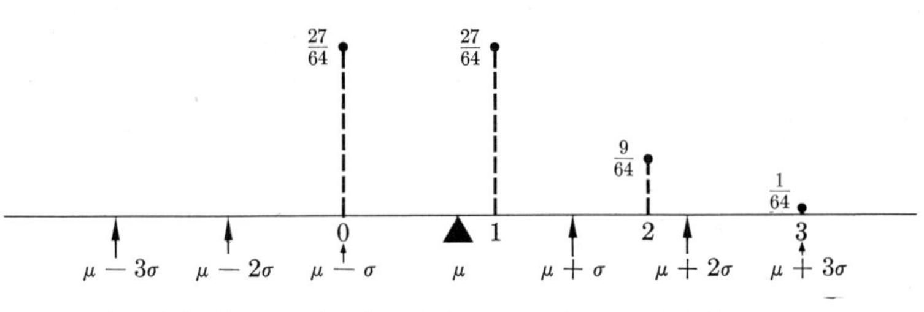

Fig. 6–2 Intervals of width 2σ, 4σ, 6σ around the mean.

Solution For the mean and standard deviation, calculations give

$$\mu = E(X) = \tfrac{3}{4}, \qquad \sigma = \tfrac{3}{4}.$$

Figure 6–2 shows a graph of the probability function. The mean, $\mu = \frac{3}{4}$, is marked with a small wedge, ▲, to suggest a fulcrum. Intervals extending 1σ, 2σ, and 3σ to the left and right of the mean are also shown, along with the corresponding probabilities.

a) The probability at or within $\pm 1\sigma$ from μ is

$$\tfrac{27}{64} + \tfrac{27}{64} = \tfrac{27}{32} \approx 0.84.$$

b) The probability at or within $\pm 2\sigma$ from μ is

$$\tfrac{27}{64} + \tfrac{27}{64} + \tfrac{9}{64} = \tfrac{63}{64} \approx 0.984.$$

c) The probability at or within $\pm 3\sigma$ from μ is

$$\tfrac{27}{64} + \tfrac{27}{64} + \tfrac{9}{64} + \tfrac{1}{64} = 1.$$

6–10 Theorem: Chebyshev's theorem

If the random variable X has finite mean μ and finite standard deviation σ, and h is any positive number, then the probability outside the closed interval $[\mu - h\sigma, \mu + h\sigma]$ is less than $1/h^2$. In symbols,

$$P(|X - \mu| > h\sigma) < 1/h^2. \tag{1a}$$

Alternatively, the probability on that interval is greater than $1 - (1/h^2)$, or, in symbols,

$$P(|X - \mu| \leq h\sigma) > 1 - (1/h^2). \tag{1b}$$

Discussion The theorem says, for example, that more than $1 - \frac{1}{4}$, or $\frac{3}{4}$, of the total probability is at or within $\pm 2\sigma$ from μ, for any random variable with finite mean μ and finite standard deviation σ. In the example above, we found that the actual probability in the band from $\mu - 2\sigma$ to $\mu + 2\sigma$, inclusive, was $\frac{63}{64}$, which is much greater than $\frac{3}{4}$. The theorem also says that more than $\frac{8}{9}$ of the total probability is assigned to the band from $\mu - 3\sigma$ to $\mu + 3\sigma$, inclusive. In the example, that band contained the total probability, 1.

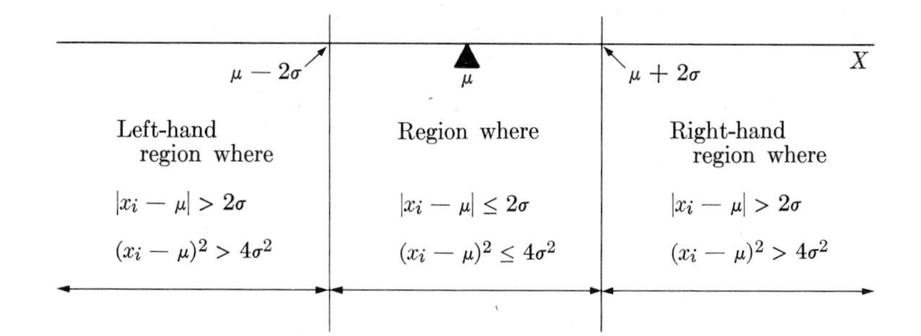

Fig. 6–3 Regions for Chebyshev's theorem, $h = 2$.

Proof of Chebyshev's theorem Suppose the random variable X has mean μ and standard deviation σ. Figure 6–3 represents the domain of its probability function. We shall first prove the theorem for the case $h = 2$, and to this end we have separated the possible values of X into two sets:

a) those on the closed interval $\mu - 2\sigma$ to $\mu + 2\sigma$, which includes any lying at the boundaries, and

b) the remainder, those lying beyond the boundaries of the interval.

We want to prove that the probability associated with values of X in the set (a) is at least $\frac{3}{4}$. For convenience, we refer to the set (a) as the values of X *on the interval* and to set (b) as the values *outside the interval.* The values outside the interval consist of those to the left of $\mu - 2\sigma$, which we call the *left-hand region,* and those to the right of $\mu + 2\sigma$, which we call the *right-hand region.*

It is clear from Fig. 6–3 that any point on the x-axis and outside the indicated middle interval is more than 2σ from the mean μ. Therefore the square of its distance from the mean is more than $4\sigma^2$.

Recall the definition of the variance:

$$\sigma^2 = E[(X - \mu)^2] = \sum_{i=1}^{t} (x_i - \mu)^2 f(x_i), \tag{2a}$$

where $x_1, x_2, \ldots, x_t$ are the possible values of X, and $f(x_1), f(x_2), \ldots, f(x_t)$ are their associated nonzero probabilities. Now the numbering of the x's (the subscripts) is completely arbitrary; so, for convenience, let $x_1, x_2, \ldots, x_r$ denote those that are *outside* the interval, if there are any outside.

Case 1. If there are no values of x_i outside the interval, all values are at or within 2σ of the mean, so the probability of set (a) is 1, and hence is more than $\frac{3}{4}$.

Case 2. If the number outside the interval is $r \geq 1$, then $\sigma^2 > 0$ and we break up the sum in Eq. (2a) into two parts:

$$\begin{aligned}\sigma^2 = {} & [(x_1 - \mu)^2 f(x_1) + (x_2 - \mu)^2 f(x_2) + \cdots + (x_r - \mu)^2 f(x_r)] \\ & + [(x_{r+1} - \mu)^2 f(x_{r+1}) + \cdots + (x_t - \mu)^2 f(x_t)].\end{aligned} \tag{2b}$$

Every squared deviation $(x_i - \mu)^2$ is positive or zero, and $f(x_i)$ is also positive. If any squared deviation is replaced by a smaller number, then the right-hand side of Eq. (2b) is reduced. We shall make such reductions and get an inequality that yields the proof.

We reduce the first r squared deviations (arising from values of x_i outside the interval) by replacing each of them by the smaller value $4\sigma^2$. Then we reduce the rest of the squared deviations (arising from values of x_i on the interval) by replacing each of them by the smaller or possibly equal value 0. When we have made these replacements, we get the inequality

$$\sigma^2 > 4\sigma^2[f(x_1) + f(x_2) + \cdots + f(x_r)]. \tag{3}$$

Because $\sigma^2 > 0$, we may divide both sides of the inequality (3) by $4\sigma^2$, and get

$$\tfrac{1}{4} > [f(x_1) + f(x_2) + \cdots + f(x_r)] = P(|X - \mu| > 2\sigma). \tag{4}$$

The last equality in (4) follows from the definition of $P(|X - \mu| > 2\sigma)$: it is the probability that X is more than 2σ from the mean μ, and this probability is the sum of the probabilities assigned to the points $x_1, x_2, \ldots, x_r$ that are outside the interval. Therefore, from the inequality (4), we see that less than $\frac{1}{4}$ of the total probability is assigned to points lying outside the interval. Hence the probability assigned to points lying at or within a distance 2σ from the mean is *more than* $1 - \frac{1}{4} = \frac{3}{4}$. This completes the proof of Chebyshev's theorem for $h = 2$.

The demonstration just given can be generalized to intervals $\mu - h\sigma$ to $\mu + h\sigma$ for any $h > 0$. We replace 2σ by $h\sigma$ and $4\sigma^2$ by $h^2\sigma^2$ throughout the argument. When these replacements are made, the inequality that replaces (4) is

$$\boxed{\frac{1}{h^2} > P(|X - \mu| > h\sigma).} \tag{5}$$

This says that the probability assigned to values of X outside the interval $\mu - h\sigma$ to $\mu + h\sigma$ is less than $1/h^2$. Hence the probability assigned to values of X at or within a distance $h\sigma$ of the mean is more than $1 - (1/h^2)$. □

Remark 1 In a binomial experiment of n trials, we know that the observed number of successes x can be any number from 0 through n. It is common to use the ratio x/n as an estimate for the underlying probability p when that probability is unknown. (For example, in the thumbtack-tossing experiment, we don't know the probability p that the tack will land point up, so we toss it a number of times n, record the number of times x that it lands point up, and use x/n as our estimate of p.) Most of us feel intuitively that the observed ratio x/n ought to lie close to the true value of p. We know, for instance, that an unlikely event such as getting ten heads in ten tosses of a coin would lead to a bad estimate for p, but such bad luck should have a very low probability when

n is large. Chebyshev's theorem can be applied to the binomial distribution to reassure us that this is so. The formal theorem is known as Bernoulli's law of large numbers. The theorem is credited to Jacob Bernoulli and was published in 1713 after his death. It is said to be the first published limit theorem of probability. [Cf. J. R. McCord and R. M. Moroney, *Introduction to Probability Theory* (Macmillan, 1964), p. 183.]

6–11 Theorem: Law of large numbers (Bernoulli)

In a binomial experiment, let $\overline{p}$ be the observed proportion of successes, p the probability of success on a single trial, n the number of independent trials. Then for any positive number d, as n tends to infinity the probability tends to 1 that the inequality $|\overline{p} - p| \leq d$ is satisfied.

Proof In order to apply Chebyshev's theorem to $\overline{p}$ we need to know the mean and the standard deviation of the random variable

$$\overline{p} = \frac{X}{n}, \tag{6}$$

given that n is a positive integer and X has mean and variance

$$E(X) = np, \qquad \text{Var}\,(X) = npq,$$

as given by Theorem 6–6. We get $\overline{p}$ from X by multiplying X by $c = 1/n$. Using Theorem 5–7 for the mean, and Theorem 6–7 for the variance, we have

$$E(\overline{p}) = \mu_{\overline{p}} = \frac{1}{n}\,E(X) = \frac{1}{n}\,(np) = p,$$

$$\text{Var}\,(\overline{p}) = \sigma_{\overline{p}}^2 = \sigma_{X/n}^2 = \frac{1}{n^2}\,\sigma_X^2 = \frac{npq}{n^2} = \frac{pq}{n}\cdot$$

More compactly, the mean and standard deviation of $\overline{p}$ are

$$\boxed{\mu_{\overline{p}} = p, \qquad \sigma_{\overline{p}} = \sqrt{\frac{pq}{n}}\,.} \tag{7}$$

We now apply Theorem 6–10, Chebyshev's theorem, to the random variable $\overline{p} = X/n$. For any positive number h the following inequalities hold:

$$1 \geq P(|\overline{p} - p| \leq h\sqrt{pq/n}\,) > 1 - \frac{1}{h^2}\cdot \tag{8}$$

The first inequality in (8) is true because every probability is less than or equal to one. The second inequality is guaranteed by Chebyshev's theorem. By proper choice of h, we can introduce the distance d of Theorem 6–11, thus:

$$d = h\sqrt{pq/n} \qquad \text{if and only if} \qquad h = d\sqrt{n/pq}. \tag{9}$$

Throughout the inequalities (8) we substitute for h its value in terms of d as given by Eqs. (9), and get

$$1 \geq P(|\overline{p} - p| \leq d) > 1 - \frac{pq}{nd^2}. \tag{10}$$

Holding d fixed and letting n increase, we see from the first and last inequalities in (10) that

$$P(|\overline{p} - p| \leq d)$$

tends to 1, as stated in Theorem 6–11. □

Remark 2 If we subtract 1 from each of the three major terms in (10), we have the equivalent statement

$$0 \geq P(|\overline{p} - p| \leq d) - 1 > -\frac{pq}{nd^2}. \tag{11a}$$

By changing signs and reversing the inequality symbols in (11a), we get

$$0 \leq 1 - P(|\overline{p} - p| \leq d) < \frac{pq}{nd^2}. \tag{11b}$$

The form (11b) is particularly useful when we wish to apply Theorem 6–11 to a problem of estimating an unknown p, as in the next example.

Example 2 It is desired to use $\overline{p}$ to estimate p, with probability greater than 0.97 that the error is no more than 0.05. How large should n be?

Solution From the statement of the problem, we know that we can take $d = 0.05$ in any of the inequalities (10) through (11b). But we don't know what value to substitute for p or for $q = 1 - p$. However, the rightmost term in (11*b*) is less than or equal to $1/(4nd^2)$ because $pq = p(1 - p) = p - p^2 = \frac{1}{4} - (p - \frac{1}{2})^2$ is less than or equal to $\frac{1}{4}$ for all values of p. By the transitive law for inequalities, we can use (11b) and $pq \leq \frac{1}{4}$ to get

$$0 \leq 1 - P(|\overline{p} - p| \leq d) < \frac{1}{4nd^2}. \tag{12}$$

To complete the solution of Example 2, we take

$$d = 0.05, \qquad \frac{1}{4nd^2} \leq 1 - 0.97 = 0.03.$$

Solving for n, we find

$$n \geq \frac{1}{(0.03)(4d^2)} \approx 3333.$$

Remark 3 The value of n obtained by applying the inequality (12) is much larger than necessary. When we study the central limit theorem in Chapter 8, we shall discover that the random variable

$$Z = \frac{\overline{p} - p}{\sqrt{pq/n}}$$

is approximately distributed according to the standard normal distribution. The theory that we shall there develop can be used to yield the number 471 as a sufficiently large value of n. (See Example 2, Section 9–2 for the complete solution along these lines.)

EXERCISES FOR SECTION 6–3

In Exercises 1 through 4, assume that $\mu_X = 0$ and $\sigma_X = 1$. Use Chebyshev's theorem to fill in the blanks with appropriate numbers.

1. $P(|X - \mu| \leq 2) >$ ______.
2. $P(-3 \leq X \leq 3) >$ ______.
3. $P(|X| > 3) <$ ______.
4. $P(|X| \leq k) > 0.96$ if $k \geq$ ______.

In Exercises 5 through 8, assume that $\mu_X = 7$ and $\sigma_X = 2$. Use Chebyshev's theorem to fill in the blanks with appropriate numbers.

5. (a) $P(3 \leq X \leq 11) >$ ______; (b) $P(1 \leq X \leq 13) >$ ______.
6. (a) $P(|X - 7| > 2) <$ ______; (b) $P(|X - 7| > 3) <$ ______.
7. $P(|X - 7| \leq 5) >$ ______.
8. $P(|X - 7| \leq k) > 0.99$ if $k \geq$ ______.
9. When $h \leq 1$, Chebyshev's theorem is useless. Why?
10. Make the required substitutions and generalize the proof of Theorem 6–10 from intervals
$$|X - \mu| \leq 2\sigma \quad \text{to} \quad |X - \mu| \leq h\sigma.$$
11. Using Chebyshev's theorem, what values of h guarantee that more than 90% of the probability is at or within $h\sigma$ of the mean? What values of h guarantee more than 99%?
12. Use Chebyshev's theorem to fill in the following blanks:
    a) $P(|X - \mu| > 2\sigma) <$ ______;
    b) $P(|X - \mu| > 3\sigma) <$ ______;
    c) $P(|X - \mu| > 5\sigma) <$ ______.
13. Under what conditions on X is its variance zero? How much probability then lies more than 0.01σ away from the mean?
14. *Alternative form of Chebyshev's theorem.* In our statement of Chebyshev's theorem, we included the endpoints $\mu - h\sigma$ and $\mu + h\sigma$ as part of the closed middle region $|X - \mu| \leq h\sigma$. We then proved that the probability *outside* that region satisfied $P(|X - \mu| > h\sigma) < 1/h^2$. Show that, if the endpoints $\mu - h\sigma$ and $\mu + h\sigma$ are *not* included in the middle region but are incorporated into the two outer regions, the corresponding probability satisfies $P(|X - \mu| \geq h\sigma) \leq 1/h^2$. In words: the

probability is less than or equal to $1/h^2$ that the random variable X takes values that differ from its mean $E(X) = \mu$ by $h\sigma$ or more.

In Exercises 15 through 22, the random variable X takes the values $-c$, 0, and $+c$, with probabilities p, $1 - 2p$, and p, respectively.

15. Find μ and σ^2.
16. Show that $c = \sigma$ when $p = \frac{1}{2}$.
17. If $c = 2\sigma$, what does p equal?
18. If $c = 3\sigma$, what does p equal?
19. In Exercise 16, what is the probability of an absolute deviation $|X - \mu|$ at least as great as one standard deviation? As great as 2σ?
20. In Exercise 17, what is the probability of an absolute deviation at least as great as 2σ? As great as 3σ?
21. In Exercise 18, what is the probability of an absolute deviation at least as great as 3σ? As 4σ?
22. By proper choice of p, can you make $c = h\sigma$, for any positive h? If so, what is the proper choice of p, in terms of h, and what is the probability of an absolute deviation at least as great as $h\sigma$?

Remark. Exercise 22 shows that if h is given, then we can find a random variable X such that the probability that X takes values at least $h\sigma$ away from its mean is $1/h^2$, the maximum allowed by the alternative form of Chebyshev's theorem given in Exercise 14. In this sense, the conclusion in Chebyshev's theorem is the best possible. But a probability distribution that has the maximum allowable probability at least $h\sigma$ away from μ for one particular value of h may not do so for a different value of h. Part of the charm of the Chebyshev theorem is that it works for all probability distributions with finite means and variances.

23. Suppose, in Example 2, that we wish the probability to be greater than 0.98 that $\bar{p}$ differs from p by no more than 0.05. How large should n be?
24. Using inequality (12), show that if X is the number of successes in a binomial experiment of n trials with probability p of success on each trial, and $\bar{p} = X/n$, then
$$P(|\bar{p} - p| \leq d) > 1 - (1/4nd^2) \qquad \text{for any} \quad d > 0.$$
25. True or false? By making n sufficiently large, we can be as sure as we please that the estimate $\bar{p} = X/n$ is within a given nonzero distance d of p:
    a) if we interpret the phrase "be as sure as we please" to mean "make the probability equal to 1".
    b) if we interpret the phrase "be as sure as we please" to mean "make the probability arbitrarily near 1, but not necessarily equal to 1".

6-4 CHEBYSHEV'S THEOREM FOR A FREQUENCY DISTRIBUTION OF MEASUREMENTS

We have seen how the standard deviation σ provides a yardstick for measuring distances from the mean of a random variable X. Chebyshev's theorem tells us that less than the fraction $1/h^2$ of the probability will be assigned to points

Table 6–4 Percentages of measurements contained in intervals about the mean

	Empirical rule	Chebyshev's theorem
Closed interval	Contains about this percentage of the measurements	Contains more than this percentage of the measurements
$\bar{x} - s$ to $\bar{x} + s$	68%	0%
$\bar{x} - 2s$ to $\bar{x} + 2s$	95%	75%
$\bar{x} - 3s$ to $\bar{x} + 3s$	99.7% (nearly all)	89%

that are more than h standard deviations away from the mean, for any positive h. We may well wonder if an analogous theorem holds for measurements or observations in a *sample*. The answer is "yes". We state the result formally as a theorem, but we do not give the proof, since it is almost identical with the proof given in Section 6–3.

6–12 Theorem: Chebyshev's theorem for measurements

More than the fraction $1 - (1/h^2)$ of the measurements in any sample lie at or within h standard deviations of the average of the measurements.

Example 1 Suppose the measurements are $-8, -1, -1, 0, 0, 0, 0, 1, 1, 8$. Verify that more than $\frac{3}{4}$ of the measurements are at or within 2 standard deviations of $\bar{x}$, and more than $\frac{8}{9}$ are at or within 3 standard deviations of $\bar{x}$.

Solution The sum of the 10 measurements is 0, hence $\bar{x} = 0$. The sum of the squares of the measurements is 132. Hence the average of the squared deviations is $s^2 = \text{Ave}(x^2) - \bar{x}^2 = 13.2 - 0 = 13.2$, and the standard deviation is $s \approx 3.6$. The interval containing all measurements at or within 2 standard deviations from the mean extends from -7.2 to $+7.2$, and contains 80% of the measurements (hence more than $\frac{3}{4}$ of them). The interval extending 3 standard deviations from the mean in both directions goes from -10.8 to $+10.8$ and contains 100% of the measurements (hence more than $\frac{8}{9}$ of them).

Remark For work with large numbers of measurements, stronger results than those given by Chebyshev's theorem usually hold. Table 6–4 gives a rough rule for the percentage of measurements usually found in intervals about the mean. The numbers given for the empirical rule agree exactly with those for the *normal probability distribution*, which we shall study in more detail in Section 7–4 and in Chapter 8.

The results of Chebyshev's theorem *guarantee* lower bounds on the percentage of measurements at or within h standard deviations of the average. They may help us (a) discover an error in calculation or (b) interpret and use the standard deviation. The middle column of Table 6–4 is even more valuable in helping to interpret a set of measurements. The figures 68%, 95%, and 99.7% are not to be taken literally. If you find 64% or 73% of the measurements within one standard deviation of the mean, you should not be startled. Indeed, it is possible to find 100% of the measurements within one standard deviation of the average.

EXERCISES FOR SECTION 6–4

Given that $\bar{x} = 0$ and $s_x = 1$ for a set of n measurements, use Chebyshev's theorem to fill in the following blanks:

1. The number of measurements that lie at or within 3 units of the mean is more than ______.
2. The number of measurements that lie between —2 and 2, or at —2 or 2, is greater than ______.
3. The number of measurements that will be greater than 2 or less than —2 is less than ______.
4. If $k \geq$ ______, then more than 96% of the measurements lie between, or at, $-k$ and k.

Let $\bar{x} = 5$ and $s_x = 2$, and use the empirical rule of Table 6–4 to answer the following:

5. About what percent of the measurements lie between 3 and 7, inclusive? Between 1 and 9, inclusive?
6. About what percent of the measurements are greater than 9 or less than 1?

Let $\bar{x} = 1$ and $s_x = 3$, and use Chebyshev's theorem to fill in the following blanks:

7. (a) The number of measurements that lie between —5 and 7, inclusive, is more than ______. (b) The number of measurements that lie between —8 and 10, inclusive, is more than ______.
8. (a) The number of measurements that are greater than 7 or less than —5 is less than ______. (b) The number of measurements that are greater than 10 or less than —8 is less than ______.
9. Seven measurements have a sample mean of 0 and a sample standard deviation of 3. Prove that a measurement of +8 could not have occurred.
10. A sample of seven measurements is said to have a mean of 10, standard deviation 3, largest and smallest measurements 16 and 4. Show that these results are inconsistent.

7 Joint distributions and continuous distributions

7–1 JOINT PROBABILITY FUNCTION OF TWO RANDOM VARIABLES

Several random variables may be associated with outcomes of the same experiment. In fact, if we wanted to study relations between weight, height, and age, we might let measures of weight, height, and age be three random variables whose values are determined by the experiment of measuring these characteristics of a person chosen at random from the population of a community. In Example 1, we look at a simpler problem involving two random variables whose values are numbers determined by tossing a coin 3 times in succession.

Table 7–1 Sample space for sequence of 3 coin tosses

Sample point	Number of heads, X	Number of runs, Y
$H\ H\ H$	3	1
$H\ H\ T$	2	2
$H\ T\ H$	2	3
$H\ T\ T$	1	2
$T\ H\ H$	2	2
$T\ H\ T$	1	3
$T\ T\ H$	1	2
$T\ T\ T$	0	1

Example 1 Let X be the number of heads and let Y be the number of runs when a coin is tossed 3 times in succession and each toss is recorded as H or T. Make tables showing possible pairs of values of X and Y and their corresponding probabilities.

Solution Table 7–1 shows a sample space for the experiment with the value of X and of Y for each point.

Table 7–2 Probabilities of paired values of X and Y from Table 7–1

		Values of X (number of heads)				
		0	1	2	3	Row totals
Values of Y (number of runs)	1	$\frac{1}{8}$			$\frac{1}{8}$	$\frac{2}{8}$
	2		$\frac{2}{8}$	$\frac{2}{8}$		$\frac{4}{8}$
	3		$\frac{1}{8}$	$\frac{1}{8}$		$\frac{2}{8}$
Column totals		$\frac{1}{8}$	$\frac{3}{8}$	$\frac{3}{8}$	$\frac{1}{8}$	1

Since X takes values 0, 1, 2, 3 and Y takes values 1, 2, 3, we set up a four-by-three array in Table 7–2 to display the probabilities of occurrences of the various combinations. We get each probability by counting the number of occurrences of that combination of values of X and Y and dividing the count by 8.

The entries in Table 7–2 give the values of the *joint probability function* of X and Y. Missing entries have the value 0. The entry $\frac{2}{8}$ that occurs in the column for $X = 1$ and the row for $Y = 2$ is the probability that $X = 1$ and, at the same time, $Y = 2$:

$$P(X = 1, Y = 2) = \tfrac{2}{8}.$$

The headings across the top of the table, together with the column totals at the bottom, provide the probability distribution for X by itself. Similarly, the probability distribution for values of Y can be obtained from the columns at the left and at the right. These are called *marginal* distributions, probably because they are found on the margins of tables. For this example, they are:

Probability distribution for X, the number of heads:

$f(x)$	$\frac{1}{8}$	$\frac{3}{8}$	$\frac{3}{8}$	$\frac{1}{8}$
x	0	1	2	3

Probability distribution for Y, the number of runs:

$g(y)$	$\frac{2}{8}$	$\frac{4}{8}$	$\frac{2}{8}$
y	1	2	3

A graph of the joint probability function of Table 7–2 is shown in Fig. 7–1.

7–1 Definition: Joint probability function

Let X and Y be two random variables. Suppose that the possible values of X are $x_1, x_2, \ldots, x_m$, and that the possible values of Y are $y_1, y_2, \ldots, y_n$.

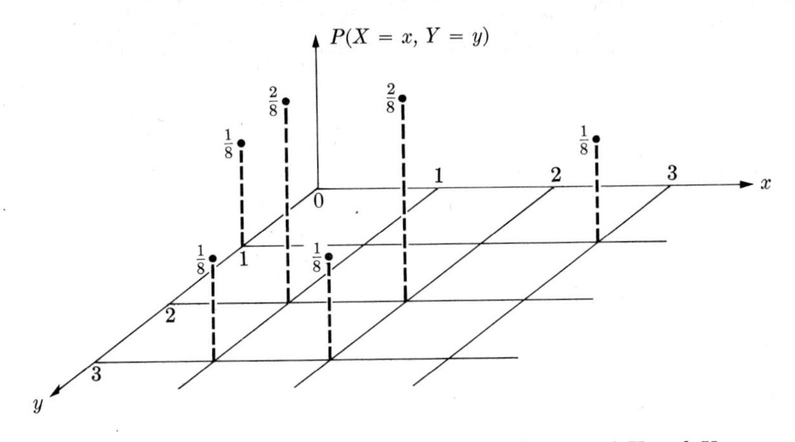

Fig. 7–1 Probability graph for joint distribution of X and Y.

For each ordered pair (x_i, y_j), let $P(x_i, y_j)$ be the probability that X takes the value x_i and Y takes the value y_j, where $i = 1, 2, \ldots, m$ and $j = 1, 2, \ldots, n$. Then $P(x_i, y_j)$ is the value of the joint probability function of X and Y at (x_i, y_j).

For the 3-coin example, the joint probability function of X and Y given by Table 7–2 tells more about the joint behavior of X and Y than we can get from their separate probability distributions. It shows that certain combinations have zero probability: for example, $X = 3$ and $Y = 3$. Suppose we let E be the event $X = 3$ and F the event $Y = 3$. Then, from Table 7–2, we see that

$$P(E \cap F) = 0,$$

while

$$P(E) = \tfrac{1}{8}, \qquad P(F) = \tfrac{2}{8}.$$

Thus

$$P(E) \cdot P(F) = \tfrac{1}{32}.$$

Therefore the events E and F are not independent. We also say that the random variables X and Y are not independent, because they fail to satisfy the general criterion for independence that we now state.

7–2 Definition: Independence

Let X and Y be two random variables over a discrete sample space S. Let the values of X be $x_1, x_2, \ldots, x_m$ and let the values of Y be $y_1, y_2, \ldots, y_n$. Then X and Y are statistically independent if and only if they satisfy the product rule:

$$\boxed{P(X = x_i \text{ and } Y = y_j) \quad = \quad P(X = x_i) \cdot P(Y = y_j)} \tag{1}$$

for all pairs of values (x_i, y_j).

Since Eq. (1) is not satisfied for $x_i = 3$, $y_j = 3$, in the 3-coin example, the random variables "number of heads" and "number of runs" are not independent.

The notion of *mathematical expectation* can be extended to functions of two or more random variables. We shall treat the expectation of a function H of two random variables X and Y. The function H is *not* ordinarily the probability function of X and Y. In particular, we need to study such functions as

$$H(X, Y) = X + Y,$$

or

$$H(X, Y) = XY,$$

or even

$$H(X, Y) = X,$$

because we shall frequently need their expectations in later chapters.

7-3 Theorem: Mathematical expectation of a function of two variables

Let $H(X, Y)$ be a function of the two variables X and Y over a discrete sample space S. Then the mathematical expectation of H, written $E[H(X, Y)]$, is equal to the sum of the products of all possible values $H(x_i, y_j)$ each multiplied by its probability $P(x_i, y_j)$:

$$\boxed{E[H(X, Y)] = \sum H(x_i, y_j) \cdot P(x_i, y_j),} \tag{2}$$

the sum being taken over all possible pairs (x_i, y_j).

Proof The proof is like that for Theorem 5-6 for a function of one random variable. Let $Z = H(X, Y)$, and note that as X ranges over its possible values $\{x_i\}$ and Y ranges over its possible values $\{y_j\}$, the set of possible values for Z is a discrete set $z_1, z_2, \ldots$ that is either finite or countably infinite. It is fair to assume that the elementary events in the original sample space consisted of the points (x_i, y_j). The sum of the probabilities of the elementary events that produce $Z = z_1$ gives the probability $P(Z = z_1)$, the sum for z_2 gives the probability $P(Z = z_2)$, and so on. Now apply the original definition of mathematical expectation, Definition 5-3, to the Z-distribution, and get

$$E(Z) = \sum_i z_i \cdot P(Z = z_i). \; \square$$

Example 2 Using the data of Table 7-2, find the mathematical expectations of (a) XY, (b) X/Y, (c) X, (d) Y, (e) $X + Y$.

Solution a) *Product* XY. Table 7-3 shows $x_i y_j P(x_i, y_j)$ in each cell where the probability is not zero. We omit the terms with zero probabilities, since they contribute nothing to the sum in Eq. (2).

We may sum the entries in Table 7-3 in any way we wish, provided we include all entries. The row sums are $\frac{3}{8}$, $\frac{12}{8}$, and $\frac{9}{8}$ and the sum of these is $\frac{24}{8}$ or 3.

Table 7–3 Values of XY and their probabilities, for data of Table 7–2 (empty cells have probability zero)

		Values of X				
		0	1	2	3	Row sums
	1	$0 \times \frac{1}{8}$			$3 \times \frac{1}{8}$	$\frac{3}{8}$
Values of Y	2		$2 \times \frac{2}{8}$	$4 \times \frac{2}{8}$		$\frac{12}{8}$
	3		$3 \times \frac{1}{8}$	$6 \times \frac{1}{8}$		$\frac{9}{8}$
Column sums		0	$\frac{7}{8}$	$\frac{14}{8}$	$\frac{3}{8}$	$\frac{24}{8} = 3$

As a check, we note that the column sums also add up to 3. Hence

$$E(XY) = 0 \times \tfrac{1}{8} + 3 \times \tfrac{1}{8} + 2 \times \tfrac{2}{8} + 4 \times \tfrac{2}{8} + 3 \times \tfrac{1}{8} + 6 \times \tfrac{1}{8}$$
$$= \tfrac{24}{8} = 3.$$

b) *Quotient* X/Y. We omit the table for values of X/Y, which, like Table 7–3, would be constructed from the data of Table 7–2. When we multiply each probability by the value of X/Y for that cell, and add, the result is

$$E(X/Y) = \tfrac{0}{1} \times \tfrac{1}{8} + \tfrac{3}{1} \times \tfrac{1}{8} + \tfrac{1}{2} \times \tfrac{2}{8} + \tfrac{2}{2} \times \tfrac{2}{8} + \tfrac{1}{3} \times \tfrac{1}{8} + \tfrac{2}{3} \times \tfrac{1}{8} = \tfrac{7}{8}.$$

c) and d) *Means of* X *and* Y. To compute $E(X)$, we may either work directly from Table 7–2 or from the probability distribution of X that we derived from it. The value of X remains constant for the cells in the same column of Table 7–2; we multiply the value of X in a cell by the probability in that cell, and then sum. The result is

$$E(X) = 0 \times \tfrac{1}{8} + 1 \times \tfrac{3}{8} + 2 \times \tfrac{3}{8} + 3 \times \tfrac{1}{8} = 1.5.$$

Similarly,

$$E(Y) = 1 \times \tfrac{2}{8} + 2 \times \tfrac{4}{8} + 3 \times \tfrac{2}{8} = 2.$$

e) *Sum* $X + Y$. We compute $E(X + Y)$ directly from Table 7–2. Again the value of $X + Y$ in each cell is multiplied by the probability in that cell, and the results are added:

$$\begin{aligned} E(X + Y) &= (0 + 1) \times \tfrac{1}{8} + (3 + 1) \times \tfrac{1}{8} \\ &+ (1 + 2) \times \tfrac{2}{8} + (2 + 2) \times \tfrac{2}{8} \\ &+ (1 + 3) \times \tfrac{1}{8} + (2 + 3) \times \tfrac{1}{8} \\ &= 3.5 = 1.5 + 2 = E(X) + E(Y). \end{aligned}$$

We might wonder whether equations like

$$E(XY) = E(X) \cdot E(Y)$$

and

$$E\left(\frac{X}{Y}\right) = \frac{E(X)}{E(Y)}$$

are true. We found that $E(XY) = 3$ and $E(X) \cdot E(Y) = 1.5 \times 2 = 3$, so it is true (in this example, *but not in general*) that $E(XY) = E(X) \cdot E(Y)$. But

$$E\left(\frac{X}{Y}\right) = \frac{7}{8} \neq \frac{1.5}{2} = \frac{E(X)}{E(Y)}.$$

The next theorem gives a general result about the mean of the sum of two random variables. The result, that the mean of the sum is the sum of the means, is not surprising; if every customer in a men's clothing store buys a hat, a suit, and a pair of shoes, then the average total price paid is the average paid for hats plus the average paid for suits plus the average paid for shoes.

7-4 Theorem: Mean of sum

Let X and Y be random variables with finite means $E(X)$ and $E(Y)$. Then the mean of their sum is the sum of their means:

$$\boxed{E(X + Y) = E(X) + E(Y).} \tag{3}$$

Proof By definition of the mean,

$$E(X + Y) = \sum(x_i + y_j) \cdot P(x_i, y_j), \tag{4a}$$

or

$$E(X + Y) = \sum x_i P(x_i, y_j) + \sum y_j P(x_i, y_j), \tag{4b}$$

the sums being taken over all pairs of values (x_i, y_j) that (X, Y) can have. In the proof, we shall find it convenient to use the marginal probability distributions

$$f(x_i) = \sum_j P(x_i, y_j) = P(X = x_i)$$

and

$$g(y_j) = \sum_i P(x_i, y_j) = P(Y = y_j).$$

For example, if X has values x_1, x_2, and Y has values y_1, y_2, y_3, then the first sum on the right side of Eq. (4b), written out in full, is

$$\begin{aligned}\sum x_i P(x_i, y_j) &= x_1 P(x_1, y_1) + x_1 P(x_1, y_2) + x_1 P(x_1, y_3)\\ &\quad + x_2 P(x_2, y_1) + x_2 P(x_2, y_2) + x_2 P(x_2, y_3)\\ &= x_1[P(x_1, y_1) + P(x_1, y_2) + P(x_1, y_3)]\\ &\quad + x_2[P(x_2, y_1) + P(x_2, y_2) + P(x_2, y_3)].\end{aligned} \tag{5}$$

The bracketed coefficient of x_1 is the sum of the probabilities of all pairs (x_1, y_j) that contain x_1. Hence this coefficient is just $P(X = x_1)$ or, more compactly, $f(x_1)$. Similarly, the bracketed coefficient of x_2 is $f(x_2)$. Therefore Eq. (5) says that

$$\Sigma x_i P(x_i, y_j) = x_1 f(x_1) + x_2 f(x_2), \tag{6a}$$

and this is $E(X)$ in the special case under discussion.

Similarly, if we expand the second term on the right of Eq. (4b), and combine terms that multiply y_1, y_2, and y_3, in order, we get

$$\Sigma y_j P(x_i, y_j) = y_1 g(y_1) + y_2 g(y_2) + y_3 g(y_3) \tag{6b}$$

and this is $E(Y)$ in the special case.

More generally, if X has m possible values $x_1, x_2, \ldots, x_m$ and Y has n possible values $y_1, y_2, \ldots, y_n$, the equations that replace Eqs. (6a, b) are

$$\Sigma x_i P(x_i, y_j) = x_1 f(x_1) + x_2 f(x_2) + \cdots + x_m f(x_m) \tag{7a}$$

and

$$\Sigma y_j P(x_i, y_j) = y_1 g(y_1) + y_2 g(y_2) + \cdots + y_n g(y_n). \tag{7b}$$

The right side of Eq. (7a) is the sum of the products of the possible values of X, each multiplied by its probability, and is therefore $E(X)$. Similarly, the right side of Eq. (7b) is $E(Y)$. When these results are introduced on the right side of Eq. (4b), we get the desired result,

$$E(X + Y) = E(X) + E(Y). \ \square$$

★Remark 1 The foregoing proof assumes that X and Y have a finite sample space. For a countably infinite sample space the sums in Eqs. (7a) and (7b) would become infinite series. The hypothesis that X has a finite mean then implies that the series of absolute values

$$\sum_{i=1}^{\infty} |x_i| \cdot f(x_i)$$

converges. A similar hypothesis applies to

$$\sum_{j=1}^{\infty} |y_j| \cdot g(y_j).$$

The convergence of these two infinite series is sufficient to guarantee convergence of the series

$$\sum |x_i + y_j| \cdot P(x_i, y_j)$$

[the sum being taken over all pairs of values (x_i, y_j) that (X, Y) can have], as well as the convergence of the two series on the right side of Eq. (4b).

Equation (3) can also be extended to the sum of three random variables X, Y, and Z:

$$E[(X+Y)+Z] = E(X+Y)+E(Z) = E(X)+E(Y)+E(Z). \tag{8}$$

More generally, the mean of the sum of any finite number of random variables is the sum of their means, provided each of the random variables in the sum has a finite mean. For future reference, we state this result formally as a corollary of Theorem 7-4.

7-5 Corollary: Mean of sums of random variables

The mean of the sum of any finite collection of random variables is the sum of their means, provided each of the random variables has a finite mean.

Remark 2 Each binomial trial of a binomial experiment produces either 0 or 1 success. Therefore each binomial trial can be thought of as producing a value of a random variable associated with that trial and taking the values 0 and 1, with probabilities q and p respectively. The several trials of a binomial experiment produce a new random variable X, the total number of successes, which is just the sum of the random variables associated with the single trials. For example, if a marksman takes 5 shots and gets two bull's-eyes, one on his third shot and one on his fifth, the numbers of successes on the 5 individual shots are 0, 0, 1, 0, 1. The number of successes on each shot is a value of a random variable that has values 0 or 1, and there are 5 such random variables here. Their sum is X, the total number of successes, which in this experiment has the value $x = 2$. We use this idea and Corollary 7-5 to give an alternative proof of Theorem 5-4. To this end, let X be the number of successes in a binomial experiment of n trials. In Theorem 5-4, we found that $E(X) = np$, by more or less direct calculations. We shall now use a less direct, but easier, way of finding $E(X)$. We introduce n random variables, called *counting variables*, one for each trial:

$$\begin{aligned}
X_1 &= \text{number of successes on trial 1}\\
X_2 &= \text{number of successes on trial 2}\\
X_3 &= \text{number of successes on trial 3}\\
&\vdots\\
X_n &= \text{number of successes on trial } n
\end{aligned}$$

What are the possible values of X_1? Obviously, only 0 or 1. If the first trial results in a success, then $X_1 = 1$. Otherwise, $X_1 = 0$. The associated probabilities are p and q, respectively. Therefore, the mean of X_1 is

$$E(X_1) = 1 \times p + 0 \times q = p.$$

The same is true of each of the other counting variables:

$$E(X_i) = p \qquad \text{for} \quad i = 1, 2, \ldots, n.$$

Now apply Corollary 7–5. The total number of successes on the n trials is exactly equal to the sum of the values of the counting variables:

$$X = X_1 + X_2 + \cdots + X_n.$$

Therefore the mean number of successes on the n trials is

$$E(X) = E(X_1 + X_2 + \cdots + X_n) = E(X_1) + E(X_2) + \cdots + E(X_n) = np.$$

Example 3 An ordinary six-sided die is thrown until each face has appeared at least once as the top face. What is the expected number of throws?

Solution We let X be the number of throws required for the six faces to appear, but represent X as the sum of six counting variables $X_1, X_2, \ldots, X_6$ as follows:

X_1 = number of throws for the first face to appear,

X_2 = additional number of throws until a face different from the first face appears (call this the second face),

X_3 = additional number of throws after the second face appears until a different face appears (call this the third face),

and so on. The first throw produces the first face, so $X_1 = 1$. Whatever that first face was, we consider a repetition as a failure (probability $\frac{1}{6}$) but any other face would represent a success (probability $p_2 = \frac{5}{6}$). Thus X_2 has a geometric distribution with $p_2 = \frac{5}{6}$. By Theorem 5–5, the mean of this distribution is

$$E(X_2) = \frac{1}{p_2} = \frac{6}{5}.$$

After the second face has appeared, the probability of getting a different face is $p_3 = \frac{4}{6}$. The random variable X_3 has a geometric distribution with mean

$$E(X_3) = \frac{1}{p_3} = \frac{6}{4}.$$

The pattern is now clear: when three different faces have appeared, the probability of getting a new face is $p_4 = \frac{3}{6}$ and the mean number of throws is

$$E(X_4) = \frac{1}{p_4} = \frac{6}{3}.$$

Similar reasoning gives the means for X_5 and X_6 as

$$E(X_5) = \frac{1}{p_5} = \frac{6}{2}, \qquad E(X_6) = \frac{1}{p_6} = \frac{6}{1}.$$

For the overall experiment, we therefore have

$$\begin{aligned} E(X) &= E(X_1 + X_2 + X_3 + X_4 + X_5 + X_6) \\ &= 1 + \tfrac{6}{5} + \tfrac{6}{4} + \tfrac{6}{3} + \tfrac{6}{2} + \tfrac{6}{1} = 14.7. \end{aligned}$$

Products The mean of the product of two random variables is not always equal to the product of their means, as the following example shows.

Example 4 *Two dependent random variables.* Suppose the joint probability function of X and Y is as follows:

		Values of Y 0	1	Row sums
Values of X	0	0	$\frac{1}{2}$	$\frac{1}{2}$
	1	$\frac{1}{2}$	0	$\frac{1}{2}$
Column sums		$\frac{1}{2}$	$\frac{1}{2}$	1

Compare $E(X \cdot Y)$ and $E(X) \cdot E(Y)$.

Solution

$$\begin{aligned}
E(XY) &= 0 \cdot 0 \cdot 0 + 0 \cdot 1 \cdot \tfrac{1}{2} + 1 \cdot 0 \cdot \tfrac{1}{2} + 1 \cdot 1 \cdot 0 = 0, \\
E(X) &= 0 \cdot \tfrac{1}{2} + 1 \cdot \tfrac{1}{2} = \tfrac{1}{2}, \\
E(Y) &= 0 \cdot \tfrac{1}{2} + 1 \cdot \tfrac{1}{2} = \tfrac{1}{2}; \\
E(XY) &= 0 \neq E(X) \cdot E(Y) = \tfrac{1}{4}.
\end{aligned}$$

We note in this example that we can predict X from Y and Y from X; for if $X = 1$ then $Y = 0$, and if $X = 0$ then $Y = 1$, and these are the only values possible. Dependence sometimes has the effect of improving the prediction of the value of one random variable from the known value of a dependent variable.

In Example 4 the random variables are dependent, and

$$E(XY) \neq E(X) \cdot E(Y).$$

In Example 2, the random variables were also dependent, but there we found that $E(XY)$ was equal to the product $E(X) \cdot E(Y)$. Thus, for dependent random variables, the mean of the product may or may not equal the product of the means. But for *independent* random variables, the following theorem is true.

7-6 Theorem: Mean of a product of independent random variables

Let X and Y be independent random variables with means $E(X)$ and $E(Y)$. Then the mean of their product is equal to the product of their means:

$$\boxed{E(XY) = E(X) \cdot E(Y).} \tag{9}$$

Proof As in the proof of Theorem 7-4, let f and g be the marginal probability functions for X and Y respectively. The condition for independence of X and Y is

$$P(x_i, y_j) = f(x_i) \cdot g(y_j), \tag{10}$$

for all pairs of values (x_i, y_j). From the definition of $E(XY)$, and Eq. (10), we have

$$E(XY) = \sum x_i y_j f(x_i) g(y_j). \tag{11}$$

Again the sum is to extend over all pairs (x_i, y_j). In particular, if x_i takes two values x_1 and x_2, and if y_j has three possible values y_1, y_2, and y_3, the sum in Eq. (11) is

$$\begin{aligned}\sum x_i y_j f(x_i) g(y_j) &= x_1 y_1 f(x_1) g(y_1) + x_2 y_1 f(x_2) g(y_1) \\ &\quad + x_1 y_2 f(x_1) g(y_2) + x_2 y_2 f(x_2) g(y_2) \\ &\quad + x_1 y_3 f(x_1) g(y_3) + x_2 y_3 f(x_2) g(y_3) \\ &= x_1 f(x_1)[y_1 g(y_1) + y_2 g(y_2) + y_3 g(y_3)] \\ &\quad + x_2 f(x_2)[y_1 g(y_1) + y_2 g(y_2) + y_3 g(y_3)]. \end{aligned} \tag{12}$$

The bracketed coefficient of $x_1 f(x_1)$ is the sum of the products of values of y_j each multiplied by its probability; by definition, this is $E(Y)$. Substituting $E(Y)$ for the two bracketed coefficients in Eq. (12), we get

$$\begin{aligned}\sum x_i y_j f(x_i) g(y_j) &= x_1 f(x_1) E(Y) + x_2 f(x_2) E(Y) \\ &= [x_1 f(x_1) + x_2 f(x_2)] E(Y). \end{aligned} \tag{13}$$

The bracketed coefficient here is $E(X)$, since we have assumed, temporarily, that X has just two values, x_1 and x_2. Thus, for the particular case where x_i has 2 values and y_j has 3 values, Eqs. (11) and (13) together give the desired result:

$$E(XY) = E(X) \cdot E(Y).$$

More generally, if x_i has m values $x_1, x_2, \ldots, x_m$, and y_j has n values $y_1, y_2, \ldots, y_n$, the expanded form of the sum in Eq. (12) contains $m \times n$ terms. But $x_1 f(x_1)$ is a common factor in n of these terms, the other factor being

$$y_1 g(y_1) + y_2 g(y_2) + \cdots + y_n g(y_n), \tag{14}$$

which equals $E(Y)$. Similarly, $x_2 f(x_2)$ is multiplied by the sum (14), and so are $x_3 f(x_3)$ and so on to $x_m f(x_m)$. Finally, the sum in Eq. (13) becomes

$$[x_1 f(x_1) + x_2 f(x_2) + \cdots + x_m f(x_m)] E(Y), \tag{15}$$

and the bracketed coefficient is $E(X)$. Hence the theorem is true in general for *independent* random variables in a finite sample space. It is also true in a countably infinite sample space, as a slight change in our proof would show. □

Table 7–4

		Values of X				
		1	2	3	4	Row totals
Values of Y	0	$\frac{1}{24}$	$\frac{1}{12}$	$\frac{1}{12}$	$\frac{1}{24}$	$\frac{1}{4}$
	1	$\frac{1}{12}$	$\frac{1}{6}$	$\frac{1}{6}$	$\frac{1}{12}$	$\frac{1}{2}$
	2	$\frac{1}{24}$	$\frac{1}{12}$	$\frac{1}{12}$	$\frac{1}{24}$	$\frac{1}{4}$
Column totals		$\frac{1}{6}$	$\frac{1}{3}$	$\frac{1}{3}$	$\frac{1}{6}$	1

Value of XY	0	1	2	3	4	6	8
Probability	$\frac{1}{4}$	$\frac{1}{12}$	$\frac{5}{24}$	$\frac{1}{6}$	$\frac{1}{6}$	$\frac{1}{12}$	$\frac{1}{24}$

Example 5 The joint probability distribution of two random variables is given in Table 7–4. Show that X and Y are independent. Compute $E(XY)$ directly and compare with $E(X) \cdot E(Y)$.

Solution Each cell entry in the main body of the table is the product of the corresponding row and column sums, so the random variables are obviously independent.

We have:

$$\begin{aligned} E(XY) &= 0 \times \tfrac{1}{4} + 1 \times \tfrac{1}{12} + 2 \times \tfrac{5}{24} + 3 \times \tfrac{1}{6} \\ &\quad + 4 \times \tfrac{1}{6} + 6 \times \tfrac{1}{12} + 8 \times \tfrac{1}{24} = \tfrac{5}{2}, \\ E(X) &= 1 \times \tfrac{1}{6} + 2 \times \tfrac{1}{3} + 3 \times \tfrac{1}{3} + 4 \times \tfrac{1}{6} = 2\tfrac{1}{2}, \\ E(Y) &= 0 \times \tfrac{1}{4} + 1 \times \tfrac{1}{2} + 2 \times \tfrac{1}{4} = 1, \end{aligned}$$

and

$$\tfrac{5}{2} = (2\tfrac{1}{2}) \times 1.$$

Example 6 A sequence of Bernoulli trials is made until the first *success* occurs, and is then continued until the first *failure* following that success occurs. Let X be the number of trials to and including the first success, and let Y be the number of trials after that first success up to and including the first failure following that success. Find the joint probability function of X and Y.

Solution Let p be the probability of a success and q the probability of a failure on each trial. Then X and Y have geometric distributions as follows:

$$\begin{aligned} P(X = x) &= q^{x-1}p, \qquad x = 1, 2, 3, 4, \ldots, \\ P(Y = y) &= p^{y-1}q, \qquad y = 1, 2, 3, 4, \ldots. \end{aligned}$$

Moreover, since the trials are independent and the outcome of the second part of the experiment, which yields a value of Y, has nothing to do with the outcome of the first part, X and Y are independent. Therefore,

$$P(X = x, Y = y) = q^{x-1}p \cdot p^{y-1}q = q^x p^y, \qquad x = 1, 2, \ldots; \quad y = 1, 2, \ldots.$$

We verify that the sum of these probabilities is 1 as follows. Holding x fixed, we sum the probabilities $q^x p^y$ on y from 1 to ∞ and get

$$\sum_{y=1}^{\infty} q^x p^y = q^x(p + p^2 + p^3 + \cdots) = q^x p/(1 - p) = pq^{x-1},$$

because the $1 - p$ in the denominator cancels one q factor in the numerator. Then we sum the new geometric progression pq^{x-1} on x from 1 to ∞ and get

$$\sum_{x=1}^{\infty} pq^{x-1} = p(1 + q + q^2 + \cdots) = p/(1 - q) = p/p = 1.$$

Therefore the sum over all possible (x, y) is 1, as it should be.

We also know from Theorem 5–5 that the mean of the X-distribution is $E(X) = 1/p$ and the mean of the Y-distribution is $E(Y) = 1/q$. Since X and Y are independent, $E(XY) = E(X) \cdot E(Y) = 1/pq$.

EXERCISES FOR SECTION 7–1

1. Refer to Table 7–4. (a) Show that $P(XY = 2) = \frac{5}{24}$ and $P(XY = 4) = \frac{1}{6}$. (b) Compute $P(X \leq 2 | Y \geq 1)$. (c) Find $P(X + Y = 3)$. (d) Compute $P(X/Y = 1)$.
2. Compare $E(XY)$ with $E(X) \cdot E(Y)$ and compare $E(X + Y)$ with $E(X) + E(Y)$ if X and Y have the following joint probability distribution:

		Values of X		
		0	1	Row totals
Values of Y	0	0	$\frac{1}{4}$	$\frac{1}{4}$
	1	$\frac{1}{3}$	$\frac{5}{12}$	$\frac{3}{4}$
Column totals		$\frac{1}{3}$	$\frac{2}{3}$	1

3. In Exercise 2 above, let $Z = XY$ and construct the probability distribution of Z. (That is, what are the values that Z may have and what are their probabilities?) Compute $E(Z)$ from this distribution.
4. Construct the probability function of $W = X + Y$ from the table in Exercise 2. Compute $E(W)$ from this function.

5. *Prediction and dependence.* Show that X and Y are dependent if they have the following joint probability distribution:

		Values of X		
		0	1	Row totals
Values of Y	0	0.1	0.3	0.4
	1	0.4	0.2	0.6
Column totals		0.5	0.5	1.0

6. Look at the marginal probability distribution of Y in Exercise 5. Since the probability of 1 is greater than the probability of 0, in the absence of any other information our prediction of Y would be 1, and we would be right 60% of the time. (a) Find the conditional probabilities

$$P(Y = 0|X = 0), \quad P(Y = 1|X = 0), \quad P(Y = 0|X = 1), \quad P(Y = 1|X = 1).$$

(b) What value of Y would you predict on the basis of these probabilities, knowing that $X = 0$? that $X = 1$? (c) Since the possible values $X = 0$, $X = 1$ occur with equal probabilities 0.5 and 0.5, what is the expected percent of correct predictions of values of Y that you would make, given X?

7. (Continuation.) In Exercise 6 above, reverse the roles of X and Y. Can you do better at predicting X, given Y, than you could without knowledge of Y? Compare the expected percents of correct predictions, knowing, and not knowing, the value of Y.

8. (Continuation.) Would there be an improvement in predicting X, given Y, if their joint distribution were as follows?

		Values of X		
		0	1	
Values of Y	0	0.4	0.3	0.7
	1	0.2	0.1	0.3
		0.6	0.4	1.0

9. (Continuation.) In Exercise 8 above, would there be an improvement in predicting Y, given X?

10. In the 3-coin example, Table 7–2 gives the frequency distribution for paired values of X and Y, number of heads and number of runs. (a) For which value (or values) of X is the corresponding value of Y completely determined? (b) Is there a value of Y for which the corresponding value of X is determined? (c) With no information about the value of Y, what is the probability that X is 2? (d) Given that

$Y = 1$, what is the probability that $X = 2$? (e) Given that $Y = 2$, what is the probability that $X = 2$?

11. A motorist tries to unlock the door of his car in the dark. He has 4 keys in his pocket, and only one of them unlocks the car. He tries the keys, selecting one after another at random without replacement. What is the expected number of keys that he will try before he gets the door unlocked?

12. Do Exercise 11 assuming that the motorist samples with replacement. (The sample space is infinite, since he might never find the key.)

13. a) In Example 6, show that $E(X + Y) = E(XY)$.
    b) For the same example, show that $E(X + Y) \geq 4$ for all p and q, $0 < p < 1$, $q = 1 - p$.

14. In the isolates problem of Example 4, Section 3–8, we obtained an estimate of the mean number of isolates. Use the theory of Section 7–1 to calculate exactly the theoretical mean number of isolates.

7–2 PROBABILITY GRAPHS FOR CONTINUOUS RANDOM VARIABLES: PROBABILITIES REPRESENTED BY AREAS

The random variables discussed up to now have had a finite or countably infinite number of possible values, each with positive or zero probability. These earlier distributions all had the features that each admissible point in the sample space had a probability either strictly positive or zero, and that the sum of the strictly positive probabilities was 1. Such distributions are called *discrete* distributions, to distinguish them from the *continuous* distributions which we now introduce.

In physical measurement problems, for example, it is convenient to think of the random variable as continuous. Length and time are typical of measurements that can, theoretically, take every possible value in an interval, however crudely we may choose to round off the values in practice. Thus we associate possible values of these measures with all the points on an interval of the real line, the continuum. Here we want to assign probabilities not in chunks, as in discrete distributions, but more smoothly. An example will show the need.

Imagine the second hand on an electric clock that stops at your blindfolded command. What is the chance that it stops *exactly* at 10 seconds after a full minute? By "exactly", we mean 10.000 . . . seconds with zeros carried on forever. Obviously, there is no chance at all. And the same thing is true for every possible number of seconds between 0 and 60. Yet the clock must stop somewhere; the probability that it stops is 1. Thus if we tried to add probabilities as usual, for one point at a time over the 60 seconds, we would have nothing but zeros to add, and they cannot add to 1, though they must.

But this difficulty is not a new one. The area of a square one unit on a side is 1. Yet the area of every interior line segment perpendicular to its base is zero. Hence we cannot get the area of the square by adding areas of line segments. Nor can we get probability over the 60-second time interval by adding probabilities of individual points.

The notion of areas signals a way around this problem. We assign probabilities to intervals, rather than to single points, and we do it by representing probabilities as areas over intervals.

Example 1 *Stopped clock.* If a clock stops at a random time, what is the probability that the hour hand stops between the numerals 1 and 5?

Solution The space between the numerals 1 and 5 is $\frac{4}{12}$ of the circumference, so most people would say the probability is $\frac{1}{3}$.

Example 2 *Cut string.* A child plays with a pair of scissors and a piece of string 8 inches long. He cuts the string in two. What is the probability that the longer piece is at least 6 inches long?

Solution We imagine the string laid out along the x-axis from 0 to 8. If the cut falls between 0 and 2 or between 6 and 8, the longer piece will be at least 6 inches long. The combined lengths of these two intervals is 4 inches, so most people would say the probability is $\frac{4}{8}$ or $\frac{1}{2}$.

Each of these examples illustrates a random variable whose possible values form a continuous interval. In the clock example, the random variable T is a number from 0 through 12 indicating the instant when the clock stopped. Any real value of T between 0 and 12 is possible. There are infinitely many possibilities; we wish to treat them all alike in assigning probabilities, but there is a difficulty that we haven't met before. We can't count equally likely cases, and probabilities have not been assigned.

However, there is a way out of the difficulty. We assign to each *interval* of values of T between 0 and 12 a probability proportional to the length of the interval. Since the entire interval from 0 to 12 has probability 1, an interval of unit length is assigned probability $\frac{1}{12}$, and an interval of length 4 is assigned probability $\frac{4}{12}$. Thus, as we said in the solution to Example 1, the probability that T will fall between 1 and 5 is $\frac{4}{12}$:

$$P(1 \leq T \leq 5) = \frac{5-1}{12-0} = \frac{4}{12}.$$

Likewise, in the string-cutting example, the random variable X represents the length of the piece cut off between the end at 0 and the cut. Again, since X is to have the same chances of falling in any two intervals of equal length, we assign to each interval a probability proportional to its length. The entire interval from 0 to 8 has probability 1; any shorter interval, say of length d, is assigned probability $d/8$. The longer piece of string is at least 6 inches long if $0 \leq X \leq 2$ or $6 \leq X \leq 8$, and since these are mutually exclusive intervals their probabilities add:

$$\begin{aligned} P(0 \leq X \leq 2 \quad \text{or} \quad 6 \leq X \leq 8) &= P(0 \leq X \leq 2) + P(6 \leq X \leq 8) \\ &= \frac{2-0}{8-0} + \frac{8-6}{8-0} = \frac{1}{2}. \end{aligned}$$

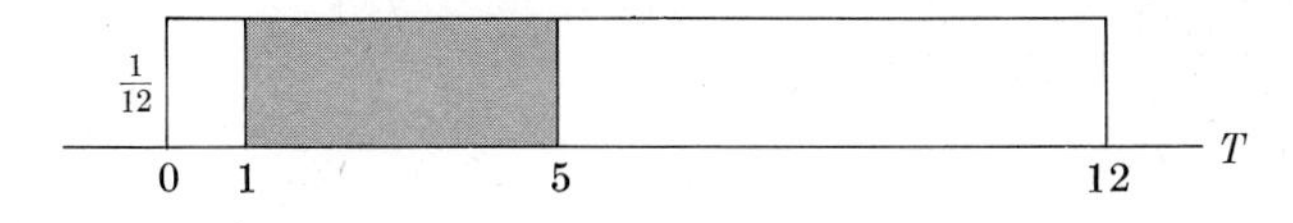

Fig. 7–2 Area probability graph for stopped clock.

We wish to represent such results graphically. We do so by representing probabilities by *areas*.

Clock example: area probability graph We return now to the clock example, and construct what we shall call its *area probability graph*. The total area under this graph is to be 1. Its base should extend from 0 to 12 along the T-axis. Because the clock can stop at random at any point between 0 and 12, we want to assign equal probabilities to any two time intervals that have equal widths. If their probabilities are to be measured by areas, the altitudes of the corresponding rectangles should have equal lengths. Thus, since the total area under it is to be 1, the area graph is bounded above by a line parallel to the T-axis and at distance $\frac{1}{12}$ above the axis. Figure 7–2 shows the area graph. The shaded area represents the probability that the clock stops between 1 and 5. The area of the rectangle is $(5 - 1)/12 = \frac{4}{12}$.

String example The area probability graph for the string-cutting example is shown in Fig. 7–3. The base extends along the X-axis from 0 to 8 and the height above the axis is $\frac{1}{8}$. Thus the total area is equal to 1. The shaded areas represent the probability that the longer piece of string is at least 6 inches long. The sum of these two areas is $\frac{1}{2}$, which is the desired probability.

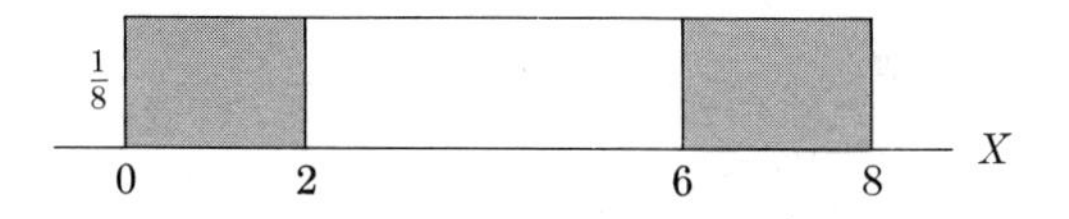

Fig. 7–3 Area probability graph for the string example.

Uniform distribution Both the clock and the cut-string examples illustrate the area probability graph of a *uniform distribution*. Continuous uniform distributions spread probability smoothly between two numbers a and b.

Example 3 *Commuter*. A commuter who drives to work from his home in the suburbs tries to reach his office by 9 o'clock. Because of fluctuations in traffic and other factors, he actually arrives between 8:45 and 9:05. The relative frequencies of his various arrival times suggest to him that the area probability graph for his time of arrival is approximated very well by an isosceles triangle. If this is true, what is the probability that he will arrive at work on time?

Solution Let his time of arrival be represented by T, a random variable. To simplify the writing, let $T = 0$ correspond to 8:55, the midpoint of his arrival times. Then his area probability graph is an isosceles triangle with base from $T = -10$ to $T = +10$, and vertex above $T = 0$ at a distance h that will make the total area of the triangle equal to 1: if $\frac{1}{2} \cdot h \cdot 20 = 1$, then $h = \frac{1}{10}$.

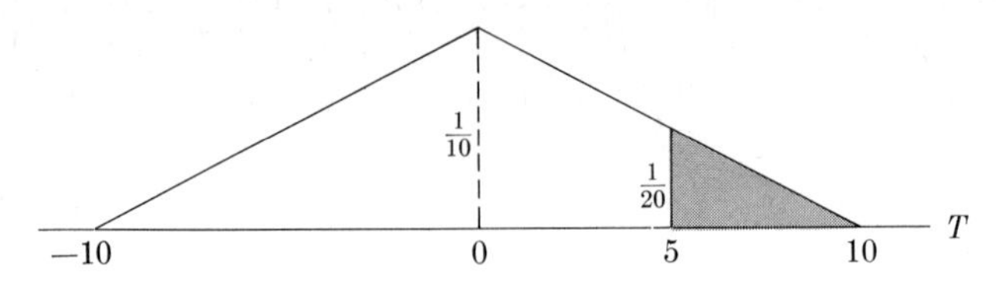

Fig. 7–4 Area probability graph for commuter's time of arrival at work.

The graph is shown in Fig. 7–4. The area of the small shaded triangle represents the probability of his arriving *after* 9 o'clock; the altitude is $\frac{1}{20}$, and the shaded area is $(\frac{1}{2})5(\frac{1}{20}) = \frac{1}{8}$. Therefore the probability that he will arrive on time is $1 - \frac{1}{8} = \frac{7}{8}$.

EXERCISES FOR SECTION 7–2

1. In the stopped-clock example, what is the probability that the clock stops between 8:13 and 9:45?
2. In the cut-string example, what is the probability that the shorter piece is at least 2 inches long? 3 inches? 3.9 inches? π inches?
3. In the cut-string example, what is the probability that the longer piece is (a) at least twice as long as the shorter? (b) At least a inches, where a is a fixed number such that $4 < a < 8$?
4. In the commuter example, what is the probability that he arrives at least 10 minutes early? Not more than 5 minutes early or late?

7–3 CUMULATIVE PROBABILITY GRAPHS

If 2 dice are thrown, what is the probability that the sum is less than or equal to 5? If 3 coins are tossed, what is the probability that at most 2 show heads? If a whole number between 1 and 10 is chosen at random, what is the probability that it has at most 3 divisors?

Questions like these have the general form: What is the probability that a random variable takes on a value less than or equal to a prescribed number x? Answers to such questions are given by the *cumulative distribution function*, which applies to continuous sample spaces as well as to those with a finite or countably infinite number of points.

7–7 Definition: Cumulative distribution function (cdf)

Let X be a random variable, x a real number, and $F(x)$ the probability that X takes on values less than or equal to x:

$$F(x) = P(X \leq x). \tag{1}$$

Then the function F defined by Eq. (1) is called the cumulative distribution function of X.

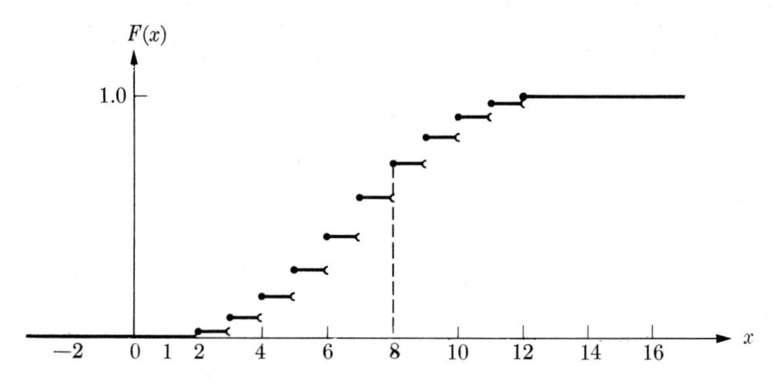

Fig. 7–5 Two dice. Graph of cumulative distribution function.

Table 7–5 Two dice: probability function f and cumulative distribution F

x	<2	2	3	4	5	6	7	8	9	10	11	12	>12
$f(x)$	0	$\frac{1}{36}$	$\frac{2}{36}$	$\frac{3}{36}$	$\frac{4}{36}$	$\frac{5}{36}$	$\frac{6}{36}$	$\frac{5}{36}$	$\frac{4}{36}$	$\frac{3}{36}$	$\frac{2}{36}$	$\frac{1}{36}$	0
$F(x)$	0	$\frac{1}{36}$	$\frac{3}{36}$	$\frac{6}{36}$	$\frac{10}{36}$	$\frac{15}{36}$	$\frac{21}{36}$	$\frac{26}{36}$	$\frac{30}{36}$	$\frac{33}{36}$	$\frac{35}{36}$	$\frac{36}{36}$	1

Example 1 *Two dice.* Table 7–5 shows the possible total scores x when 2 dice are thrown, together with their probabilities $f(x)$ and the cumulative probabilities $F(x)$ as given by Eq. (1). The cumulative distribution function has a value, defined by Eq. (1), for every real number x, not just those listed in Table 7–5. For example,

$$F(3.7) = P(X \leq 3.7) = f(2) + f(3)$$
$$= F(3) = \tfrac{3}{36}.$$

In fact, if x is any number greater than or equal to 3 and less than 4, $3 \leq x < 4$, then $F(x) = F(3) = \frac{3}{36}$. The graph of F, shown in Fig. 7–5, consists of a sequence of horizontal line segments and two rays. One ray coincides with that part of the x-axis to the left of $x = 2$, because $P(X \leq x)$ is 0 if $x < 2$. There is a jump in the graph at $x = 2$ because $F(x) = 0$ to the left of 2 and $F(2) = \frac{1}{36}$. The graph has constant height $F(x) = \frac{1}{36}$ for $2 \leq x < 3$. At $x = 3$, there is a jump amounting to $f(3) = \frac{2}{36}$ and bringing $F(x)$ up to $F(3) = \frac{3}{36}$. As we proceed along the graph from left to right, we find a jump at each integer $x = 2, 3, 4, 5, 6, \ldots, 12$. For $x \geq 12$, the graph has another ray extending indefinitely to the right because $F(x) = P(X \leq x)$ is 1 if x is any real number greater than or equal to 12.

The heavy dots on the horizontal line segments, above $x = 2, 3, 4$ and so on, indicate that these line segments are closed at their left ends. The small

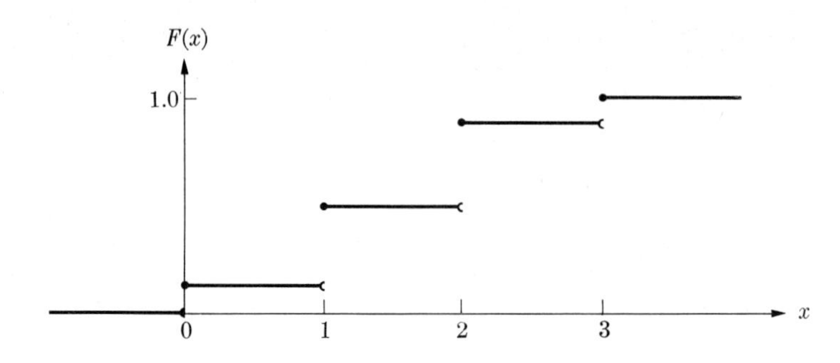

Fig. 7-6 Three coins. Cumulative distribution function for number of heads.

Table 7-6 Three coins: number of heads, probability function, and cumulative distribution function

x	<0	0	1	2	3	>3
$f(x)$	0	$\frac{1}{8}$	$\frac{3}{8}$	$\frac{3}{8}$	$\frac{1}{8}$	0
$F(x)$	0	$\frac{1}{8}$	$\frac{4}{8}$	$\frac{7}{8}$	1	1

open crescents at the right ends of the horizontal segments indicate that the segments are open at their right ends—they do not contain their right endpoints. The vertical dashed line through $x = 8$ indicates that the value of $F(8)$ is to be read from the left end of the higher horizontal segment where the heavy dot appears.

Example 2 *Three coins.* If three coins are tossed, the number of heads that show is a random variable X. Its probability function and cumulative distribution function are shown in Table 7-6.

The graph of the cumulative distribution function is shown in Fig. 7-6. Again we see that jumps occur at $x = 0$, 1, 2, and 3. The jump at $x = 0$ is $f(0) = \frac{1}{8}$, at 1 it is $f(1) = \frac{3}{8}$, at 2 it is $f(2) = \frac{3}{8}$, and at 3 it is $f(3) = \frac{1}{8}$. For negative x's, $F(x) = 0$; for $x > 3$, $F(x) = 1$.

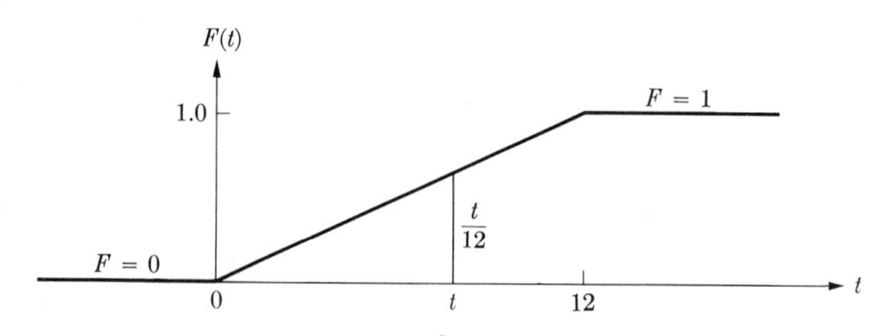

Fig. 7-7 Stopped clock. Cumulative distribution function for T.

Example 3 *Stopped clock.* The area probability graph is shown in Fig. 7–2. The cumulative distribution function is defined by

$$F(t) = P(T \leq t). \tag{2}$$

Discussion The probability given by Eq. (2) is 0 for $t < 0$ and 1 for $t > 12$. For t between 0 and 12, the *probability* is given by the area of a rectangle of base from 0 to t and of altitude $\frac{1}{12}$; hence the area is $t/12$. Therefore F, the cumulative distribution function, may be described as follows:

$$F(t) = \begin{cases} 0, & \text{if} \quad t < 0, \\ t/12, & \text{if} \quad 0 \leq t \leq 12, \\ 1, & \text{if} \quad t > 12. \end{cases} \tag{3}$$

The graph of F is shown in Fig. 7–7. To find the probability that the clock stopped between 0 and 3, we use $F(3) = \frac{3}{12}$. The probability that it stopped between 1 and 5 is $F(5) - F(1) = \frac{5}{12} - \frac{1}{12} = \frac{1}{3}$.

Example 4 *Commuter.* Figure 7–8 shows the area probability graph for the time before or after 8:55 that Mr. Commuter arrives at work. We now study the cumulative distribution function F defined by

$$F(t) = P(T \leq t). \tag{4}$$

We recall that T here represents the number of minutes from 8:45 until Mr. C. arrives at work.

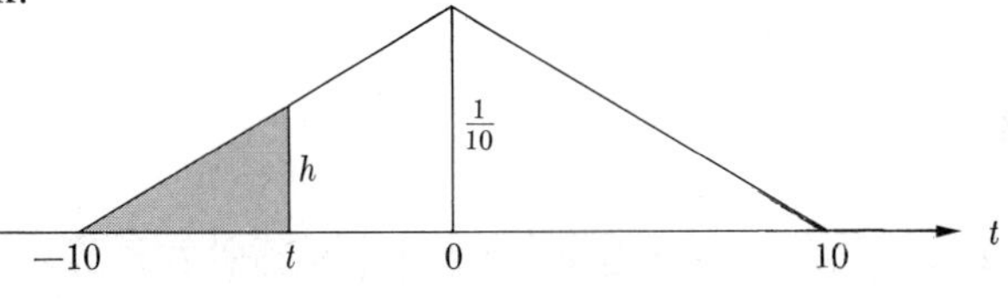

Fig. 7–8 Commuter example. Area probability graph.

For t less than -10, the probability given by (4) is zero. For a value of t between -10 and 0, the probability is given by the area of a triangle like the one shown shaded in Fig. 7–8. From similar triangles, we have

$$\frac{h}{t + 10} = \frac{1/10}{10} = \frac{1}{100}.$$

Hence

$$h = \frac{t + 10}{100}. \tag{5}$$

Check: If $t = -10$, Eq. (5) gives $h = 0$; if $t = 0$, $h = \frac{1}{10}$. These results agree with what we know is correct, and since (5) is linear, it is also correct for

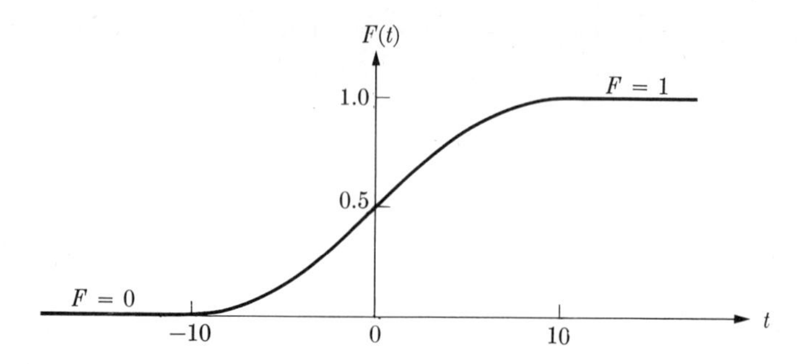

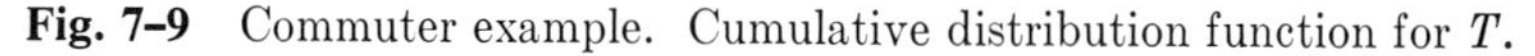
Fig. 7–9 Commuter example. Cumulative distribution function for T.

$-10 < t < 0$. Therefore the area of the shaded triangle is $\frac{1}{2}(t+10)h$, or

$$F(t) = \frac{\frac{1}{2}(t+10)(t+10)}{100} = \frac{(t+10)^2}{200}, \qquad -10 \le t \le 0. \tag{6}$$

For t between 0 and 10, compute the area of the small triangle to the right of t, and subtract the result from 1. We omit the calculation, which again makes use of similar triangles. The result is

$$F(t) = 1 - \frac{(10-t)^2}{200}, \qquad 0 \le t \le 10. \tag{7}$$

For $t > 10$, $F(T) = 1$. Combining the several bits of information about $F(t)$, we have the following description of F:

$$F(t) = \begin{cases} 0, & \text{for } t < -10, \\ \dfrac{(t+10)^2}{200}, & \text{for } -10 \le t \le 0, \\ 1 - \dfrac{(10-t)^2}{200}, & \text{for } 0 \le t \le 10, \\ 1, & \text{for } t > 10. \end{cases} \tag{8}$$

The graph of F is shown in Fig. 7–9.

For example, $F(5) = 1 - (10-5)^2/200 = 1 - \frac{1}{8} = \frac{7}{8}$, and this is the probability that he arrives at a time $T \le +5$, i.e., by 9 o'clock. Similarly, $F(10) - F(0) = 1 - \frac{1}{2} = \frac{1}{2}$ is the probability that he arrives between 8:55 and 9:05.

EXERCISES FOR SECTION 7–3

1. In Example 2, for 3 coins, what is the probability that the number of heads is between $\sqrt{2}$ and π? Is it equal to $F(\pi) - F(\sqrt{2})$?

2. In Example 1, for 2 dice, what is the probability that the sum is between 2.3 and 9.4? Is it equal to $F(9.4) - F(2.3)$?

3. Let X be a random variable whose values are all between 0 and 10 and let F be its cumulative distribution function. What is the numerical value of $F(11) - F(-1)$? Why?
4. Sketch the graph of the cdf for the number of divisors of an integer selected at random from 1 through 10. (See Table 5-5.) Where does it have jumps, and how big are they?
5. For its area probability graph, a random variable X has a right triangle of base 5 running from $X = 2$ to $X = 7$. The hypotenuse slopes upward to the right. Sketch its graph and find (a) $P(X \leq 4)$, (b) $P(X > 5)$, and (c) $F(x) = P(X \leq x)$ in these three cases:

$$\text{i) } x < 2, \qquad \text{ii) } 2 \leq x \leq 7, \qquad \text{iii) } x > 7.$$

7-4 THE NORMAL CURVE AND THE NORMAL PROBABILITY DISTRIBUTION

The most important probability distribution in the whole field of probability and statistics is the normal probability distribution. In the present section, we shall describe what this normal probability distribution is, how we use it, and why it is so important.

What is the normal probability distribution? It is a special way of assigning probabilities to intervals of real numbers associated with continuous random variables. These probabilities are assigned by means of a special curve, called the *normal curve*, and are related to a special kind of random variable, called the *standard normal random variable*. The procedure is clarified by the following definition.

7-8 Definition: Standard normal random variable

Let X be a random variable whose possible values are the real numbers between $-\infty$ and $+\infty$. Then X is called a *standard normal random variable* if the probability assigned to the interval from a to b is the area from a to b between the x-axis and the *normal curve*, whose equation is

$$y = \frac{1}{\sqrt{2\pi}} e^{-x^2/2}. \tag{1}$$

Graph of the standard normal curve The graph of Eq. (1) can be obtained in the usual way by constructing a table of values. We recognize π as an old friend:

$$\pi = 3.14159 \ldots.$$

The number e (like π, an important irrational number) is the base of the natural logarithms:

$$e = 2.71828 \ldots.$$

Table 7–7 Coordinates of points on the standard normal curve,

$$y = \frac{1}{\sqrt{2\pi}} e^{-x^2/2}$$

x	y	x	y	x	y
0.00	.3989	±1.50	.1295	±3.00	.0044
±0.25	.3867	±1.75	.0863	±3.25	.0020
±0.50	.3521	±2.00	.0540	±3.50	.0009
±0.75	.3011	±2.25	.0317	±3.75	.0004
±1.00	.2420	±2.50	.0175	±4.00	.0001
±1.25	.1826	±2.75	.0091		

With these facts in mind, a table of logarithms is sufficient to enable us to compute the y that matches a given value of x. Table 7–7 gives coordinates of points on the normal curve for values of x between -4.00 and $+4.00$, at intervals of 0.25.

If we plot the points (x, y) whose coordinates are given in Table 7–7 for values of x between -3.00 and $+3.00$ and then draw a smooth curve through them, we get the normal curve shown in Fig. 7–10.

Properties of the standard normal curve From an inspection of the graph in Fig. 7–10, we note that the normal curve:

a) is symmetric about the y-axis;
b) has its highest point at $(0, 1/\sqrt{2\pi})$, where $1/\sqrt{2\pi} \approx 0.40$;
c) is concave downward between $x = -1$ and $x = +1$, and concave upward for values of x outside that interval;
d) extends without limit to the left and to the right, and approaches the x-axis very rapidly as we move away from $x = 0$ in either direction.

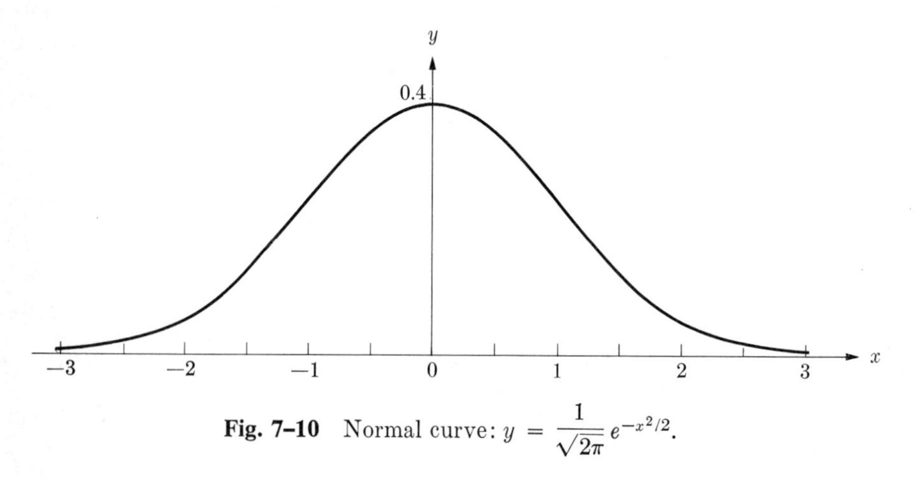

Fig. 7–10 Normal curve: $y = \dfrac{1}{\sqrt{2\pi}} e^{-x^2/2}$.

We shall accept on faith a further property of the normal curve that can be proved by advanced calculus:

e) the total area under the curve and above the x-axis equals 1.

Property (e) corresponds to the fact that 1 is the probability that the standard normal random variable X takes a value between $-\infty$ and $+\infty$. Properties (a) and (e) imply that the area below the curve and to the left of the y-axis is $\frac{1}{2}$, and so is the area to the right of the y-axis.

How do we use the normal probability distribution? The definition tells us what probabilities are to be assigned to the standard normal random variable. For example, the probability that X takes a value in the interval from $x = a$ to $x = b$ is equal to the measure of the area bounded by the normal curve, the x-axis, and the vertical lines $x = a$ and $x = b$ as indicated by the shaded region in Fig. 7–11.

How do we determine the measures of such areas? Recall that the total area under the normal curve is 1. Partial areas of the type shown in Fig. 7–11 can be approximated by rectangles. However, in practice we shall use Table 7–8 (which gives areas from 0 to x at intervals of 0.1, where $0 \leq x \leq 4.0$) or the larger Table III at the back of the book.

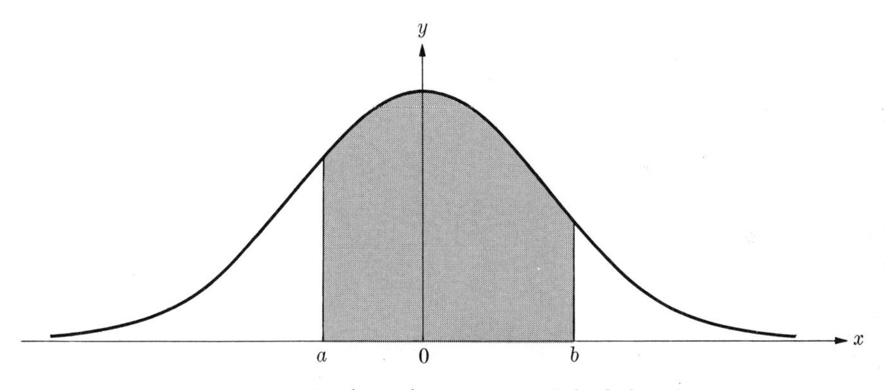

Fig. 7–11 $P(a \leq X \leq b)$ = area of shaded region.

The following examples show how Table 7–8 is used to find probabilities.

Example 1 What is the probability that a standard normal random variable takes a value between 0 and 1?

Solution $P(0 < X < 1) = A(1) = 0.3413$. About 34% of the total probability is between 0 and 1 and, by symmetry, about 68% is between -1 and $+1$.

Example 2 What is the probability that a standard normal random variable takes a value between -2 and $+2$?

Table 7–8 Area under the standard normal curve from 0 to x, shown shaded, is $A(x)$

$$y = \frac{1}{\sqrt{2\pi}} e^{-x^2/2}$$

x	Area, $A(x)$	x	Area, $A(x)$
0.0	.0000	2.1	.4821
0.1	.0398	2.2	.4861
0.2	.0793	2.3	.4893
0.3	.1179	2.4	.4918
0.4	.1554	2.5	.4938
0.5	.1915	2.6	.4953
0.6	.2257	2.7	.4965
0.7	.2580	2.8	.4974
0.8	.2881	2.9	.4981
0.9	.3159	3.0	.4987
1.0	.3413	3.1	.4990
1.1	.3643	3.2	.4993
1.2	.3849	3.3	.4995
1.3	.4032	3.4	.4997
1.4	.4192	3.5	.4998
1.5	.4332	3.6	.4998
1.6	.4452	3.7	.4999
1.7	.4554	3.8	.4999
1.8	.4641	3.9	.5000
1.9	.4713	4.0	.5000
2.0	.4772		

Solution Since the normal curve is symmetric about the y-axis, the area from -2 to $+2$ is twice the area from 0 to 2:

$$\begin{aligned} P(-2 < X < 2) &= (\text{area from } x = -2 \text{ to } x = +2) \\ &= 2(\text{area from } x = 0 \text{ to } x = +2) \\ &= 2A(2) \\ &= 2(0.4772) = 0.9544. \end{aligned}$$

Slightly more than 95% of the total area lies over the interval between -2 and $+2$.

Example 3 What is the probability that a standard normal random variable takes a value between 0.3 and 3.2?

Solution The required probability is the area between 0.3 and 3.2, and this is the area from 0 to 3.2 minus the area from 0 to 0.3. That is,

$$\begin{aligned} P(0.3 \leq X \leq 3.2) &= A(3.2) - A(0.3) \\ &= 0.4993 - 0.1179 = 0.3814. \end{aligned}$$

Example 4 Find $P(-0.3 \leq X \leq 3.2)$.

Solution The desired probability is the area from -0.3 to 0 plus the area from 0 to 3.2. The area from -0.3 to 0 is the same as the area from 0 to $+0.3$, $A(0.3)$. Hence

$$\begin{aligned} P(-0.3 \leq X \leq 3.2) &= A(0.3) + A(3.2) \\ &= 0.1179 + 0.4993 = 0.6172. \end{aligned}$$

Example 5 Find $P(-3.2 \leq X \leq -0.3)$.

Solution Since the normal curve is symmetric about the y-axis, the area from -3.2 to -0.3 is the same as the area from $+0.3$ to $+3.2$, found in Example 3. Hence

$$\begin{aligned} P(-3.2 \leq X \leq -0.3) &= P(+0.3 \leq X \leq +3.2) \\ &= A(3.2) - A(0.3) \\ &= 0.3814. \end{aligned}$$

Example 6 Find

a) $P(X > 0.3)$, b) $P(X > -0.3)$, c) $P(|X| > 0.3)$.

Solutions a) The area to the right of 0.3 is 0.5000 minus the area from 0 to 0.3:

$$\begin{aligned} P(X > 0.3) &= 0.5000 - P(0 \leq X \leq 0.3) \\ &= 0.5000 - 0.1179 = 0.3821. \end{aligned}$$

b) The area to the right of -0.3 is equal to the area from -0.3 to 0 plus 0.5000; hence

$$\begin{aligned} P(X > -0.3) &= P(-0.3 < X \leq 0) + 0.5 \\ &= P(0 \leq X < 0.3) + 0.5000 \\ &= 0.1179 + 0.5000 = 0.6179. \end{aligned}$$

c)

$$\begin{aligned} P(|X| > 0.3) &= P(X < -0.3) + P(X > 0.3) \\ &= 2P(X > 0.3) \\ &= 2(0.3821) = 0.7642. \end{aligned}$$

Why is the normal probability distribution important? In the first place, measurements of many things give rise to distributions that are approximately normal. For example, the Pearson-Lee data on father-son statures provides the data used in plotting Fig. 7–12 showing the distribution of heights, in inches, of 1078 sons.

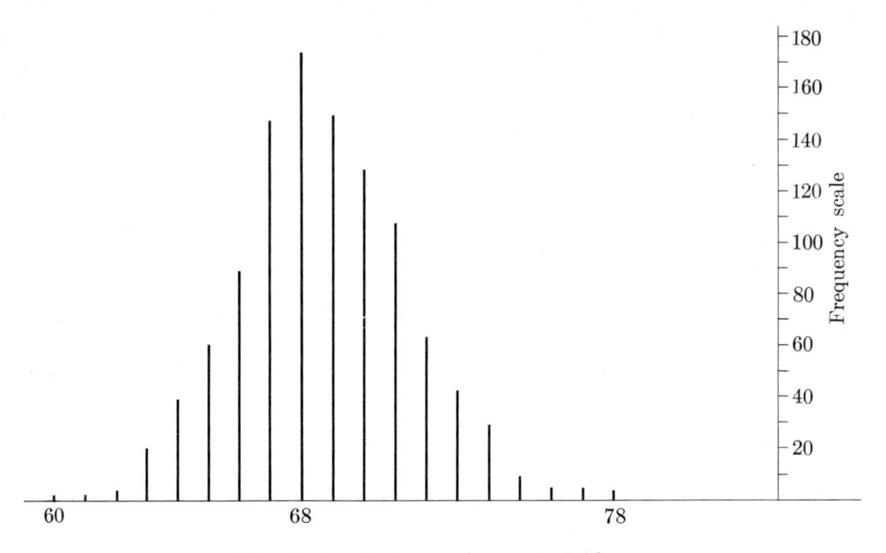

Fig. 7–12 Statures of sons, in inches.

The distribution of heights has a bell-shaped appearance that reminds us of the normal curve. Other examples of measurements that have approximately normal distributions include diameters of machined parts, lengths of tobacco leaves, IQ scores, and College Board aptitude and achievement test scores.

Since probabilities are readily available for the normal distribution, we have at hand a convenient method for obtaining approximations for probabilities in many other distributions. Moreover, some random variables, although not normally distributed, can be subjected to transformations that yield new random variables which are exactly or approximately normally distributed. The result is a new ease of calculation of probabilities. We shall see an important example of this feature when we study the binomial probability distribution in Chapter 8.

In the second place (and this is one of the most important reasons), under very general conditions, sums of random variables, normal or nonnormal, are approximately normally distributed. For instance, if the random variables are independent and if we know only their means and standard deviations, it is possible to make accurate probability calculations for the distribution of their sum, from the normal table. We return to this feature in Chapters 8 and 10.

Mean and variance The *mean* of the standard normal distribution is 0, and its variance and standard deviation are both 1:

$$\mu = 0, \qquad \sigma^2 = 1, \qquad \sigma = 1. \tag{2}$$

Note that these values for the mean and variance imply that the random variable is a standard deviate in the language introduced at the close of Section 6–1, and we naturally speak here of a *standard normal deviate*. The symmetry of the normal curve shows that, if a mean exists, this mean is zero, but it is not obvious

Table 7–9 Probability distribution of discrete random variable Z approximating the normal

Values, z	Probabilities, $P(z) = A(z + \frac{1}{4}) - A(z - \frac{1}{4})$	z^2	$z^2P(z)$
0	2(.0987 − .0000) = .1974	0	0
± .50	.2734 − .0987 = .1747	.25	.0437
±1.00	.3944 − .2734 = .1210	1.00	.1210
±1.50	.4599 − .3944 = .0655	2.25	.1474
±2.00	.4878 − .4599 = .0279	4.00	.1116
±2.50	.4970 − .4878 = .0092	6.25	.0575
±3.00	.4994 − .4970 = .0024	9.00	.0216
±3.50	.4999 − .4994 = .0005	12.25	.0061
±4.00	.5000 − .4999 = .0001	16.00	.0016
		Total	.5105

$$\sigma_Z^2 \approx 2(\text{Total}) = 2(.5105) = 1.021$$

that the variance is 1. We need calculus to prove that $\sigma^2 = 1$, but we can approximate σ^2 in the following way.

We approximate the continuous normal random variable X by a random variable Z that takes on only the 17 values

$$-\tfrac{8}{2}, \ldots, -\tfrac{3}{2}, -\tfrac{2}{2}, -\tfrac{1}{2}, 0, \tfrac{1}{2}, \tfrac{2}{2}, \tfrac{3}{2}, \ldots, \tfrac{8}{2}.$$

These values are the multiples of $\frac{1}{2}$ between -4 and $+4$. Since practically all of the probability of X is between -4 and $+4$, we stop at these boundaries for Z. With each value of Z we associate the probability under the normal curve in a band of width one-half centered at the value. For example, with the value $z = \frac{1}{2}$ we associate the probability of the standard normal distribution from $x = \frac{1}{4}$ to $x = \frac{3}{4}$, namely 0.1747. See Table 7–9.

Now, $\text{Var}(Z) = E(Z - \mu_Z)^2 = E(Z^2)$, since $\mu_Z = 0$. If we square each positive value of z, multiply by its probability, add these products, and then double that result to take account of the corresponding negative values of z, we get the variance of Z:

$$\text{Var}(Z) \approx 1.021. \tag{3}$$

Naturally we don't expect the variances to be exactly the same for the discrete random variable Z and the continuous normal random variable X. Equation (3) does, however, suggest that the variance of X should be near 1; it is, in fact, exactly 1.

Other normal random variables We have thus far discussed the *standard* normal random variable with mean equal to 0 and standard deviation equal to 1. Questions about a normal random variable Y, with mean μ_Y and standard deviation σ_Y, can be answered by translating them into questions about the related random variable

$$X = \frac{Y - \mu_Y}{\sigma_Y}.$$

The mean and variance formulas of Chapters 5 and 6 hold for continuous random variables as well as for those with a finite number of values. Consequently this new random variable has mean 0, because

$$\mu_X = E\left(\frac{Y - \mu_Y}{\sigma_Y}\right) = \frac{1}{\sigma_Y} E(Y - \mu_Y) = 0.$$

The variance of X is

$$\sigma_X^2 = \text{Var}\left[\frac{Y - \mu_Y}{\sigma_Y}\right] = \frac{1}{\sigma_Y^2} \text{Var}\,(Y - \mu_Y) = \frac{1}{\sigma_Y^2} \cdot \sigma_Y^2 = 1.$$

Moreover, X is a normal random variable, too, and since its mean is 0 and its variance is 1, it is a *standard* normal random variable.

We say that there is a *family* of normal distributions. Which particular member of the family we have is determined by μ and σ.

WARNING!

The word "normal" in normal distribution must not be thought of in its everyday sense of "usual" or "to be expected". Distributions other than the normal are not abnormal. When we say standard normal distribution, we mean specifically the distribution having ordinates given by Eq. (1), no more and no less. Sometimes the normal distribution is called Gaussian, especially in engineering and physics. In France it is called Laplacean. These names are probably used because the distribution was invented by de Moivre.

Figure 7-13 shows probability curves for normal random variables:

a) with standard deviation 1 and mean μ—the standard normal curve shifted to be symmetric about a vertical line through μ;
b) with mean 0 and standard deviation 2—the standard normal curve flattened some near its peak and raised some in its tails;
c) with arbitrary mean μ and standard deviation σ.

The following example shows how the standard normal table is used to answer questions about nonstandard normal random variables.

Example 7 Mr. Commuter has a statistician friend who believes that Mr. Commuter's time of arrival, measured in minutes after 8:55, is approximately a normal random variable T, with mean 0 and standard deviation 2.5. If this approximation is valid, what is the probability that Mr. Commuter arrives between 8:50 and 9:00 o'clock?

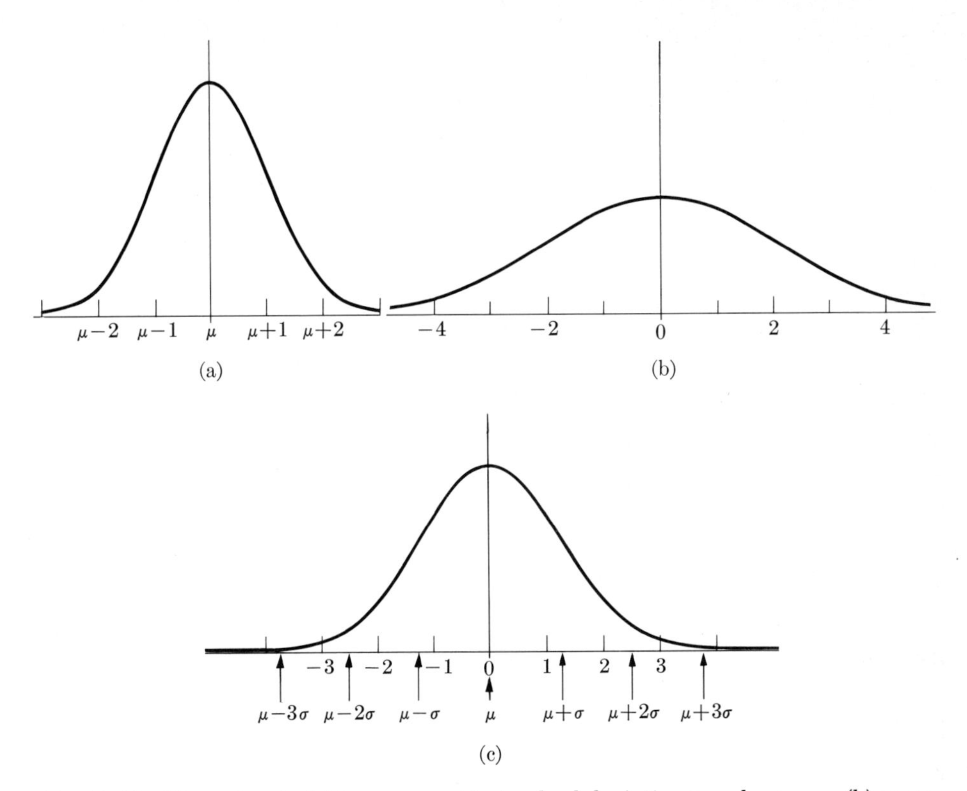

Fig. 7–13 Normal probability curves: (a) standard deviation 1 and mean μ, (b) mean 0 and standard deviation 2, (c) arbitrary mean μ and standard deviation σ.

Solution The given clock times correspond respectively to $T = -5$ and $T = +5$. The probability that T lies between -5 and $+5$ is

$$P(-5 \leq T \leq 5).$$

The statement

$$-5 \leq T \leq 5$$

is equivalent to

$$\frac{-5-0}{2.5} \leq \frac{T-0}{2.5} \leq \frac{5-0}{2.5}$$

or to

$$-2 \leq X \leq +2,$$

where

$$X = \frac{T-0}{2.5}.$$

Since T is normal with mean 0 and standard deviation 2.5, X is normal with mean 0 and standard deviation 1, i.e., X is a *standard* normal random variable.

From Table 7–8,

$$P(-2 \leq X \leq +2) = 2(0.4772) = 0.9544.$$

Hence, under the assumptions, Mr. Commuter arrives at work between 8:50 and 9:00 o'clock about 95% of the time.

In what follows, we shall use the normal distribution (a continuous distribution) to approximate discrete distributions. On the other hand, approximations go both ways. Discrete distributions on finite sample spaces can ordinarily approximate continuous distributions quite closely, if the discrete sample points are sufficiently numerous and appropriately placed. The word "finite" does not restrict the number of possible sample points to 10, 10^{10}, or even $10^{10^{10}}$. Therefore we do not hesitate to apply our discrete theory to continuous measurement problems. We are justified in three ways: (a) discrete distributions can be made to approximate continuous ones for most properties that we study; (b) many of our results for discrete variables hold exactly for continuous variables as well, or they have suitable analogues; and (c) practical measurement is ordinarily discrete rather than continuous.

EXERCISES FOR SECTION 7–4

1. X is a standard normal random variable. Find
   a) $P(X < 0.3)$, b) $P(X < -0.3)$,
   c) $P(1.2 < X < 3.6)$, d) $P(-1 < X < 2)$.
2. X is a standard normal random variable. Find k such that $P(X < k)$ is
   a) 0.4, b) 0.05, c) 0.95, d) 0.75.
3. X is a standard normal random variable. Find k such that $P(|X| < k)$ is
   a) 0.4, b) 0.8, c) 0.9, d) 0.05.
4. The College Entrance Examination Board test scores are scaled to approximate a normal distribution with mean $\mu = 500$ and standard deviation $\sigma = 100$. (a) What is the probability that a randomly selected student will score 700 or more? 580 or less? (b) What is the probability that 3 randomly selected students will all score 700 or more? That at least 2 of the 3 will score less than 700?
5. Make up three problems about CEEB scores, similar to those in Exercise 4, and solve them.
6. Use Table 7–8 to show that the "empirical rule" entries in Table 6–3 apply to distributions that are approximately normal.
7. Check that $1/\sqrt{2\pi}$ is close to 0.4. Note that the ordinate of the normal at $x = 0$ is $1/\sqrt{2\pi} \approx 0.4$.
8. From your normal tables find the area above the interval from 0 to h for $h = 0.01$, 0.1, 0.2, 0.3, 0.4, 0.5. Compare these values with $0.4h$. How big must h be before the difference is 10%?
9. If h is very small, the area under the standard normal curve from a to $a + h$ may be approximated by the area of a rectangle of base h and altitude equal to the

normal ordinate at $x = a$. For example (from Table 7–7), the ordinate at $x = 1.00$ is 0.2420. Compute $P(1.00 \leq X \leq 1.00 + h)$ directly from Table 7–8 and compare the results with $0.2420h$ for (a) $h = 0.1$, (b) $h = 0.2$, (c) $h = 0.5$, (d) $h = 1.0$.

10. Although we can't find the area under the normal curve without advanced calculus methods, we can approximate it closely by summing areas of rectangles. Consider a grid on the x-axis:

$$\ldots, -\tfrac{5}{2}, -\tfrac{4}{2}, -\tfrac{3}{2}, -\tfrac{2}{2}, -\tfrac{1}{2}, 0, \tfrac{1}{2}, \tfrac{2}{2}, \tfrac{3}{2}, \ldots,$$

or, more generally, with points at $n/2$, where n is any integer—positive, negative, or zero. On the interval between two adjacent grid points erect a rectangle whose height is the value of the normal ordinate at the midpoint. Then if $f(x)$ is a normal ordinate, the sum of the areas of the rectangles is

$$A = \sum_{n=1}^{\infty} f(n/4), \qquad \text{where } n \text{ is an odd integer.}$$

a) Explain why we need sum only over positive values of n.
b) Look up the ordinates at $n/4$, $n = 1, 3, 5, \ldots$, as far as your table goes, add them, and compare the sum with 1, the true normal area.

11. Do Exercise 10 with a grid on the x-axis at the integers

$$\ldots, -4, -3, -2, -1, 0, 1, 2, 3, 4, \ldots.$$

12. Use a table of logarithms to compute the ordinate y of the point on the normal curve

$$y = \frac{1}{\sqrt{2\pi}} e^{-x^2/2}$$

at $x = 0$, at $x = 1$, and at $x = 2$.

8 Approximating the binomial distribution by the normal: the central limit theorem

8–1 PROPERTIES OF THE BINOMIAL DISTRIBUTION

In this section we study the shapes of graphs of binomial distributions produced under two conditions: (1) for a fixed number of trials n, but different values of p; and (2) for a fixed value of p, but different values of n. We study especially how the graphs change shape as n grows large. Such a study helps us understand the family of binomial distributions, and it also helps us understand other sequences of probability distribution functions, because the changes within the binomial family resemble the changes within many other families of distributions.

Some properties are merely stated and illustrated without proof. The binomial tables at the back of the book can provide further numerical illustrations. In the discussion of figures, we sometimes abbreviate the notation for the binomial ordinate at x, which is $b(x; n, p)$, to $b(x)$.

1) Fixed *n*, varying *p* As p varies, the shape of the graph of the binomial distribution changes. Figure 8–1(a) through (i) illustrates this for $n = 5$. For p near zero or near one (Fig. 8–1a, b), the probability spikes up at $x = 0$ and $x = n$, respectively. The abscissa corresponding to the largest ordinate is called *the mode.* For p more centrally located, $b(x)$ increases with each successive x until the largest $b(x)$ is achieved (Fig. 8–1c, d, g, h) and then, except possibly for a tie at $x + 1$ (Fig. 8–1e, f, i), $b(x)$ decreases as x continues to increase. Thus, unless two adjacent ordinates are tied in value, there is just one largest ordinate, and the ordinates decrease steadily as we move to the right or to the left from the mode. The proof would divert us, but it is an exercise in the manipulation of inequalities that is within the range of an enthusiastic student.

When $p = \frac{1}{2}$ (Fig. 8–1i), the distribution is symmetric about $n/2$; if n is odd, two central values of x have equal ordinates (Fig. 8–1i); if n is even, the ordinate at the middle value of x is the largest (Fig. 8–2). If $p \neq \frac{1}{2}$, the distribution is asymmetric.

In Figs. 8–1 and 8–2 the fulcrum ▲ on the horizontal axis shows the mean, μ, for each distribution. You can see that the means of these binomial distribu-

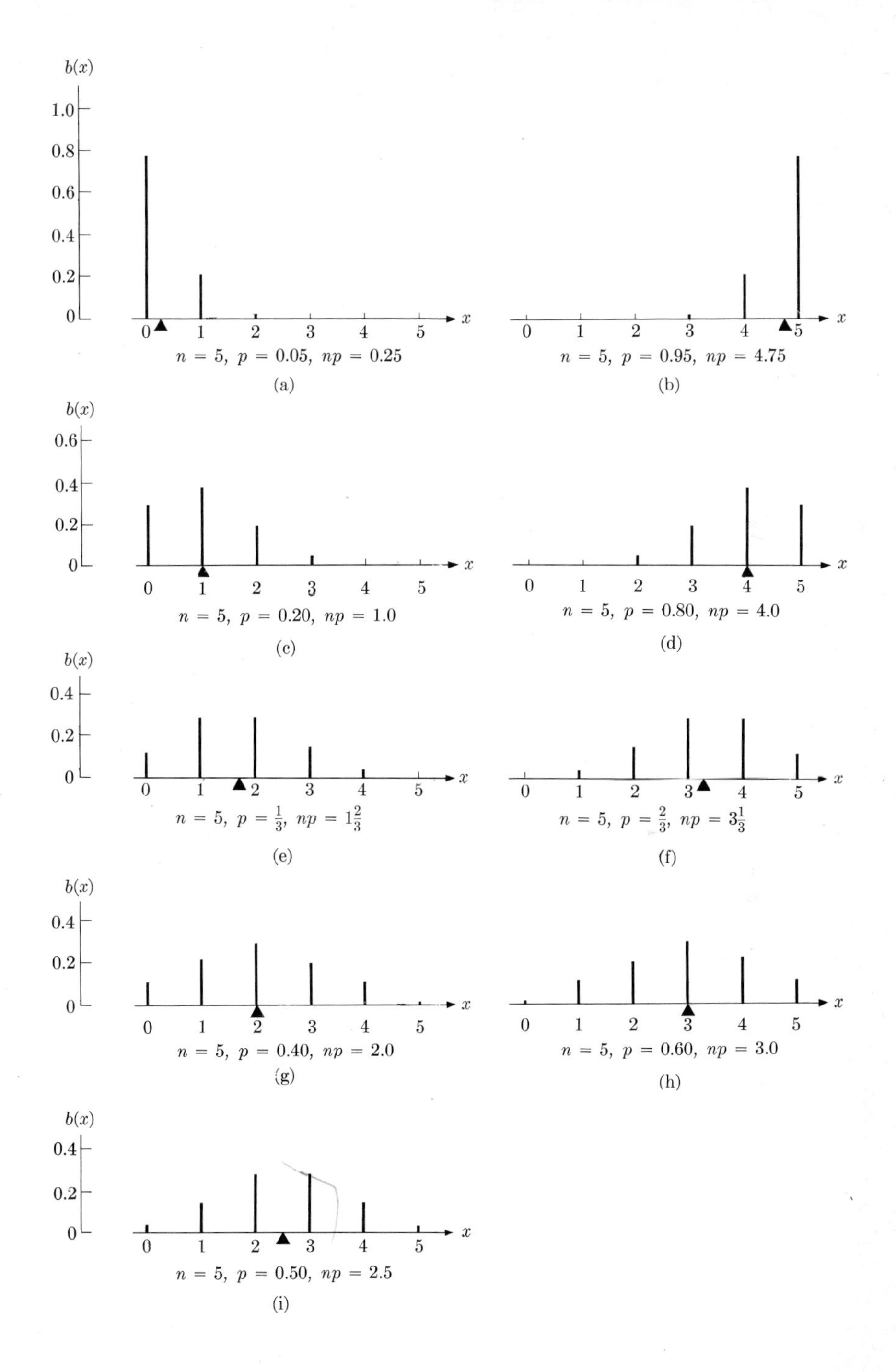

Fig. 8–1 Binomial distributions for $n = 5$, displaying the change in form as p varies; x = number of successes, $b(x)$ = probability of exactly x successes.

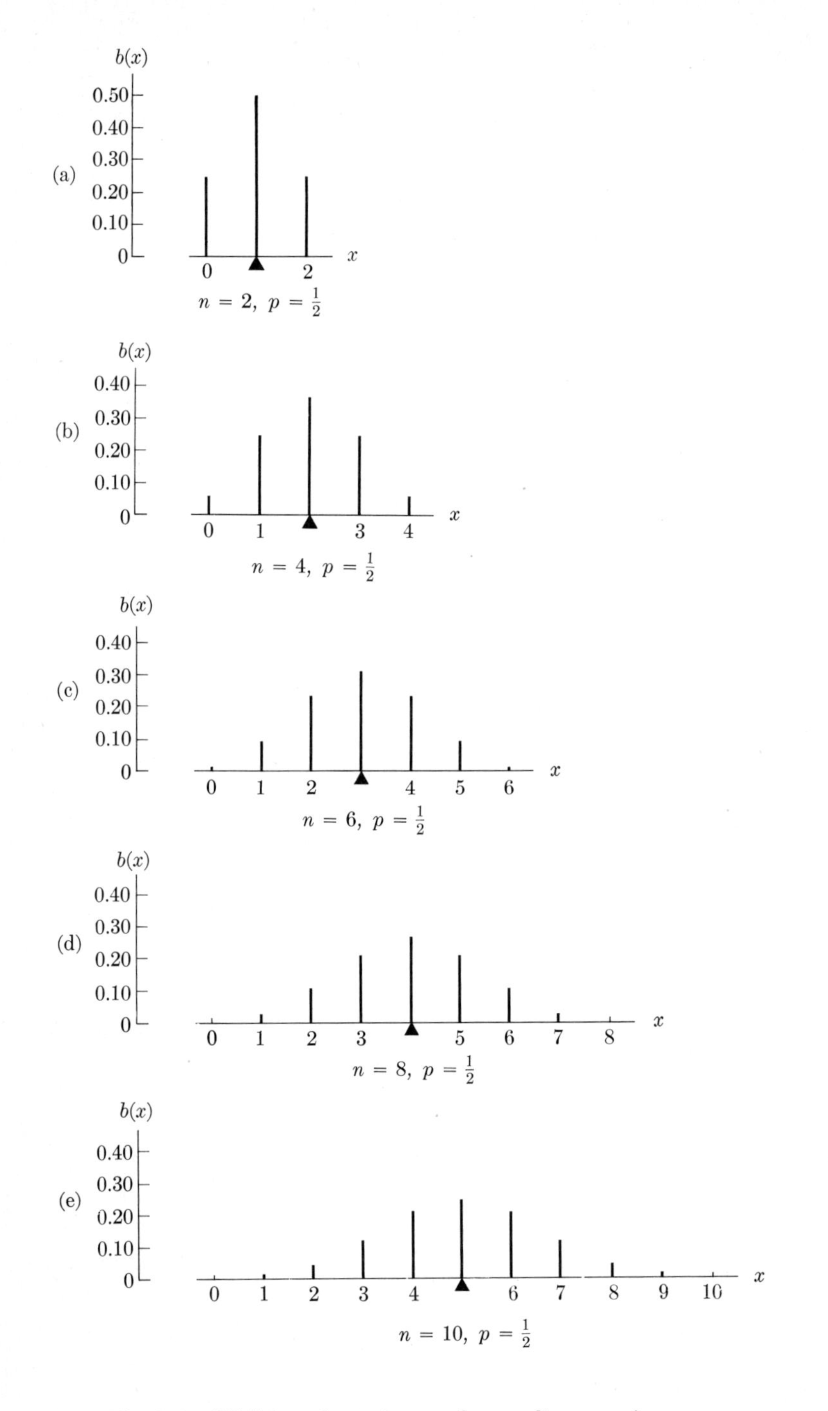

Fig. 8–2 Walking, flattening, and spreading as n increases.

tions are within one unit of the abscissa with largest probability (the mode). In binomial distributions, the mean and mode are *always* within one unit of each other. Furthermore, if np is an integer, the mode and mean are identical. We shall not prove these facts.

2) Fixed p, increasing n As n increases, the successive binomial distributions (a) "walk" to the right, (b) flatten, and (c) "spread". We discuss these features in turn.

a) *"Walking"*. As n increases, the mean μ moves to the right a distance p for each unit increase in n because $\mu = np$. The mode and the other large ordinates are near the mean, so the central mass of the distribution also "walks" to the right as n increases.

Later we shall try to obtain a limiting shape for the binomial distribution as n grows large, and to achieve this end we must prevent the distribution from walking off. We can do this by replacing the random variable X by a new variable $X - np$. For if X is replaced by $X - np$, then μ is replaced by $\mu - np$, or zero. Hence this adjustment keeps the successive distributions centered at the origin and prevents walk-off.

b) *Flattening*. Consider further the unadjusted random variable X. As the means walk, the distributions flatten (Fig. 8–2a through e). We wish to study the rate of flattening. It can be proved that, for large n, the sizes of the central ordinates are inversely proportional to $\sqrt{n}$. We shall illustrate this fact graphically. To do this, let us first recall that $y = mx$ is an equation of a straight line through the origin with slope m. When y is a constant times x, y varies directly as x. If $x = 1/\sqrt{n}$, then we usually say that y varies inversely as $\sqrt{n}$. But we can also say that y varies directly as $1/\sqrt{n}$; and when we plot y against $1/\sqrt{n}$, we get a straight line through the origin. The point is that we have a linear relation if we regard $1/\sqrt{n}$ as an independent variable. In other words, one way to show that y is inversely proportional to $\sqrt{n}$ is to show that y is directly proportional to $1/\sqrt{n}$. We use this idea in Fig. 8–3.

Mode proportional to $1/\sqrt{n}$ To return to the main discussion, Fig. 8–3 shows how the middle ordinates of symmetric binomial distributions ($p = \frac{1}{2}$) decrease as n grows. The relation is smooth when n is taken as even. (A similar smooth relation holds for n odd.) The modal value of x is $n/2$. When we choose the horizontal axis as the axis of $1/\sqrt{n}$, we see that, as n grows, P(mode) decreases, following a curve that is almost a straight line through the origin. (A scale of values of n is marked below the axis.) The points on the curve for $p = \frac{1}{2}$ have coordinates $\left(1/\sqrt{n}, b(\frac{1}{2}n)\right)$, n even. Our binomial table IV–A can be used to check a point on the curve. For $n = 24$, $b(12) \approx 0.161$ and $1/\sqrt{24} \approx 0.204$.

Similarly, the relation between P(mode) and $1/\sqrt{n}$ is approximated by a straight line through the origin for binomial distributions with $p = \frac{1}{5}$ (for smoothness, we have chosen values of n that are multiples of 5, and then the mode is $n/5$). Our binomial table IV–A can be used to check a point for $n = 25$.

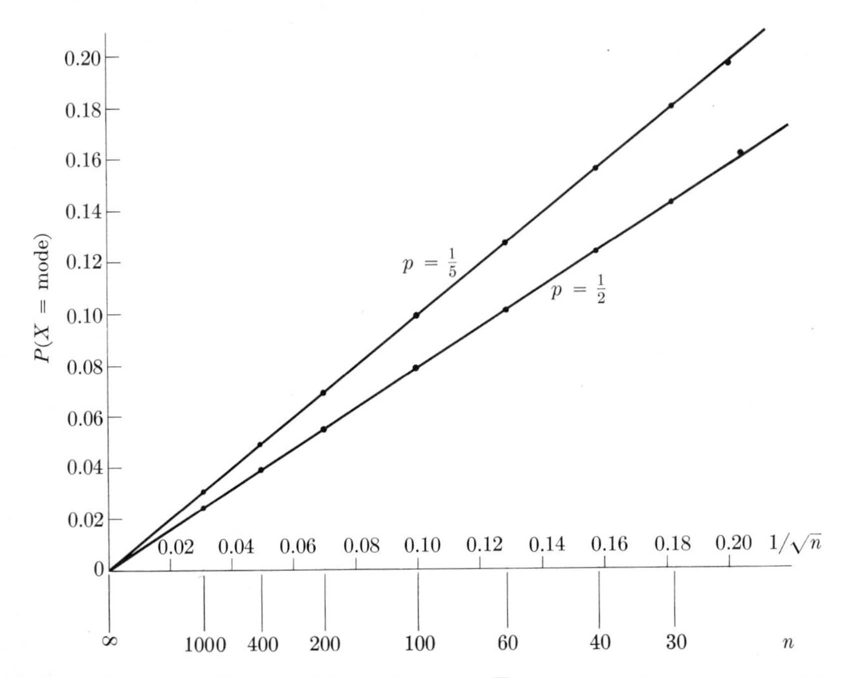

Fig. 8-3 Plot of $P(X = \text{mode})$ against $1/\sqrt{n}$ to show the nearly straight-line relationship for $p = \frac{1}{2}$ and $p = \frac{1}{5}$.

Then $1/\sqrt{n} = 0.20$, and $b(5) \approx 0.196$. This graph is adequate to illustrate the approximation: *the modal ordinate and its neighbors decrease inversely as* $\sqrt{n}$.

When we study the limiting distribution of X as n grows, we shall need to prevent the binomial distributions with large n from collapsing onto the x-axis. As Fig. 8-3 illustrates, as n grows, $1/\sqrt{n}$ tends to 0, and the largest ordinate tends to zero.

c) *Spreading.* We recall that the sum of the ordinates is always 1. Naturally, if the distributions flatten as n increases, and the total probability must remain constant, successive distributions must spread out. They spread at a rate proportional to $\sqrt{n}$; for the standard deviation is a measure of spread, and its value for a binomial distribution, $\sqrt{npq}$, is proportional to $\sqrt{n}$ when p is constant. (We derived this standard deviation in proving Theorem 6-6.)

The *total* range of the binomial is from 0 to n, and increases at a rate proportional to n. But we know from Chebyshev's theorem that there is very little probability near the ends of the distribution compared with the amount within a few standard deviations of the mean. The standard deviation is sensitive to the rate at which the central mass of the distribution spreads (the 75% or 99% near the mean), and it is this central mass that we want to study.

To summarize: as n grows, (a) successive binomial distributions walk to the right at a rate proportional to n; (b) the modal ordinates flatten at a rate proportional to $1/\sqrt{n}$; and (c) the distributions spread out, i.e., their standard deviations increase in proportion to $\sqrt{n}$.

EXERCISES FOR SECTION 8–1

All problems in this set refer to binomial distributions; m and n are integers.

1. Find the mean and standard deviation of the binomial distribution with (a) $n = 4$, $p = \frac{1}{2}$, (b) $n = 10$, $p = \frac{1}{5}$.
2. If $\mu = 45$, $\sigma = \sqrt{npq} = 6$, find n and p.
3. If $\mu = 10$, $\sigma = \sqrt{npq} = 3$, find n and p.
4. Show that if $p = \frac{1}{2}$ and $n = 2m - 1$, then $b(m - 1) = b(m)$.
5. Show that if $p = \frac{1}{2}$ and $n = 2m$, then $b(m)$ is larger than $b(m - 1)$ or $b(m + 1)$.
6. (Continuation.) Let $p = \frac{1}{2}$, $n = 2m$, and $m \geq 2$. Show that $b(r - 1) \leq b(r)$ for $1 \leq r \leq m$.
7. If $p = \frac{1}{3}$, how fast does the sequence of means of binomial distributions walk to the right, per unit increase in n?
8. Make a graph like that of Fig. 8–3 for $p = 0.4$, using your binomial tables. Choose n's that are multiples of 5 to smooth the plotting. Do not forget to label the axes. Assume that the curve through the points passes through the origin.
9. It is desired to show that points like those in Fig. 8–3 do not lie along a straight line. It is convenient to choose values of n that are perfect squares, say 1, 9, 25. If $p = \frac{1}{2}$, the modal ordinates are $b(0; 1, \frac{1}{2})$, $b(4; 9, \frac{1}{2})$, $b(12; 25, \frac{1}{2})$ at $1/\sqrt{1}$, $1/\sqrt{9}$, $1/\sqrt{25}$, respectively. Show that the slopes of the chords connecting the points are not equal. Use tables.
10. For $n = 2$, find the values of p for which the three ordinates $b(x; 2, p)$ lie on a straight line when a binomial probability graph like Fig. 8–1 is made.
11. For $n = 4$, $p = 0.2$, use your binomial tables to assist in plotting the graph of the probability function, as in Fig. 8–1. Be sure to indicate the mean.
12. For $n = 25$, $p = \frac{1}{2}$, use your tables to find the probabilities contained within σ of the mean μ, within 2σ of the mean, and within 3σ of the mean. Compare these results with those given by the Chebyshev inequality and with those given by the empirical rule of Table 6–4, Section 6–4.
13. Assuming that Fig. 8–3 is correct and that the curve for any p passes through the origin, show that, for a fixed p ($\neq 0$ or 1), as n approaches infinity $b(x; n, p)$ approaches zero, and therefore in the limit, the sequence of binomial distributions does collapse onto the x-axis. [*Puzzle:* The total probability had to add to 1; where did it go?]

8–2 TOOLS FOR STUDYING THE LIMIT OF THE BINOMIAL DISTRIBUTION

When n is large, the binomial distribution can be adjusted so that it is closely approximated by the standard normal distribution. We shall study the appropriate adjustments that lead to this approximation. Although the formula for the standard normal distribution looks a bit unfriendly, bristling as it does with roots, exponents, and transcendental numbers like π and e, these are not important features. For our purposes, the important features are that tables of the normal are widely available, and that the transition from a binomial probability problem to a normal probability problem is easy to make once one knows how.

With the exact binomial distribution, we are somewhat hampered by the extent of the available tables. As we have seen, even the largest tables extend only to $n = 1000$ and run to hundreds of pages, with large gaps in the list of values for n. But a page or two of normal tables can usually provide results with sufficient accuracy for most binomial problems. Naturally, we appreciate having binomial tables, but we also appreciate a fine approximation.

In addition, the normal distribution formula sometimes offers a more manageable expression for a binomial probability than does a complicated summation.

We recall (Section 7–1, Remark 2) that the random variable X, the total number of successes in n binomial trials, is the sum of n random variables, one variable for each binomial trial, each having the possible values 0 and 1. Sums of several random variables under general conditions are approximately normally distributed, but we cannot prove this fact here. However, a graphical demonstration of the way the binomial can be approximated by the normal distribution illustrates how the distribution of sums can be approximated by the normal distribution as n grows.

We need three pieces of equipment to help us study the limiting behavior of the binomial family:

1. Because the binomial is a discrete distribution and the normal a continuous one, the probabilities represented by binomial ordinates need to be replaced by areas, since areas are used to represent probabilities in continuous distributions.
2. We need a change of scale for X that prevents both walk-off and flattening.
3. We need to know how to approximate binomial probabilities by areas under the normal curve.

1) Ordinates and areas In Section 7–2 we found, for a continuous random variable, how to represent probabilities by areas contained between the graph of the probability function $f(x)$ and the x-axis. Figure 8–4 illustrates this point. The total probability (area) contained between the curve and the x-axis is 1. The probability that the random variable X takes a value between a and b is given by the *area* of the shaded part of Fig. 8–4. Note that $P(X = a) = 0$, because the area over a is just that of a line segment $f(a)$ units long, but of zero width. On the other hand, the *ordinates* of discrete distributions represent probabilities, and $P(X = 3)$, for example, need not be zero. If we wish to fit a binomial distribution with a continuous probability function, the following is a simple way to use areas to represent the probabilities usually given by ordinates.

We replace each ordinate of a binomial distribution by centering at x a rectangle whose width is one unit and whose height equals that of the original

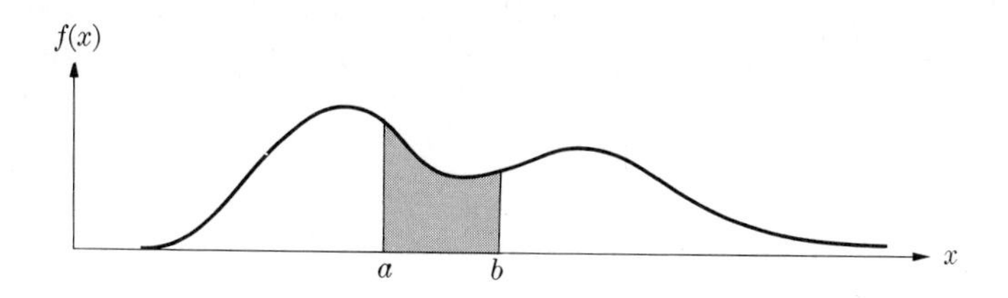

Fig. 8–4 Continuous probability density function; shaded area gives $P(a \leq X \leq b)$.

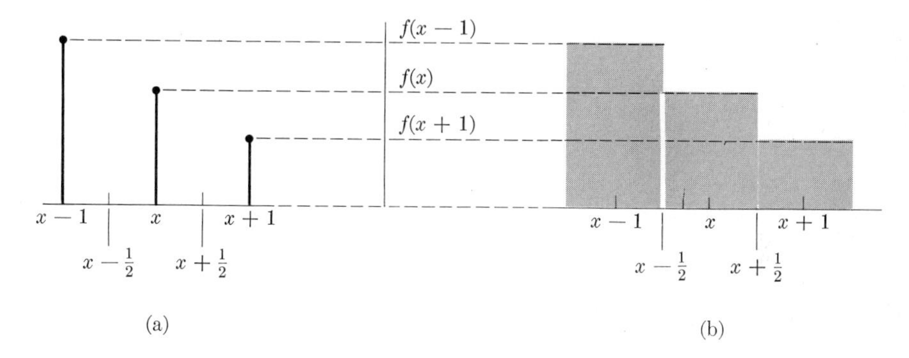

Fig. 8–5 (a) Probabilities as ordinates. (b) Probabilities as areas.

binomial ordinate. The area of the rectangle then has the same numerical measure as the height of the ordinate. To illustrate, the area over the interval from $x - \frac{1}{2}$ to $x + \frac{1}{2}$ in Fig. 8–5(b) has the same numerical value as the height of the ordinate at x in Fig. 8–5(a). Thus the probabilities given as ordinates in Fig. 8–5(a) are represented as areas in Fig. 8–5(b).

2) Changing scale To prevent "walk-off", we change the random variable X by subtracting the mean and creating the new variable $Y = X - np$. The distribution of the new variable Y has mean zero, and thus is centered at the origin, like the standard normal distribution. The standard deviation of Y is the same as that of X: namely, $\sigma = \sqrt{npq}$.

The standard normal distribution has mean zero and standard deviation 1. Our new random variable Y has mean zero, and a further adjustment is needed to get a variable with $\sigma = 1$. We divide Y by $\sqrt{npq}$, if $npq > 0$, to get a new random variable Z whose standard deviation is 1, like that of the standard normal distribution. Let

$$Z = \frac{Y}{\sqrt{npq}} = \frac{X - np}{\sqrt{npq}}.$$

The variance of Z is 1 because (Theorem 6–7)

$$\operatorname{Var} Z = \operatorname{Var}\left(\frac{Y}{\sqrt{npq}}\right) = \operatorname{Var}\left(\frac{1}{\sqrt{npq}}\, Y\right) = \frac{1}{npq} \operatorname{Var} Y = \frac{npq}{npq} = 1,$$

$$\sigma_Z = \sqrt{\operatorname{Var} Z} = \sqrt{1} = 1.$$

Thus the variable Z has $\mu_Z = 0$, $\sigma_Z = 1$, just as does the standard normal distribution. Furthermore, since the standard deviation of Z is a constant, it does not depend on n. Therefore, with Z, we do not have the problem of spreading that afflicted X.

Now that walk-off and spreading are taken care of, what about flattening? When we changed to the variable Z, we adjusted the scale of X. This adjustment changed the width of the rectangles we used in the areal representation of probabilities. We make a compensating change in their heights as discussed in the next example.

Example 1 Let $n = 8$, $p = \frac{1}{3}$; then $\mu_X = np = \frac{8}{3}$, and $\sigma_X = \sqrt{npq} = \sqrt{8(\frac{1}{3})(\frac{2}{3})} = \frac{4}{3}$. The rectangles for the area graph (see Fig. 8–5b) of the distribution of X have abscissa boundaries $x \pm \frac{1}{2}$, $x = 0, 1, \ldots, 8$. Discuss the area graph for the new random variable Z.

Discussion Let us carry through our transformation for the boundaries of the rectangles centered at 0, 1, 2, and 3.

Boundaries for X	Boundaries for $Y = X - \frac{8}{3}$	Boundaries for $Z = 3Y/4$
$-\frac{1}{2}$	$-\frac{19}{6}$	$-\frac{19}{8}$
$\frac{1}{2}$	$-\frac{13}{6}$	$-\frac{13}{8}$
$\frac{3}{2}$	$-\frac{7}{6}$	$-\frac{7}{8}$
$\frac{5}{2}$	$-\frac{1}{6}$	$-\frac{1}{8}$
$\frac{7}{2}$	$+\frac{5}{6}$	$+\frac{5}{8}$

The successive boundaries for the X-rectangles are 1 unit apart, but the corresponding boundaries for the Z-rectangles are $\frac{3}{4}$ unit apart. We want the probabilities represented by the areas in the Z-scale to be identical with those in the X-scale. To preserve areas we have to increase the height of the rectangles on the Z-scale by multiplying by $\frac{4}{3}$. The area in the X-scale was, for a rectangle centered at x,

$$\text{base} \times \text{height} = 1 \times b(x; 8, \tfrac{1}{3}) = b(x; 8, \tfrac{1}{3}).$$

The corresponding rectangle on the Z-scale, with height $\frac{4}{3}$ that of the rectangle on the X-scale, has the same area:

$$\text{base} \times \text{height} = \tfrac{3}{4} \times \tfrac{4}{3} b(x; 8, \tfrac{1}{3}) = b(x; 8, \tfrac{1}{3}).$$

More generally, for any binomial distribution, the rectangle centered at x on the X-scale has area

$$\text{base} \times \text{height} = 1 \times b(x; n, p) = b(x; n, p).$$

The corresponding rectangle on the Z-scale, with height equal to that of the rectangle on the X-scale multiplied by $\sqrt{npq}$, has the same area:

$$\text{base} \times \text{height} = \frac{1}{\sqrt{npq}} \times [\sqrt{npq}\, b(x; n, p)] = b(x; n, p),$$

as we require.

Note that the height of the binomial ordinate has been multiplied by $\sqrt{npq}$, a quantity proportional to $\sqrt{n}$. Recall that the flattening of the central binomial ordinates was *inversely* proportional to $\sqrt{n}$. This means that the distribution of Z *does not have the flattening feature* because $\sqrt{n} \times (1/\sqrt{n}) = 1$.

To summarize: The random variable

$$\boxed{Z = \frac{X - np}{\sqrt{npq}}}$$

has $\mu_Z = 0$, $\sigma_Z = 1$ and these values equal the mean and standard deviation of the standard normal distribution. Furthermore, as n grows large the central ordinates (heights of rectangles) of the area probability graph of Z do not flatten. Since $\mu_Z = 0$, walk-off does not occur; and since $\sigma_Z = 1$, spreading does not occur for the central mass of the distribution.

In Section 8-3, we shall study the behavior of the probability function of the new random variable Z as n increases and show its relation to the normal distribution. But before we do that, we shall see how to use the table of areas under the normal curve to approximate binomial probabilities.

3) Using the normal tables for binomial problems Given the binomial distribution with $n = 8$, $p = \frac{1}{2}$, we find

$$\mu_X = np = 8(\tfrac{1}{2}) = 4, \qquad \sigma_X = \sqrt{npq} = \sqrt{8(\tfrac{1}{2})(\tfrac{1}{2})} = \sqrt{2} \approx 1.41.$$

The graph with areas as probabilities for this distribution is shown in Fig. 8-6.

If we use the areas of Fig. 8-6 to evaluate the probability of 3 or more successes in this example, we want to include all the area in the rectangles above the x-axis to the right of $x = 2\frac{1}{2}$. If we used only the area to the right of $x = 3$, we would leave half of $P(3)$ behind. We therefore take as the left-hand boundary for x the value $2\frac{1}{2}$.

We use a standard normal table to obtain an approximation of the area to the right of $x = 2\frac{1}{2}$. To do this, we need to change from $P(X \geq 2\frac{1}{2})$ to $P(Z \geq z)$, and we have for z, the left-hand boundary value of Z,

$$z = \frac{x - np}{\sqrt{npq}} = \frac{2\frac{1}{2} - 4}{\sqrt{2}} \approx -1.06.$$

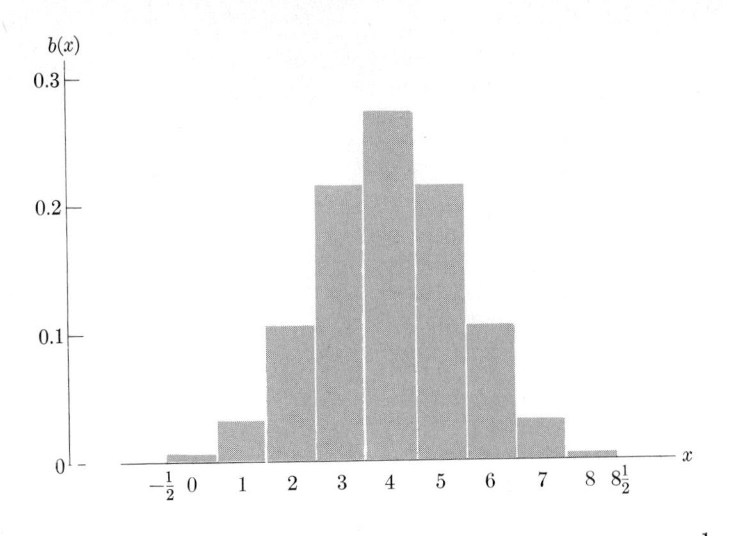

Fig. 8-6 Area graph for the binomial distribution $n = 8$, $p = \frac{1}{2}$.

Therefore $P(X \geq 2\frac{1}{2}) \approx P(Z \geq -1.06)$. We approximate this by the area under the standard normal curve to the right of -1.06, obtained with the help of Table III. The area from $z = 0$ to $z = +1.06$ is the same as the area from $z = -1.06$ to $z = 0$ or, from the tables, 0.3554. The area to the right of 0 is $\frac{1}{2}$. So the total probability is $0.3554 + 0.5000 = 0.8554$. This compares well with the true binomial four-place answer 0.8555.

Note We included the tail area of the normal that goes to infinity; we did not stop with the z corresponding to $x = 8\frac{1}{2}$, the right-hand bound of the rightmost rectangle. It is customary to regard the small area under this long right-hand tail of the normal as part of the area corresponding to the rightmost rectangle. A similar remark applies to leftmost rectangles.

Example 2 *Extreme ordinate.* For $n = 8$, $p = \frac{1}{2}$, approximate $P(8)$, using the normal table. The left boundary of the rectangle centered at $x = 8$ is $7\frac{1}{2}$. Therefore the corresponding boundary value of Z is

$$z = \frac{7\frac{1}{2} - 4}{\sqrt{2}} \approx 2.47.$$

The normal area to the right of 2.47 is 0.0068. The exact value is $P(8) = (\frac{1}{2})^8 = \frac{1}{256} \approx 0.0039$. The absolute magnitude of the error is not large, but the percentage error is nearly 75%.

Example 3 *Central ordinate.* For $n = 8$, $p = \frac{1}{2}$, approximate $P(4)$, using the normal table. The rectangle boundaries are $x = 3\frac{1}{2}$ and $4\frac{1}{2}$, so the left and right boundaries for Z are

$$z_1 = \frac{3\frac{1}{2} - 4}{\sqrt{2}} \approx -0.35, \qquad z_2 = \frac{4\frac{1}{2} - 4}{\sqrt{2}} \approx +0.35.$$

The area from 0 to +0.35 is 0.1368, and we need to double this to get $P(4) \approx 0.2736$. The true binomial value is

$$P(4) = \binom{8}{4}\left(\frac{1}{2}\right)^8 \approx 0.2734.$$

Note that both the absolute error and the percentage error are less in this example for a central ordinate than for the tail ordinate in the previous example. The method ordinarily approximates central ordinates better than tail ordinates.

In the next section we show that the foregoing tools do the job; they enable us to show that for large n the adjusted binomial distribution is approximated by the standard normal distribution.

EXERCISES FOR SECTION 8–2

1. Why do we want to study the normal approximation to the binomial distribution? Give three reasons.
2. (a) Plot side by side for $n = 4$, $p = 0.3$, the ordinary binomial probability function (see Fig. 8–1, for example) and the corresponding representation by areas (see Fig. 8–5b). (b) If you find the area to the left of $x = 0$ in the area diagram, what value do you get? (c) If you find the area to the left of $x = \frac{1}{2}$, what value do you get? (d) Is $b(0; 4, 0.3)$ the answer to (b) or to (c)? Explain what the other answer is.
3. (Continuation.) (a) Over what interval would the area be taken to find $P(2 \leq X \leq 3)$? (b) Over what interval would the area be taken to find $P(0 \leq X \leq 2)$? Would other intervals do? How about the interval with left boundary at -87? How far to the left can you place the left-hand boundary?
4. (a) For $n = 4$, $p = \frac{1}{5}$, make an area chart for the binomial. Don't forget to label the axes and put on the scales. (b) Calculate μ and σ for this distribution. (c) Now make an area chart for Y which is $X - \mu$, again labeling axes and putting on the scales. (d) Now change the scale to Z, which is $(X - \mu)/\sigma$, and make an area chart for Z, labeling axes and scaling them. Don't forget to adjust the heights of the rectangles.
5. (Continuation.) For $n = 4$, $p = \frac{1}{5}$, use the data from Exercise 4 and your normal tables to compute the normal approximation to.

   a) $P(X = 2)$, b) $P(X = 0)$, c) $P(0 \leq X \leq 2)$,
   d) $P(X \geq 2)$, e) $P(0 \leq X \leq 4)$.

8–3 AREAS FOR BINOMIAL DISTRIBUTIONS TEND TO AREAS UNDER THE NORMAL AS *n* GROWS: THE CENTRAL LIMIT THEOREM FOR THE BINOMIAL

In this section we illustrate how the area graphs for adjusted binomial distributions tend to the shape of the normal distribution. We shall also numerically compare probabilities obtained from the binomial tables with those obtained from their normal approximations. We study a sequence of binomial distributions for a fixed p. To keep track of the value of n, it is convenient to make n a

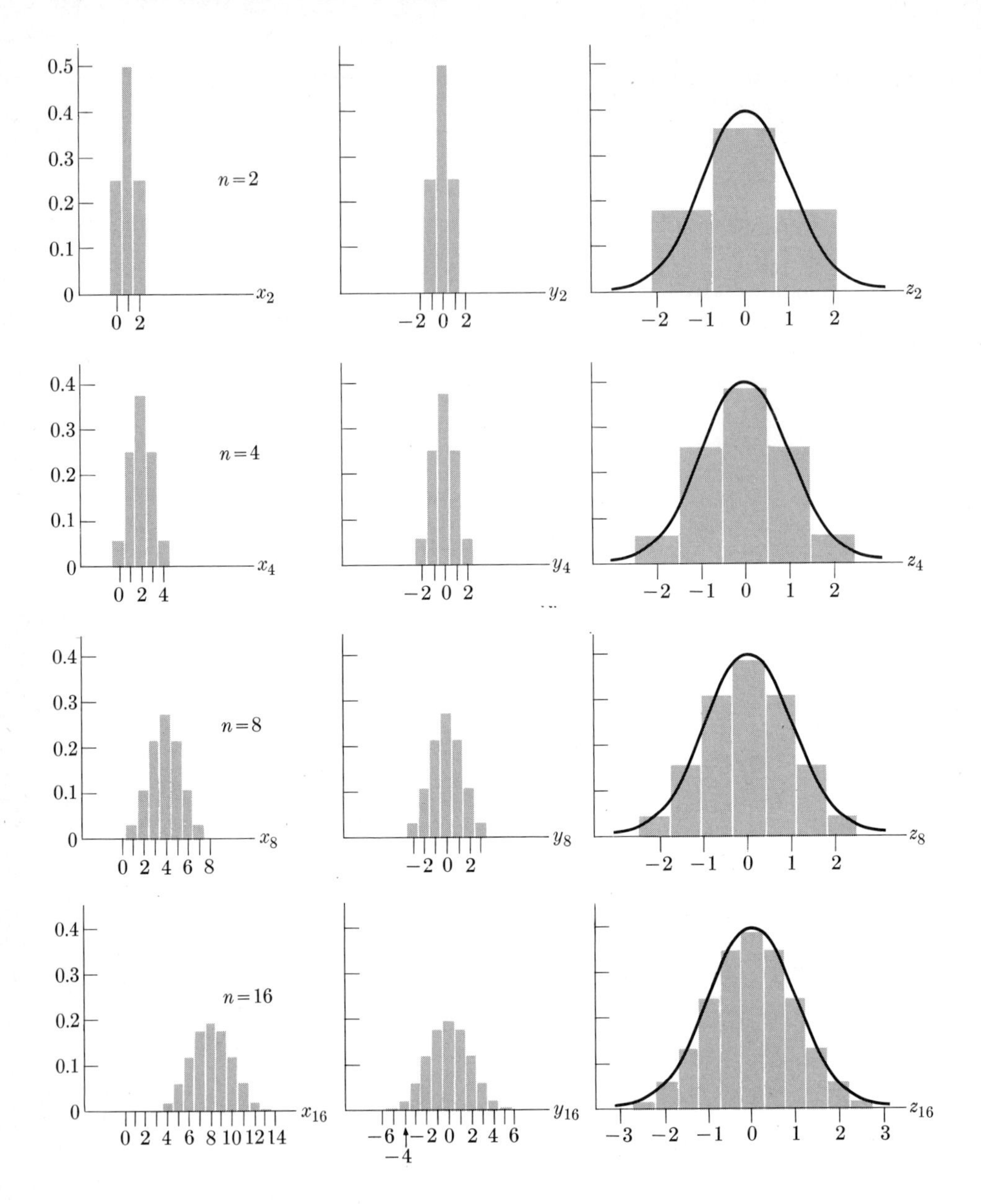

Fig. 8–7 Area graphs for binomial approaching normal, $p = \frac{1}{2}$. Horizontal scale for z_n differs from that for x_n and y_n.

subscript on any variable arising from a binomial experiment with number of trials n. Thus X_n is the number of successes in a binomial experiment composed of n trials. We review the adjustments of Section 8–2 briefly.

We have seen that the distributions of the variables X_n walk off and flatten as n increases. To prevent walk-off, a new set of variables $Y_n = X_n - np$ is

Table 8–1 Tabular values for single binomial terms and for cumulative binomials for $p = \frac{1}{2}$, together with their normal approximations, corresponding to the graphs of Fig. 8–7

	x	$b(x; n, \frac{1}{2})$	Normal approximation to $b(x)$	$P(X \geq x)$	Normal approximation to $P(X \geq x)$
$n = 2$:	0	.250	.240	1.000	1.000
	1	.500	.520	.750	.760
	2	.250	.240	.250	.240
$n = 4$:	0	.062	.067	1.000	1.000
	1	.250	.242	.938	.933
	2	.375	.383	.688	.691
	3	.250	.242	.312	.309
	4	.062	.067	.062	.067
$n = 8$:	0	.004	.007	1.000	1.000
	1	.031	.032	.996	.993
	2	.109	.106	.965	.961
	3	.219	.217	.855	.856
	4	.273	.277	.637	.638
	5	.219	.217	.363	.362
	6	.109	.106	.145	.144
	7	.031	.032	.035	.039
	8	.004	.007	.004	.007
$n = 16$:	0	.000	.000	1.000	1.000
	1	.000	.000	1.000	1.000
	2	.002	.002	1.000	.999
	3	.009	.009	.998	.997
	4	.028	.028	.989	.988
	5	.067	.066	.962	.960
	6	.122	.121	.895	.894
	7	.175	.175	.773	.773
	8	.196	.197	.598	.599
	9	.175	.175	.402	.401
	10	.122	.121	.227	.227
	11	.067	.066	.105	.106
	12	.028	.028	.038	.040
	13	.009	.009	.011	.012
	14	.002	.002	.002	.003
	15	.000	.000	.000	.001

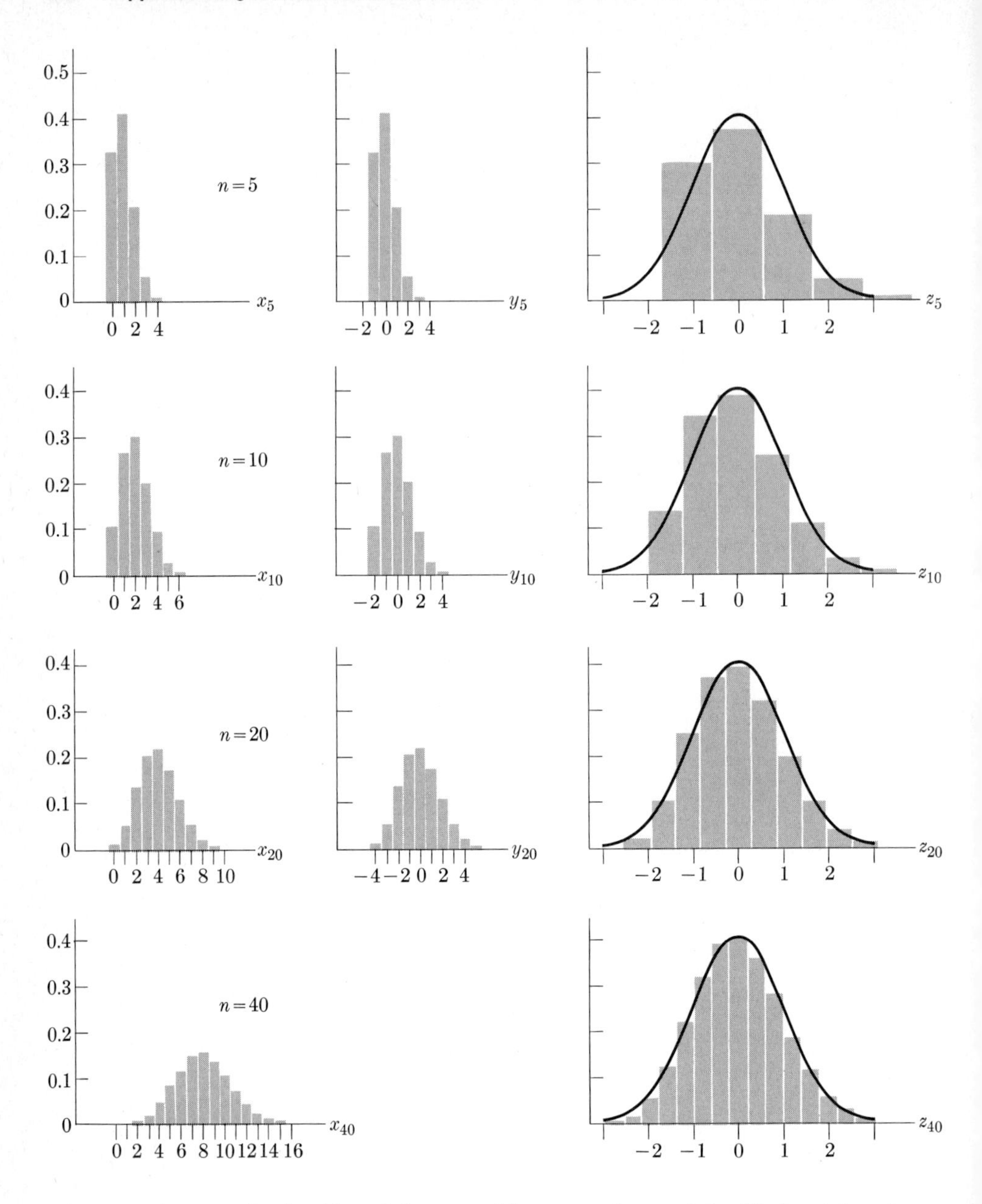

Fig. 8–8 Area graphs for binomial approaching normal, $p = 0.2$. Horizontal scale for z_n differs from that for x_n and y_n.

introduced. The mean of the distribution of X_n is np, and the standard deviation is $\sqrt{npq}$. Consequently, the mean of Y_n is 0, and the standard deviation is $\sqrt{npq}$. Thus each Y_n has the same mean as the standard normal distribution but, in general, does not have the same standard deviation.

The sequence of distributions of Y_n still flattens as n increases, but this can be prevented by introducing a third set of variables $Z_n = Y_n/\sqrt{npq}$. Note

Table 8–2 Normal approximations to binomial probabilities for $p = \frac{1}{5}$, corresponding to Fig. 8–8

	x	$b(x; n, \frac{1}{5})$	Normal approximation to $b(x)$	$P(X \geq x)$	Normal approximation to $P(X \geq x)$
$n = 5$:	0	.328	.288	1.000	1.000
	1	.410	.424	.672	.712
	2	.205	.241	.263	.288
	3	.051	.044	.058	.047
	4	.006	.003	.007	.003
	5	.000	.000	.000	.000
$n = 10$:	0	.107	.118	1.000	1.000
	1	.268	.229	.893	.882
	2	.302	.307	.624	.654
	3	.201	.229	.322	.346
	4	.088	.094	.121	.118
	5	.026	.021	.033	.024
	6	.006	.003	.006	.003
	7	.001	.000	.001	.000
	8	.000	.000	.000	.000
$n = 20$:	0	.012	.025	1.000	1.000
	1	.058	.056	.988	.975
	2	.137	.120	.931	.919
	3	.205	.189	.794	.799
	4	.218	.221	.589	.610
	5	.175	.189	.370	.390
	6	.109	.120	.196	.201
	7	.055	.056	.087	.081
	8	.022	.019	.032	.025
	9	.007	.005	.010	.006
	10	.002	.001	.003	.001
	11	.000	.000	.001	.000
	12	.000	.000	.000	.000
$n = 40$:	0	.000	.002	1.000	1.000
	1	.001	.004	1.000	.998
	2	.006	.010	.999	.995
	3	.021	.023	.992	.985
	4	.047	.046	.972	.962
	5	.085	.078	.924	.917
	6	.125	.115	.839	.838
	7	.151	.145	.714	.723
	8	.156	.157	.563	.578
	9	.139	.145	.407	.422
	10	.107	.115	.268	.277
	11	.073	.078	.161	.162
	12	.044	.046	.088	.083
	13	.024	.023	.043	.038
	14	.011	.010	.019	.015
	15	005	.004	.008	.005
	16	.002	.001	.003	.002
	17	.001	.000	.001	.000
	18	.000	.000	.000	.000

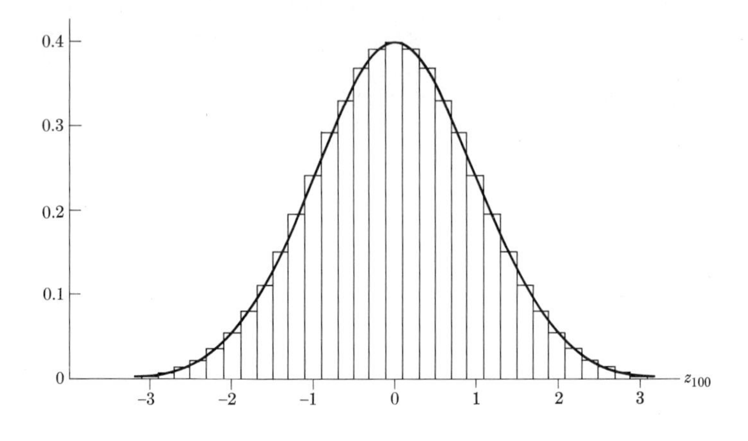

Fig. 8-9 Area graphs for binomial approaching normal, $p = \frac{1}{2}$, $n = 100$.

that we divided Y_n by its standard deviation. This choice gives an adjustment inversely proportional to $\sqrt{n}$, which, from the work in Sections 8-1 and 8-2, we know we need. Furthermore, the choice of the standard deviation as the divisor makes each Z_n have standard deviation 1. Thus the distribution of each Z_n has mean 0 and standard deviation 1, just as the standard normal distribution does. These agreements make the use of the approximation seem promising.

We wish to discuss the sequence of probability area graphs corresponding to $Z_1, Z_2, \ldots, Z_n, \ldots$ for a given p. Recall that the probability ordinates for Z_n are the ordinates for X_n multiplied by $\sqrt{npq}$ to get the correct ordinates for the area probability graph. We are especially interested in large values of n, but we shall illustrate for sequences of distributions where n is of moderate size.

In Fig. 8-7 we show, in parallel columns, area graphs of the distributions for X_n, Y_n, and Z_n for $p = \frac{1}{2}$ and $n = 2, 4, 8$, and 16. On the area graph for Z_n a standard normal curve is superimposed so that you can judge by eye the agreement between these area graphs and the normal curve.

We want a numerical as well as a visual assessment of the agreement between the graphs of the normal and the adjusted binomials. Table 8-1 shows a comparison between the binomial values from Tables IV-A and IV-B and the probabilities approximated by normal areas by the methods described in Section 8-2. Comparisons are given both for single binomial terms and for cumulatives. You can readily see that the results are in close agreement, and that they improve as n increases.

The asymmetric binomials tend to the normal more slowly than do the symmetric ones, so we also show in Fig. 8-8 a sequence for $p = 0.2$, $n = 5, 10, 20$, and 40. Table 8-2 shows the corresponding numerical comparisons between binomial probabilities and their normal approximations. By $n = 20$, the maximum error in a single term is 0.017, and in the cumulative it is 0.021.

In Figs. 8-9 and 8-10 we illustrate the comparison between the standard normal and the area graph of Z_n for $n = 100$, $p = \frac{1}{2}$, and $p = \frac{1}{5}$. Tables 8-3

Table 8–3 Numerical values corresponding to Fig. 8–9 ($n = 100$, $p = \frac{1}{2}$)

x	$b(x; 100, \frac{1}{2})$	Normal approximation to $b(x)$	$P(X \geq x)$	Normal approximation to $P(X \geq x)$
34	.000	.000	1.000	1.000
35	.001	.001	.999	.999
36	.002	.002	.998	.998
37	.003	.003	.997	.997
38	.004	.005	.994	.994
39	.007	.007	.990	.989
40	.011	.011	.982	.982
41	.016	.016	.972	.971
42	.022	.022	.956	.955
43	.030	.030	.933	.933
44	.039	.039	.903	.903
45	.048	.048	.864	.864
46	.058	.058	.816	.816
47	.067	.067	.758	.758
48	.074	.074	.691	.691
49	.078	.078	.618	.618
50	.080	.080	.540	.540
51	.078	.078	.460	.460
52	.074	.074	.382	.382
53	.067	.067	.309	.309
54	.058	.058	.242	.242
55	.048	.048	.184	.184
56	.039	.039	.136	.136
57	.030	.030	.097	.097
58	.022	.022	.067	.067
59	.016	.016	.044	.045
60	.011	.011	.028	.029
61	.007	.007	.018	.018
62	.004	.005	.010	.011
63	.003	.003	.006	.006
64	.002	.002	.003	.003
65	.001	.001	.002	.002
66	.000	.000	.001	.001
67	.000	.000	.000	.000

and 8–4 give the corresponding numerical comparisons. By $n = 100$, the agreement is quite close—within 0.001 in the cumulative for $p = \frac{1}{2}$, and within 0.010 for $p = \frac{1}{5}$.

These graphical and numerical results are intended to illustrate, though not to prove, that the limit of a sequence of adjusted area graphs of binomial distributions with fixed p and increasing n is a normal distribution. The result is put more precisely in the following theorem that we state and use but do not prove.

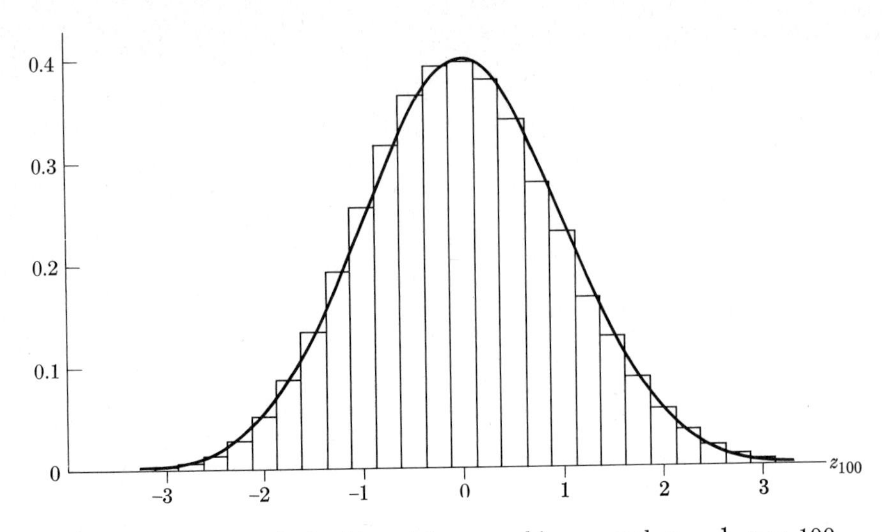

Fig. 8-10 Area graphs for binomial approaching normal, $p = \frac{1}{5}$, $n = 100$.

8-1 Theorem: De Moivre-Laplace Theorem

Let $X_1, X_2, \ldots, X_n, \ldots$ be a sequence of random variables, where X_n is the number of successes in a binomial experiment with n trials, each with probability of success p, $0 < p < 1$. Let Z_n, $n = 1, 2, \ldots$, be the corresponding sequence of adjusted random variables, where

$$\boxed{Z_n = \frac{X_n - np}{\sqrt{npq}},}$$

and let z be a constant. Then, as n approaches infinity, $P(Z_n \geq z)$ approaches the area to the right of z for the standard normal distribution.

Remark In practice, the result of the theorem says that for large values of n

$$P(X_n \geq x) = P(X_n > x - \tfrac{1}{2}) = P\left(\frac{X_n - np}{\sqrt{npq}} > \frac{x - \frac{1}{2} - np}{\sqrt{npq}}\right)$$

$$= P\left(Z_n > \frac{x - \frac{1}{2} - np}{\sqrt{npq}}\right) \approx P(Z > z),$$

where Z is a standard normal random variable and

$$z = \frac{x - \frac{1}{2} - np}{\sqrt{npq}}. \tag{1}$$

Example 1 For $n = 100$, $p = \frac{1}{5}$, find $P(X_{100} \geq 24)$.

Table 8-4 Numerical values corresponding to Fig. 8-10 ($n = 100$, $p = \frac{1}{5}$)

x	$b(x; 100, \frac{1}{5})$	Normal approximation to $b(x)$	$P(X \geq x)$	Normal approximation to $P(X \geq x)$
6	.000	.000	1.000	1.000
7	.000	.001	1.000	1.000
8	.001	.001	1.000	.999
9	.001	.002	.999	.998
10	.003	.004	.998	.996
11	.007	.008	.994	.991
12	.013	.014	.987	.983
13	.022	.022	.975	.970
14	.034	.032	.953	.948
15	.048	.046	.920	.915
16	.064	.060	.871	.870
17	.079	.075	.808	.809
18	.091	.088	.729	.734
19	.098	.096	.638	.646
20	.099	.099	.540	.550
21	.095	.096	.441	.450
22	.085	.088	.346	.354
23	.072	.075	.261	.266
24	.058	.060	.189	.191
25	.044	.046	.131	.130
26	.032	.032	.087	.085
27	.022	.022	.056	.052
28	.014	.014	.034	.030
29	.009	.008	.020	.017
30	.005	.004	.011	.009
31	.003	.002	.006	.004
32	.002	.001	.003	.002
33	.001	.001	.002	.001
34	.000	.000	.001	.000
35	.000	.000	.000	.000

Solution We use Eq. (1) to find the boundary value

$$z = \frac{x - \frac{1}{2} - np}{\sqrt{npq}} = \frac{24 - \frac{1}{2} - 20}{\sqrt{100(\frac{1}{5})(\frac{4}{5})}} = +0.875.$$

The area to the right of 0.875 in the normal table is 0.191, close to the true three-place value 0.189 obtained from large binomial tables.

The $\frac{1}{2}$ adjustment in the numerator of the expression for z in Eq. (1) is the same adjustment that we used in our normal calculations at the close of Section 8-2, so as not to omit half the probability at x. (See Fig. 8-5b.)

The De Moivre–Laplace Theorem is one form of a quite general set of "central limit theorems". These theorems treat the limiting distributions of sums of random variables, and these limiting distributions are ordinarily normal. The value of the theorems is that they enable us to compute approximate probabilities for sums using the normal distribution without ever knowing the exact distribution of the sum. Exact distributions are often hard to get, so we are grateful for such approximations.

In this section we have illustrated the approach to normality of a sequence of adjusted binomial distributions. Since the number of successes, X, in a binomial experiment is an example of a random variable which is itself the sum of several independent random variables, we have also illustrated the more general idea that sums of independent random variables, suitably adjusted, tend to be normally distributed, under quite general conditions.

Note *Accuracy of the approximation.* Although the De Moivre–Laplace Theorem is an exciting mathematical result, its practical value is the suggestion that the normal approximation may fit the binomial well even for moderate values of n. Mathematical investigations of the error in the normal approximation are extremely difficult, and their results are not easy to report here. However, extensive empirical investigation suggests that when the mean μ is "far" from 0 and n, the extreme values of X, the approximation is quite good. If μ is at least 3σ from both 0 and n, then it appears that the maximum error in evaluating a single term is at most 0.011, and in the cumulative at most 0.025.

It will be observed in Table 8–1 that for $n = 8$, $p = \frac{1}{2}$ the maximum errors in the individual terms and in the cumulative are both 0.004. Yet $n = 8$ does not quite put μ a distance of 3σ from the extremes. For $n = 40$, $p = \frac{1}{5}$, μ is more than 3σ from both extremes, and the maximum errors are 0.010 and 0.015 for individual terms and for cumulatives, respectively.

EXERCISES FOR SECTION 8–3

1. From Table 8–1, compute the maximum error in the approximation for a single term for $n = 2, 4, 8, 16$. Plot these maximum errors against a scale of $1/n$, and from the approximate linearity see that the maximum error is roughly proportional to $1/n$.

2. Compute the maximum error in the approximation for the cumulatives for $n = 2, 4, 8, 16$ from Table 8–1. Plot these maximum errors against $1/n$. From the approximate linearity, what do you conclude?

3. (a) Carry out the graphical work of Exercise 1 for Table 8–2. Note that the result for $n = 5$ does not line up with the rest of the points. (b) From Table 8–4, obtain the maximum error for $n = 100$ for individual terms. Drop $n = 5$, enlarge the scale for $1/n$, and plot the points for $n = 10, 20, 40, 100$, and note that the points fall close to a line through the origin. [*Remark:* We do not have tables extensive enough to study the corresponding result for the cumulative for Table 8–2.]

4. For $n = 8$, $p = \frac{1}{2}$, plot the area chart of Z carefully on graph paper to large scale and get a larger version of the graph shown in Fig. 8–7. Then from the table of normal ordinates, Table 7–7, plot points for the normal ordinates at intervals of $\frac{1}{2}$ and sketch in the normal curve.
5. For $n = 9$, $p = \frac{1}{2}$, do the graphical work corresponding to Exercise 4.
6. A professional basketball player sinks 60% of his foul shots, in the long run. During a season he gets 100 tries. Approximately, what is the probability that he sinks 70 or more of these? What is the probability that he makes exactly 60?
7. If 1000 coins are tossed, the most likely outcome is 500 heads. Use the normal approximation to find how likely that outcome is. Compare your numerical answer with a reading from Fig. 8–3.
8. (Continuation.) The area of a rectangle is base $\times$ height. We could approximate the result in Exercise 7 by multiplying the height of the normal ordinate at $z = 0$ by $1/\sigma$, because the width of the rectangle is 1 in the x-units, and $1/\sigma$ in the z-units. Make the calculation and compare it with the result obtained in Exercise 7.

9 | Some statistical applications of probability

9–1 ESTIMATION AND THE TESTING OF HYPOTHESES

Our work in the general theory of probability and our detailed study of the family of binomial distributions are applied in this chapter to a few problems in statistical inference. We focus attention on two related problems: *estimation* and *hypothesis testing.* Each is studied first without the use of prior information, as it would ordinarily be treated by probabilists of the objective school, and then with the use of prior information, in the manner of the personalistic school.

Starting with our experiments in Chapter 1, we have used observed averages to estimate population means and observed proportions to estimate p, the binomial probability of a success. These estimates are familiar and natural. If a professional basketball player sinks 65 foul shots out of 100, we estimate the value of p, the probability that he sinks a foul, to be $\frac{65}{100} = 0.65$. Assuming that the outcomes of successive shots are independent (not too safe an assumption in this example), how sure are we that $0.6 \leq p \leq 0.7$? The method of *confidence limits* presented in this chapter gives one way of making probabilistic statements about such an *interval estimate.*

Suppose that the basketball player has a long history of foul-shooting, with an average success of 0.54. Is there good reason to suppose that his new performance of 0.65 represents a change in his probability of making a successful foul shot? This kind of question is treated in the statistical testing of hypotheses, discussed later in this chapter. Similar problems arise when a new medication for relief from headaches is proposed. Does the new medication relieve more headaches than the usual remedy? Does a new dust for the disease *Botrytis* reduce the number of affected plants, as compared with no treatment?

If prior information is available, both the theory of estimation and that of hypothesis testing can be extended by the use of Bayes's Theorem. In Sections 9–6 and 9–7 we discuss examples of such extensions within the limitation of discrete probability distributions.

9-2 ESTIMATING *p*, THE BINOMIAL PROBABILITY OF SUCCESS

An estimate of *p* To establish our notation, recall the usual way of obtaining a numerical estimate of p, the probability of success for a single trial of a binomial experiment. We execute n trials, count the number of successes x, and compute $x/n = \overline{p}$ to obtain a value for the estimate of p.

If team A beats team B 4 times in 20, we estimate A's probability of beating team B as $\frac{4}{20} = 0.2$. Now we wish to discuss the properties of the estimate $\overline{p}$, giving special attention to its variability. What is its mean and what is its standard deviation? Answers to these questions follow easily from our work in Chapters 7 and 8.

The usual estimate of p is

$$\overline{p} = \frac{X}{n},$$

where X is the number of successes. Since X is a random variable, so is $\overline{p}$. Thus the value of $\overline{p}$ changes from one binomial experiment of size n to another. The possible values of X are $x = 0, 1, \ldots, n$, and so the possible values of $\overline{p}$ are $x/n = 0, 1/n, 2/n, \ldots, (n - 1)/n, 1$.

Mean and standard deviation of $\overline{p}$ We recall from Section 6-3, Eqs. (7), that

$$\mu_{\overline{p}} = p, \qquad \sigma_{\overline{p}} = \sqrt{\frac{pq}{n}}. \tag{1}$$

Thus "on the average" $\overline{p}$ gives the correct value of p, which is the result most people expect. In taking the expectation of $\overline{p}$, it may be worthwhile to rewrite $E(\overline{p}) = p$ in a notation which makes it more explicit that p and n are the quantities that determined the binomial distribution. Thus, to indicate that p is the expectation of $\overline{p}$, *given n and p*, we write

$$E(\overline{p}; n, p) = p.$$

This feature has a name: we say that $\overline{p}$ is an *unbiased estimate* of p. Lack of bias gives some backing for the use of $\overline{p}$ as an estimate of p. The everyday use of the term unbiased means "without prejudice". Since $\overline{p}$ neither averages to more than p nor to less, it can be thought of in that way. But the warm glow we reserve for the man who is unprejudiced need not be retained for unbiased statistics, as we now explain.

Closeness That an estimate has a long-run mean with value p is not an adequate basis for its use. We ought at least to know that $\overline{p}$ is often close to p. To illustrate this need, consider the outcome of a binomial experiment of size $n = 100$. Let us deliberately throw away the results of the last 99 trials and estimate the probability of success to be 1 if the first trial resulted in success, and 0 otherwise. This method of estimating is also unbiased because it corresponds to B/n, with $n = 1$ and B the number of successes on the *first* trial, and here again

$$E\left(\frac{B}{n}\right) = \frac{1}{n}\,E(B) = p.$$

Yet we do not like this estimate very well. Indeed we could easily be persuaded to use, instead, $X/99$ or $X/101$, with X the number of successes on the 100 trials, even though these estimates are biased, since

$$E\left(\frac{X}{99}\right) = \frac{100}{99}\,p, \qquad E\left(\frac{X}{101}\right) = \frac{100}{101}\,p.$$

What is missing from our discussion is some notion of closeness. We want an estimate that is more likely than other estimates to be close, in some sense, to the true value of p.

As just stated, the notion of closeness is rather vague, and we shall not pursue it. But we can find from binomial tables how often $\bar{p}$ is within a given distance of p. In addition, we can tell from the form of $\sigma_{\bar{p}}$ in Eq. (1) that when the sample size n grows large, the standard deviation becomes small. And the law of large numbers, Theorem 6–11, gives us bounds on the chances that $\bar{p}$ can depart from p by as much as d.

Recall that $\bar{p}$ is the *average* of the values of n random variables, each of which has value 0 or 1, while X is the *sum* of these same n random variables. The results for σ_X and $\sigma_{\bar{p}}$ are symptomatic of a more general fact: *sums of independent random variables ordinarily vary more than their component random variables, while averages vary less.*

Example 1 In a biological study, 1000 fruit flies are to be counted for the presence or absence of a certain characteristic. If $\bar{p}$ is used to estimate p, the probability that a fruit fly has the characteristic, how near p will the value of $\bar{p}$ be?

Solution *Conservative normal approach.* If we are confident that np is at least $3\sqrt{npq}$ from both 0 and n (see the Note on accuracy of the normal approximation to the binomial, Section 8–3), we are justified in using the theory of the normal distribution. We know that $Z = (X - np)/\sqrt{npq}$ is approximately normally distributed. If we divide both numerator and denominator on the right by n, and recall that $\bar{p} = X/n$, we have

$$Z = \frac{\bar{p} - p}{\sqrt{pq/n}}.$$

Now, since Z is approximately distributed according to the standard normal distribution, we can say that the probability is approximately 0.95 that

$$-2 \le Z \le 2 \qquad \text{or} \qquad -2 \le \frac{\bar{p} - p}{\sqrt{pq/n}} \le 2, \tag{2}$$

where the 2's represent 2 standard deviations, to approximate the more precise 1.96 from the normal table.

We now multiply all terms of the right-hand expression of inequality (2) by $\sqrt{pq/n}$, and get

$$-2\sqrt{pq/n} \leq \bar{p} - p \leq 2\sqrt{pq/n}, \tag{3a}$$

or

$$|\bar{p} - p| \leq 2\sqrt{pq/n}. \tag{3b}$$

One difficulty is that we don't know $\sigma_{\bar{p}}$ because we don't know the value of p. However, we can find the value of p that maximizes $\sigma_{\bar{p}}^2 = pq/n$. Since the graph of $pq = p(1 - p)$ is a parabola that is symmetrical about the line $p = \frac{1}{2}$, the maximum value of pq is attained when $p = q = \frac{1}{2}$. Therefore the maximum value of pq is $\frac{1}{2} \cdot \frac{1}{2} = \frac{1}{4}$, and

$$\sigma_{\bar{p}} = \sqrt{\frac{pq}{n}} \leq \sqrt{\frac{1}{4n}} = \frac{1}{\sqrt{4n}}.$$

Maximizing pq as $pq = \frac{1}{4}$, we find from the normal distribution that the probability is approximately 0.95 for p near $\frac{1}{2}$, and larger otherwise, that

$$|\bar{p} - p| \leq \frac{2}{\sqrt{4n}} = \frac{1}{\sqrt{n}}. \tag{4}$$

Since we have replaced $\sigma_{\bar{p}}$ by something larger, the interval has been extended in both directions and therefore the probability should be slightly larger than 0.95. The new, extended interval from $p - 1/\sqrt{n}$ to $p + 1/\sqrt{n}$ contains all the probability that was associated with the interval from $p - 2\sigma_{\bar{p}}$ to $p + 2\sigma_{\bar{p}}$.

If we choose h standard deviations instead of 2, the appropriate probability is obtained from our normal table, Table III. (Throughout this derivation, we have ignored the $\frac{1}{2}$ correction for continuity that we used in Chapter 8 to adjust our rectangular graphs for the binomial.)

To return to the foregoing example, the conservative normal approximation gives the probability as at least 0.95 that

$$|\bar{p} - p| \leq \frac{1}{\sqrt{n}} = \frac{1}{\sqrt{1000}} \approx 0.032.$$

The approximation 0.95 from the normal distribution compares with the number 0.75 that we would have gotten had we limped along with Chebyshev's Theorem.

The example and demonstration above illustrate the following theorem about binomial experiments. The theorem uses the following notation: X is the number of successes, p is the probability of success on a single trial, n is the number of trials, and $\bar{p} = X/n$ is the estimate of p.

9-1 Theorem: Large sample distribution of Z

In a binomial experiment, for fixed p and large n,

$$Z = \frac{\overline{p} - p}{\sqrt{pq/n}}$$

is approximately distributed according to the standard normal distribution.

Discussion Theorem 9-1 is equivalent to Theorem 8-1. We gave illustrations for the distribution of X, but we cannot prove the theorem without much more advanced mathematics. Nevertheless, we shall use it to compute approximate probabilities.

Example 2 It is desired to use $\overline{p}$ to estimate p, with probability 0.97 or higher that $\overline{p}$ is within 0.05 of p. How large should n be?

Solution Since n is evidently large, we apply Theorem 9-1:

$$Z = \frac{\overline{p} - p}{\sqrt{pq/n}}$$

is approximately distributed according to the standard normal distribution. The condition

$$|\overline{p} - p| \leq 0.05 \tag{5a}$$

is equivalent to

$$|Z| \leq \frac{0.05}{\sqrt{pq/n}}. \tag{5b}$$

The probability that inequality (5b) is true is to be 0.97; this means that the area under the standard normal curve from $-0.05/\sqrt{pq/n}$ to $+0.05/\sqrt{pq/n}$ is to be 0.97. Hence the area from 0 to $0.05/\sqrt{pq/n}$ is $\frac{1}{2}(0.97) = 0.485$; so, from Table III,

$$\frac{0.05}{\sqrt{pq/n}} \approx 2.17;$$

whence

$$\sqrt{n/pq} \approx 43.4,$$
$$n \approx (43.4)^2 pq \approx 1884pq.$$

Although p is unknown, the maximum value of $pq = p(1 - p)$ is $\frac{1}{4}$; hence

$$n \approx \tfrac{1884}{4} = 471$$

should be sufficiently large. The contrast with the result $n \approx 3333$ obtained in Example 2 of Section 6-3 by applying Chebyshev's Theorem is noteworthy.

EXERCISES FOR SECTION 9-2

Many of these problems require the use of Table III, IV-A, or IV-B.

1. What additional considerations, besides lack of bias, influence the choice of a method of estimation?
2. Rework Example 1, given that 400 fruit flies are used in the study.
3. Rework Example 1, given that 900 fruit flies are used in the study and that we choose 3 standard deviations instead of 2.
4. To have the probability be 0.95 or higher that $\bar{p}$ is within 0.1 of p, how large should n be?
5. Use your binomial tables for $n = 20$, $p = 0.01$, to find $P(|\bar{p} - p| \leq 0.1)$.

Fill in the missing cells in the following table:

	n	p	$\sigma_{\bar{p}}$
6.	4	$\frac{1}{2}$	
7.	9	0.2	
8.	100		0.05
9.		0.1	0.1
10.		p	p
11.	1000		0.01

12. Find the maximum value of $\sigma_{\bar{p}}$ when $n = 4$, 100, and 1000.
13. For $n = 5$, $p = 0.2$, compute $\sigma_{\bar{p}}$ directly from values in Table IV-A.

Find the estimates based on the normal approximation for $P(|\bar{p} - p| \leq d)$ when

14. $d = 0.1$, $n = 1000$
15. $d = 0.05$, $n = 100$
16. $d = 0.2$, $n = 10$
17. $d = 0.01$, $n = 400$

Use the conservative normal approximation to fill in the missing cells in the following:

	$P(\|\bar{p} - p\| \leq d)$	d	n
18.	0.05		49
19.	0.10	0.01	
20.	0.10	0.05	
21.	0.20		16
22.	0.50		100
23.		0.10	81
24.		0.05	9

25. The unknown size of a total population of animals is x. From this population, m are captured at random, marked, and released. On a second occasion, n are captured, of which r are found to be marked. Suggest an estimate for x.

9-3 CONFIDENCE LIMITS FOR p WITH LARGE n

In addition to reporting a value of $\overline{p}$ as an estimate of p, it may be helpful to make a statement to summarize our knowledge and our uncertainty about an interval in which p lies. The method of confidence limits offers a way to do this. In this method, we make a statement based on the result of the experiment. For example, with $n = 100$, $x = 40$, and the value of $\overline{p} = 0.4$, we can report with about 95% confidence that the statement $0.3 \leq p \leq 0.5$ is true. Each performance of an experiment gives rise to a statement that p is contained in a specific interval. (Just how this is done will be explained in the next paragraph.) Some statements will be true, some false, but 95% confidence means that, in the long run, 95% or more will be true.

The results of Section 9-2 enable us to construct such statements. We illustrate first with limits of 2 standard deviations that will give us 95% confidence. The approximate normality of the distribution of

$$Z = \frac{\overline{p} - p}{\sqrt{pq/n}}$$

for large n allowed us to write from inequality (4) of the preceding section that the probability is at least* 0.95 that

$$|\overline{p} - p| \leq \frac{2}{\sqrt{4n}} = \frac{1}{\sqrt{n}}. \tag{1}$$

An alternative way to write inequality (1) is

$$\frac{-1}{\sqrt{n}} \leq \overline{p} - p \leq \frac{1}{\sqrt{n}}. \tag{2}$$

Since we want a statement about p, we wish p were alone as the middle member of an inequality. We achieve this in two steps.

Step 1. Add $-\overline{p}$ to every member of the inequality (2) to get

$$-\overline{p} - \frac{1}{\sqrt{n}} \leq -p \leq -\overline{p} + \frac{1}{\sqrt{n}}. \tag{3}$$

Step 2. To change the sign of $-p$: multiply every member of inequality (3) by -1 and reverse the signs of inequality. The result is

$$\boxed{\overline{p} + \frac{1}{\sqrt{n}} \geq p \geq \overline{p} - \frac{1}{\sqrt{n}}.} \tag{4}$$

---

* Since an approximation is involved, we cannot absolutely guarantee the "at least", but the result is usually conservative; the use of 2 standard deviations instead of 1.96 makes the "at least" even more forceful.

For example, with $n = 100$ and the value of $\bar{p} = 0.4$, inequality (4) gives

$$0.5 \geq p \geq 0.3,$$

and we are 0.95 confident or 95% confident that the statement is true.

For each experiment, we substitute for n and $\bar{p}$ in (4) and get a 95% confidence statement about p.

Example 1 *Bowling.* In a league season, a good bowler bowled 400 frames and got 120 strikes. Set conservative 95% confidence limits on p, which measures his probability of a strike. (Assume independence between frames.)

Solution The value of $\bar{p}$ is $120/400 = 0.3$. From inequality (4) the limits are $0.30 \pm 1/\sqrt{400} = 0.30 \pm 1/20$, and so the interval runs from 0.25 and 0.35. Thus, we say $0.25 \leq p \leq 0.35$, with 95% confidence.

Generalization If, in inequality (4) of Section 9–2, we use h rather than 2 for our multiplier of $1/\sqrt{4n}$, the conservative estimate of $\sigma_{\bar{p}}$, we have for $h > 0$

$$|\bar{p} - p| \leq \frac{h}{\sqrt{4n}}. \tag{5}$$

The probability that the statement is true, called the *confidence coefficient,* is at least the area under the standard normal distribution from $-h$ to $+h$. The "at least" again is the consequence of replacing $\sigma_{\bar{p}}$ by $1/\sqrt{4n}$. The same steps that led to inequality (4) lead to the *confidence statement*

$$\boxed{\bar{p} + \frac{h}{\sqrt{4n}} \geq p \geq \bar{p} - \frac{h}{\sqrt{4n}},} \tag{6}$$

or, "p is in the interval $\bar{p} \pm h/\sqrt{4n}$". The interval $\bar{p} \pm h/\sqrt{4n}$ is called a *confidence interval.* The numbers $\bar{p} + h/\sqrt{4n}$ and $\bar{p} - h/\sqrt{4n}$ are called the *upper* and *lower confidence limits*, respectively.

Example 2 In the bowling example, what confidence statement would be used for a 0.50 confidence coefficient?

Solution From the standard normal tables, we find that the interval from -0.67 to $+0.67$ gives the desired area of 0.50. Therefore we choose $h = 0.67$. For the bowling example, we get

$$\bar{p} \pm \frac{h}{\sqrt{4n}} \quad \text{as} \quad 0.300 \pm \frac{0.67}{\sqrt{4(400)}} \approx 0.300 \pm 0.017,$$

and 50% lower and upper confidence limits are 0.283 and 0.317, respectively.

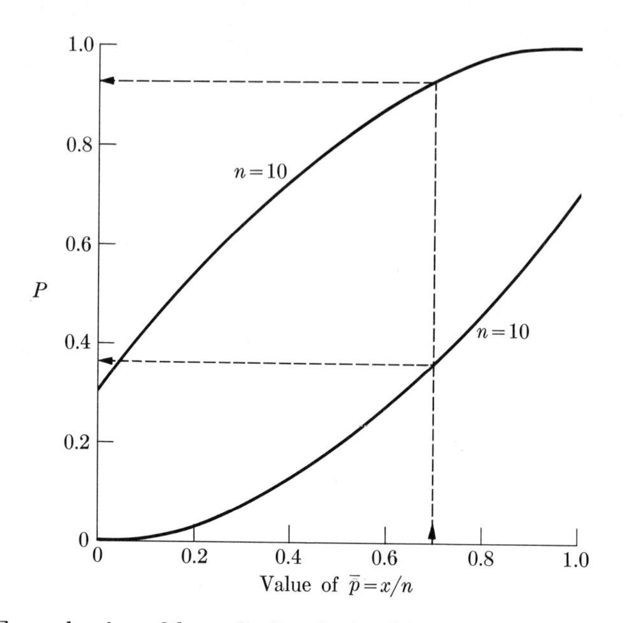

Fig. 9–1 Example of confidence limits obtained from curves given in Chart I. If a binomial experiment gives 7 successes in 10 trials, enter the horizontal axis with $\overline{p} = 0.7$, erect perpendicular to curves for $n = 10$, and read lower and upper confidence limits as 0.36 and 0.93.

Chart for confidence limits: Chart I By rather heavy algebraic methods it is possible to compute limits that are not as conservative as these, and by using extensive tables and computations we can compute limits for small as well as large values of n. We shall not develop such methods here, but we have made a chart that is easy to use for 0.95 confidence limits, and we present it as Chart I at the back of the book.

We enter the chart with the observed value $\overline{p}$ on the horizontal axis, and erect a perpendicular from that point. The perpendicular crosses the two curves for the given value of n in two points. Using the vertical scale, we read the upper and lower confidence limits for p as the heights of the points.

Example 3 If a binomial experiment produces 7 successes in 10 trials, set 95% confidence limits on p, the probability of success on a single trial.

Solution Figure 9–1 reproduces the two curves for $n = 10$ from Chart I. We enter with $\overline{p} = 0.7$ and read approximately the values 0.36 and 0.93. Having the curve is merely a convenience for the eye, because, of course, $\overline{p}$ can only take a value from the set of tenths 0, 0.1, 0.2, 0.3, . . . , 0.9, 1.0 in this example.

Example 4 Use the curves in Chart I to get 95% confidence limits for the bowling example, Example 1.

Solution Entering with $\overline{p} = 0.3$ and interpolating roughly between the curves for $n = 250$ and $n = 1000$, we find 0.26 and 0.35 as the lower and upper confidence limits. Our own more conservative limits were 0.25 and 0.35.

Optimistic approach to confidence limits If $\overline{p}$ is not close to $\frac{1}{2}$ and n is large, the limits obtained by the conservative approach of substituting $p = \frac{1}{2}$ into inequality (3a) of Section 9–2 may be rather broad. A more optimistic value for the appropriate confidence interval could be obtained by setting $p = \overline{p}$ under the square roots to give the general limits

$$\boxed{\overline{p} - h\sqrt{\overline{p}\overline{q}/n} \leq p \leq \overline{p} + h\sqrt{\overline{p}\overline{q}/n},} \tag{7}$$

where $\overline{q} = 1 - \overline{p}$. As usual, h is the number of standard deviations, and the confidence level is obtained from the normal table. These intervals are too short for small values of $\overline{p}$ or of $1 - \overline{p}$ (consider $\overline{p} = 0$ or $\overline{p} = 1$), but they are fairly satisfactory over most of the range of $\overline{p}$.

The "conservative versus optimistic" approach hinges on the standard deviation of $\overline{p}$:

$$\sigma_{\overline{p}} = \sqrt{pq/n}.$$

In an experiment where we are trying to *estimate* p, we don't really know the value of $pq = p(1 - p)$. In the *conservative* approach we play it safe by taking the maximum value for pq, which is $\frac{1}{4}$. This broadens the confidence interval and puts us on the safe side in saying that the probability that the corresponding interval contains the true value of p is *at least* so much. In the *optimistic* approach we use the estimate

$$\overline{p}\overline{q} \approx pq$$

to give an estimated value of $\sigma_{\overline{p}}$ as

$$\sigma_{\overline{p}} \approx \sqrt{\overline{p}\overline{q}/n}.$$

Unless $\overline{p} = \frac{1}{2}$, using this optimistic estimate of $\sigma_{\overline{p}}$ produces a narrower confidence interval than we get from the conservative approach. The narrower confidence interval, in effect, claims to give a somewhat more precise estimate of p.

Choosing the method For $\overline{p}$ near 0 or 1, say between 0 and .2 or between .8 and 1.0, the chart is probably wisest, because the largest corrections occur near the extremes. When $\overline{p}$ is near 0.5, the optimistic limits are quite accurate as soon as n is at least 20. And, of course, this formula is best when some confidence level other than 0.95 is used. The chart is good for any 0.95 level unless the interpolation is difficult.

EXERCISES FOR SECTION 9–3

All the following problems deal with binomial distributions.

1. A random sample of 25 households from a large town shows that 10 buy newspaper A. (a) Use inequality (4) to set 95% confidence limits on the proportion p in the town who buy newspaper A. (b) Use inequality (6) to set 50% confidence limits. (c) Set 95% confidence limits, using Chart I.

2. In a sample of 20 students drawn from a large population, 16 recalled recently learned material better immediately after sleeping 8 hours than after 8 hours awake. Set 80% confidence limits on the population proportion p who would have performed better after sleeping 8 hours, had all been tested.
3. A random sample of 50 families from Cambridge, Mass., showed 10 families with incomes over $10,000 during 1969. Set a 90% confidence interval on p, the percent of families with incomes over $10,000.
4. On 100 different local telephone calls, a secretary fails to complete 25 at the first attempt. Use Chart I to set 95% confidence limits on p, her long-run proportion completed on the first attempt.
5. Some parents of the fifth-grade pupils in a large school system complained that their children could not read clear handwriting. A lengthy test on handwritten material showed that 225 out of a random sample of 250 pupils did read the handwriting ($\bar{p} = 0.90$). Use Chart I to set 95% confidence limits on the population proportion p reading handwriting.
6. Use Chart I to set 95% confidence limits on p, the proportion of defective teacups produced if a random sample of 25 had no defectives.
7. Use inequality (4) to decide on the sample size required to set a 95% confidence limit of total length less than or equal to 0.04 on a binomial p.
8. Use Chart I to estimate the sample size required to set a 95% confidence limit on p of total length less than or equal to 0.10.
9. *Operative mortality* (L. B. Ellis and D. E. Harken). Among 211 cardiac invalids in a series having heart operations, the operative mortality was 18%. Set 95% confidence limits on the operative mortality for this operation on this kind of patient.

9–4 TESTING OF A BINOMIAL STATISTICAL HYPOTHESIS

Sometimes we want to decide whether the performance of a binomial process is consistent with the assumption that the probability of success has a given value, p_0.

Example 1 *Methods of memorizing.* Two methods of memorizing some difficult material are being tried to see which gives better retention. Pairs of students are matched for both IQ and academic performance and are then taught the same material, one member of a pair using method A, the other member using method B. After scoring tests for recall, we know which method worked better for each matched pair of students. We want to use these experimental data to decide which method works better for most students in a larger population from which this sample was drawn.

Discussion Suppose that the investigation had 25 pairs of students. If method A was better for x of them, then $x/25$ estimates p, the proportion of such pairs from a large population that would recall better using method A. If we merely want to reach a decision, a natural way is to choose method A if $x \geq 13$, and method B if $x < 13$. This is an obvious device that is fair to both methods.

Suppose now that the outcome was $x = 13$. Although favoring method A, our evidence is weak. We can readily imagine that a repetition of this experiment could come out favoring method B. When the evidence in favor of some hypothesis—here, "method A is better"—is weak, we may prefer to act as if there were no difference between the methods. Then we may reserve the decision "method A is better" or "the methods differ in their efficiency" for occasions when the evidence is stronger. Just how we act and when we choose one way or another of making the decision are practical matters that depend upon the specific problem. Sometimes the issues and the answers are obvious from the context; but sometimes they aren't, and then we require more information. Let us spell out three possibilities here.

1. *No initial preference between A and B; problem not important.* In a training handbook we want to recommend one method for rote learning, either A or B, but not both, and we know that both methods do quite well. Then whichever method our experiment of $n = 25$ pairs points to is likely the one that we will choose to put in the handbook.

2. *Result of great importance or controversial; no initial preference.* If the method chosen is to be used very extensively or if our evaluation of an important scientific principle pivots on which method performs better, then we may be unwilling to choose method A on the basis of weak evidence such as $x = 13$. Imagine signing your name to the statement "Method A is better!" in a full-page ad in *The New York Times* with a "P.S. I found that method A was better in 13 and worse in 12 out of 25 trials." In these important situations with weak evidence, we usually do not want to make a choice. Instead, we say something like "The evidence is that the methods are about equal in merit." But when the evidence is extreme, perhaps $x \geq 21$, we may still be willing to make the decision. If we *have* to *choose* between the methods, then, other things like costs being equal, we would choose the method with the larger number of successes.

3. *New untried method vs. the standard tried-and-true method.* Suppose that method B has been used with consistent success in the past and that method A is proposed to replace it. Then method A has to prove that it is definitely better than method B or we will not give up the latter. Changing over is usually a costly process, and we know that new methods in practice often reveal unexpected difficulties and frequently do not perform as well in the field as they did in the laboratory. Again, we are not inclined to give up method B for weak evidence in favor of A. How strong the evidence will need to be will depend upon the importance of the problem and upon costs. Let us note, too, that in this case no one cares much about the hypothesis "method A is *worse* than method B", for only a *better* method can be of value.

These three possibilities show that in some situations, (1) we are willing to reach a decision on the basis of modest evidence and we act whichever way the data point; in others, (2) we regard two options or hypotheses as having equivalent merit unless extreme evidence in favor of one or the other is available; and in still others, (3) we are initially in favor of one hypothesis, and we will reject it only if an extreme result favors its denial. We shall not discuss possibility (1) much further. Possibility (2) divides into two cases: (a) a two-sided two-decision problem where we conclude either "the methods differ" or "we detect no difference"; and (b) a three-decision problem where we conclude "method A is better", "method B is better", or "we detect no difference". Usually formal work in statistics discusses the two-sided two-decision problem, but in practice the worker often acts as if he had a three-decision problem. The two-decision problem is simpler to explain, and in this book we choose this easier course; but we could develop our discussion for the three-decision problem if we wished. Possibility (3) fits well with one-sided tests. We do not wish to imply that these three exhaust the possibilities, but they set the stage adequately for our work in this section.

Sign test Example 1 illustrates the ideas behind the "sign test". A binomial experiment is performed, and each outcome is scored $+$ or $-$ according as it is favorable or unfavorable to a hypothesis: for example, "method A is better". The total number of $+$'s is used for a decision. We need a decision rule. In our Example 1, possibility 2, let us agree that if 18 or more pairs favor method A, we decide in favor of A; and, symmetrically, if 7 or fewer cases favor A (18 or more favor B), we decide in favor of B. Otherwise we accept the hypothesis, "The methods are about equally good".

How would this rule work? If the methods are equally good, we say the *null hypothesis* is true. For the binomial, this means $p = \frac{1}{2}$. (The word "null" is assigned to any special hypothesis that we choose to treat as a standard. Here it is $p = \frac{1}{2}$, but we might in another problem have chosen $p = 0.1$. For example, if the cure rate for a disease has been 10% and we want to test a new treatment, the null value of p would ordinarily be taken as $p = 0.1$. We want to see if the new method has a higher value of p.) The probability that we correctly decide "The methods are about equally good" using the suggested decision rule when $p = \frac{1}{2}$ and $n = 25$ can be computed from Table IV–B to be 0.956. And so our rule "accepts" the null hypothesis more than 95% of the time, when the null hypothesis is true. What if it is false? The answer depends on the degree of falseness. If $p = 0.60$, the probability of deciding "A is better" is

$$P(X \geq 18 | n = 25, p = 0.6) = 0.154;$$

but

$$P(X \geq 18 | n = 25, p = 0.8) = 0.891,$$

and so the larger p is, the more likely we are to "reject" the null hypothesis of equality in favor of method A. Symmetrical results in favor of B hold when p is smaller than $\frac{1}{2}$.

Language of testing statistical hypotheses We discuss a two-decision rule, since this is simple and frequently useful. The decisions are

1) accept the null hypothesis,
2) reject the null hypothesis.

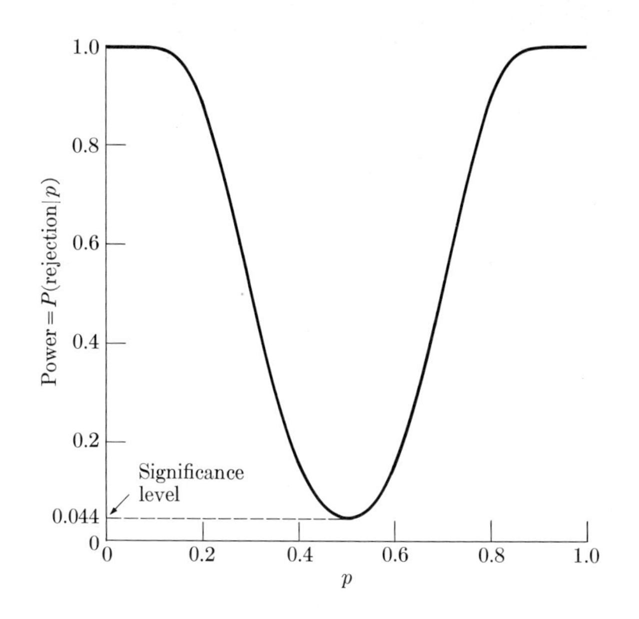

Fig. 9–2 Power of the sign test for testing the null hypothesis $p = \frac{1}{2}$ when $n = 25$, if we accept when $8 \leq x \leq 17$, reject otherwise.

We have a *test statistic*, here the count x. We have an *acceptance region* for the statistic. In this example it is $8 \leq x \leq 17$. We have a *rejection region*, here $0 \leq x \leq 7$ and $18 \leq x \leq 25$. We have a *significance level*, which is the probability of rejecting the null hypothesis when it is true: here $1 - 0.956 = 0.044$. The test has *power*, which is the probability of rejecting the null hypothesis when it is false. Each degree of falsity has its own power. Here, when $p = 0.6$, the power (= probability of rejecting the null hypothesis) consists of

$$P(0 \leq x \leq 7) + P(18 \leq x \leq 25) = 0.001 + 0.154 = 0.155.$$

It may seem silly to add in the probability of rejecting when the test went in the wrong direction, but in the two-decision problem that is the standard procedure. Note that it contributes only 0.001 to the power. When we compute the power for every possible alternative hypothesis, here every value of p, we have a *power function* whose graph is a *power curve*. Such a power curve is illustrated for our example in Fig. 9–2.

Meaning of "accept" and "reject" Just as "success" and "failure" for the two outcomes of a binomial trial are merely labels with no meaning in themselves, so in statistical decision theory the words "accept" and "reject" are neutral labels standing for the actions that are taken. In many scientific investigations, the decision to "accept" the null hypothesis means that one will continue to believe that the particular value of the parameter is one that may be near the true value. Rejecting in Example 1 might mean that we believe method A is better and that we are going to give it further tests on other kinds of material. It could even mean that we believe method A is better and yet that we won't use it because it is too costly for the observed gain. If we "reject" a lot of material because too many defectives are found in a sample, this rejection may call for any one of a number of possible actions: the lot may be (1) reworked, or (2) 100% inspected, or (3) destroyed, or (4) sold at reduced price; or perhaps (5) a slip of paper will be sent to a committee reporting the rejection. It is especially important not to count on the words "accept" and "reject" to have their everyday meanings.

Let us take a second example of the use of the two-sided sign test. On the basis of a sample from a population, one wants to decide whether some percentile of the population, for example the median, is equal to a known standard or not. The method proceeds by translating measurements in the sample into a + or − according as the measurement is above or below the standard. If the standard equals the population percentile, then binomial theory applies to the number of +'s and −'s, as we describe in the following examples.

Example 2 *A middling class?* National results on a standardized achievement test are scored so that half of all students score 100 or over, and half score less (100 is the population median). A teacher wonders whether his class differs from this standard. Of his class of 20, 16 scored higher than 100, and 4 lower. He sees at once that more than half of his class is above the standard. But he may also ask, "Considering sampling fluctuations, is it reasonable that my class is a sample from a large population of students half of whom score 100 or more, and half less?" Specifically, he visualizes the population from which his students are drawn as the students attending his school over a number of years, and he is willing to regard this class as a random sample from such a population.

Discussion In this question, the teacher visualizes a population in which the probability of scoring 100 or more is $p_0 = \frac{1}{2}$. We call p_0 his *null hypothesis*, or standard. The word "null", in this setting, means that the population is an average one. As *alternative hypotheses* he considers that the true p for the population of students from which his school draws has $p \neq \frac{1}{2}$. He thinks that his school's population may be lower or higher than the national average.

One way to test such questions is to use the sample to set confidence limits on the unknown p. If the confidence interval contains the value $p = \frac{1}{2}$, we accept the possibility that the null hypothesis is true, otherwise we reject the null hypothesis in favor of the alternatives. Recall that in a testing problem, we speak of the *significance level* of the test. It measures the probability of rejecting the null hypothesis when it is true. It is the complement of the confidence level, if a confidence interval is used to make the test. Thus if the confidence level is chosen as 0.95, the significance level is 0.05.

In Example 2, the 95% confidence limits for p, when 16 successes are observed out of 20, can be read from Chart I. We get 0.57 and 0.94 as lower and upper confidence limits for p. Since $p = \frac{1}{2}$ is not in the interval, at the 5%

level we reject the null hypothesis that $p = \frac{1}{2}$ for the population of students from which the class is drawn.

The device of using confidence regions to make statistical tests has broad applicability.

On rejecting the null hypothesis some authors say that "the result is statistically significant at the 0.05 level". Unfortunately, the word "significant" in everyday use means not only "suggestive" but also "important" and "weighty". We do not carry these everyday meanings over into statistics. Such usage would horrify readers with our heartless report that "the doubling of the death rate was not significant", or baffle them with our excitement over small changes when we say that "the half-pound increase in the weight of the adult pigs is significant". An important difference may or may not be statistically significant, and a statistically significant result may or may not be important. In reporting to laymen, everyone is helped if words like significant or significance are avoided entirely. A statistically significant result is one that the data support as showing a real effect, as opposed to a result that might readily arise from sampling variation.

One-sided tests In the test just described, the approach was two-sided, because the teacher thought both $p > \frac{1}{2}$ and $p < \frac{1}{2}$ were possibilities. He might have formulated the problem with the null hypothesis $p \leq \frac{1}{2}$ and the alternative $p > \frac{1}{2}$. Then he would reject the null hypothesis only for large values of X.

The worst value of the null hypothesis, from the point of view of being able to distinguish it from the alternative hypothesis, is $p = \frac{1}{2}$. Assume, temporarily, that $p = \frac{1}{2}$, and compute the probability of observing a result as extreme as, or more extreme than, the one observed in the direction of the alternative (16 or more above 100 in our example). The resulting probability is called a *descriptive level of significance* or a *nominal level of significance.* If it is as small as or smaller than the level of significance the investigator would use in the problem, he rejects the null hypothesis in favor of the alternative. Our descriptive level of significance computed on the basis of the largest null-hypothesis value of p (from Table IV–B) is $P(X \geq 16) = 0.006$, and we again would reject at the 0.05 level of significance.

Giving the descriptive level of significance has the advantage of offering a valuable report about the outcome without committing us to a two-decision approach. Just as we can profitably compute and report an observed proportion without making a test, we can compute and report a descriptive level of significance under a relevant null hypothesis and use it as an aid in interpreting the data. One way of thinking about such a descriptive level of significance is that it expresses the proper amount of surprise to attach to the observed result if the null hypothesis is true. The smaller the descriptive level, the greater the surprise, and therefore the greater one's skepticism about the chosen null hypothesis.

Other null hypotheses The sorts of tests just described are not limited to null hypotheses related to $p = \frac{1}{2}$.

Example 3 *New medication vs. a standard.* A standard medication reduces reports of post-operative pain in 80% of patients treated. A new medication for the same purpose produces 90 patients relieved among the first 100 tested. What is an appropriate test of significance?

Solution If the new medication is better than the old—few are—we want to detect it. We take as the null hypothesis $p \leq 0.8$, and as the alternative $p > 0.8$. We compute $P(X \geq 90)$ for $n = 100$, $p = 0.8$, because 0.8 is the standard the new medication ought to exceed if it is to replace the old tried-and-true medication. For the normal approximation we compute

$$z = \frac{90 - \frac{1}{2} - (0.8)100}{\sqrt{(0.8)(0.2)100}} \approx 2.37.$$

The descriptive level of significance is approximately

$$P(Z > 2.37) \approx 0.0089.$$

If the investigator is using the 5% or the 1% level of significance he rejects the null hypothesis in favor of the alternative. This means that he believes that the new medication relieves more patients than the old.

Accepting or rejecting the null hypothesis Why do we reject the null hypothesis, or accept it? As you must have observed, we reject it when the probability of the occurrence of the observed event, or more extreme ones, is small. But that alone is not the reason. We reject it because the data do not support it, *and* because we think that the alternative hypothesis is tenable and that the data do support the alternative. The teacher knows that there is variation in teaching ability and in school systems. His students were not randomly drawn from the national population, but from a special neighborhood. And the doctor looking for a new medication knows that medications better than the standard ones are found from time to time (after all, the standard was once unknown) but not often, so he will be very cautious about replacing a standard. In most such examples the decisions are not final; new data can overthrow them.

What does it mean to "accept" the null hypothesis? Suppose the teacher had observed 12 students with scores of 100 or over and 8 with lower scores, and had tested the null hypothesis $p = \frac{1}{2}$ against the alternatives $p \neq \frac{1}{2}$. At usual levels of significance he accepts the null hypothesis. However, he does not believe therefore that $p = \frac{1}{2}$ exactly. A sample of 20 carries practically no information for discriminating $p = 0.500$ from $p = 0.501$. All the teacher accepts is that p is near $\frac{1}{2}$. Furthermore, he reserves the right to change his mind if he gets more data that conflict with the hypothesis that p is near $\frac{1}{2}$. Obviously in the real world a point null hypothesis, like $p = \frac{1}{2}$ or (in continuous measurement problems) $\mu = 0$, cannot be exactly true because there are infinitely many competing hypotheses in any neighborhood of it. We cannot detect departures in the far-out decimal places, even though they are likely to be present.

Example 4 *Acceptance sampling or quality control.* After the bugs are ironed out of a production process, its output is stable and it is said to be "in control". Keeping it in control is a major statistical and engineering task. A production process has been in control for some time with proportion defectives $p_0 \approx 0.05$. Samples of 25 are inspected, and if 4 or more defectives are observed the process is regarded as "out of control"; otherwise the process is accepted as "in control". For various values of the *true* proportion defective p, how likely is this criterion to accept the process as "in control"?

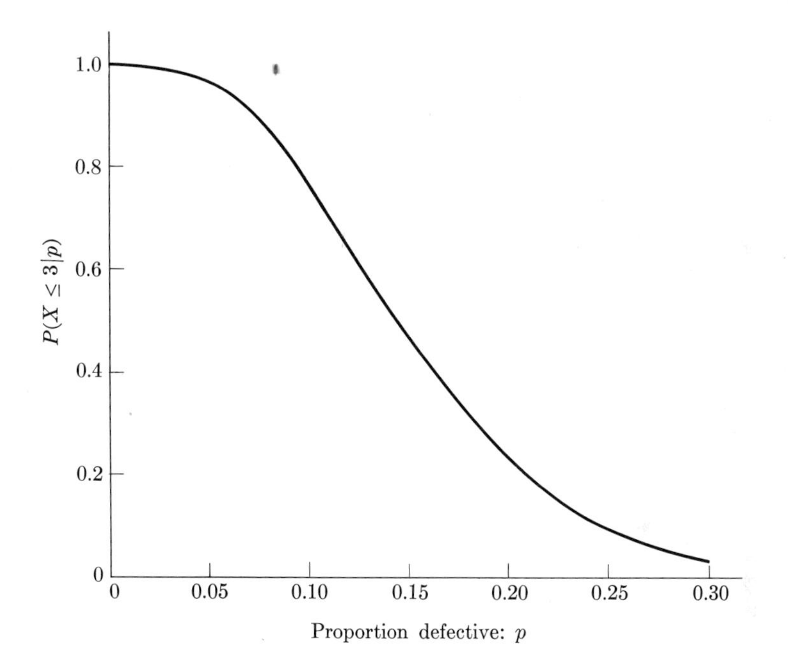

Fig. 9–3 Operating characteristic: $n = 25$, rejection number 4, acceptance number 3.

Solution Let the random variable X be the number of defectives in the sample. From Table IV–B, the probability of accepting the process as "in control", $P(X \leq 3)$, can be readily computed for any given proportion of defectives, p. Figure 9–3 shows $P(X \leq 3)$ for different proportions of defectives p, $0 \leq p \leq 0.30$. The ordinate gives the probability of accepting the process as "in control" for the value of p shown in the abscissa. The graph is called the *operating characteristic* of the test. The ordinate of the operating characteristic of a test is the complement of the ordinate of the power function, and so for every value of p,

$$\begin{pmatrix}\text{ordinate of operating}\\ \text{characteristic}\end{pmatrix} = 1 - \begin{pmatrix}\text{ordinate of}\\ \text{power function}\end{pmatrix}.$$

When the process is operating at a level of 5% or fewer defectives, the sample rarely gives the judgment "out of control". The graph shows the probability of accepting as greater than 0.95. On the other hand, if p is large compared with 0.05, say $p = 0.15$, 4 or more defectives occur more than half the time, and the sampling process is likely to detect the change soon, if not immediately. If p is very large, say 0.25, the sample is almost certain to detect this at once, and appropriate action will be taken.

When more than 4 of 25 items are defective, the quality control man rejects the hypothesis that the process is "in control" in favor of the hypothesis that the process is "out of control" because he knows there is a good chance it may be, since machine settings, inattention, and new raw materials are all common sources of trouble. Therefore, when he sees a high number of defectives, he would rather assume that something has gone wrong and look for it, than merely assume that a very unusual sample has occurred in a process that is in control.

For $p = 0.15$, Fig. 9-3 shows that about half the time the fact that the process is "out of control" will not be detected by the sample. The manufacturer may wish to steepen the curve so as to discriminate more immediately and sharply between a process producing 5% defectives and, say, 15%. By increas-

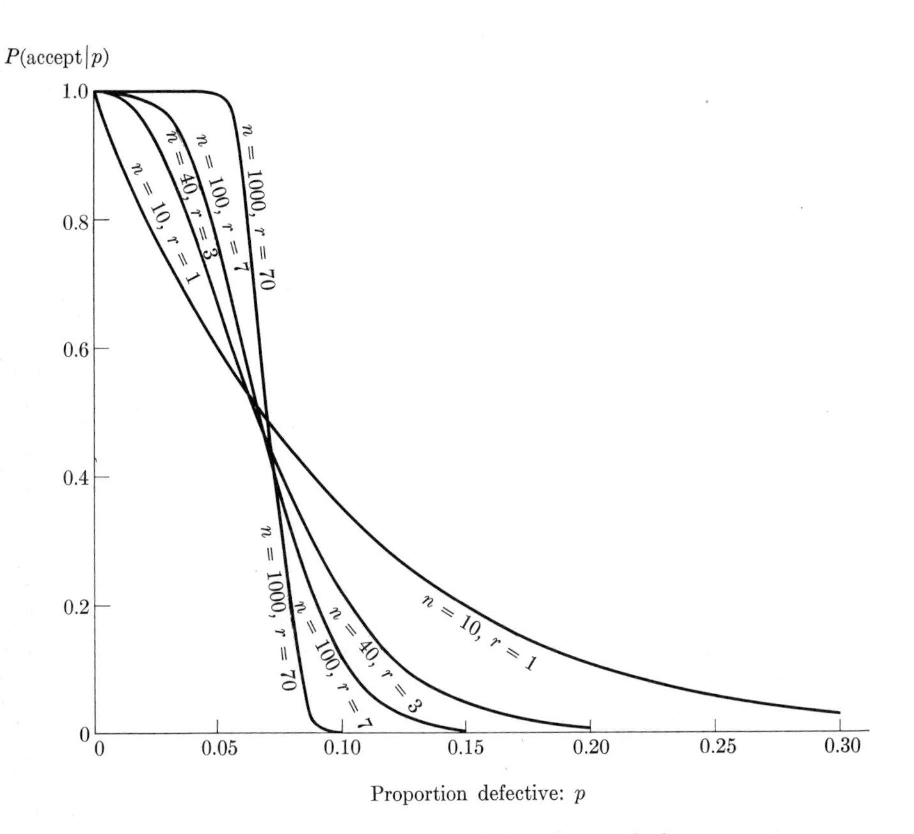

Fig. 9-4 Operating characteristics of several plans.

ing the sample size n and changing the rejection number r ($r = 4$ in the example), the shape of the operating characteristic can be changed.

Figure 9–4 shows a set of operating characteristics for several acceptance sampling plans, all designed to have probability of about $\frac{1}{2}$ of rejecting the process when $p = 0.07$. As n increases, r increases, and the operating characteristic becomes steeper at $p = 0.07$. The curves were made with the aid of large binomial tables, but the normal approximation could have been used for the larger values of n.

Such plans are used to help control a process or to help a buyer decide, by inspecting a random sample, whether the lot of material he purchased has the quality the seller claims.

Example 5 *Biased dice.* Are inexpensive plastic dice whose pips are drilled out biased in favor of even numbers because there are $6 + 4 + 2 = 12$ pips drilled from these, but only $5 + 3 + 1 = 9$ drilled from the odd numbers? The drilling means that even sides may be slightly lighter on the average. Over a period of time, Willard H. Longcor tossed such dice 1,160,000 times and got "evens" 588,410 times. Test whether the dice are biased in favor of evens.

Solution Under the null hypothesis evens and odds are equally likely. The expected number of "evens" is 580,000. The standard deviation is

$$\sigma_X = \sqrt{1{,}160{,}000(\tfrac{1}{2} \times \tfrac{1}{2})} = 538.$$

The standard deviate (see end of Section 6–1 just after Corollary 6–8) is

$$z = \frac{588{,}410 - 580{,}000}{538} = \frac{8410}{538} = 15.6.$$

Since the total is more than 15 standard deviations from the null value, the inexpensive dice are definitely biased toward "evens". Indeed, our normal table goes only as far as 3.09 standard deviations because the tail probability is by then down to 0.001. Deviations more extreme than this correspond to levels of significance smaller than any in common use, and so correspond to strong rejection of the null hypothesis.

Choice of significance level The significance levels in most common use are 5%, 1%, and 10%. Generally speaking, tests based upon smaller samples should use larger levels of significance. For example, if one were two-sidedly testing $p = \frac{1}{2}$ against $p \neq \frac{1}{2}$ for a binomial sample of size 4, taking the extreme values $x = 0$ and $x = 4$ as the rejection region gives a significance level of $2(1/2^4) = 1/8$, which already exceeds 10%. Therefore, if one is unwilling to accept a level larger than 10% in this example, he should abandon the investigation or design a larger study. One of the main merits of the study of significance levels and power is that they may reveal that a proposed investigation should not be carried out. The anticipated effects may be too small to have a chance of being detected by the proposed investigation.

Hypotheses suggested by the data: problems of multiplicity In discussing testing methods, we have implicitly assumed that the hypotheses were set up for testing in advance of the investigation. When he is analyzing a body of data, hypotheses come to the experimenter's mind on the basis of the results actually found. He notices a large difference and asks whether it should be regarded as significant. Suppose that he makes a test of significance and finds that the difference is significant at one of the usual levels—perhaps it is 2.5 standard deviations out. What should he make of it? One possibility is to regard it as a good suggestion to be checked on a further sample. Thus he may not take the result of such a test as seriously as one planned in advance. Why should he take one attitude toward a difference that he has chosen to look at in advance and another toward one that arises from detective work? After all, it is the same number, and conceivably two investigators working independently on the same data could be in different positions on the same number, one having a hypothesis in advance, the other just exploring his data. Partly it is a matter of bookkeeping. If we plan to test 100 hypotheses at the 5% level, then from sampling fluctuations we expect 5 plus or minus a few to be significant even if there are no differences. If about this many are found, then one suspects that none of the differences is large and that the observed differences are primarily due to sampling variation.

Example 6 In studying Longcor's data on runs in dice tossing, 225 observed relations were compared with theoretical values using a 5% level. Sampling variation would have forecast 11 significant differences, and so when only 13 were found, the whole set of observed relations was regarded as close to the theoretical. (These were for high-precision dice.)

When a difference is suggested by the data, we have no bookkeeping available comparable to that in the example. We do not know how many hypotheses have been implicitly tested and ignored—in a large body, thousands may be secretly and unwittingly examined by eye. Since we have no way of telling how many have been looked at, we therefore have no way to appraise the total outcome.

This idea must not be carried to extremes. When data point toward a very significant difference—say, one of 5 or more standard deviations—then one should take it seriously. Differences of such a size usually hold up under repeated experiments even when many hypotheses are examined, whereas differences of the order of 1 and 2 standard deviations frequently do not. The reason is that achieving a difference of 5 standard deviations by sampling variation is so rare an event that even thousands of comparisons are not likely to yield one.

EXERCISES FOR SECTION 9-4

In the following exercises, if a significance level has not been chosen, choose one that you think might be appropriate and state what it is.

1. From Fig. 9–3, what is the probability that the process is judged "in control" if $p = 0.10$? If $p = 0.30$? What is the probability it is judged "out of control" if $p = 0.20$?
2. From Fig. 9–3, what proportions of defectives lead to judgments of "in control" 80% of the time? 10% of the time?
3. Check the answers to Exercise 2, using binomial tables.
4. Make an operating characteristic like Fig. 9–3 for the plan: samples of size 2, reject if 1 or more defectives are found. Compare with Fig. 9–3.
5. (Continuation.) Suppose 2 items are randomly drawn without replacement from 10 items, and the lot of 10 is rejected if the sample has one or more defectives. Show on the same graph as that of Exercise 4 the probabilities of accepting the lot of 10 for each fraction defective in the lot. (Thus you are in a position to compare the exact probabilities of this exercise with the binomial probabilities of Exercise 4 as an approximation. Binomial calculations are often made *as if* sampling were done with replacement, as an approximation for calculations for sampling without replacement.)
6. From Fig. 9–4, find for the four plans the values of p for which lots are accepted 95% of the time.
7. Use your binomial tables to find a plan (sample size n and rejection number r) that accepts about 85% of lots or processes with $p = 0.05$, and rejects about 90% of lots with $p = 0.20$.

For Exercises 8 and 9, consider the plan of Fig. 9–3 ($n = 25$, $r = 4$), and suppose that 100 large lots have 1% defectives, 100 have 5% defectives, and 100 have 20% defectives.

8. Find the expected number of lots accepted by the plan.
9. Find the expected number of defective items accepted if each of the 300 lots has 1000 items. Compute the percent of defective items accepted, and compare it with the percent in the original 300 lots.
10. Suppose a buyer uses one of the sampling plans described in this chapter. Suppose the seller always persuades the buyer to give any rejected lot a "second chance", using the same plan. What is the relationship between the actual operating characteristic for the new procedure and the operating characteristic of the original plan? (A second chance means that the buyer draws a new sample.)
11. An accepted lot is worth about \$500 to a manufacturer. A rejected one costs \$200 for reworking, and so has a net worth of about \$300. He can produce at $p = 0.10$ at no additional cost, and he can reduce p by an additional $0.01x$ for $5x$ dollars. If the plan of Fig. 9–3 is used, the probabilities of rejection for various values of p are as follows:

p	.10	.09	.08	.07	.06	.05	.04	.03	.02	.01
$P(\text{rejection} \mid p)$	.236	.183	.135	.094	.060	.034	.017	.006	.0015	.0001

At what p should he operate to maximize expected profit (or to minimize expected losses compared with the \$500 value for an accepted lot)?

12. Madame X says that she can tell by taste whether tea has been made with tea bags or with bulk tea. What null hypothesis would you choose if you wished to test her claim? She sips from 10 pairs of cups, one with each kind of tea, and correctly identifies 9 of the pairs. What descriptive level of significance would you attach to this experiment?
13. For the data of Exercise 12, use Chart I to set 95% confidence limits on p, the probability of correctly identifying a pair of cups. Then, at the 5% significance level, reject the null hypothesis, $p = \frac{1}{2}$, in favor of the alternatives, $p \neq \frac{1}{2}$, if $p = \frac{1}{2}$ is outside the confidence interval.
14. Fred has a die he believes may be loaded in favor of the side marked "six". He tosses it 4 times and gets three "sixes". Using the 5% level of significance, do these results cause you to reject the null hypothesis $p = \frac{1}{6}$?
15. Mr. Williams played 5 hands of bridge one evening and got no aces 4 times. He complained of poor shuffling. Given good shuffling, the probability p of getting at least one ace, on any one deal, is 0.7 (approximately). Are 4 no-ace hands out of 5 hands enough to reject the null hypothesis $p = 0.7$ at the 5% level of significance? (Use the binomial formula.)
16. A manufacturer of light bulbs says that only 10% of the frosted bulbs he manufactures have defective frosting, and that these defective bulbs occur at random during manufacture. A carton of 4 of his bulbs was purchased and 2 of these had defective frosting. Would you reject his claim at the 1% level of significance?
17. A patient suffering from chronic headaches has had 60% of a large number of headaches relieved by standard medication. A new component is added to his medication, and 17 of his next 20 headaches are relieved. Would you reject, at 5% level, the null hypothesis $p = 0.6$? Criticize the application of the binomial distribution to this experiment.
18. Find a 95% confidence interval for p in Exercise 17.
19. Five items are drawn from a large lot. If two or fewer are defective, the lot is accepted. Compute roughly the operating characteristic of this test, graph it, and tell for what percent defective half the lots will be accepted and half rejected.
20. To decide whether a coin is unbiased, one man flips the coin 4 times. If it comes up heads on all 4 flips, he rejects the hypothesis that it is unbiased. A second man mixes the balls in an urn containing 15 white balls and 1 red one, all alike except for color, and draws out 1 ball. If it is red, he rejects the hypothesis that the coin is unbiased, otherwise he assumes it is unbiased. (a) What is the null hypothesis for each man? (b) What is the significance level for each test? (c) What are the circumstances under which the first man's test is more powerful than the second man's?
21. Fertilizers A and B are used on 5 pairs of adjacent (randomly selected) plots of cabbage. The differences in yield in hundreds of pounds ($A - B$) are 6, 4, 2, 2, 1. Use the sign test to decide at the 10% level whether the fertilizers are equally likely to provide high yields.
22. In the Weldon dice experiment, 12 dice were thrown 26,306 times and the appearance of a 5 or a 6 was considered a success. The mean number of successes observed was 4.052383. Is this result significantly different from the expected average number of successes, 4? (Use normal approximation.)

23. In a psycho-physical experiment, a subject has a 30% detection rate for a signal, established by thousands of trials. After a vacation he returns to the laboratory and detects the signal on only one of the first 20 trials. The experimenter wonders whether the equipment and/or the subject have changed, or whether this large a deviation from the 30% rate is a frequent occurrence under sampling variation. Advise him and state your assumptions.
24. In a coffee-tasting experiment a subject tastes each of 10 pairs of cups of coffee and decides for each pair which cup contains the instant rather than the percolated coffee. The experimenter decides to call a person a "taster" if he decides correctly in at least 8 out of 10 pairs; otherwise he is called a "non-taster". Regarding this operation as a test of significance: (a) What is the null hypothesis? (b) What are the alternative hypotheses? (c) What is the level of significance? (d) If a subject has probability 0.8 of correctly calling a pair, what is the chance he will be called a "taster"?
25. To test the precision of an ordinary six-sided die, it is to be tossed 10,000 times and rejected if even numbers occur more than 5100 or fewer than 4900 times. (a) What is the significance level of the test? (b) If $P(\text{even}) = 0.51$, what is the power of the test?
26. To test whether the probability of an event is about 0.2, as opposed to being much higher, an experiment consisting of 10 independent trials is to be performed. If 0, 1, 2, 3, or 4 successes are observed, the null hypothesis of $p = 0.2$ is to be accepted, but if 5 or more successes are observed the hypothesis is rejected. (a) Find the significance level of the test. (b) If p is actually 0.5, what is the power of this test and what is its operating characteristic?
27. (Continuation.) Use your binomial tables to help you make a graph of the power of the test in the previous problem for all values of p.

9-5 CONTINGENCY TABLES; DIFFERENCES OF p's

When individuals or events have more than one attribute, it is convenient to cross-classify the sample or the population according to these attributes, as we did in the tables of Chapter 4. When the properties in question consist of two classes for each of two variables, the cross classification forms a 2×2 (pronounced "two by two") table which, together with its counts, proportions, or probabilities, is called a contingency table. A frequent first question posed is whether the data in the table could have arisen "by chance", meaning by random allocation to the categories, considering the totals in each class and allowing for a reasonable amount of sampling variation. This question harks back to the discussion of the hypergeometric distribution of Section 3-5, as the following two examples illustrate. The method is called Fisher's exact test.

Example 1 *Designing a test for classifying individuals.* A clinical psychologist has the impression that he can tell from a man's writings, such as business letters, whether the individual is the firstborn son or a later son in the family. To test his impressions he asks a colleague to prepare $2n$ sets of materials written by men from families containing two boys, half from older and half from younger sons. He plans to sort these 50–50 into two categories, and then to appraise his performance in comparison with the null hypothesis of random sorting. If he wants to allow himself the possibility of making two wrong choices and still reject the null hypothesis at the 0.05 level, then how big must n be?

Solution The tables show the arrangements that the psychologist wants to count as significant at the 0.05 level.

		All correct Actual Older	All correct Actual Younger		Two wrong Actual Older	Two wrong Actual Younger	
Psychologist's analysis	Older	n	0	n	$n-1$	1	n
	Younger	0	n	n	1	$n-1$	n
		n	n	$2n$	n	n	$2n$

Let's review the reading of these tables. In the first table, reading across the first row we note from the margin that the psychologist ascribed n letters to the category of older brother. In actuality all were written by older brothers, and so he had 0 mistakes on that line. Similarly he had 0 mistakes on the second line. If instead the result comes out like the second table, then the first line says that of n letters ascribed to older brothers, 1 was actually written by a younger brother, and $n - 1$ by an older brother. And so this line contains 1 mistake, as does the second line, for a total of two. Alternatively, we could get the same information by reading down columns. In this second table, reading down the first column we see that of the n writings by older brothers, 1 was ascribed erroneously to a younger brother.

The *null hypothesis* for this test assumes that, with probability $\binom{n}{x}\binom{n}{n-x}/\binom{2n}{n}$, the psychologist would get results like the following table.

		Actual Older	Actual Younger	Row totals
Psychologist's analysis	Older	$n-x$	x	n
	Younger	x	$n-x$	n
Column totals		n	n	$2n$

Because row totals and column totals are fixed, all entries of the table are fixed as soon as one cell is. The entry in the upper right-hand corner, labeled x, can be any number from 0 through n. The probability of any particular value is given by the hypergeometric distribution of Section 3–5. If $x = 0$ or 1, the psychologist plans to reject the null hypothesis (randomness) in favor of the alternative that he has ability to pick correctly. A one-sided test seems appropriate. The probabilities (under the null hypothesis of randomness) are computed from the hypergeometric distribution of Section 3–5 as

$$P(2n \text{ or } 2n - 2 \text{ correct}) = \frac{\binom{n}{n}\binom{n}{0}}{\binom{2n}{n}} + \frac{\binom{n}{n-1}\binom{n}{1}}{\binom{2n}{n}} = \frac{(1+n^2)(n!)^2}{(2n)!}.$$

When we evaluate this function for trial values of n, we find that the smallest n that gives a value less than 0.05 is $n = 6$, for which $P \approx 0.04$. If his test rejects the null hypothesis, the psychologist will believe that his impression is correct, and no doubt then he will try to find out what clues in the letters are tipping him off.

If he uses this test for $n = 6$, to reject he must assign at least $\frac{5}{6}$ of letters by older sons to the older son category and no more than $\frac{1}{6}$ of those by younger sons to the older category. Thus his difference of observed proportions ascribed to older sons must be at least $\frac{5}{6} - \frac{1}{6} = \frac{2}{3}$ to get a difference significant at the 0.05 level.

In applications that arise more frequently than the kind illustrated by Example 1, the marginal totals for the two groups are not all fixed in advance. For example, the colleague might have constructed materials for 6 older and 6 younger sons, but the psychologist might allow himself to classify freely. In small samples, we still would make the analysis with the hypergeometric distribution essentially asking the question, "Given the margins, is it reasonable that the individuals could have been randomly assigned to the cells?" The next example illustrates the more usual situations.

Frequently we wish to compare two groups or two processes when their performance is expressed as the number of successes and failures each has achieved under comparable conditions. We use chance as a baseline to help make the comparison. We note how many successes and failures have occurred altogether in the two groups and ask whether *random* allocation of the same total numbers of successes and failures to the two groups would likely have resulted in an outcome at least as extreme as the one we have observed. If so, we "accept" the hypothesis that the groups perform about equally well; if not, we reject this hypothesis. Let us illustrate with a classical medical example.

Example 2 *Sterile bandages.* After surgeons began using clean instead of dirty bandages for wounds, the surgeon Lister went further and tried bandaging with antiseptic methods, which we shall call sterile bandaging. Data like those shown in Table 9–1 were found. How strong is the evidence in favor of the sterile bandaging in these data?

Solution We ask what the chances are of getting a table at least as favorable to sterile bandages as this table if the 3 cases of blood poisoning are distributed

Table 9–1 Illustrative data for sterile bandaging example

	Bandaging		
Patient	Sterile	Ordinary	Total
Well	4	3	7
Blood poisoning	1	2	3
Total	5	5	10

by chance. Again we use the hypergeometric of Section 3–5. Recall that in a table with fixed margins from random allocation, the probability of the given cells is

$$P(x \text{ in upper left cell}) = \frac{\binom{m}{x}\binom{w}{r-x}}{\binom{n}{r}},$$

where the notation is given by:

	Sterile	Ordinary	Total
Well	x	$r - x$	r
Blood poisoning	$m - x$	$w - r + x$	$n - r$
Total	m	w	n

The four possible tables with margins $m = w = 5$, $r = 7$, $n - r = 3$ are:

Table more favorable to "sterile"	"Observed" table	Tables less favorable to "sterile"	
5 2 0 3	4 3 1 2	3 4 2 1	2 5 3 0

The probabilities for these tables are

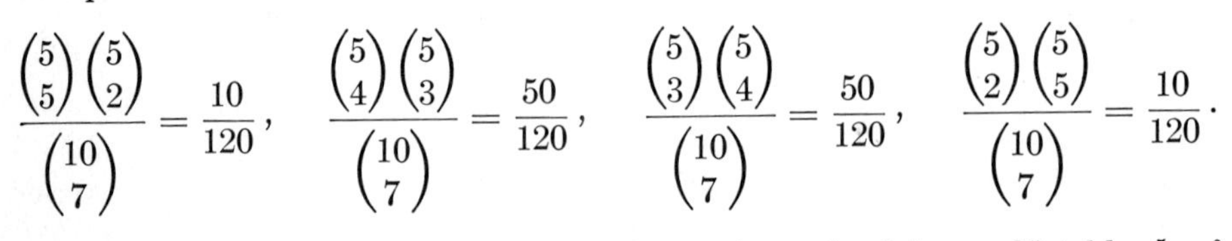

$$\frac{\binom{5}{5}\binom{5}{2}}{\binom{10}{7}} = \frac{10}{120}, \quad \frac{\binom{5}{4}\binom{5}{3}}{\binom{10}{7}} = \frac{50}{120}, \quad \frac{\binom{5}{3}\binom{5}{4}}{\binom{10}{7}} = \frac{50}{120}, \quad \frac{\binom{5}{2}\binom{5}{5}}{\binom{10}{7}} = \frac{10}{120}.$$

We see that chance assignments would have given the "observed" table $\frac{5}{12}$ of the time and a table more favorable to "sterile" $\frac{1}{12}$ of the time. And so chance would have given results at least as favorable to sterile bandaging as we found half the time. This doesn't say anything against sterile bandaging, but it doesn't give strong evidence that it is preferable to ordinary bandaging.

Suppose that the data had been considerably stronger, say:

	Sterile	Ordinary
Well	14	9
Blood poisoning	1	6

Then the only table more favorable to sterile bandages would be

$$\begin{array}{|cc|} 15 & 8 \\ 0 & 7 \end{array}\,.$$

The probability of it together with the observed table is

$$\frac{\binom{15}{15}\binom{15}{8}}{\binom{30}{23}} + \frac{\binom{15}{14}\binom{15}{9}}{\binom{30}{23}} \approx 0.040.$$

Thus under chance allocation of the seven blood-poisoning cases to the two treatments, the probability of a result at least as favorable to sterile bandages as observed would be about 0.040, and from such evidence we would be inclined to reject the hypothesis of a random distribution of blood poisoning in favor of the effectiveness of the sterile bandages.

Computation of the hypergeometric function is simplified by the existence of an extensive table: G. J. Lieberman and D. B. Owen, *Tables of the Hypergeometric Probability Distribution* (Stanford University Press, 1961).

Using the hypergeometric distribution as described in these examples offers us a general method for testing the significance of difference in performance when two groups are to be compared with respect to a dichotomy (Fisher's exact test).

Large samples For large samples, the hypergeometric approach to the comparison in contingency tables becomes computationally awkward and we take a different tack. Let us continue to illustrate with the bandage problem.

We have the idea that a particular class of wounds and a particular type of bandaging give a fixed probability of not producing a serious infection. Let p_S be the probability that a patient gets "well" if treated with sterile bandaging, p_O that he gets "well" if treated with ordinary bandaging. Then we ask: Is $p_S > p_O$ or is $p_S \leq p_O$? If $p_S \leq p_O$ then the sterile bandages are of no extra help, while if $p_S > p_O$ they are. (It is easy by hindsight to say that sterile bandages could not have been worse than ordinary, but Lister found plenty of opposition in his time, and except for his ultimate success, we cannot be sure that something in his procedure might not have been bad for the healing process. One of the techniques he recommended was to keep the bandage wet with carbolic acid.)

We regard $p_S \leq p_O$ as the set of null hypotheses and $p_S > p_O$ as the set of alternatives. To test whether $p_S > p_O$ we naturally look at the difference in observed proportions $\overline{p}_S - \overline{p}_O$. If the difference is reliably large and positive, we decide in favor of sterile bandages, otherwise not.

To set up this test we use two facts about differences. The distribution of the difference between two independent normally distributed random variables is itself normally distributed, with mean equal to the difference of the component means and variance equal to the *sum* of the component variances. Let us apply these facts to the present problem.

If the sample sizes are large, we know from Chapter 8 that $\overline{p}_S$ and $\overline{p}_O$ are approximately normally distributed, and from Chapter 6 that they have means p_S and p_O and variances $\sigma_S^2 = p_S(1 - p_S)/n_S$ and $\sigma_O^2 = p_O(1 - p_O)/n_O$, where n_S and n_O are the sample sizes associated with sterile and with ordinary bandages, respectively. The distribution of $\overline{p}_S - \overline{p}_O$ is approximately normal with mean $p_S - p_O$ and variance $\sigma_S^2 + \sigma_O^2$.

If we reject the hypothesis $p_S = p_O$ in favor of $p_S > p_O$, then we also reject all values of p_S where $p_S < p_O$. Therefore, we choose the borderline null hypothesis $p_S = p_O$ as a base for making the test. We need to know whether sterile bandages do better than ordinary. In this extreme null hypothesis $p_S - p_O = 0$. We pretend that $p_S = p_O$ and then use a one-sided test to see whether the assumption is reasonable. Let's suppose that the common value of p_S and p_O is p. Then for the case of equal p's

$$\begin{aligned}\text{Var}(\overline{p}_S - \overline{p}_O) &= \frac{p(1-p)}{n_S} + \frac{p(1-p)}{n_O} \\ &= p(1-p)\left(\frac{1}{n_S} + \frac{1}{n_O}\right). \end{aligned} \tag{1}$$

As we did earlier in setting confidence limits, we estimate p from the observed numbers, here x_S and x_O, of well patients,

$$\overline{p} = \frac{x_S + x_O}{n_S + n_O}. \tag{2}$$

Finally we use as our test statistic the approximately standard normal deviate

$$z = \frac{\overline{p}_S - \overline{p}_O}{\sqrt{\overline{p}(1 - \overline{p})\left(\dfrac{1}{n_S} + \dfrac{1}{n_O}\right)}}, \tag{3}$$

and ask whether it is significantly large.

Example 3 *Large sample difference of p's.* Suppose that in Lister's time a lengthy investigation gave:

	Sterile	Ordinary	Total
Well	80	120	200
Blood poisoning	20	80	100
Total	100	200	300

Test whether sterile bandaging performs better than ordinary bandaging.

Solution The necessary data for the test are

$$\overline{p}_S = \frac{80}{100} = 0.8, \qquad \overline{p}_O = \frac{120}{200} = 0.6, \qquad \overline{p} = \frac{80 + 120}{100 + 200} = \frac{2}{3} \approx 0.67,$$

$$\sqrt{\overline{p}(1 - \overline{p})\left(\frac{1}{n_S} + \frac{1}{n_O}\right)} = \sqrt{\frac{2}{3}\left(\frac{1}{3}\right)\left(\frac{1}{100} + \frac{1}{200}\right)} = 0.0577.$$

And so the test statistic gives

$$z = \frac{0.8 - 0.6}{0.0577} = \frac{0.2}{0.0577} = 3.47.$$

The normal table shows that the probability of results at least this favorable to sterile bandages under chance allocation is very tiny, less than .001. And so we would reject the hypothesis that sterile bandages were no better than ordinary ones. If we wish to be formal about it, we would set some significance level beyond which we plan to reject, say, .01. Our value of z gives a tail probability less than .001, and because it is less than .01 we reject the null hypothesis. Informally the question is would you rather believe sampling error produced a probability smaller than .001 or that one of many alternatives did. Most of us will prefer the latter.

More generally, if p_1 and p_2 are binomial proportions, n_1 and n_2 are large sample sizes, and x_1 and x_2 are the numbers of successes associated with p_1 and p_2, then to test whether $p_1 > p_2$ as opposed to the null hypotheses $p_1 \leq p_2$, we use the test statistic

$$z = \frac{\overline{p}_1 - \overline{p}_2}{\sqrt{\overline{p}(1 - \overline{p})\left(\frac{1}{n_1} + \frac{1}{n_2}\right)}}, \tag{4}$$

where $\overline{p}_i = x_i/n_i$, $i = 1, 2$, and $\overline{p} = (x_1 + x_2)/(n_1 + n_2)$, and refer z to a normal table. We *reject the null hypothesis at the significance level* α if

$$P(Z \geq z) = \text{area to the right of } z$$

is less than α.

Example 4 *Unemployment ratio.* Two independent labor force surveys having samples of 25,000, taken one year apart, show a change from 5% to 4% unemployed. Do we have strong grounds for believing the unemployment rate has gone down?

Solution We estimate $\bar{p} = 0.045$. Eq. (4) gives

$$z = \frac{.05 - .04}{\sqrt{\dfrac{.045(.955)2}{25{,}000}}} = 5.39.$$

The result is highly significant, and firmly establishes that there has been a reduction in the proportion of unemployed. Ordinarily, results no further from the null value than 2 or 3 standard deviations are regarded as the extremes to be accepted as reasonably due to sampling variation. Slight changes in the definition of unemployment could cause a change in rate.

Confidence limits on $p_1 - p_2$ We may have no reason to suppose that two proportions are equal, and we may want to report their observed difference together with some notion of its unreliability. Confidence limits give us a way to do this.

Example 5 *Opinion changes.* A polling agency found that in two independent samples of 2000 taken 3 months apart, the percentage responding favorably to Great Britain in an attitude question increased from 50 to 60. Set confidence limits on the true difference in proportions.

Solution Let p_1 and p_2 be the true population proportions favorable to Great Britain on the first and second occasions. We know from the theorem mentioned at the beginning of this section that if $\bar{p}_1$ and $\bar{p}_2$ are the observed proportions, then

$$\text{Var}\,(\bar{p}_1 - \bar{p}_2) = \frac{p_1(1 - p_1)}{n_1} + \frac{p_2(1 - p_2)}{n_2}. \tag{5}$$

We do not assume $p_1 = p_2$, but we shall need to estimate p_1 and p_2. We can use the observed proportions. Then upper and lower 95% confidence limits would be given by

$$\bar{p}_1 - \bar{p}_2 \pm 1.96\sqrt{\text{Var}\,(\bar{p}_1 - \bar{p}_2)}. \tag{6}$$

Other confidence coefficients could be obtained by replacing 1.96 by other numbers from the normal table.

Substituting the numbers from our numerical example in expression (6) gives

$$0.50 - 0.60 \pm 1.96\sqrt{\frac{.5(.5)}{2000} + \frac{.4(.6)}{2000}} = -0.100 \pm 1.96(0.0157)$$
$$= -0.100 \pm 0.031 = -.069, -.131.$$

And so the lower and upper 95% confidence limits on the difference are -7%

and -13%. The shift seems real, but even these large samples allow a substantial range in its actual value.

Expression (6) and similar ones obtained by replacing 1.96 by alternative values from the normal table produce confidence limits on differences that have other confidence coefficients.

Example 6 *Job stability.* In a newly industrialized country, a small study of factories showed that of 100 newly employed from the rural area, 50 left the job within 3 months, while of 200 newly employed from urban areas, 76 left within 3 months. Set 95% confidence limits on the difference in proportions leaving within 3 months.

Solution Let p_1 be the proportion of rural and p_2 the proportion of urban dwellers leaving within 3 months. Then we get $\overline{p}_1 = 0.50$, $\overline{p}_2 = 0.38$, and the limits are

$$0.50 - 0.38 \pm 1.96\sqrt{\frac{.50(.50)}{100} + \frac{(.38)(.62)}{200}} = 0.00,\ 0.24.$$

We have 95% confidence that the difference in leaving rates is between 0 and 24%.

Caution In the solution we have behaved as if the individual worker's decisions were independent of one another. If a strike, for example, or the closing of a factory made many leave at one time, then the assumption of independence would fail and our estimate of variability of the $\overline{p}$'s would be too small. Indeed, we could not then readily apply the methods described here. They would exaggerate the statistical significance of the difference. A large difference may be a message that the assumptions have broken down rather than that there is a difference in the p's. In many investigations, lack of independence does increase the actual variance while the investigator does not take account of it. He will then be surer of a difference than he ought to be, and more likely to be surprised if the effect does not hold up under repeated experimentation. Proper handling of such difficulties goes beyond the scope of this book.

In some problems, considering the proper unit of measurement is a help. For example, in a test of skill, suppose one person, Jones, using method A succeeded in 90 trials of 100, while 100 persons independently using method B once each succeeded in 40 trials. On the face of it, method A performed better, but we are suspicious because Jones did the whole 100 trials. If Jones's 100 trials are independent of one another, what we may have found is that Jones is better than the people using method B, rather than that method A is better than method B.

Although we would like our tests to focus only on the differences or features that interest us, we may actually be testing the entire probabilistic model. The feature causing significance or insignificance of results may not be the one we are trying to study. Formulas are never the whole answer; one must consider the assumptions.

EXERCISES FOR SECTION 9-5

1. In an experiment a new method, A, was successful in 3 trials out of 3, while the old method, B, succeeded once in 4 trials. Assuming a random allocation of successes, what is the probability that method A does this well?

2. If method A succeeds n times out of n and method B fails n times out of n, what is the smallest value of n that will make a one-sided test that method A is more successful than method B significant at the 5% level?

3. Method A succeeded once and failed once. Method B succeeded n times out of n. Find the probability of getting a result at least as extreme as this in favor of method B if the failures are randomly allocated. How big must n be if method B is to appear significantly better than method A at the 5% level?

4. One day out of 4, Mr. A works at the cash register, and on the other days Mr. B works at it. During 100 days the cash count at the close of the day had not checked on 3 days, and 2 of these were Mr. A's working days. If Mr. A and Mr. B were equally accurate people, how unusual is it that Mr. A would have at least 2 of the 3 errors?

5. An investigator comparing the effects of two treatments, an experimental and a control, may get 0 successes in 10 trials from the controls. Assuming that the controls do give 0 successes, how many successes would he need in an experimental group of size 5 to reach the one-sided 5% level of significance in favor of the hypothesis that the experimental group differed from the controls?

6. (Continuation.) Suppose that a control group of size n has no successes and an experimental group of size $n/2$ has 1 success. Would it have increased the significance of the difference between control and experimental results if the sample size of the experimental group had been doubled to n and the number of successes had also doubled to 2?

7. *Spinning pennies.* The children of R. E. Beckwith spun, rather than tossed, unworn pennies on a smooth surface and observed the following numbers of heads when the coins came to rest: for a 1955 Denver penny, 50 heads in 100 spins; for a 1964 Denver penny, 14 heads in 100 spins. Test these results to see whether pennies are going out of balance.

8. *The bibliophilic French.* In a sample of 1000 Frenchmen, 59% own television and of these 29% buy books. Of those without television 22% buy books. Test for a significant difference in the percentages buying books.

9. *Air Force accidents (Webb and Jones).* The table below shows for one period of time the numbers of U.S. Air Force pilots having accidents, by type of duty. Test whether Bomber service and Transport service have equality of safety.

	Bomber	Transport
No accident	26,307	36,244
Accident	291	389
Total	26,598	36,633

10. *Differences in operative procedure* (*L. B. Ellis and D. E. Harken*). Of 36 cardiac patients reoperated on by the "early" technic, 64% were improved at the end of the first year, while of 43 cardiac patients reoperated on by the Ivalon-tunnel technic 74% were improved at that time. Find the two-sided descriptive level of significance for the difference in reoperative improvement under the two technics.

11. *Hail suppression* (*Paul Schmid*). In the Swiss Hail Suppression experiment, seeding was intended to *reduce* the amount and frequency of hail. Seeding was randomly carried out on storm days. The following table shows the observations made on days with maximum wind velocity between 40 and 80 km/h.

	Days without seeding	Days with seeding
Hail	5	15
No hail	55	57
Total	60	72

Test the hypothesis of equal frequencies of hail days, with and without seeding. Discuss the outcome as opposed to the intention of the seeding.

12. Use the data of Exercise 11 to set 95% confidence limits on the difference in the proportion of days with hail under the two methods.

13. Use the data of Exercise 10 to set 90% confidence limits on the difference in success rate for the two methods.

14. Use the data of Exercise 9 to set 80% confidence limits on the differences between the two accident rates.

15. *Lister experiment.* In 1882, Donald MacAlister asked in a British paper, "Of 10 cases treated by Lister's method, 7 did well and 3 suffered from blood-poisoning; of 14 cases treated with ordinary dressings, 9 did well and 5 had blood-poisoning; what are the odds that the success of Lister's method was due to chance?" This led to a lively mathematical controversy in the columns of the paper (see C. P. Winsor, "Probability and Listerism", *Human Biology*, September 1948, Vol. 20, No. 3, pp. 161–169). In this section, we have not asked the question just this way. We asked, "If chance is at work, what are the chances of a result at least this favorable to Listerism?" A Bayesian development would be required to answer MacAlister's question. Analyze the data using the methods of this section.

Accuracy of economic forecasts. The following data relate economic forecasts by reputable forecasters to the actual results in the economy. In Table A, by Victor Zarnowitz, the abbreviation TP stands for "turning point", either actual or forecast. An actual turning point means that the economy changed direction; no turning point means that it continued to rise or fall as it had been doing. Table B, by Geoffrey H. Moore, relates forecasts of rises and falls in the market to actual movements. The data are from 1947–1965.

16. Use the data of Table A and the hypergeometric distribution to test the significance of the relation between actual and predicted turning points.

Table A Forecast made near middle of preceding year

		Forecast	
		No TP	TP
Actual	No TP	6	2
	TP	3	3

Table B Forecasts about 2 to 4 quarters (6–12 months) beyond the present quarter

		Forecast	
		Rise	Fall
Actual	Rise	23	10
	Fall	7	5

17. Use the data of Table B to test the relation between the forecasts and the performance of the economy.

9-6 BAYESIAN APPROACH TO ESTIMATION

Statistical inference If we could answer the question "What probability distribution generated my observations?", we would have answered many of the questions of statistical inference. The process takes several steps. First, we assume that some family of distributions (for example, normal or binomial) is, or approximates, the one we are sampling from. This family often depends on a few parameters (such as μ and σ for the normal, or p and n for the binomial). In this section we treat only the family of binomial distributions. Then we discuss parameters as a shorthand notation for the distribution, though it is usually the distribution that we make inferences about.

Two of the classical methods, confidence intervals and Bayesian inference, take steps toward answering the original question by indicating that some sets of parameter values are more likely to include the true value than are others. Here "likely" is intended as a vague term to be made specific for each method, as we have already made it specific for confidence intervals. Bayesian inference gets its name from Bayes's Theorem, which plays a central role in the calculations.

The Bayesian approach leads us to probability distributions for our degree of belief about the value of the parameter.

Our emphasis on beliefs rather than on the actual state of the world comes from a somewhat deterministic view of the universe. We usually think of the properties of the world, such as the probability that a coin, even a biased coin, comes up heads, as fixed. We think of there being a distribution of possible values of p for the set of coins we have drawn this one from. We think of drawing a coin, and thereby its probability p of coming up heads. We think of that probability as fixed, as we have throughout the book. But our beliefs about it start with our information about the distribution of p's for the coins we might have chosen, and these beliefs are altered to new beliefs as experimental information about the number of heads obtained in actual tosses from *this* coin is gathered. As the data grow, the belief distribution will concentrate its probability in a very small neighborhood around the true p, which in the case of a fair coin would be $\frac{1}{2}$.

Strengths The strengths of the Bayesian approach come from its unified treatment of all problems of uncertainty and from the directness of the inference from the sample to the belief distribution.

Difficulties To get started on a Bayesian inference, one must choose an initial probability distribution, called a prior distribution, and this choice is an easy target for criticism, because one man's choice may not match another's. Choosing a family of data distributions that includes ours, here the binomial, is also a difficult and important decision. Because we have studied so few families of distributions in this book, we might not appreciate that enormous numbers of them exist and that few have formulas we can work with easily. This in turn means that we shall have to make approximations, as we also do for confidence limits. We discuss some simple discrete problems that approximate continuous ones.

In setting confidence limits, we give some notion of the unreliability of our estimate. Sometimes we may have prior information, and then we may try to combine it with experimental information to get an improved estimate of p. Bayes's Theorem, Theorem 4–10, offers some assistance. Since p is a continuous variable, it is natural to use calculus methods in this approach. To avoid the calculus, we shall act as if our beliefs about p had a discrete distribution.

Example 1 A manufacturing process has a machine that inspects every item for internal flaws. Over a long period, lots have had the following relative frequencies of percent of defective items.

Table 9–2 Prior distribution of percent defectives

Relative frequency of lot	0.6	0.3	0.1
Percent defective	1	5	10

Thus 60% of the lots are classified as in the 1% defective class, 30% in the 5% defective class, and 10% in the 10% defective class.

The usual inspecting machine is broken, but the rest of the production process is working and a new lot is to be inspected by another, more expensive operation. A sample of 20 items is drawn from a large lot and none have internal flaws. What can we say of p, the proportion of defectives, for this lot?

Solution We regard the discrete distribution of this example as approximately the prior probability distribution for the random variable $100p$, the percent of defectives in a random lot chosen from this process. The new lot (the population from which the sample of 20 items is drawn) has some unknown percent of defectives, say $100p_0$. On the basis of the past data, we want to use the sample information that 0 defectives were observed in 20 trials to find a posterior distri-

Table 9–3 Calculation of posterior probabilities

Lot composition (percent defective)	Probability of 0 defectives in 20 trials	Posterior probability
1	$.6(.99)^{20} \approx .6(.818) = .4908$	.804
5	$.3(.95)^{20} \approx .3(.358) = .1074$	.176
10	$.1(.90)^{20} \approx .1(.122) = .0122$	.020
	.6104	1.000

bution for (our beliefs about) $100p_0$. As usual in applying Bayes's Theorem, we set up (in Table 9–3) the probabilities of getting this sample from each of the 3 possible compositions of the lot. We get the probabilities of 0 defectives in 20 trials from our binomial Table IV–A.

For example, the prior probability that the new lot is in the 5% defective class is 0.3. Given that the new lot is in this class, the probability of 0 defectives in 20 items is $0.95^{20} \approx 0.358$. Consequently the probability that the new lot is in the 5% class and produces 0 defectives in the sample is $0.3(0.358) = 0.1074$. The posterior probability for the 5% class is the conditional probability that the lot is in the 5% class, given the sample outcome, or $0.1074/0.6104 \approx 0.176$ (see Table 9–3).

The table of posterior probabilities suggests that the odds are 4 to 1 that $100p_0$ is in the class symbolized by 1%, about 1 to 5 that it is in the class symbolized by 5%, and that the chance is very small that it is in the class symbolized by 10%. One estimate of $100p_0$ could be obtained by computing the mean of the posterior distribution (we round to two decimals):

$$\mu = 1(0.80) + 5(0.18) + 10(0.02) = 1.90.$$

One advantage of the Bayes approach is that the probabilities derived apply to the class of lots with this prior and this sample outcome, whereas the confidence coefficient of a confidence limit statement applies to a long sequence of confidence statements. A difficulty is to supply a prior distribution for p, with the view that the p for the new lot is drawn from that distribution. Practical problems in the use of prior distributions are currently being studied by experts in probability and statistics. For an extensive application see F. Mosteller and D. L. Wallace, *Inference and Disputed Authorship: The Federalist* (Addison-Wesley, 1964). In this work Bayesian and other methods are used to solve the historical puzzle of the authorship of 12 Federalist Papers. Both Alexander Hamilton and James Madison claimed to be their sole author. The Bayesian analysis showed that Madison remembered correctly.

Discrete uniform prior Often we have no prior data and have no preference among the various possible values of the parameter to be estimated. In bi-

Table 9–4 Discrete uniform prior on the tenths (11 points)

p:	$0 = \frac{0}{10}$	$\frac{1}{10}$	$\frac{2}{10}$	$\frac{3}{10}$	$\frac{4}{10}$	$\frac{5}{10}$	$\frac{6}{10}$	$\frac{7}{10}$	$\frac{8}{10}$	$\frac{9}{10}$	$\frac{10}{10} = 1$
$P(p)$:	$\frac{1}{11}$	$\frac{1}{11}$	$\frac{1}{11}$	$\frac{1}{11}$	$\frac{1}{11}$	$\frac{1}{11}$	$\frac{1}{11}$	$\frac{1}{11}$	$\frac{1}{11}$	$\frac{1}{11}$	$\frac{1}{11}$

nomial problems, the prior distribution for p that is used to express lack of preference among intervals of p of equal lengths is the uniform distribution that we met in the "stopped clock" and the "cut string" examples of Section 7–2. Applied to p, the probability of a defective item, this uniform distribution should be continuous on the interval from 0 to 1. We shall approximate this by the discrete uniform with probability $\frac{1}{11}$ at each whole tenth on the scale, as shown in Table 9–4.

Example 2 *Two nondefective items.* Two items are randomly drawn from a large new lot, and both are nondefective. Let us assume a uniform prior for p, the probability of a defective, and approximate this by the distribution shown in Table 9–4. Compute the posterior distribution of p, find the posterior probability that $p \leq 0.2$, and find the posterior mean value of p.

Solution Let A be the event NN (two nondefective items drawn) and B the event $p = i/10$. Then, from the meaning of conditional probability, we can compute

$$P(A \cap B) = P(B)P(A|B).$$

We know

$$P(B) = P\left(p = \frac{i}{10}\right) = \frac{1}{11},$$

$$P(A|B) = (1 - p)^2 = \left(1 - \frac{i}{10}\right)^2.$$

Writing out the 11 possibilities for $P(A \cap B)$ as i runs from 0 through 11, we find

$$P\left(p = \frac{i}{10} \cap NN\right) = \frac{(10 - i)^2}{1100}, \qquad i = 0, 1, \ldots, 10.$$

Their sum is $P(A)$, which we determine numerically to be 385/1100. Finally, the posterior probability distribution for p, given NN, is

$$P\left(p = \frac{i}{10}\middle| NN\right) = \frac{1}{P(A)} \cdot P\left(p = \frac{i}{10} \cap NN\right).$$

Thus, writing $f(p|NN)$ for the posterior probability, we have

$$f(p|NN) = \frac{(10 - i)^2}{385}, \qquad i = 0, 1, \ldots, 10.$$

The numerical results are displayed in Table 9–5.

The probability that $p \leq 0.2$ is $0.260 + 0.210 + .166 = 0.636$, a considerable increase from $3/11 \approx 0.273$ for the uniform prior.

Table 9–5 Probabilities with discrete uniform prior for two nondefectives

p:	$\frac{0}{10}$	$\frac{1}{10}$	$\frac{2}{10}$	$\frac{3}{10}$	$\frac{4}{10}$	$\frac{5}{10}$	$\frac{6}{10}$	$\frac{7}{10}$	$\frac{8}{10}$	$\frac{9}{10}$	$\frac{10}{10}$	Total
$1100P(A \cap B)$:	100	81	64	49	36	25	16	9	4	1	0	385
$f(p\|NN)$:	.260	.210	.166	.127	.094	.065	.042	.023	.010	.003	.000	1.000

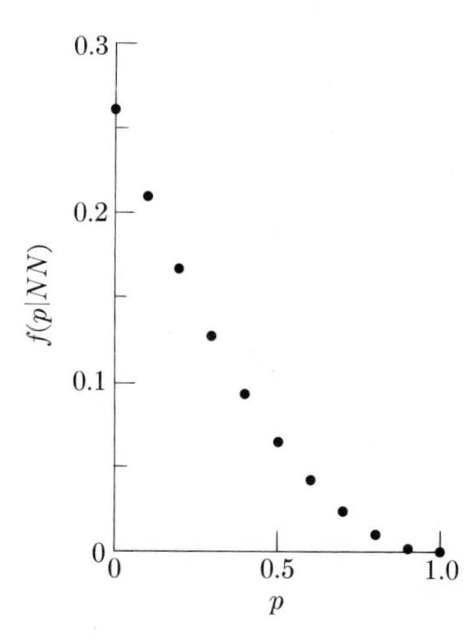

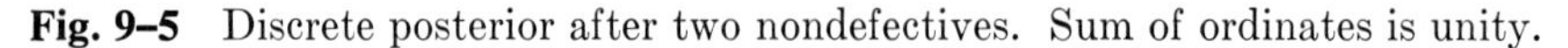

Fig. 9–5 Discrete posterior after two nondefectives. Sum of ordinates is unity.

To get the posterior mean for p we must compute

$$\begin{aligned}\mu_{\text{post},p} &= \sum pP_{\text{post}}(p) = \sum pf(p|NN) \\ &= \sum_{i=0}^{10} \frac{i}{10} \frac{(10-i)^2}{385} = \frac{0 \times 100 + 1 \times 81 + \cdots + 10 \times 0}{3850} \\ &= \frac{825}{3850} \approx 0.214.\end{aligned}$$

Hence, on the basis of two nondefective items, the mean of our belief distribution for p has reduced from 0.5 to 0.214.

The dots of Fig. 9–5, whose heights are tabled in Table 9–5, show the shape of the discrete posterior.

EXERCISES FOR SECTION 9–6

1. In the textual Example 1, replace the prior relative frequencies of lots 0.6, 0.3, 0.1, by 0.7, 0.3, 0, respectively. Find the posterior probabilities and the mean, μ, for the posterior distribution.

2. In the textual Example 1, replace the prior relative frequencies 0.6, 0.3, 0.1, by 0.4, 0.4, 0.2, respectively. Find the posterior probabilities and the mean, μ, for the posterior distribution.
3. In the textual Example 1, refer to Table 9–3. If the manufacturer uses the lot, he makes a profit of \$100 if the lot is in the 1% class, loses \$10 if the lot is in the 5% class, and loses \$1000 if the lot is in the 10% class. On the basis of the posterior distribution, should he use the lot? That is, is his expected profit positive?
4. In the textual Example 1, change the sample size to 25 and the number of defectives observed to 4. Find the posterior distribution and the mean, μ, for this distribution. (Strictly, p^4q^{21} is the probability of the sample, given p, but since $b(4; 25, p)$ is tabled and is proportional to p^4q^{21}, we use the binomial probability in the calculation of the posterior probabilities.)
5. In the textual Example 2, suppose that only one item had been drawn and it were found to be nondefective. Find the posterior distribution of p. Make a probability graph of your results.
6. (Continuation.) Use the results of Exercise 5 to compute the mean of the posterior distribution of p when one defective has been drawn.
7. In the textual Example 2, suppose that 3 items were drawn and found to be nondefective. Find $P(p = 0.4|NNN)$.

9–7 BAYESIAN INFERENCE WITH PERSONAL PROBABILITIES

Hypotheses are rejected because we believe their alternatives have a high chance of being true in the light of the total evidence. If we can quantify our prior beliefs, then Bayes's Theorem can aid us in problems like those treated in Section 9–4. Many people are unwilling to make such personalistic quantifications, but the following example illustrates how the personalistic approach through probability as degree of belief would work. The example is one about which you may have views.

Extrasensory perception Art claims that he has extrasensory perception (ESP). He says that if Bob conceals a red card in one hand and a black card in the other, he can tell which hand holds the red card. Bob doesn't believe it. Art admits he can't do this all the time, just "pretty often".

First, we give an approach like that of Section 9–4. In this form the problem is appropriate for a significance test with null hypothesis $p = \frac{1}{2}$ (Art has no ESP, and performs by guessing), alternative hypothesis $p > \frac{1}{2}$ (Art has some ESP). Bob can give Art a test consisting of a number of trials and a criterion for passing such that Art has a good chance of passing the test if he has a noticeable amount of ESP, say $p = 0.7$, and not much chance of passing if he doesn't ($p = \frac{1}{2}$). Let X be the number of successes in n trials. Then we read from our binomial tables that, with 25 trials, if $p = \frac{1}{2}$, $P(X \geq 16) \approx 0.115$, and if $p = 0.7$, $P(X \geq 16) \approx 0.811$. One test is to try 25 times and be right 16 or more times. Thus if Art has no ESP he will fail in about 89% of such tests, and if he has ESP amounting to $p = 0.7$ he will pass in about 81%. Tests with more trials can reduce the risk of passing or failing Art erroneously, and improve his chance of passing if he has some ESP but less than $p = 0.7$. Let us turn now to the personalistic approach.

To give a numerical value for Bob's disbelief in Art's ability, we would have to ask Bob what he thinks the chances are that Art has no ability ($p = 0.5$), or that he has probability $p = 0.6$, $p = 0.7$, or so on, of passing. (Properly treated there would be a continuous probability distribution for p, but we shall treat the prior distribution with a discrete approximation, as we did in Section 9–6.) For simplicity, suppose Bob's degrees of belief in Art's ability are as follows:

Hypothesis	Art's ability	Bob's prior degree of belief
H_1:	$p = 0.5$	0.98
H_2:	$p = 0.7$	0.02
		1.00

If n trials are performed and Art has x successes, then the following table helps in calculating the posterior degrees of belief from Bayes's Theorem.

	Probability of outcome
$p = 0.5$	$0.98(0.5)^x(0.5)^{n-x} = P(\text{Bob has } p = 0.5)P(\text{sample}\mid\text{Bob has } p = 0.5)$
$p = 0.7$	$0.02(0.7)^x(0.3)^{n-x} = P(\text{Bob has } p = 0.7)P(\text{sample}\mid\text{Bob has } p = 0.7)$

There are no binomial coefficients because we compute the probability of the particular sample in the order of occurrence of the calls. Had we introduced binomial coefficients, they would have dropped out of the later calculations. Therefore we may use them if there is any advantage in calculation.

In advance, Bob's odds (ratio of degrees of belief) were 0.98 to 0.02 or 49 to 1 against Art having ESP. After the experiment, Bob's odds against Art having ability are given by the ratio

$$\frac{0.98(0.5)^x(0.5)^{n-x}}{0.02(0.7)^x(0.3)^{n-x}} = \frac{49(0.5)^x(0.5)^{n-x}}{(0.7)^x(0.3)^{n-x}}.$$

Example 1 Suppose $n = 25$, $x = 17$. Find the posterior odds.

Solution From our Table IV–A, we get

$$\frac{49b(17; 25, 0.5)}{b(17; 25, 0.7)} \approx \frac{49(0.032)}{0.165} \approx 9.5.$$

Thus if Art correctly identifies 17 cards out of 25, Bob's odds have gone down from 49 to 1 to 9.5 to 1 against Art having the ability. He still doesn't believe much in Art's ability, but he has weakened by a factor of 5.

Additional data can readily be used, if available, to further modify Bob's odds.

Example 2 Starting with 49-to-1 odds, how big a sample do we need to be 95% sure that the odds are reversed (1 to 49), if Art has ESP to the extent $p = 0.7$?

Solution We want to find x and n that satisfy two properties: first,

$$49 \frac{(0.5)^x(0.5)^{n-x}}{(0.7)^x(0.3)^{n-x}} = \frac{1}{49}. \tag{1}$$

Taking three-place logarithms to the base 10 and simplifying gives

$$0.368x - 0.222n \approx 3.380. \tag{2}$$

The second property is that if $p = 0.7$ we want to find n and x so that

$$P(X \geq x) = 0.95.$$

The normal Table III shows that $z = -1.65$ is exceeded 95% of the time. Applying the central limit theorem approximation (Theorem 8–1) gives

$$\frac{x - \frac{1}{2} - 0.7n}{\sqrt{n(0.7)(0.3)}} = -1.65. \tag{3}$$

Let us pick a few convenient values of n, find the x's from Eq. (2), and see how well they satisfy Eq. (3).

n	x	Left side Eq. (3)	Normal table probability
100	70	−0.11	0.54
200	130	−1.62	0.95
300	190	−2.58	0.995

Consequently, a sample of $n = 200$ gives Art a probability of about 0.95 of reversing the odds if he has an ability $p = 0.7$.

A difficulty in this approach is the initial assessment of degree of belief; Bob might have said 999 to 1 or even 999,999 to 1. But this uncertainty is inherent in the problem. In ordinary hypothesis testing and confidence limit work there is a parallel uncertainty about appropriate choice of confidence level, significance level, and sample size.

Example 3 *Uniform discrete prior.* Suppose that we had started our ESP problem with the uniform discrete prior distribution of Section 9–6 and that Art made 20 guesses and got 12 correct and 8 wrong. Find the posterior distribution, and its mean and variance.

Table 9–6 Posterior distribution, $f(p|x = 12, n = 20)$, and calculations for its mean and variance

	$b(12; 20, p)$	$f = b/.475$	pf	p^2f
$p = 0$	.000	.000	.0000	
.1	.000	.000	.0000	
.2	.000	.000	.0000	
.3	.004	.008	.0024	.00072
.4	.035	.074	.0296	.01184
.5	.120	.253	.1265	.06325
.6	.180	.379	.2274	.13644
.7	.114	.240	.1680	.11760
.8	.022	.046	.0368	.02944
.9	.000	.000	.0000	
1.0	.000	.000	.0000	
	.475	1.000	.5907 = posterior mean	.35929 = $E(p^2)$

Solution Table 9–6 shows the work. A constant factor equal to 1/11, the prior probability associated with each discrete value of p listed in the first column of the table, has been omitted. If included, it would multiply each entry in the second column, but would not affect entries in the third column. Figure 9–6 shows the discrete posterior that approximates the correct continuous one. Notice that it spikes up near $p = 0.6$. All but about 5% of the probability is associated with $p = 0.4, 0.5, 0.6, 0.7$. The posterior mean of p is 0.59 and the posterior variance $0.3593 - .5907^2 \approx 0.0104$, giving a standard deviation of about 0.1.

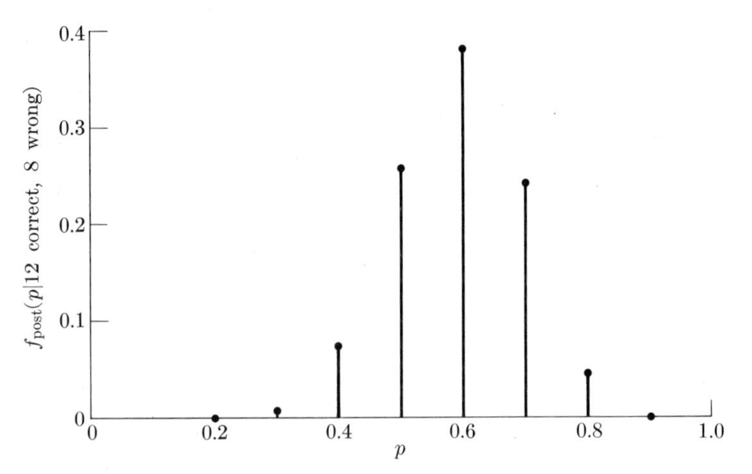

Fig. 9–6 Posterior distribution after 12 correct and 8 wrong (based upon a uniform prior distribution).

Including more evidence When additional data arise, we can readily include them by taking over the current posterior distribution as the new prior and including the data through a further application of Bayes's Theorem. You may like to know, without proof, that for a *continuous uniform* prior distribution for p, the parameter of a binomial probability distribution, the mean and variance of the posterior distribution of p are especially simple:

$$\mu_{\text{post}}(p) = \frac{x+1}{n+2}, \qquad \sigma^2_{\text{post}}(p) = \frac{\mu_{\text{post}}(1-\mu_{\text{post}})}{n+3}, \tag{4}$$

where x is the number of successes observed so far and n is the sample size. These formulas parallel the more familiar $\bar{p} = x/n$ and $\sigma^2 = p(1-p)/n$.

Bayesian confidence limits The curves of Chart I at the back of the book are based upon a continuous uniform prior distribution for p. A sample of size n with observed proportion $\bar{p}$ produces a posterior distribution for p with the mean and variance given in Eqs. (4). By taking the shortest interval containing a particular amount of probability, such as 98% for $n = 10$ for each possible $\bar{p}$, we could connect the resulting points with the curves shown for $n = 10$ in Chart I. Similar calculations are made for other values of n. The curves themselves are a convenience for the eye—the posterior distributions occur only for $\bar{p} = x/n$, $x = 0, 1, \ldots, n$. See Explanation of Chart I at the back of the book.

EXERCISES FOR SECTION 9–7

The first three of the following exercises assume that Bob's odds against Art are 9.5 to 1, and are concerned with the effects of further data on the odds. Start each problem at the 9.5-to-1 odds.

1. If Art's next 3 trials are successes, what are the new odds?
2. If Art's next 3 trials are failures, what are the new odds?
3. If Art has 3 successes and 3 failures, what are the new odds?
4. Find the relation between n and x that leaves the odds approximately constant in the ESP problem.
5. Starting with the initial 49-to-1 odds, what size n guarantees Art a 0.5 chance of at least reversing the odds if his $p = 0.7$?
6. Starting with 49-to-1 odds, if a test has $n = 200$ trials and Art's $p = 0.5$, what is the chance the odds against him go up to at least 100 to 1?
7. If Art gets exactly half right and half wrong in a test of size n, what value of n yields odds of about 1000 to 1 against him, starting with 49 to 1 against?
8. Bob holds prior odds of 1 to 1 that Madame X of Exercise 12 in Section 9–4 has $p = 0.5$ or $p = 0.7$ of being able to distinguish the pairs of teacups. After she gets 9 out of 10 right, what are Bob's posterior odds?
9. Use a 6-point uniform prior on the *even* tenths (0, .2, .4, .6, .8, 1.0) for p, the probability of success. A 3-trial experiment produces 2 successes and 1 failure. Find (a) the posterior distribution, (b) the posterior mean, (c) the posterior variance.

10. Interpret the mean and variance of the posterior distribution for p starting from a continuous uniform prior when the sample is 0. [Ans: These numbers $\frac{1}{2}$ and $\frac{1}{12}$ are just the mean and variance of the original uniform distribution.]

11. Use the facts about the posterior mean and variance for p based on a continuous uniform prior to set 95% probability limits on p if the observed number of successes in 100 trials is 30. (Assume that the posterior distribution of p is approximately normal.)

MISCELLANEOUS EXERCISES FOR CHAPTER 9

1. Bright red clothing has regularly been used by hunters for visual protection. (Recent experiments have assessed goldenrod yellow for this purpose.) Suppose we know that under standard conditions red is sighted 80% of the time. Four hundred trials with yellow are tried. (a) What would you take as the null hypothesis? (b) Using a level of significance of 0.05, set a criterion number of successes at or above which you would regard yellow as better than red. (c) If $p = 0.9$ for sighting with yellow, what is the probability that yellow does not reach the criterion?

2. Suppose that method A is known to yield 50% successes. It is desired to detect the superiority of method B if it yields as few as 55% successes ($p = 0.55$). A significance level of 0.025 is to be used. How big a sample is required to be 95% sure that a $p = 0.55$ will be detected as better? [*Note:* You need the normal approximation. First set up the criterion, using the largest null-hypothesis value of p, then find out how method B will perform against this criterion for a general value of n.]

3. The success rate in selecting applicants for a position is 60%. A new selection method may yield a different rate (not necessarily higher). Use the method of confidence limits to set up a two-sided test using a 10% level of significance, $n = 50$. If $p = 0.50$ for the new method, what is the probability that you decide the success rate is different from 0.60?

Data for Exercises 4, 5, 6. Of two brands of "fireproof" glass ovenware, a wholesaler wants to choose the one that withstands a greater sudden change of temperature. In testing the brands, he uses an oven and a tub of icewater. He tests a brand A piece and a brand B piece simultaneously as follows. He transfers the two pieces from the tub to the oven, which is set at 300°F. If neither breaks, both are returned to the tub. If neither breaks now, both are returned to the oven, in which the temperature has been advanced to 350°F. This process continues with 50° increases in oven temperature until one piece breaks. (Forget about both breaking at the same time.) The piece that breaks first is regarded as poorer, and its brand is regarded as poorer on that trial.

4. In 10 trials, brand A broke first 9 times. Is this often enough to reject at the 5% level the hypothesis that the two brands are of equal quality?

5. Another wholesaler ran a series of trials. He reported that brand A always broke first in his trials and that he rejected the hypothesis of the equality of the brands at the 5% level. What is the smallest number of trials he could have run?

6. A third wholesaler said he had run a lot of trials, and brand A had always broken first in his, too. He had concluded that brand A would always break first. In answer to a question, he said he supposed the probability of A's breaking first was 1.00. Without using pencil or paper, tell how many times in 1000 trials brand B would have to break first to reject this supposition.

Data for Exercises 7, 8, 9. A manufacturer produces capsules to be filled with medicinal powder. These are inexpensive and are sold in large numbers, in boxes of 100 capsules. One segment of the drug industry demands capsules with no discernible surface defects. Ten percent of our manufacturer's capsules have surface defects. In the handling and boxing operations the capsules get so thoroughly mixed around and stirred together that the 100 going into any box are for all practical purposes selected at random.

7. What fraction of the boxes will have just 10 defectives in each? [*Hint:* In the appropriate normal approximation, what fraction of the area lies between 9.5 and 10.5?]

8. A box of 100 with just 10 defectives in it is received by a drug concern which employs acceptance sampling procedures. Their plan is to examine two ($n = 2$) of the capsules at random and if either of them is defective ($r = 1$), reject the box. What is the probability that the box will be accepted?

9. Another druggist does his acceptance sampling by dumping 10 boxes together into one lot of 1000 with 100 defectives in it. He draws two capsules at random from the lot. He, too, rejects the lot if either of the two selected capsules is defective. How does his probability of rejecting the 1000 capsules compare with the first druggist's probability of rejecting a lot of size 100?

10. Cartons of eight 60-watt lamps are called lots. Each carton is inspected by testing 2 of the lamps, selected at random. The acceptance rule is that if both the tested lamps light, the carton is accepted; if either fails, the carton is rejected. A customer who has adopted this acceptance plan picks up a carton which happens to contain 2 defective lamps and 6 nondefective ones. What is the probability that he will accept the carton?

10 Theory of sampling. Variances of sums and of averages

10–1 CALCULATION OF THE DISTRIBUTION OF A SUM

Precise measurements When precise results are needed, an experimenter may repeat a measurement several times and compute the arithmetic average, $\bar{x}$, to estimate the required value. Although there are many good reasons for taking repeated measurements, the one dealt with in this chapter is that arithmetic averages of several measurements ordinarily vary less from one determination to another than do single measurements from one to another.

An average is obtained by summing the measurements and then dividing by the number of measurements. After we develop the theory of the variability of sums, that for averages follows easily.

Sums of many variables If 100 dice are rolled, what is the probability that the sum of the dots on their top faces exceeds 325? Here 100 independent random variables are being summed, and we are asked to compute a probability associated with the distribution of that sum. The methods of the next example can be extended to give an exact answer to the question. But the work required is long and tedious compared with the worth of the exact answer; consequently we later develop an approximate method based on the central limit theorem.

Example 1 *Sum of three independent random variables.* A coin is tossed once, and the random variable X is the number of heads; a thumbtack, with probability $\frac{1}{3}$ of landing point up, is tossed once, and the random variable Y is the number of thumbtacks landing point up; an engineer's ruler with sides numbered 1, 2, and 3 is rolled, and the random variable Z is the number on the face-down side. Find the exact distribution and the mean and variance of the random variable U, where $U = X + Y + Z$.

Solution We find the distribution in two stages: first we get the probability function of the sum of X and Y; then we combine $X + Y$ and Z. This technique can be continued for sums of still more independent random variables, say $X + Y + Z + W$, and so on.

Table 10–1 Sum of two independent random variables

		Value of Y: 1	0	$P(X = x)$
Value of X	1	$2, \frac{1}{6}$	$1, \frac{2}{6}$	$\frac{1}{2}$
	0	$1, \frac{1}{6}$	$0, \frac{2}{6}$	$\frac{1}{2}$
$P(Y = y)$		$\frac{1}{3}$	$\frac{2}{3}$	1

Probability	$\frac{2}{6}$	$\frac{3}{6}$	$\frac{1}{6}$
Value of $X + Y$	0	1	2

The entry in each cell in the top array in Table 10–1 shows, as the first entry, the value of the sum $x + y$ in the cell and, as the second entry, the probability of the cell. Probabilities are computed under the assumption of independence. The lower array of Table 10–1 shows the probability function of $X + Y$.

In Appendix 2 we show that when X, Y, and Z are independent random variables, then $X + Y$ and Z are also independent. We use this result to fill in the probabilities in the cells of the upper array in Table 10–2. The first entry in each cell is the value of the sum $(x + y) + z$ for that cell, and the second entry is the probability, i.e., the product of the row total and the column total. Collecting probabilities of the cells where $(x + y) + z = 3$, for example, we get $\frac{1}{18} + \frac{3}{18} + \frac{2}{18} = \frac{6}{18}$ for the corresponding entry in the lower array of Table 10–2. Other entries are obtained in the same way.

Table 10–2 Joint distribution of $X + Y$ and Z, together with their sum (first cell entry)

		Value of Z: 1	2	3	Probability of $X + Y$
Value of $X + Y$	0	$1, \frac{2}{18}$	$2, \frac{2}{18}$	$3, \frac{2}{18}$	$\frac{2}{6}$
	1	$2, \frac{3}{18}$	$3, \frac{3}{18}$	$4, \frac{3}{18}$	$\frac{3}{6}$
	2	$3, \frac{1}{18}$	$4, \frac{1}{18}$	$5, \frac{1}{18}$	$\frac{1}{6}$
Probability of Z		$\frac{1}{3}$	$\frac{1}{3}$	$\frac{1}{3}$	1

Probability	$\frac{2}{18}$	$\frac{5}{18}$	$\frac{6}{18}$	$\frac{4}{18}$	$\frac{1}{18}$
Value of $u = x + y + z$	1	2	3	4	5

The mean of the sum is the sum of the means (Corollary 7–5, Section 7–1). Consequently, since

$$\mu_X = \tfrac{1}{2}, \qquad \mu_Y = \tfrac{1}{3}, \qquad \mu_Z = 2,$$

then

$$\mu_U = \mu_X + \mu_Y + \mu_Z = 2\tfrac{5}{6}.$$

The latter value can be checked by computing the mean for the lower array of Table 10–2.

In the next sections, we show that the variance of a sum of *independent* random variables is the sum of their variances. Let us check the statement on the present example. Easy calculations give

$$\sigma_X^2 = \tfrac{1}{4}, \qquad \sigma_Y^2 = \tfrac{2}{9}, \qquad \sigma_Z^2 = \tfrac{2}{3}.$$

Their sum is $\frac{41}{36}$. Direct calculation of σ_U^2 from the lower array of Table 10–2 also gives $\frac{41}{36}$, so the statement checks in this example.

Fortunately, under quite general conditions, sums of many independent random variables are almost normally distributed. This information makes it easy to get an answer to a question like "What is the probability that the total score on 100 dice exceeds 325?" with enough accuracy for most purposes, without going through the step-by-step construction illustrated in Example 1. But before we can use this approximation, we must know the mean and variance of the distribution of the sum in terms of the means and variances of the variables being summed.

EXERCISES FOR SECTION 10–1

In Exercises 1 through 4 use the data of Example 1 and Tables 10–1 and 10–2.

1. Graph the probability functions of
   a) X, b) Y, c) $X + Y$,
   d) Z, e) $X + Y + Z$.

2. Use the probability functions of Exercise 1 to compute the means and variances of
   a) X, b) Y, c) Z,
   d) $X + Y$, e) $X + Y + Z$.

3. If each value of X is increased by 1 and each value of Y is decreased by 1, what is the effect on the probability function of $X + Y$?

4. Let $U = X + Y + Z$. If each value of X is decreased by $\frac{1}{2}$ $(= \mu_X)$, each value of Y is decreased by $\frac{1}{3}$ $(= \mu_Y)$, and each value of Z is decreased by 2 $(= \mu_Z)$, (a) what is the probability function of the new U which is the sum of the adjusted values of X, Y, and Z? (b) What is the mean of the new U? (c) What is the variance of the new U? Is it the same as that of the old U?

In Exercises 5 through 7, X, Y, and Z are independent and have the following probability functions:

Probability	$\frac{1}{4}$	$\frac{3}{4}$
Value of x	0	1

Probability	$\frac{1}{3}$	$\frac{1}{3}$	$\frac{1}{3}$
Value of y	-1	0	1

Probability	0.1	0.4	0.2	0.3
Value of z	1	2	3	4

5. Construct the probability functions of $X + Y$ and $X + Y + Z$.
6. Show that $P(X + Y = 1 \text{ and } Z = 2) = P(X + Y = 1) \cdot P(Z = 2)$.
7. Graph the probability functions of X, $X + Y$, and $X + Y + Z$.
8. The random variables X and Y are independent, and both take on the values 0, 1, 2 with probabilities $\frac{1}{3}$, $\frac{1}{3}$, $\frac{1}{3}$. Set up the probability function of $U = X + Y$ and compute σ_U^2 from it. Is $\sigma_U^2 = \sigma_X^2 + \sigma_Y^2$?
9. (Continuation.) Suppose that Z is another random variable with the same probability function as X and Y of Exercise 8. Find the probability function of $W = X + Y + Z$. Compute $E(W)$ and σ_W^2 and compare the results with $E(X)$ and σ_X^2.
10. X and Y independently take on the values 0, 1, 2, with probabilities $\frac{1}{4}$, $\frac{1}{2}$, $\frac{1}{4}$, respectively. Find the probability function of $U = X + Y$ and sketch its graph. Compute $E(U)$ and Var (U) from the probability function of U. Compare the results with $E(X)$ and Var (X).
11. The joint probability function of X and Y is as follows:

		Value of Y			
		-1	0	1	$P(X = x)$
Value of X	-1	0.1	0.1	0.2	0.4
	1	0.1	0.2	0.3	0.6
$P(Y = y)$		0.2	0.3	0.5	1.0

Find $E(X)$, E(Y), $E(X + Y)$, Var (X), Var (Y), Var $(X + Y)$. Is the variance of $X + Y$ equal to the sum of the variances of X and Y in this example? Are X and Y independent?

12. Three engineer's rulers with numbers 1, 2, 3 on the faces are tossed. Find the probability function of the *median* of the numbers on their lower faces. (In this exercise, if there is no tie, 2 is the median; if there is a tie, the tied value is the median.)

10-2 THE VARIANCE OF THE DISTRIBUTION OF THE SUM OF TWO INDEPENDENT RANDOM VARIABLES

To derive the variance of the sum of two independent random variables, we let

$$U = X + Y, \tag{1}$$

where X and Y are the independent variables and $\mu_X, \mu_Y, \sigma_X^2, \sigma_Y^2$ are their respective means and variances. Equation (1) defines a new random variable U whose variance, like that of any random variable, is the mean of its square minus the square of its mean:

$$\sigma_U^2 = E(U^2) - \mu_U^2. \tag{2}$$

Hence, by substitution from Eq. (1),

$$\sigma_U^2 = E(X + Y)^2 - \mu_{X+Y}^2. \tag{3}$$

We know from Theorem 7-4 that

$$\mu_{X+Y} = \mu_X + \mu_Y. \tag{4}$$

Furthermore,

$$E(X + Y)^2 = E(X^2 + 2XY + Y^2). \tag{5}$$

Now X^2, $2XY$, and Y^2 are also random variables and, by Corollary 7-5, the mean of their sum is the sum of their means:

$$\begin{aligned} E(X^2 + 2XY + Y^2) &= E(X^2) + E(2XY) + E(Y^2) \\ &= E(X^2) + 2E(XY) + E(Y^2). \end{aligned} \tag{6}$$

The middle term of the last line of Eq. (6) follows from the property that the mean of a constant times a random variable is that constant times the mean:

$$E(2XY) = 2E(XY).$$

So far, we have not used the assumption that X and Y are independent. We now make use of their independence, and Theorem 7-6, to write

$$E(XY) = E(X) \cdot E(Y) = \mu_X \mu_Y. \tag{7}$$

We now substitute from Eq. (7) into Eq. (6), then from (6) into (5), and the result, together with Eq. (4), into Eq. (3), to get

$$\begin{aligned} \sigma_U^2 &= E(X^2) + 2\mu_X\mu_Y + E(Y^2) - \mu_X^2 - 2\mu_X\mu_Y - \mu_Y^2 \\ &= E(X^2) - \mu_X^2 + E(Y^2) - \mu_Y^2 \\ &= \sigma_X^2 + \sigma_Y^2. \end{aligned}$$

We have therefore established the following theorem:

10–1 Theorem: σ^2_{X+Y} for independent variables

If $U = X + Y$ and if X, Y, U have variances σ^2_X, σ^2_Y, and σ^2_U, and if X and Y are independent, then

$$\boxed{\sigma^2_U = \sigma^2_X + \sigma^2_Y.} \tag{8}$$

Example 1 A machine makes small round discs of average thickness 0.5 inch, but with standard deviation of thickness 0.003 inch. Assemblies of two such discs lie one on top of the other (like a king in checkers). What is the standard deviation of the heights of finished assemblies?

Solution Let X stand for the thickness of the bottom disc, Y for that of the top, and U $(= X + Y)$ for the thickness of the assembly. We assume that X and Y are independent. Then $\sigma^2_U = 0.003^2 + 0.003^2$, $\sigma_U \approx 0.0042$.

Example 2 *Measurements.* A boy saws boards in lengths of about 3 feet, and the standard deviation of the lengths is 0.2 inch. To check his precision the boy measures the lengths of the boards and obtains a distribution of measurements with standard deviation 0.25 inch. What is the standard deviation of the distribution of his errors of measurement?

Solution Let the random variable X represent the actual length of a board, and let Y be the *error* of measurement; then the measurement $U = X + Y$. Assume that X and Y are independent. Then $\sigma_X = 0.2$, $\sigma_U = 0.25$. To find σ_Y, we substitute in $\sigma^2_U = \sigma^2_X + \sigma^2_Y$ and get

$$0.0625 = 0.04 + \sigma^2_Y.$$

Solving for σ^2_Y gives

$$\sigma^2_Y = 0.0225, \qquad \sigma_Y = 0.15.$$

Thus the standard deviation of the distribution of measurement errors is 0.15 inch, or roughly the size of the standard deviation of the distribution of actual lengths. The boy's sawing is nearly as accurate as his measuring.

Example 3 *Disparate standard deviations.* Let X have standard deviation σ and Y standard deviation $k\sigma$, where k is large compared with 1. Find approximately the standard deviation of $U = X + Y$, when X and Y are independent.

Solution By formula (8),

$$\sigma^2_U = \sigma^2_X + \sigma^2_Y = \sigma^2 + k^2\sigma^2 = (k^2 + 1)\sigma^2,$$

$$\sigma_U = \sqrt{(k^2 + 1)\sigma^2} = k\sigma\sqrt{1 + \frac{1}{k^2}} \approx k\sigma.$$

Roughly, then $\sigma_U \approx k\sigma = \sigma_Y$, the larger standard deviation.

Remark The moral of Example 3 is that in trying to reduce the variability of a measurement that is composed as the sum of several independent components, reducing the variability of the component with largest variance will reduce the variability of the total substantially. But the reduction or even the elimination of the variability of a component with small variance will do little good for the total. Unfortunately, often one knows how to reduce the variance of a component already having small variance, but has no way to get better control over the one making the large contribution. Time and money should not be wasted reducing the small variance.

Use of normal tables When data are approximately normally distributed and we know their mean and variance, we can use normal tables (Table III) to compute approximate probabilities.

Example 4 Performance on an algebra test and on a test of dramatic ability are approximately independent, and the distribution of the sum of their scores is approximately normal. If the distribution of each set of test scores has mean 50 and standard deviation 10, what proportion of examinees scored a total of 125 points or more?

Solution Let X be the algebra score, Y the dramatic score, U their total. Then $\mu_U = 50 + 50 = 100$, and

$$\sigma_U^2 \approx 10^2 + 10^2 = 200, \qquad \sigma_U \approx \sqrt{200} \approx 14.1.$$

If we let

$$Z = \frac{U - 100}{\sqrt{200}},$$

then Z is approximately normal and has mean 0 and standard deviation 1. We can find $P(Z \geq a)$, approximately, from the normal table, Table III. The event $U \geq 125$ is equivalent to the event

$$Z \geq \frac{125 - 100}{\sqrt{200}} \approx 1.77.$$

Therefore

$$P(U \geq 125) \approx P(Z \geq 1.77)$$

and, from Table III, we find

$$P(Z \geq 1.77) \approx 0.0384.$$

Thus about 4% of the examinees score 125 or higher.

10-2 Theorem: Weighted sums of measurements

Let measurements X and Y be independently drawn from distributions with means μ_X, μ_Y and variances σ_X, σ_Y. Let their weighted sum, with weights a and b, be a new random variable Z

$$\boxed{Z = aX + bY.} \tag{9}$$

Then

$$\boxed{\mu_Z = a\mu_X + b\mu_Y} \tag{10}$$

and

$$\boxed{\sigma_Z^2 = a^2\sigma_X^2 + b^2\sigma_Y^2.} \tag{11}$$

Proof You are asked to prove this theorem in Exercise 16.

Example 5 *Prices of alloys.* Manufactured blocks of expensive alloys are sold by weight. Blocks of alloy A have standard deviation 3 pounds and cost \$100 per pound. Blocks of alloy B have standard deviation 4 pounds and cost \$50 per pound. In repeated orders of two blocks (one A and one B), what is the standard deviation of the total price if the blocks are independently assembled to fill the order?

Solution Let X be the weight in pounds of a block of alloy A, Y that for alloy B. Then the total price in dollars is

$$Z = 100X + 50Y.$$

By Theorem 10–2,

$$\sigma_Z^2 = 100^2\sigma_X^2 + 50^2\sigma_Y^2 = 130{,}000, \qquad \sigma_Z \approx 361, \quad \text{in dollars.}$$

10–3 Corollary: Differences

If X and Y are independent, with means μ_X, μ_Y and variances σ_X^2, σ_Y^2, then the distribution of their difference D,

$$\boxed{D = X - Y,} \tag{12}$$

has mean

$$\boxed{\mu_D = \mu_X - \mu_Y} \tag{13}$$

and variance

$$\boxed{\sigma_D^2 = \sigma_X^2 + \sigma_Y^2.} \tag{14}$$

Proof Theorem 10–2 applies to this corollary when we take $a = 1$, $b = -1$, $a^2 = 1$, $b^2 = 1$. Substituting these values in Eqs. (10) and (11) yields Eqs. (13) and (14). □

Example 6 *Rods with washers.* Rods (circular in cross section) have outside diameters that are normally distributed with mean 1.0 inch and standard deviation 0.003 inch. Washers (with holes circular in cross section) have inside diameters also normally distributed, with mean 1.005 inch and standard deviation 0.004 inch. When rods and washers are randomly paired, in what percentage of pairs are washers too small to fit on their rods? (Assume that the difference of two independent normally distributed random variables is normally distributed. This is true although we don't prove it in this book.)

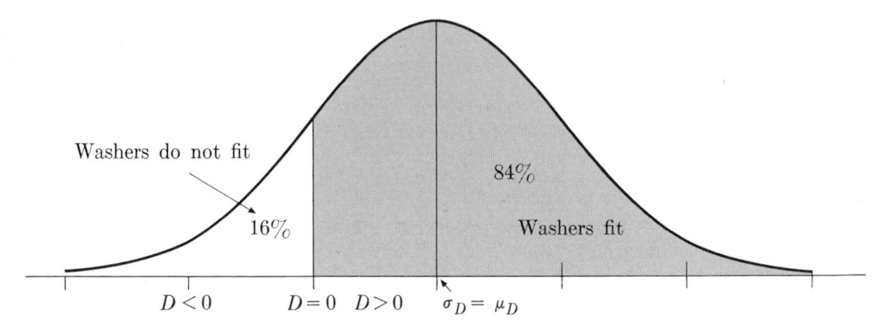

Fig. 10–1 Distribution of difference of washer and rod diameters.

Solution Let X measure the inside diameter of a washer in inches, Y the outside diameter of a rod. Let $D = X - Y$ measure the difference in these diameters. If $D > 0$, the rod fits in the hole, otherwise not. By Corollary 10–3, $\mu_D = 1.005 - 1.000 = 0.005$, in inches, and $\sigma_D = \sqrt{0.004^2 + 0.003^2} = 0.005$, in inches. Since $\mu_D = \sigma_D$, the percent of washers too small to fit is equal to the probability to the left of -1 for a standard normal random variable Z, where

$$Z = \frac{D - \mu_D}{\sigma_D}.$$

If $D < 0$, then $Z < -1$, and

$$P(\text{washer does not fit}) = P(D < 0) = P(Z < -1).$$

From Table III,

$$P(Z < -1) = P(Z > 1) = 0.5000 - P(0 \leq Z \leq 1) = 0.1587.$$

Thus about 16% of the washers are too small to fit their rods (Fig. 10–1).

EXERCISES FOR SECTION 10–2

In Exercises 1 through 5, X and Y are independent, $U = X + Y$.

1. The variables X and Y both take values -1, 0, 1 with equal probabilities $\frac{1}{3}, \frac{1}{3}, \frac{1}{3}$. Find σ_U^2.

2. Find the missing entries in the following table:

	σ_X	σ_Y	σ_U
a)	3	4	
b)	7		25
c)	9	40	
d)		8	10
e)	8		17
f)		12	13
g)	1		2
h)	1	1	

3. If $\sigma_U^2 = 8$, $\sigma_Y^2 = 8$, what can you tell about the distribution of X?

4. What theorem from plane geometry does Eq. (8) look like? What do σ_X, σ_Y, and σ_U correspond to in that theorem from geometry?

5. If $\sigma_X = \sigma_Y$, show that $\sigma_U = \sqrt{2}\sigma_X$. (Why not $-\sqrt{2}\sigma_X$?)

6. (Continuation.) Use the result of Exercise 5 to solve the disc assembly example, Example 1 of the text.

7. When a die is rolled, a man receives \$4 if an ace appears, but loses \$1 otherwise. On this basis, if a die is rolled twice, what are the mean and the standard deviation of the distribution of his net receipts?

8. (Continuation.) Suppose in the game of Exercise 7 the die is rolled a third time. What are the mean and the standard deviation of the distribution of the man's net receipts on the three payments? [*Hint:* Let the outcome of the first two rolls be X.]

9. The scores of College Board examinations have a mean of 500 and a standard deviation of 100. If two students are drawn at random from College Board examinees, approximately what are the mean and the standard deviation of the distribution of the sum of their mathematics scores?

10. (Continuation.) Consider College Board examinees who have taken both an English and a Mathematics examination. Find the mean and the standard deviation of the distribution of the sum of the two scores, or explain why you cannot.

11. In a large chemical operation, a man ladles out material with two scoops, one with capacity one pound, and the other, two pounds. For precise work, he uses the one-pound scoop twice to put two pounds of material into the mix. The standard deviation of the distribution of weight for the two-pound scoop is 0.5 ounce, for a single one-pound scoop 0.3 ounce. Should he have used the two-pound scoop for more precision?

12. Two tests with independent scores X and Y are given. They have standard deviations of 7 and 24. For each student we take the sum of his scores on the tests, $X + Y$. Show that the standard deviation of the distribution of these $X + Y$ scores is 25.

13. The distributions of lengths of two kinds of wooden parts A and B are approximately normal, with means $\mu_A = 2$ inches and $\mu_B = 4$ inches, and standard deviations $\sigma_A = 0.009$ inch, $\sigma_B = 0.040$ inch. An A part and a B part are ran-

domly assembled and laid end to end to form a length about 6 inches long. If an assembly is to fit, it must be between 5.92 and 6.08 inches long. What percentage of random assemblies fail to fit? (Assume that the distribution of the sum of two independent normally distributed random variables is normal.)

14. Two measurements X and Y are independently drawn from the same distribution with mean μ and variance σ^2, and a weighted sum $S = wX + (1 - w)Y$ is computed. (a) Find μ_S. (b) Find σ_S^2. (c) Find the value of w that minimizes σ_S^2. (d) Find the minimum value of σ_S^2. [*Remark:* For any w, S is called an *unbiased estimate* of μ because $\mu_S = \mu$, and with $w = \frac{1}{2}$, S is called the *minimum variance unbiased estimate* of μ.]

15. An assembly is made by putting two washerlike objects face-to-face on an axle. If the total thickness of the two objects is between 0.549 and 0.551 inch inclusive, the assembly is satisfactory—otherwise not. The objects are randomly assembled from an approximately normal population with mean thickness 0.275 inch and standard deviation 0.0006 inch. What percent of the assemblies is unsatisfactory? (Assume that the distribution of the sum of two independent normal random variables is normal.)

16. Prove Theorem 10–2 (concerning weighted sums of measurements).

In Exercises 17 through 19, X, Y, Z are independent random variables taking the following values, each with probability $\frac{1}{4}$:

Values of X: $-4,\ -1,\ 2,\ 3$

Values of Y: $-3,\ -1,\ 2,\ 3$

Values of Z: $-2,\ -1,\ 0,\ 3$

The ★'s in Exercises 18 and 19 mean that the second part of each is lengthy rather than difficult.

17. Calculate σ_{Y+Z}^2 by definition, and by formula.
18. Calculate σ_{Y+Z+X}^2 by formula and (★) by definition.
19. Calculate $\sigma_{3X+2Y-6Z}^2$ by formula and (★) by definition.

10–3 VARIANCE OF THE SUM AND OF THE AVERAGE OF SEVERAL VARIABLES

To solve the problem of the distribution of sample averages for the 100-dice problem given at the beginning of this chapter, we need first the variance of the sum of many independent random variables, not just two as dealt with in Section 10–2. As soon as we have the variance of the sum we get the variance of sample averages by a trivial operation.

Subscripts for random variables Note that previously we have used subscripts mainly on the values of the random variables, such as x_i, not on the random variables themselves, because there were only one or two of them. Now we study many random variables, so we need subscripts for them. Thus in a theorem about the sum of n random variables, we denote the random variables by $X_1, X_2, \ldots, X_n$. But if we have only a few random variables we shall continue to denote them by X, Y, Z.

To extend the theorem on the variance of a sum of two random variables, we need to show that we can add one more variable, and then another, and so on. For example, if X, Y, and Z are three independent random variables, we know that the variance of $U = X + Y$ is $\sigma_X^2 + \sigma_Y^2$. If Z is independent of X and of Y, we naturally expect it to be independent of their sum U. If Z is independent of U, then we know that $W = U + Z$ has variance $\sigma_W^2 = \sigma_U^2 + \sigma_Z^2 = \sigma_X^2 + \sigma_Y^2 + \sigma_Z^2$. Provided Z is independent of U, this argument is enough to show that we have a general method for adding one more variable. Thus we can extend the theorem from 3 independent random variables to 4, then from 4 to 5, and so on to n variables. Of course a rigorous development requires a slightly more formal induction argument.

The entire argument above depends on the intuitively obvious but somewhat subtle fact that if X, Y, and Z are jointly independent, then the two variables Z and U $(= X + Y)$ are independent. The statement is true, but its proof, while easy, requires extra notation that we have not developed. A proof is given in Appendix 2. This proof, together with the previous arguments, completes the demonstration of the first part of the following desired theorem:

10-4 Theorem: Variance of sums and weighted sums of independent random variables

If $X_1, X_2, \ldots, X_n$ are independent random variables with means $\mu_1, \mu_2, \ldots, \mu_n$ and variances $\sigma_1^2, \sigma_2^2, \ldots, \sigma_n^2$ and

$$T = X_1 + X_2 + \cdots + X_n,$$

then the mean and variance of T are

$$\boxed{\begin{aligned} \mu_T &= \mu_1 + \mu_2 + \cdots + \mu_n, \\ \sigma_T^2 &= \sigma_1^2 + \sigma_2^2 + \cdots + \sigma_n^2. \end{aligned}} \tag{1a}$$

Also, given constant weights $w_1, w_2, \ldots, w_n$, the weighted sum

$$W = w_1X_1 + w_2X_2 + \cdots + w_nX_n$$

has mean and variance

$$\boxed{\begin{aligned} \mu_W &= w_1\mu_1 + w_2\mu_2 + \cdots + w_n\mu_n, \\ \sigma_W^2 &= w_1^2\sigma_1^2 + w_2^2\sigma_2^2 + \cdots + w_n^2\sigma_n^2. \end{aligned}} \tag{1b}$$

Proof for weighted sum W If random variables X, Y, and Z are independent, then so are the new random variables L, M, N, where $L = aX$, $M = bY$, $N = cZ$, obtained by multiplying X, Y, and Z by the constants a, b, and c, respectively (see Appendix 2). We obtain the result on weighted sums by applying the first part of the theorem to new random variables $W_i = w_i X_i$, $i = 1, 2, \ldots, n$, and then noticing that $\text{Var}(W_i) = w_i^2 \sigma_i^2$. □

Example 1 *A three-unit assembly.* In an electrical circuit, resistances in series form a resistance equal to the sum of their resistances. A 10,000-ohm, a 20,000-ohm, and a 50,000-ohm resistance are each drawn from a large stock to form an 80,000-ohm resistance. The standard deviations of these three kinds are 30, 60, and 150 ohms, respectively. Find the standard deviation of the distribution of 80,000-ohm resistances formed in this manner.

Solution If the resistances are randomly assembled, then

$$\sigma^2 = (30)^2 + (60)^2 + (150)^2 = 27{,}000,$$
$$\sigma = \sqrt{27{,}000} \approx 164.$$

Note Two or more random variables are said to be *identically distributed* if their probability functions are the same. Such identically distributed random variables arise from sampling with replacement or from repeated measurements.

10-5 Corollary: Variance of identically distributed variables — sampling theory

If $X_1, X_2, \ldots, X_n$ are independently and identically distributed random variables with means μ and variances σ^2, and if

$$T = X_1 + X_2 + \cdots + X_n,$$

then

$$\boxed{\mu_T = n\mu} \tag{2}$$

and

$$\boxed{\sigma_T^2 = n\sigma^2.} \tag{3}$$

Proof Equation (2) follows at once from Corollary 7-5, Section 7-1. Equation (3) is obtained by substituting σ^2 for each σ_i^2 in Eq. (1a) of Theorem 10-4 above. □

We can use Corollary 10-5 to give an easy proof that the variance of the binomial distribution is

$$\sigma^2 = npq = np(1 - p).$$

10-6 Corollary: Variance of a binomial distribution

Let p be the probability of success on a single binomial trial, and let X be the total number of successes in n such trials. Then the variance of X is

$$\boxed{\sigma_X^2 = np(1 - p).} \tag{4}$$

Proof The distribution of the number of successes, B, on *one* binomial trial is

Probability	p	$1 - p$
Number of successes	1	0

Consequently, the mean number of successes on a single trial is

$$\mu_B = p \cdot 1 + (1 - p) \cdot 0 = p.$$

The variance of the number of successes on a single trial is

$$\sigma_B^2 = E(B^2) - \mu_B^2 = 1^2 p + 0^2(1 - p) - p^2 = p - p^2 = p(1 - p).$$

The total number of successes on n trials is the sum of n independent random variables like B, one for each trial, and each having mean p and variance $p(1 - p)$. Therefore by Corollary 10–5 the mean, the variance, and the standard deviation for the number of successes in n trials are given by the formulas:

$$\boxed{\begin{aligned} &\text{a)}\ \mu_X = np, \\ &\text{b)}\ \sigma_X^2 = np(1 - p), \\ &\text{c)}\ \sigma_X = \sqrt{np(1 - p)}. \end{aligned}} \quad \square \qquad (5)$$

Example 2 *1000 thumbtacks.* If thumbtacks have a probability $p = 0.3$ of landing point up, what is the probability that at least 320 out of 1000 tossed land point up?

Solution Let X be the number landing point up. Then by Eq. (5), the mean of X is $(0.3)(1000) = 300$, and the standard deviation is

$$\sigma_X = \sqrt{1000(0.3)(0.7)} \approx 14.5.$$

Thus if 320 land point up, the number of "ups" in excess of the mean is $320 - 300 = 20$, and the number of standard deviations from the mean is $20/14.5 \approx 1.38$. From Table III the probability in excess of 1.38 standard deviations is approximately 0.0838, or about 8%.

Central limit theorems In Chapter 8 we studied the behavior of the binomial distribution as n increases. We found that if X is the number of successes in n independent binomial trials, then the related random variable

$$Z = \frac{X - np}{\sqrt{npq}}$$

has a distribution that is closely approximated by the standard normal distribution if n is large. That result is a special case of a more general *central limit theorem* which we now state, and use, without proof.

10–7 Theorem: Central limit theorem

Let $X_1, X_2, \ldots, X_n, \ldots$ be a sequence of identically distributed independent random variables, each with mean μ and variance σ^2. Let

$$T_n = X_1 + X_2 + \cdots + X_n.$$

Then, for each fixed value of z, as n tends to infinity,

$$P\left(\frac{T_n - n\mu}{\sigma\sqrt{n}} > z\right)$$

approaches the probability that the standard normal random variable Z exceeds z.

Remark By subtracting $E(T_n) = n\mu$ from T_n and then dividing by $\sigma\sqrt{n} = \sqrt{n\sigma^2} = \sqrt{\text{Var}\ (T_n)}$, we have obtained a new random variable whose mean is zero and whose standard deviation is 1, as are those of the standard normal. In more advanced work in probability, it is proved that the distribution of this new random variable approaches that of the standard normal as n tends to infinity. In practical terms, this means that if n is large we may use the standard normal tables to answer such questions as the one asked in Example 3 of this section.

More general central limit theorem Under broad conditions, for large values of n when $X_1, X_2, \ldots, X_n$ are independent random variables with means $\mu_1, \mu_2, \ldots, \mu_n$ and variances $\sigma_1^2, \sigma_2^2, \ldots, \sigma_n^2$, the quantity

$$Z_n = \frac{\sum X_i - \sum \mu_i}{\sqrt{\sum \sigma_i^2}}$$

is approximately normally distributed with zero mean and unit variance. The essential feature is that each of the variables X_i makes a modest contribution to the total variance $\sum \sigma_i^2$. This means that one, two, or a few random variables should not swamp the variability of the others, as happened in Example 3 of Section 10–2. If one has 10 variables, 9 with variance 1 and the other with variance 100, then the distribution of the latter will largely shape the distribution of their sum.

To illustrate that approximate normality comes soon, Fig. 10–2 shows the distribution of the sum of 5 random variables all having distribution

$f(x)$:	.1	.2	.3	.4
x:	1	2	3	4

together with its normal approximation. Since the original distribution starts out looking like a right triangle, it has no symmetry to speed the sum's approach to normality.

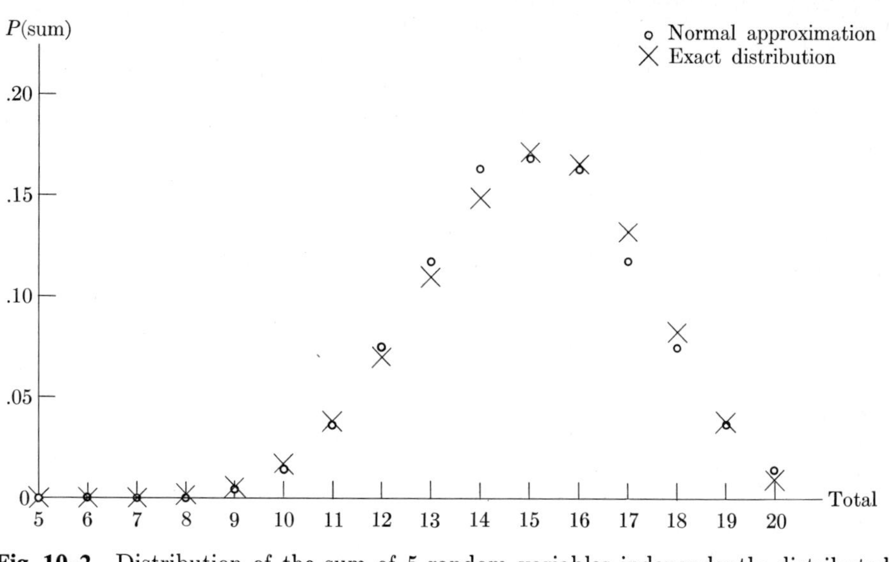

Fig. 10–2 Distribution of the sum of 5 random variables independently distributed according to

$f(x)$:	.1	.2	.3	.4
x:	1	2	3	4

Sums and differences of independent normal random variables We shall not prove the theorem that the distribution of the sum of two independent normally distributed random variables is itself normally distributed, but it is true and will be of frequent use to us. Furthermore, if the X_i's of Theorem 10–4 are independent and *normally* distributed, then the sum T and the weighted sum W are again *normally* distributed with the means and variances given by formulas (1a) and (1b). In particular, this implies that the difference of two independent normally distributed variables is normally distributed.

Example 3 *100 dice.* Find the probability that when 100 dice are rolled the sum of the dots on their topmost faces exceeds 325.

Solution For the roll of one die we found the mean and variance of the number of dots on a face in Example 3, Section 6–4, to be

$$\mu = 3.5, \qquad \sigma^2 = \tfrac{35}{12}.$$

Applying Corollary 10–5 with $n = 100$, we find that the mean, variance, and standard deviation of the sum of 100 rolls are

$$\mu_T = 100(3.5) = 350, \qquad \sigma_T^2 = 100(\tfrac{35}{12}), \qquad \sigma_T \approx 17.1.$$

Thus $Z = (T_n - \mu_T)/\sigma_T = (325 - 350)/17.1 \approx -1.46$. This means that the value 325 is 1.46 standard deviations to the left of the mean. Now

$$P(T > 325) = P\big((T_n - \mu_T)/\sigma_T > -1.46\big) = P(Z > -1.46).$$

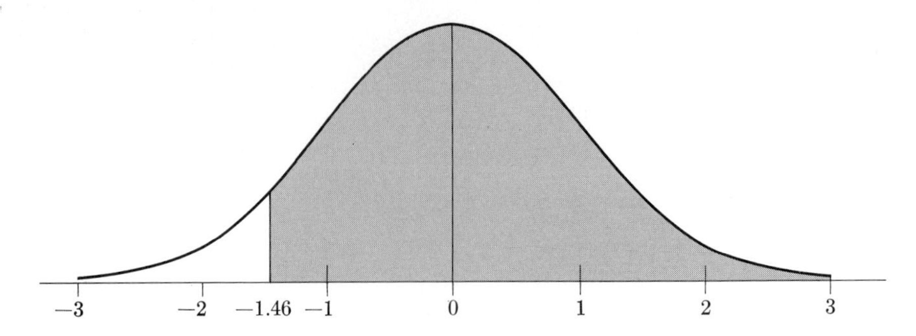

Fig. 10–3 Shaded area gives probability of total score on 100 dice exceeding 325.

With the aid of Table III we find that the probability to the right of −1.46 is 0.9279 for the standard normal, or all but about 7%. (See Fig. 10–3.)

Going from sums to averages If we divide a sum of measurements by the number of them, we obtain an average. Therefore an average is a trivial adjustment of a sum. Theorem 10–7 completes the information we need about the distribution of a sum. Now we quickly derive the mean and variance of the *distribution of sample averages.* These results will confirm the intuitive notion that sample averages are more stable than single measurements.

As in Theorem 10–4, let

$$T = X_1 + X_2 + \cdots + X_n.$$

Then the sample average is

$$\overline{X} = \frac{T}{n}.$$

We know that when we multiply a random variable by a constant we multiply its mean, or expected value, by that same constant and we multiply its variance by the square of the constant. Therefore

$$\mu_{\bar{X}} = \frac{1}{n}\,\mu_T, \qquad \sigma_{\bar{X}}^2 = \frac{1}{n^2}\,\sigma_T^2.$$

Of course, since the mean of a sum is the sum of the means,

$$\mu_T = \mu_1 + \mu_2 + \cdots + \mu_n,$$

and by Theorem 10–4,

$$\sigma_T^2 = \sigma_1^2 + \sigma_2^2 + \cdots + \sigma_n^2.$$

Thus we have proved the following theorem.

10–8 Theorem: Means and variances of sample averages

Let the random variables $X_1, X_2, \ldots, X_n$ be independent, with means $\mu_1, \mu_2, \ldots, \mu_n$ and variances $\sigma_1^2, \sigma_2^2, \ldots, \sigma_n^2$. Let the average of these variables be $\overline{X}$, where

$$\overline{X} = \frac{1}{n}(X_1 + X_2 + \cdots + X_n).$$

Then $\overline{X}$ has a distribution with mean

$$\boxed{\mu_{\overline{X}} = \frac{1}{n}(\mu_1 + \mu_2 + \cdots + \mu_n)} \qquad (6)$$

and variance

$$\boxed{\sigma^2_{\overline{X}} = \frac{1}{n^2}(\sigma_1^2 + \sigma_2^2 + \cdots + \sigma_n^2).} \qquad (7)$$

10–9 Corollary: Averages of independent random variables having identical means and variances

Let the independent variables $X_1, X_2, \ldots, X_n$ have identical means, μ, and identical variances, σ^2, and let their average be

$$\overline{X} = \frac{1}{n}(X_1 + X_2 + \cdots + X_n).$$

Then

$$\boxed{\mu_{\overline{X}} = \mu} \qquad (8)$$

and

$$\boxed{\sigma^2_{\overline{X}} = \frac{\sigma^2}{n}, \qquad \sigma_{\overline{X}} = \frac{\sigma}{\sqrt{n}}.} \qquad (9)$$

Proof Substitute into Eqs. (6) and (7) of Theorem 10–8. □

Sampling with replacement The most important application of Corollary 10–9 is to *sampling with replacement* from a finite population, or sampling from an infinite population. Let X_1 represent some measured characteristic of the population element that is drawn *first* in the sample, X_2 that of the element drawn *second*, and so on. In sampling with replacement the probability functions of the random variables $X_1, X_2, \ldots, X_n$ are the same; the variables are *identically distributed.* Thus the observed measurements in such samples are values of random variables with equal means and variances, so Corollary 10–9 applies.

In words, Eqs. (8) and (9) say that the *expected value* of the *average* of n measurements is the population mean μ, and the standard deviation of the averages from one set of measurements to another is inversely proportional to the square root of the number of measurements. Thus averages of 4 independent measurements drawn from the same population have a standard deviation equal to $\frac{1}{2}$ the standard deviation of single measurements; and averages of 100 independent measurements drawn from the same population have a standard deviation equal to $\frac{1}{10}$ the standard deviation of single measurements. This shrinking of the standard deviation as n increases causes a tightening up of the probability distribution of $\overline{X}$ around the population mean and practically guarantees that the sample average lies close to the population mean when n is sufficiently large.

Bias, or systematic error In an actual measuring procedure, there may be a systematic error. For example, one may consistently tend to read too high. Such a systematic error is not reduced by taking the average of repeated measurements.

Example 4 *Average for 100 dice.* Find the probability that when 100 dice are rolled their sample average exceeds 3.7.

Solution For a single die, $\mu = 3.5, \sigma^2 = \frac{35}{12}$. By Corollary 10–9, $\mu_{\overline{X}} = \mu = 3.5$, $\sigma^2_{\overline{X}} = \sigma^2/n = \frac{35}{1200} \approx 0.0292$, and $\sigma_{\overline{X}} \approx 0.171$. If the sample average exceeds 3.7, then it exceeds the population mean 3.5 by at least 0.2 ($= 3.7 - 3.5$), or by $0.2/0.171 \approx 1.17$ standard deviations. From the normal tables, the probability in excess of 1.17 standard deviations is 0.121. There is less than 1 chance in 8 that the sample average for the 100 dice exceeds 3.7.

10–10 Corollary: Mean proportion of successes for the binomial

Consider a binomial experiment composed of n binomial trials each with probability p of success and with total number of successes X. Let $\overline{p} = X/n$ be the proportion of successes. Then the mean and variance of $\overline{p}$ are

$$\mu_{\overline{p}} = p \qquad \text{and} \qquad \sigma^2_{\overline{p}} = \frac{p(1-p)}{n} = \frac{pq}{n}. \tag{10}$$

Proof The proof for the mean was given in Theorem 5–4; the variance follows from Corollary 10–6, since $\sigma^2_{\overline{p}} = (1/n^2)\sigma^2_X$. □

Example 5 *Voting.* If 60% of a large population favors a certain candidate, what is the probability that in a random sample of 100 voters the proportion in favor of the candidate is under 50%?

Solution $\mu_{\overline{p}} = p = 0.6$, $\sigma^2_{\overline{p}} = (0.6)(0.4)/100 = 0.0024$, $\sigma_{\overline{p}} \approx 0.049$. The difference $0.5 - 0.6 = -0.1$ is $-0.1/0.049 \approx -2$ standard deviations, or 2 standard deviations below the mean. Using the normal approximation, we find that the probability is about 0.025, or about 1 chance in 40.

EXERCISES FOR SECTION 10–3

In these exercises, samples are drawn independently with replacement.

In Exercises 1 through 10, the population has mean $\mu = 6$ and standard deviation $\sigma = 10$.

1. What is the variance?
2. What is the standard deviation of the averages of samples of size 4 drawn from this population?
3. What standard deviate, $x' = (x - \mu)/\sigma$, does the observation 16 correspond to?
4. A new population is formed by adding 3 to every observation in the given population. Find the mean and standard deviation of the new population.
5. A new population is formed by multiplying every observation in the given population by 3. Find the mean and standard deviation of this new population.
6. Find the mean and standard deviation of the distribution formed by taking the sum of the observations of every sample of size 4 drawn from the original population with replacement.
7. Chebyshev's Theorem guarantees that at least 75% of the original population lies between what two numbers?
8. If the original population is normally distributed, what proportion of it lies to the right of the mean?
9. If the original population is normally distributed, what proportion of it has values larger than 11?
10. If the original population is normally distributed, what is the value exceeded by 80% of the population?

11. The standard deviation of a population of scores is 36. What is the standard deviation of the distribution of sample averages for samples of size 16 drawn from this population?
12. A large population of measurements has mean $\mu = 20$ and standard deviation $\sigma = 5$. Consider samples of 4 measurements, each randomly drawn from the original population. What is the expected value of the sample average $\overline{X}$ for such samples? What is the standard deviation of the distribution of $\overline{X}$?
13. The distribution of weights (in pounds) of a large group of equipped army recruits is very closely approximated by a normal curve with mean 185.0 pounds and standard deviation 15.0 pounds. (a) If two recruits are picked at random from this group, what is the probability that *both* their weights are between 170.0 and 200.0 pounds? (b) If 81 recruits from the large group are to enplane with an allowance of 190.0 pounds per man, what is the chance that the transport will be overloaded?
14. A certain mental test yields scores in months of mental age. The errors of measurement in this test average zero in the long run, with a standard deviation σ of 2 months. A fifth-grade class of 36 students took this test. What is the probability that the average *score for the class* is in error by 1 month or more (either too high or too low)?
15. Given a large population of test scores with mean 20 and variance 9, what is the standard deviation of the sampling distribution of averages of samples of size 25 drawn from this population?
16. In crossing two pink flowers of a certain variety the resulting flowers are either white, red, or pink, and the probabilities that attach to these various outcomes are $\frac{1}{4}$, $\frac{1}{4}$, and $\frac{1}{2}$, respectively. If 300 flowers are obtained by crossing pink flowers of this variety, what is the probability that 90 or more of these flowers are white?

17. In a certain large society the standard deviation of the number of children in a family is 1.5. If an anthropologist wants the standard deviation of $\overline{X}$, his estimate of the mean number of children per family, to be 0.1, how many families should be in his random sample?
18. The standard deviation of the distribution of sample averages in samples of size 9 is 4. What is the standard deviation of the population from which the sample is drawn?
19. A mass-produced object has 2 mirrored faces. The percentages of such objects with 2, 1, or 0 marred mirrored faces are 1, 1, 98, respectively. (a) If four objects are randomly drawn from a very large lot with this composition, compute the exact probability of 2 or fewer faces being unmarred all told (compute as if the sampling were done with replacement). (b) If 900 objects are randomly drawn, compute the approximate probability that there are 1775 or fewer unmarred faces. [*Hint:* First compute the mean and standard deviation for the number of marred faces on one object for part (b).]
20. The listening time per week to the musical programs of a radio station by a certain large group of people is approximately normally distributed, with mean $\mu = 4$ hours and standard deviation $\sigma = 1$ hour. A sample of 16 people to be drawn from this group has average listening time $\overline{X}$. What is the probability that $\overline{X}$ will differ from the group mean by more than half an hour?
21. In a before-and-after experiment, the mean difference in the population is 1.00, and the standard deviation of the differences for individuals in the population is 2.00. Find the probability of a positive difference for a randomly selected individual.
22. From a distribution with variance $\sigma^2 = 1$, two independent random samples are drawn as in the following table:

	Sample size	Sample average
Sample 1	10	$\overline{X}_1$
Sample 2	5	$\overline{X}_2$

To estimate the population mean:

a) One man weights the sample averages in proportion to their sample sizes and claims that the sample variance of such a weighted estimate, $\frac{2}{3}\overline{X}_1 + \frac{1}{3}\overline{X}_2$, is $\frac{1}{15}$. Justify this result.
b) A second man merely averages the sample averages and uses $\frac{1}{2}(\overline{X}_1 + \overline{X}_2)$. Show that the sampling variance of such an estimate is $\frac{3}{40}$.
c) Show that both methods are unbiased (have mean equal to the true mean μ).
d) Explain (a) from the point of view of one sample of size 15.

23. A population of 500 is composed of two subpopulations or strata. Stratum 1 has 400 members and stratum 2 has 100 members. A random sample of size 20 is taken from each stratum. The average of the sample of stratum 1 is 4.0 and the standard deviation of this sample is 1.0. The average of the sample of stratum 2 is 8.0 and the standard deviation of the sample is 2.0. (a) Estimate the grand mean of the total population. (b) Estimate the standard deviation of the estimate of the grand mean.
24. Which of the following choices is true? In sampling, as the size of the sample increases, the standard deviation of the theoretical distribution of sample averages

(a) decreases in value, approaching 0; (b) increases in value, growing without bound; (c) does not necessarily increase or decrease, but approaches the value of the true (universe) standard deviation.

25. A fair coin is tossed n times. Each time a head appears a is *added* to the score, each time a tail appears b is *subtracted* from the score. Determine the mean and variance of the distribution of the score after n throws.

26. Suppose that the birth-weights of children are normally distributed with mean 7 pounds and standard deviation $\frac{1}{2}$ pound. Suppose that the two sexes are equally likely and that birth-weight is independent of sex. (a) What is the probability that four children in a family are all males? (b) What is the probability that the four children, whatever their sex, have an average birth-weight over 7.5 pounds? (c) What is the probability that the children are all males *and* have an average birth-weight over 7.5 pounds? (d) Are you using an assumption not stated? If so, what?

27. In a motor-skill experiment, 100 subjects perform a task in 25 randomly selected groups of 4, but each subject has a separate cubicle, so that he does not influence other members of the group. From the frequency distribution of the scores of these 100 subjects, the experimenter finds the mean to be 30 points and the standard deviation 10 points. Then the experimenter obtains for each of the 25 groups the *sum* of the scores of the 4 subjects. He wants to know

    a) the value of the mean of the 25 sums,
    b) the value of the standard deviation of the 25 sums.

    Tell him the answer to (a) exactly. Using the data given and your knowledge of the theory in this chapter, make a good estimate of the result for (b).

    On hearing the result for the standard deviation of the sums for the groups, the experimenter is surprised. He says "The law of averages should have made the standard deviation for the groups smaller, not larger than the standard deviation of the individual scores". Comment crisply on his misapprehension.

10–4 DEPENDENT RANDOM VARIABLES: COVARIANCE AND CORRELATION

In the earlier sections of this chapter, we found that the variance of a sum of *independent* random variables is the sum of their variances. What if the variables are dependent? The result, derived in this section, introduces two important new concepts: *covariance* and *correlation.*

Having in mind some analogies between the variance of the sum of two random variables, on the one hand, and the theorem of Pythagoras and its generalization, the law of cosines, on the other hand, may help us to keep our bearings as we go through the algebra. Table 10–3 shows the two theorems for triangles and the corresponding results for random variables. The new symbols in the table are Cov (X, Y), read "*covariance* of X and Y", and ρ (Greek letter "rho"), which represents the *correlation* between X and Y.

Two random variables can be uncorrelated without being independent, but all independent random variables are uncorrelated. And so lack of correlation is a weaker requirement than independence.

Table 10–3 Analogous theorems for triangles and for random variables

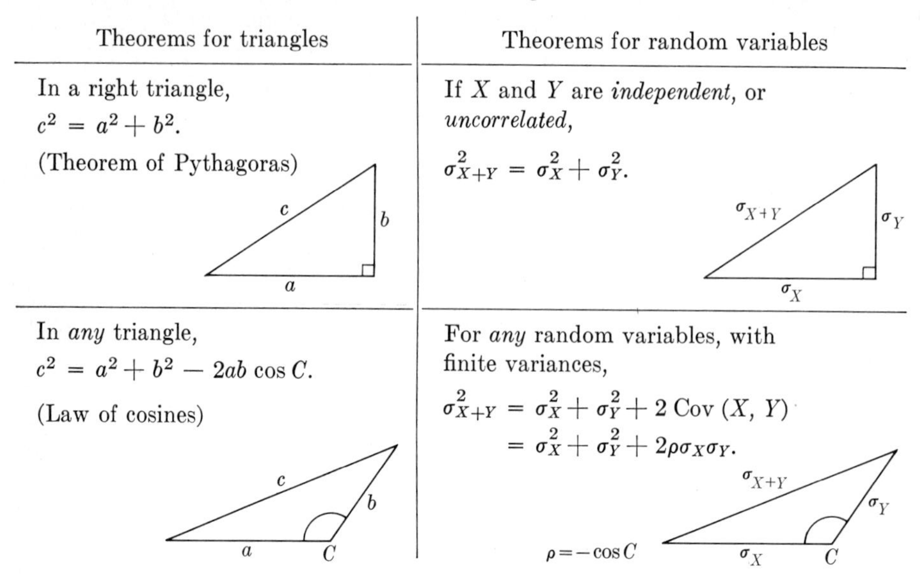

Theorems for triangles	Theorems for random variables
In a right triangle, $c^2 = a^2 + b^2$. (Theorem of Pythagoras)	If X and Y are *independent*, or *uncorrelated*, $\sigma_{X+Y}^2 = \sigma_X^2 + \sigma_Y^2$.
In *any* triangle, $c^2 = a^2 + b^2 - 2ab \cos C$. (Law of cosines)	For *any* random variables, with finite variances, $\sigma_{X+Y}^2 = \sigma_X^2 + \sigma_Y^2 + 2 \operatorname{Cov}(X, Y) = \sigma_X^2 + \sigma_Y^2 + 2\rho\sigma_X\sigma_Y$.

Variance of $X + Y$ To compute the variance of $X + Y$, we apply the definitional formula for variance,

$$\sigma_{X+Y}^2 = E(X + Y - \mu_{X+Y})^2. \tag{1}$$

Whether X and Y are independent or not, the mean of $X + Y$ is the mean of X plus the mean of Y:

$$\mu_{X+Y} = \mu_X + \mu_Y. \tag{2}$$

We now substitute from Eq. (2) into the right side of Eq. (1), group X-terms and Y-terms, square, and get

$$\sigma_{X+Y}^2 = E[(X - \mu_X)^2 + 2(X - \mu_X)(Y - \mu_Y) + (Y - \mu_Y)^2]. \tag{3}$$

Using the fact that the expected value of a sum is the sum of the expected values, and that

$$E[2(X - \mu_X)(Y - \mu_Y)] = 2E[(X - \mu_X)(Y - \mu_Y)],$$

we expand the right side of Eq. (3), rearrange terms, and get

$$\sigma_{X+Y}^2 = E(X - \mu_X)^2 + E(Y - \mu_Y)^2 + 2E[(X - \mu_X)(Y - \mu_Y)]. \tag{4}$$

We recognize the first two terms on the right side of Eq. (4) as σ_X^2 and σ_Y^2 from their definitions:

$$E(X - \mu_X)^2 = \sigma_X^2, \qquad E(Y - \mu_Y)^2 = \sigma_Y^2. \tag{5}$$

Thus Eq. (4) agrees with one of the entries for σ_{X+Y}^2 in Table 10–3 provided we define Cov (X, Y) as follows.

10–11 Definition: Covariance

The *covariance* of two random variables is the mean value of the product of their paired deviations from their own means. In symbols:

$$\boxed{\operatorname{Cov}(X, Y) = E[(X - \mu_X)(Y - \mu_Y)].} \tag{6}$$

It is evident from Eq. (6) that the *commutative law for covariance* holds:

$$\operatorname{Cov}(X, Y) = \operatorname{Cov}(Y, X).$$

It is also clear that the covariance of a random variable with itself is its variance, because

$$\operatorname{Cov}(X, X) = E[(X - \mu_X)(X - \mu_X)] = \operatorname{Var}(X) = \sigma_X^2.$$

If we substitute from Eqs. (5) and (6) into Eq. (4), we see that we have proved the following theorem:

10–12 Theorem: Variance of a sum of dependent or independent random variables

Let X and Y be random variables with finite variances σ_X^2 and σ_Y^2, and with covariance as defined above. Then the variance of their sum is given by

$$\boxed{\sigma_{X+Y}^2 = \sigma_X^2 + \sigma_Y^2 + 2\operatorname{Cov}(X, Y).} \tag{7}$$

Example 1 *Means zero.* Let X and Y have the joint probability function shown in the table below. Find the variances of X and Y, their covariance, and the variance of $X + Y$.

		Value of Y		
		-1	1	$P(X = x)$
Value of X	-2	0.4	0.1	0.5
	2	0.1	0.4	0.5
	$P(Y = y)$	0.5	0.5	1.0

Solution Since the means are $\mu_X = \mu_Y = 0$,

$$\begin{aligned}
\sigma_X^2 &= E(X - \mu_X)^2 = E(X^2) = 0.5(2)^2 + 0.5(-2)^2 = 4,\\
\sigma_Y^2 &= E(Y - \mu_Y)^2 = E(Y^2) = 0.5(1)^2 + 0.5(-1)^2 = 1,\\
\operatorname{Cov}(X, Y) &= E[(X - \mu_X)(Y - \mu_Y)] = E(XY)\\
&= 0.4(2)(1) + 0.1(2)(-1) + 0.1(-2)(1) + 0.4(-2)(-1)\\
&= 1.2,\\
\sigma_{X+Y}^2 &= \sigma_X^2 + \sigma_Y^2 + 2\operatorname{Cov}(X, Y)\\
&= 4 + 1 + 2(1.2) = 7.4.
\end{aligned}$$

Note that the covariance term increases the sum of the variances of X and Y by nearly 50% to give the variance of $X + Y$.

Computational formula When neither mean is zero, the covariance can be computed from a convenient formula derived from Eq. (6):

$$\text{Cov}\,(X, Y) = E[(X - \mu_X)(Y - \mu_Y)] = E[XY - \mu_X Y - \mu_Y X + \mu_X \mu_Y].$$

The expectation of a sum is the sum of the expectations, so we expand the right-hand term:

$$\text{Cov}\,(X, Y) = E(XY) - E(\mu_X Y) - E(\mu_Y X) + E(\mu_X \mu_Y).$$

Recall that $E(cX) = cE(X)$, and $E(c) = c$. Since μ_X and μ_Y are constants, we can write

$$\text{Cov}\,(X, Y) = E(XY) - \mu_X E(Y) - \mu_Y E(X) + \mu_X \mu_Y.$$

We now substitute for $E(Y)$ and $E(X)$ their values μ_Y and μ_X and get

$$\boxed{\text{Cov}\,(X, Y) = E(XY) - \mu_X \mu_Y.} \tag{8}$$

Example 2 Let X, Y have the joint probability function shown in the table below. Find the variances of X and Y, their covariance, and the variance of $X + Y$.

x	y	$P(x, y)$
1	1	0.1
1	−1	0.5
−1	1	0.1
−1	−1	0.3

Solution

$\mu_X = 0.2$
$\mu_Y = -0.6$
$\sigma_X^2 = 0.96$
$\sigma_Y^2 = 0.64$

$$\begin{aligned}\text{Cov}\,(X, Y) &= 0.1(1)(1) + 0.5(1)(-1) + 0.1(-1)(1) \\ &\quad + 0.3(-1)(-1) - (0.2)(-0.6) = -0.08, \\ \sigma_{X+Y}^2 &= \sigma_X^2 + \sigma_Y^2 + 2\,\text{Cov}\,(X, Y) \\ &= 0.96 + 0.64 + 2(-0.08) = 1.44.\end{aligned}$$

Here the covariance has the effect of decreasing the variance of the sum 10% below the value $\sigma_X^2 + \sigma_Y^2 = 1.60$ that would result if X and Y were independent.

Independence From Theorem 7–6, Section 7–1, we know that if X and Y are independent, then

$$E(XY) = E(X)E(Y) = \mu_X \mu_Y.$$

Therefore we find that if X and Y are independent their covariance is zero, because then

$$\text{Cov}(X, Y) = E(XY) - \mu_X\mu_Y = \mu_X\mu_Y - \mu_X\mu_Y = 0.$$

Thus the following theorem is true.

10–13 Theorem

If X and Y are independent, then

$$\boxed{\text{Cov}(X, Y) = 0.} \tag{9}$$

The population correlation In some problems, it is convenient to have a measure of the degree of linear relationship between X and Y that does not depend upon their units of measurement. The covariance does depend upon the units of measurement, as the following theorem shows.

10–14 Theorem

Let X and Y be random variables with means μ_X, μ_Y and variances σ_X^2, σ_Y^2. Form the *standardized* variables

$$X' = \frac{X - \mu_X}{\sigma_X}, \qquad Y' = \frac{Y - \mu_Y}{\sigma_Y}. \tag{10}$$

Then

$$\boxed{\text{Cov}(X, Y) = \sigma_X\sigma_Y \text{Cov}(X', Y').} \tag{11}$$

Proof Both X' and Y' are constructed to have means 0 (and standard deviations 1). Their covariance is therefore

$$\text{Cov}(X', Y') = E(X'Y') = E\left[\frac{X - \mu_X}{\sigma_X} \cdot \frac{Y - \mu_Y}{\sigma_Y}\right].$$

By factoring the constant $1/\sigma_X\sigma_Y$ outside the expectation, we find

$$\text{Cov}(X', Y') = \frac{E[(X - \mu_X)(Y - \mu_Y)]}{\sigma_X\sigma_Y} = \frac{\text{Cov}(X, Y)}{\sigma_X\sigma_Y}.$$

Multiplying through by $\sigma_X\sigma_Y$ completes the proof of Eq. (11). □

Remark A shift of origin of X leaves the numerator and denominator of X' unchanged on the right in Eq. (10). A change in scale, or units, e.g., from feet to yards or from pounds to ounces, multiplies both the numerator and denominator by the same constant, and therefore leaves X' unchanged. Thus any combination of these transformations leaves X' and Y' unchanged. Therefore, $\text{Cov}(X', Y')$ is independent of the origin and units of measurement of X and Y and provides a measure of linear relationship between them with the desired properties.

10-15 Definition: Correlation

Let X and Y be random variables with positive standard deviations σ_X and σ_Y. Then the *correlation* between X and Y, denoted by ρ, is defined by

$$\rho = \text{Cov}(X', Y'),$$

where

$$X' = \frac{X - \mu_X}{\sigma_X}, \qquad Y' = \frac{Y - \mu_Y}{\sigma_Y}. \tag{12a}$$

Alternatively, from Eqs. (11) and (12a):

$$\rho = \frac{\text{Cov}(X, Y)}{\sigma_X \sigma_Y}, \qquad \text{Cov}(X, Y) = \rho\sigma_X\sigma_Y. \tag{12b}$$

When X and Y are independent, Cov $(X, Y) = 0$, and therefore $\rho = 0$. But the converse is false: $\rho = 0$ does not necessarily imply that X and Y are independent, as shown by the three-coin example, Table 7-2. When $\rho = 0$, we say X and Y are *uncorrelated*.

Computational formula for correlation If we use Eq. (8) to evaluate Cov (X, Y) in the numerator of Eq. (12b), we get an alternative formula, often useful for calculations:

$$\rho = \frac{E(XY) - \mu_X\mu_Y}{\sigma_X\sigma_Y}. \tag{13}$$

Example 3 For the population of Example 1, compute ρ.

Solution The substitution into Eq. (12b) of the values shown in the solution to Example 1 gives

$$\rho = \frac{1.2}{\sqrt{4 \times 1}} = 0.6.$$

Example 4 For the population of Example 2, find the correlation between X and Y.

Solution The substitution into Eq. (12b) of the values found in the solution of Example 2 gives

$$\rho = \frac{-0.08}{\sqrt{0.96(0.64)}} \approx \frac{-0.08}{0.784} \approx -0.10.$$

Range of ρ Values of ρ can range from -1 to $+1$ inclusive. For discrete probability functions, $\rho = 1$ or -1 if and only if all (X, Y) pairs with non-zero probability lie along a straight line. (If the line is horizontal or vertical, then σ_X or σ_Y is zero, and ρ is undefined.) The slope of the line determines the sign of ρ. We shall not prove these facts. However, if we replace Cov (X, Y) by $\rho\sigma_X\sigma_Y$ in Eq. (7), Theorem 10–12, we get the following theorem, which bears a striking resemblance to the law of cosines, with ρ playing the role of $\cos(180° - C)$.

10–16 Theorem: Variance of a sum of correlated random variables

Let X and Y be random variables with finite variances σ_X^2, σ_Y^2 and correlation ρ. Then the variance of their sum is

$$\boxed{\sigma_{X+Y}^2 = \sigma_X^2 + \sigma_Y^2 + 2\rho\sigma_X\sigma_Y.} \tag{14}$$

Note that *zero correlation* gives the result

$$\sigma_{X+Y}^2 = \sigma_X^2 + \sigma_Y^2.$$

Sample correlation coefficient For a set of pairs of measurements (x_i, y_i), $i = 1, 2, \ldots, n$, there is a *sample correlation coefficient*, r, defined by analogy with ρ as

$$\boxed{r = \frac{\sum(x_i - \bar{x})(y_i - \bar{y})}{ns_xs_y},} \tag{15}$$

where s_x and s_y are the sample standard deviations as defined in Section 6–2. Two additional forms of r are

$$r = \frac{\sum x_iy_i - n\bar{x}\bar{y}}{ns_xs_y}, \tag{16}$$

$$r = \frac{\sum x_iy_i/n - \bar{x}\bar{y}}{s_xs_y}. \tag{17}$$

We shall study r in Chapter 11.

Variance of the sum of two or more correlated random variables Sometimes we deal with sums of several correlated random variables. For example, students may take several tests in the same course. Let the score on the first test be X_1; on the second, X_2; on the third, X_3; and so on. What can we say of the mean and variance of the distribution of the sums of the scores? We answer this question by extending the results of Theorems 10–12 and 10–16 to sums of more than two random variables. First we extend the results algebraically to three variables; the results for n variables can be proved similarly, but we shall merely state the general result.

Variance of the sum of three variables Let X_1, X_2, and X_3 have means μ_1, μ_2, μ_3 and variances σ_1^2, σ_2^2, σ_3^2, respectively. Let

$$Z = X_1 + X_2 + X_3.$$

Then, by definition,

$$\begin{aligned}\sigma_Z^2 &= E(Z - \mu_Z)^2 \\ &= E(X_1 + X_2 + X_3 - \mu_1 - \mu_2 - \mu_3)^2 \\ &= E[(X_1 - \mu_1) + (X_2 - \mu_2) + (X_3 - \mu_3)]^2.\end{aligned}$$

Recall that

$$(a + b + c)^2 = a^2 + b^2 + c^2 + 2ab + 2ac + 2bc.$$

Therefore

$$\begin{aligned}\sigma_Z^2 = E[&(X_1 - \mu_1)^2 + (X_2 - \mu_2)^2 + (X_3 - \mu_3)^2 \\ &+ 2(X_1 - \mu_1)(X_2 - \mu_2) + 2(X_1 - \mu_1)(X_3 - \mu_3) \\ &+ 2(X_2 - \mu_2)(X_3 - \mu_3)].\end{aligned}$$

The expectation of a sum is the sum of the expectations. Therefore

$$\boxed{\begin{aligned}\sigma_Z^2 = \sigma_1^2 + \sigma_2^2 + \sigma_3^2 &+ 2 \operatorname{Cov}(X_1, X_2) + 2 \operatorname{Cov}(X_1, X_3) \\ &+ 2 \operatorname{Cov}(X_2, X_3).\end{aligned}} \tag{18}$$

In words, *the variance of a sum of several random variables is the sum of their variances plus twice the sum of the covariances of all possible pairs of random variables.* And this statement is true for many variables as well as for three. In Eq. (18), the 2's that are coefficients of the covariances can be viewed as coming from the squaring operation, or equivalently from

$$\operatorname{Cov}(X_i, X_j) + \operatorname{Cov}(X_j, X_i) = 2 \operatorname{Cov}(X_i, X_j), \qquad i \neq j.$$

10-17 Theorem: Variance of a sum

Let the variables $X_1, X_2, \ldots, X_n$ have variances $\sigma_1^2, \sigma_2^2, \ldots, \sigma_n^2$ and covariances $\operatorname{Cov}(X_i, X_j)$, $i \neq j$, and let

$$Z = X_1 + X_2 + \cdots + X_n.$$

Then

$$\boxed{\sigma_Z^2 = \sum_{i=1}^{n} \sigma_i^2 + 2 \sum \operatorname{Cov}(X_i, X_j),} \tag{19}$$

where the sum of the covariance terms runs over all possible pairs with $i < j$, $i = 1, 2, \ldots, n$; $j = 1, 2, \ldots, n$.

Note that n variances are added and that $n(n - 1)/2$ covariances are added because there are $\binom{n}{2} = n(n - 1)/2$ selections of 2 variables from n variables, disregarding order. If $n = 3$, $n(n - 1)/2 = 3$, which checks with our result for three variables.

Sometimes we introduce ρ_{ij} as the correlation between X_i and X_j, and then the covariance can be written as

$$\text{Cov}(X_i, X_j) = \rho_{ij}\sigma_i\sigma_j. \tag{20}$$

We shall not explicitly substitute (20) into Eq. (19), but in the special case where all ρ_{ij} equal ρ and all σ_i equal σ, Eq. (19) can be rewritten as

$$\begin{aligned}\sigma_Z^2 &= n\sigma^2 + n(n-1)\,\text{Cov}(X_i, X_j)\\ &= n\sigma^2 + n(n-1)\rho\sigma^2 = n\sigma^2[1 + (n-1)\rho].\end{aligned} \tag{21}$$

Example 5 *Simplified Army alpha test.* A test is composed of four parts: (1) arithmetic, (2) number completion, (3) word analogies, and (4) information. Let X_1 stand for the score on arithmetic, X_2 for that on number completion, and so on. Scores on each part have mean 50 and standard deviation 10, and the variables are correlated as follows (after J. P. Guilford, *Psychometric Methods*, McGraw-Hill, 1936, p. 491):

	(1)	(2)	(3)	(4)	Sum
(1)		.49	.09	.31	.89
(2)			.40	.16	.56
(3)				.29	.29
					1.74

Find the mean and standard deviation of Z, the sum of the scores of the four parts.

Solution The mean of the sum is the sum of the means, or $4(50) = 200$. Since all four variables have the same standard deviation, 10, Eq. (19) reduces to

$$\sigma_Z^2 = 100[4 + 2\,(\text{sum of correlations})] = 100[4 + 2(1.74)] = 748.$$

Therefore $\sigma_Z = \sqrt{748} \approx 27.3$. The sums for such tests are usually approximately normally distributed, so we know that about $\frac{2}{3}$ of the scores are within one standard deviation of the mean, i.e., between 173 and 227.

Example 6 *Matching.* A deck of cards numbered 1, 2, . . . , n is shuffled and the cards are dealt one at a time. If card number i is the ith card dealt, we count it as a *match*, otherwise not. Find the mean and the variance of the total number of matches.

Solution We introduce n random variables X_i, $i = 1, 2, \ldots, n$; one for each card dealt. If the ith card dealt has the number i on it, then $X_i = 1$; otherwise $X_i = 0$. The total number of matches is

$$T = X_1 + X_2 + \cdots + X_n. \tag{22}$$

We want to find the mean and variance of T. To this end, we need to find the mean and variance of each X_i and the covariances of all pairs X_i and X_j.

For any X_i, we have

$$\mu_i = E(X_i) = 1 \cdot P(X_i = 1) + 0 \cdot P(X_i = 0) = P(X_i = 1) = 1/n,$$

because there is just one favorable card, i, among the n cards. Since all $\mu_i = 1/n$,

$$\mu_T = \sum \mu_i = n\left(\frac{1}{n}\right) = 1.$$

It is easy to get Var (X_i), because $X_i = 1$ with probability $1/n$, and 0 with probability $1 - 1/n$. Therefore

$$\text{Var}\,(X_i) = E(X_i^2) - \mu_i^2 = 1^2\left(\frac{1}{n}\right) + 0^2\left(1 - \frac{1}{n}\right) - \frac{1}{n^2} = \frac{n-1}{n^2}. \tag{23}$$

To find the covariance of X_i and X_j, we apply the computational formula, Eq. (8):

$$\text{Cov}\,(X_i, X_j) = E(X_iX_j) - \mu_i\mu_j.$$

The possible values of X_iX_j are 0 and 1, and

$$\begin{aligned} E(X_iX_j) &= 1 \cdot P(X_iX_j = 1) + 0 \cdot P(X_iX_j = 0) = P(X_iX_j = 1) \\ &= \frac{(n-2)!}{n!} = \frac{1}{n(n-1)}, \end{aligned}$$

since $X_iX_j = 1$ if and only if both of the cards numbered i and j are in their respective matching places, and there are $(n-2)!$ permutations of the remaining $n - 2$ cards that correspond to this event. Therefore, if $i \neq j$,

$$\text{Cov}\,(X_i, X_j) = \frac{1}{n(n-1)} - \frac{1}{n} \cdot \frac{1}{n} = \frac{1}{n^2(n-1)}. \tag{24}$$

We are now ready to substitute from Eqs. (23) and (24) into the first line of Eq. (21). The result is

$$\begin{aligned} \sigma_T^2 = \text{Var}\,(T) &= n \cdot \frac{n-1}{n^2} + n(n-1) \cdot \frac{1}{n^2(n-1)} \\ &= 1 - \frac{1}{n} + \frac{1}{n} = 1. \end{aligned} \tag{25}$$

The results are striking—the variance and the mean are both equal to one. Obviously, the result applies to the number of matches when a bridge deck is shuffled and then compared card by card with a previously laid out target deck. Although any number of matches between 0 and 52 is possible, the mean number of matches is only 1, and the standard deviation is also 1, which is very small compared with the number of cards in the deck, 52. By Chebyshev's Theorem, we do not expect large fluctuations about the mean in repeated experiments of this type.

EXERCISES FOR SECTION 10-4

1. Let X and Y have the following joint probability function:

x	y	$P(x, y)$
0	1	0.4
0	-1	0.3
1	0	0.2
1	1	0.1

   Find (a) $E(X)$, (b) $E(Y)$, (c) σ_X^2, (d) σ_Y^2, (e) $E(XY)$, (f) Cov (X, Y), (g) ρ, the correlation between X and Y.

2. If $\sigma_X^2 = 0$ or $\sigma_Y^2 = 0$, what can you say about the covariance of X and Y? Explain.
3. If $Y = kX$, for some constant k, how is Cov (X, Y) related to Var (X)?
4. Prove that Cov $(aX, bY) = ab$ Cov (X, Y).
5. What is the correlation between aX and X, if (a) $a = 1$? (b) $a = -1$? (c) $a = 5$? (d) $a = -5$?
6. What is the correlation between X and Y if $Y = aX + b$?
7. On the first quiz in a mathematics class, the class average was 85 and the standard deviation was 5. On the second quiz, the class average was 80 and the standard deviation was 7. The same 30 students took both tests. The correlation between grades on the two tests was 0.7. Let X represent the grade on the first test, for a randomly chosen student, and let Y represent his grade on the second test. Find (a) μ_{X+Y}, (b) σ_{X+Y}^2. (c) Assuming that the probability that the variable $\overline{X} = \frac{1}{2}(X + Y)$ takes a value greater than k can be approximated by a normal random variable with mean equal to $\mu_{\overline{X}}$ and variance equal to $\sigma_{\overline{X}}^2$, find (approximately) the number of students whose average scores, on the two tests, were 75 or better.
8. Prove the "distributive law":

$$\text{Cov}\,(X + Y, Z) = \text{Cov}\,(X, Z) + \text{Cov}\,(Y, Z).$$

9. By repeated applications of the commutative law,

$$\text{Cov}\,(Y, X) = \text{Cov}\,(X, Y),$$

   and the distributive law,

$$\text{Cov}\,(X + Y, Z) = \text{Cov}\,(X, Z) + \text{Cov}\,(Y, Z),$$

   show that

$$\text{Cov}\,(X + Y, X + Y) = \text{Cov}\,(X, X) + 2\,\text{Cov}\,(X, Y) + \text{Cov}\,(Y, Y).$$

   To what previously known result does this correspond?

10. (Continuation.) Show how the method of Exercise 9 can be generalized to produce formula (18) for the variance of a sum of three random variables.

11. The 3 random variables X, Y, Z all have means equal to zero and variances equal to 1, and any two of them have correlation $\rho = 0.5$. Find (a) the mean of $X + Y + Z$, (b) the variance of $X + Y + Z$, (c) the probability that $U = X + Y + Z$ exceeds 2, assuming that U is normally distributed.

12. If the variance of $X_1 + X_2 + \cdots + X_n$ in Eq. (21) is zero, but $\sigma^2 \neq 0$, what can you say about the correlation ρ? Note that this is the smallest value ρ can have.

13. Twenty married couples attend a masquerade ball. The men form one line and the women form another, each in random order. What is the probability that the 4th man in line is the husband of the 4th woman in line? That the 5th man and woman are husband and wife, and that the 6th man and woman are also husband and wife?

14. In Exercise 13, suppose the ith man in line dances with the ith woman in line, $i = 1, 2, \ldots, 20$. What is the expected number of couples with husband and wife dancing together? What is the standard deviation of that number?

15. A secretary addresses 40 envelopes and types 40 individual letters to be mailed in them. While she is out of the office for lunch, a practical joker puts the letters into the envelopes at random and mails them. What is the probability that both Mr. Jones and Miss Smith receive the letters intended for them? That Mr. Jones, Miss Smith, and Mr. Thornton all receive their proper letters? What is the expected number of persons who receive the letters meant for them? Using Chebyshev's Theorem, state an upper bound for the probability that 25% or more of the letters go to their intended recipients.

16. Take the clubs and diamonds from an ordinary bridge deck and lay them out in order: Club ace, king, ..., two; Diamond ace, ..., two. Shuffle the hearts and spades together, then lay them out one after another beside the other cards: first card beside the Club ace, second beside the Club king, and so on. Whenever the card from the shuffled half-deck matches the target card in color and denomination, score a 1; otherwise don't count it. Count and record the total number of matches. Perform this experiment 10 times and make a frequency distribution of the numbers of matches in the 10 performances. Find the mean and variance of this frequency distribution and compare them with the theoretical values.

17. Each of n random variables has variance σ^2, and each pair has correlation ρ. Find the standard deviation of their average.

18. Each of n questions on an examination is scored so that its mean is zero and its variance σ^2. The total score for the examination, T, is the sum of the scores of the individual test questions. The correlation between the scores of any pair of questions is $\rho = 1/(n - 1)$, $n \geq 2$. Find the variance of T.

19. Let X and Y be independent random variables with zero means and variances σ_X^2, σ_Y^2. Find the correlation between X and Z, where $Z = X + Y$.

20. Let $X_1, X_2, \ldots, X_k$ be independent random variables with zero means and equal variances σ^2. Let $a_1, a_2, \ldots, a_k$ and $b_1, b_2, \ldots, b_k$ be numbers such that

$$\sum_{i=1}^{k} a_i b_i = 0.$$

If $U = a_1X_1 + a_2X_2 + \cdots + a_kX_k$, and $V = b_1X_1 + b_2X_2 + \cdots + b_kX_k$, prove that U and V are uncorrelated.

21. Students applying for admission to a program are required to take an aptitude test and an achievement test. For this particular student population, the scores X and Y on the two tests have means 50, standard deviations 10, and correlation $\rho = 0.62$. Only students who get a combined score $X + Y$ of at least 136 points are admitted to the program. Assume that the combined scores are normally distributed. (a) Find the mean and variance of the distribution of $X + Y$. (b) What percent of the applicants are admitted? (c) If the director wants to admit only the top 25% of the applicants, based on this score, what new cutoff score would you suggest?

22. There are $2n$ items on a test. Each item is scored so that its mean is zero and its variance is 1. The correlation between any pair of items is ρ. Let the score on the first n items be Y, where

$$Y = X_1 + X_2 + \cdots + X_n,$$

and let the score on the second n items be Z, where

$$Z = X_{n+1} + X_{n+2} + \cdots + X_{2n},$$

and where X_i is the score on the ith item. Show that the correlation between Y and Z is

$$\rho(Y, Z) = \frac{n\rho}{1 + (n - 1)\rho}.$$

Discuss the result for the two special cases (a) $n = 1$, (b) $n\rho = 100$.

23. An experimenter measured two characteristics on each of $n = 25$ animals, thus obtaining 25 pairs of measurements (x_i, y_i), $i = 1, 2, \ldots, 25$. From these data, he computed

$$\sum x_i = 125, \quad \sum y_i = 50, \quad \sum x_i^2 = 725,$$
$$\sum y_i^2 = 325, \quad \sum x_i y_i = 350.$$

Find $\bar{x}$, $\bar{y}$, s_x, s_y, and the correlation, r, between x and y.

For Exercises 24 through 31, an ordinary six-sided die is thrown n times in succession. Let $X_1 = 1$, and let X_i be 1 if the ith throw produces a result different from those obtained on the first $i - 1$ throws. Otherwise $X_i = 0$. Thus $Y = X_1 + X_2 + \cdots + X_n$ is the number of different faces of the die that have appeared on top in the n throws.

24. Find (a) $P(X_2 = 1)$, (b) $E(X_2)$, (c) Var (X_2).
25. Find Cov (X_1, X_2).
26. Find $\rho(X_1, X_2)$.
27. Find (a) $E(X_1 + X_2)$, (b) Var $(X_1 + X_2)$.
28. Find (a) $P(X_3 = 1)$, (b) $E(X_3)$, (c) Var (X_3).
29. Find (a) Cov (X_1, X_3), (b) Cov (X_2, X_3).
30. Find (a) $E(X_1 + X_2 + X_3)$, (b) Var $(X_1 + X_2 + X_3)$.

☠31. For fixed $n \geq 3$, and $k = 1, 2, \ldots, n-1$, find:

a) $P(X_n = 1)$, $E(X_n)$, and $\text{Var}(X_n)$,
b) $\text{Cov}(X_k, X_n)$,
c) $E(X_1 + X_2 + \cdots + X_n)$.

In Exercises 32 through 34, two ordinary dice are thrown once. Let X be the maximum of the scores on their topmost faces, and let Y be the absolute value of the difference between those scores.

32. Write out a sample space for the experiment, together with the paired values (X, Y) for the sample points.
33. Obtain, from the sample space of Exercise 32, the joint probability function of X and Y.
34. From the result of Exercise 33, compute the covariance and the correlation of X and Y.
35. *Effect of correlation on* $\sigma_{\bar{X}}$. If n measurements are drawn from populations with identical means and variances, μ and σ^2, and all pairs of observations are equally correlated ρ, compare $\sigma_{\bar{X}}^2$ to its value when $\rho = 0$.

★10-5 SAMPLING WITHOUT REPLACEMENT FROM A FINITE POPULATION

The key results of this section are: (1) the mean of a random sample drawn without replacement from a finite population is an unbiased estimate of the population mean; (2) the standard deviation of the sample mean is almost exactly $(\sigma^2/n)(1 - f)$, where σ^2 is the population variance, n is the sample size, and f the fraction n/N, where N is the population size. Unless f is substantial, say more than 0.1, the reduction in standard deviation owing to finiteness of the population is negligible. The surprising result is that for the same accuracy in a sample of people, we need nearly the same sample size from the population of a small town as we do from the United States of America. (See Exercise 22 for Section 10-5.)

Our development of properties of distributions of sample averages has been restricted to infinite populations, or to sampling *with* replacement from a finite population. In many practical sampling problems the sampling is done without replacement, and the reader may wish to know the mean and variance of the distribution of sample averages from such populations, and to compare the results with the corresponding formulas for sampling with replacement. For example, if you plan to ask opinions on a school issue from a random sample of 10 students drawn from a class of 50, you wish to know what accuracy to expect.

Example 1 For a population of size $N = 4$ with measurements $x_1 = 0$, $x_2 = 3$, $x_3 = 2$, $x_4 = 3$, find the mean and variance of the distribution of sample averages for random samples of size $n = 2$, drawn without replacement.

Table 10–4 Samples (X_1, X_2) and sample averages $\bar{x}_j$

		X_2, second sample element			
		$x_1 = 0$	$x_2 = 3$	$x_3 = 2$	$x_4 = 3$
X_1, first sample element	$x_1 = 0$		1.5	1.0	1.5
	$x_2 = 3$	1.5		2.5	3.0
	$x_3 = 2$	1.0	2.5		2.5
	$x_4 = 3$	1.5	3.0	2.5	

$$\sum \bar{x}_j = 24.0, \qquad \sum \bar{x}_j^2 = 54.00$$

Solution Using the sampling notation introduced in Section 10–4, we let X_1 represent the measurement of the population element drawn first, X_2 that of the element drawn second. There are $4 \cdot 3 = 12$ possible samples of individuals, as shown in Table 10–4 together with the sample averages $\bar{x}_j$, $j = 1, \ldots, 12$. The table is symmetrical about the main diagonal because each pair can be reversed, but the diagonal cells are empty because the sampling is without replacement.

The probability function of the random variable $\overline{X}$, whose values are the $\bar{x}$'s, is given by the probability table

$f(\bar{x})$	$\frac{2}{12}$	$\frac{4}{12}$	$\frac{4}{12}$	$\frac{2}{12}$
$\bar{x}$	1.0	1.5	2.5	3.0

We can compute the mean and variance of $\overline{X}$ either from Table 10–4 or from the probability table. Using Table 10–4, we get

$$\mu_{\overline{X}} = E(\overline{X}) = \sum_{j=1}^{12} \frac{\bar{x}_j}{12} = \frac{24}{12} = 2,$$

$$\sigma_{\overline{X}}^2 = \sum_{j=1}^{12} \frac{\bar{x}_j^2}{12} - [E(\overline{X})]^2 = \frac{54}{12} - 2^2 = \frac{1}{2}.$$

This example illustrates the distribution of sample averages and the calculation of $E(\overline{X})$ and $\sigma_{\overline{X}}^2$ for a small sample from a small population. We now generalize the results.

Let the ith individual in a population of size N have a measured characteristic x_i, $i = 1, 2, \ldots, N$; then the population has mean

$$\mu = \sum_{i=1}^{N} \frac{x_i}{N} \tag{1}$$

and variance

$$\sigma^2 = \sum_{i=1}^{N} \frac{(x_i - \mu)^2}{N}. \tag{2}$$

In the derivation of the mean and variance of averages for random samples drawn from this population, the algebra is simplified if we translate the origin of the x_i-measurements to their mean. Assume that this has been done; then Eq. (1) implies that

$$\text{a) } \mu = 0, \qquad \text{b) } \sum_{i=1}^{N} x_i = 0, \qquad \text{c) } \sigma^2 = \sum_{i=1}^{N} \frac{x_i^2}{N}. \tag{3}$$

Suppose a random sample of n is drawn, without replacement, from this population. We introduce n random variables, $X_1, X_2, \ldots, X_n$, one for each element in the *sample.* The value of X_1 is the measurement of the characteristic of the particular *population member* drawn *first,* the value of X_2 is the measurement of the population member drawn *second,* and so on, X_j being the measurement of the population member drawn as the jth element of the sample. Observe that order counts in this setup; we distinguish between the sample of 2 elements in which $X_1 = x_2$, $X_2 = x_5$, say, and the sample in which $X_1 = x_5$, $X_2 = x_2$.

We want the mean and variance of the sum

$$Z = X_1 + X_2 + \cdots + X_n \tag{4}$$

and of the sample average

$$\overline{X} = \frac{Z}{n} = \frac{X_1 + X_2 + \cdots + X_n}{n}. \tag{5}$$

We therefore need to compute the means, μ_j, variances, σ_j^2, and covariances, Cov (X_j, X_k).

Mean and variance of X_j The population measurements $x_1, x_2, \ldots, x_N$ have equal probabilities of being the jth in the sequence $X_1, X_2, \ldots, X_n$. Hence the probability function of X_j is given by

$$P(X_j = x_1) = P(X_j = x_2) = \cdots = P(X_j = x_N) = \frac{1}{N}.$$

Therefore

$$\mu_j = E(X_j) = \sum_{i=1}^{N} \frac{1}{N} \cdot x_i = \sum_{i=1}^{N} \frac{x_i}{N} = \mu = 0,$$

where the result $\mu = 0$ follows from Eq. (3a).

Since $\mu_j = 0$, the variance of X_j is given by

$$\sigma_j^2 = E(X_j^2) = \sum_{i=1}^{N} \frac{1}{N} \cdot x_i^2 = \sigma^2.$$

To summarize, the implications for a population with general mean μ and variance σ^2 are

$$\mu_1 = \mu_2 = \cdots = \mu_n = \mu \qquad \text{and} \qquad \sigma_1^2 = \sigma_2^2 = \cdots = \sigma_n^2 = \sigma^2.$$

By applying the result of Corollary 7-5 of Section 7-1, we find the following:

$$E(Z) = E(\sum X_j) = \sum E(X_j) = \sum \mu = n\mu, \tag{6}$$

where the summations extend from $j = 1$ to n, and

$$E(\overline{X}) = E(Z/n) = E(Z)/n = n\mu/n = \mu. \tag{7}$$

So the mean of sample averages is the population mean.

Covariance of X_j and X_k We lean again on the fact that the origin of measurements has been chosen to make the population mean equal to zero. Since X_j and X_k also have zero means,

$$\text{Cov}\,(X_j, X_k) = E(X_j X_k). \tag{8}$$

To evaluate this expectation, we need the joint probability function of X_j and X_k. When $j \neq k$, the ordered pair (X_j, X_k) is a sample of 2 different individuals from the population, say (x_r, x_s). There are $N(N - 1)$ such ordered pairs of individuals from the population, as in Table 10-4 where $N = 4$. These pairs are equally likely, so each has probability $1/N(N - 1)$. Therefore

$$P(X_j = x_r, X_k = x_s) = \frac{1}{N(N - 1)}, \tag{9}$$

where Eq. (9) is true for all pairs of different random variables and all pairs of different individuals in the population.

To compute $E(X_j X_k)$, for fixed $j \neq k$, observe that this product can take every value $x_r x_s$, with r and s different. Therefore

$$\begin{aligned} E(X_j X_k) &= \sum_{r \neq s} \frac{1}{N(N - 1)} x_r x_s \\ &= \frac{1}{N(N - 1)} \sum_{r \neq s} x_r x_s. \end{aligned} \tag{10}$$

The final sum in Eq. (10) is most easily evaluated as a part of the square of $x_1 + x_2 + \cdots + x_N$, because

$$(x_1 + x_2 + \cdots + x_N)^2 = x_1^2 + x_2^2 + \cdots + x_N^2 + \sum_{r \neq s} x_r x_s, \tag{11}$$

where the final sum in Eq. (11) is the same as that in Eq. (10). But, from Eq. (3b),

$$\sum_{i=1}^{N} x_i = 0,$$

and therefore the left side of Eq. (11) is 0^2, or 0. Hence

$$\sum_{r \neq s} x_r x_s = -(x_1^2 + x_2^2 + \cdots + x_N^2) = -\sum_{i=1}^{N} x_i^2 = -N\sigma^2. \tag{12}$$

The final result follows from Eq. (3c).

Substituting from Eq. (12) into Eqs. (10) and (8), we find

$$\operatorname{Cov}(X_j, X_k) = -\frac{\sigma^2}{N-1}. \tag{13}$$

Variance of $X_1 + X_2 + \cdots + X_n$ It is now an easy matter to find the variance of the sum Z:

$$Z = X_1 + X_2 + \cdots + X_n.$$

The variances of all the X_j's are equal, and the covariances of all pairs X_j, X_k are also equal. Therefore Eq. (21) of Section 10–4 applies, with the following result:

$$\sigma_Z^2 = n\sigma^2 - \frac{n(n-1)}{N-1}\sigma^2$$

$$= n\sigma^2\left[1 - \frac{n-1}{N-1}\right]$$

or

$$\boxed{\sigma_Z^2 = n\sigma^2 \frac{N-n}{N-1}.} \tag{14}$$

To go from the variance of the sum Z to that of the sample average $\overline{X} = Z/n$, we recall that

$$\sigma_{cZ}^2 = c^2\sigma_Z^2. \tag{15}$$

Taking $c = 1/n$ in Eq. (15) and substituting for σ_Z^2 from Eq. (14), we get

$$\boxed{\sigma_{\overline{X}}^2 = \frac{\sigma^2}{n}\left(\frac{N-n}{N-1}\right).} \tag{16}$$

Note that the variance of $\overline{X}$ is zero when the sample size n is equal to the population size N, because then each sample is the whole population and there is no variation of the sample average from one sample to another; all have sample averages $\overline{X}$ equal to μ. When $n = 1$, we are drawing samples of size 1, and $\overline{X} = x_i$, where the ith individual is the one chosen in the sample. Naturally, for $n = 1$, $\sigma_{\overline{X}}^2 = \sigma^2$, because σ^2 is the variance of the population of individuals. These two special cases ($n = N$ and $n = 1$) may give us extra confidence that Eq. (16) is correct in other cases too.

Example 2 Calculate $\mu_{\overline{X}}$ and $\sigma_{\overline{X}}^2$ for the data of Example 1, using formulas (7) and (16).

Solution We need μ and σ^2 for the population:

$$\mu = \sum_{i=1}^{4} \frac{x_i}{4} = \frac{8}{4} = 2, \qquad \sigma^2 = \sum_{i=1}^{4} \frac{x_i^2}{4} - \mu^2 = \frac{22}{4} - 4 = \frac{3}{2}.$$

We substitute these results into formulas (7) and (16), with $N = 4$, $n = 2$, to get

$$\mu_{\bar{X}} = \mu = 2, \qquad \sigma_{\bar{X}}^2 = \frac{\sigma^2}{n}\left(\frac{N-n}{N-1}\right) = \frac{3/2}{2}\left(\frac{4-2}{4-1}\right) = \frac{1}{2}.$$

These values agree with the calculations of Example 1.

Example 3 Among 50 students, a fraction p hold a favorable opinion on an issue, and a fraction $q = 1 - p$ are against it. In random samples of size 10, find the mean and variance of the proportion $\bar{p}$ who hold a favorable opinion.

Solution Let a student with a favorable opinion have the score 1, a student with an unfavorable opinion, the score 0. For the population, there are $50p$ members with scores 1, and $50q$ with scores 0. Therefore, for the population of 50 students,

$$\mu = \frac{(50p \times 1) + (50q \times 0)}{50} = p,$$

$$\sigma^2 = \frac{(50p \times 1^2) + (50q \times 0^2)}{50} - p^2 = p - p^2 = p(1-p) = pq.$$

We apply formulas (7) and (16), with $N = 50$, $n = 10$, to get

$$\mu_{\bar{p}} = \mu = p,$$

$$\sigma_{\bar{p}}^2 = \frac{pq}{10}\left(\frac{50-10}{50-1}\right) = \frac{4pq}{49} = \frac{pq}{12.25}. \tag{17}$$

Discussion Had the samples been drawn with replacement, we know from Corollary 10–10, Section 10–3, that we would then have $\mu_{\bar{p}} = p$ and

$$\sigma_{\bar{p}}^2 = \frac{\sigma^2}{n} = \frac{pq}{n}. \tag{18}$$

Compare the denominators 12.25 and n of the rightmost terms of Eqs. (17) and (18). In this problem, if we measure accuracy by the size of $\sigma_{\bar{p}}^2$, a sample of 10 drawn without replacement is worth a little more than a sample of 12 drawn with replacement.

When n is small compared with N, however, the factor $(N - n)/(N - 1)$ in Eq. (16) is near 1, and there is little difference between the values of $\sigma_{\bar{p}}^2$ computed according to Eq. (16) and according to the formula $\sigma_{\bar{p}}^2 = \sigma^2/n$ corresponding to sampling with replacement.

EXERCISES FOR SECTION 10–5

1. Three similar tags, numbered 1, 2, and 3, are put into a bowl. Two tags are drawn without replacement. Set up a sample space of possible samples of size $n = 2$. Compute the probability function for sample averages of the numbers on the tags

and find its mean and variance. Compare these with the mean and variance of the original measurements.

2. Measurements associated with four objects are —3, —1, 1, and 3. (a) Compute the mean and variance of this set of measurements. (b) Form the probability function of averages of samples of size $n = 1$ drawn without replacement from this population and give the mean and variance of this distribution.
3. Same as Exercise 2(b) but for $n = 2$.
4. Same as Exercise 2(b) but for $n = 3$.
5. Same as Exercise 2(b) but for $n = 4$.
6. Compare the means obtained in Exercises 2(a), 2(b), 3, 4, 5.
7. Compare the variances obtained in Exercises 2(a), 2(b), 3, 4, 5.

In Exercises 8 through 14, the objects in the population are a nickel, dime, quarter, half-dollar, and dollar.

8. Set up a sample space for samples of size $n = 2$ drawn without replacement from this population.
9. Find the probability function of the amount of money contained in the samples of Exercise 8.
10. Find the mean and variance of the probability function of Exercise 9.
11. Compute the mean and variance of the original set of five amounts.
12. Use the results of your work in Exercises 9 and 11 to compute the probability function of the average amount of money in a sample. Find the mean and variance of this distribution.
13. Compare the means obtained in Exercises 10, 11, and 12.
14. Compare the variances found in Exercises 10, 11, and 12.

In Exercises 15 and 16 the population consists of 4 students who are carrying the following amounts of money: 40¢, 80¢, \$1.00, and \$1.60. A sample of 2 students is randomly drawn without replacement.

15. Find the mean and variance of the amounts of money carried by the 4 students.
16. Set up a sample space for sums of money carried by the 2 students in the samples. Find the mean and variance of these sums and compare them with the population mean and variance. Check your answers by comparing them with Eqs. (6) and (14).
17. In a population of 8 seedlings, at the end of three weeks 4 had no true leaves, 2 had 2 true leaves, and 2 had 4 true leaves. Compute the mean and variance of the random variable: number of true leaves on a seedling drawn at random from this population.
18. (Continuation.) A sample of size $n = 3$ is drawn without replacement from the population of Exercise 17. The average number of true leaves per seedling is the new random variable to be considered. Find its mean and variance.
19. In a lottery, 1000 tickets are sold. There is 1 first prize ticket worth \$50, there are 4 second prize tickets worth \$1 each, and the rest are worthless. If two tickets are drawn from this population, what is their total worth, on the average? What is the standard deviation of their total worth?

20. A school with 1500 students plans to draw a random sample of 100 students to estimate the average expenditure for school supplies. If the standard deviation of the expenditures measured in dollars is actually $\sigma = 2$, what is the standard deviation of sample averages?

21. (Continuation.) Suppose the sample size in Exercise 20 is n (not necessarily 100). If it is desired to make the standard deviation of sample averages 0.1 or less, how large a sample (approximately) should be drawn?

22. Let $f = n/N$ represent the fraction that the sample size n is of the population size N. Then if N is large compared with 1, show that, approximately,

$$\sigma_{\bar{X}}^2 = (\sigma^2/n)(1 - f).$$

Use Eq. (16) in the following Exercises 23 through 28:

23. If n and N are fixed, show that $\sigma_{\bar{X}}^2$ increases when σ^2 increases.

24. For fixed N and σ^2, show that $\sigma_{\bar{X}}^2$ decreases as n increases.

25. For fixed σ^2 and n, show that $\sigma_{\bar{X}}^2$ increases with N (if $n > 1$), but is always less than or equal to σ^2/n.

26. Show that the sample size n_1 required to give a *standard deviation* for sample averages half as big as that for samples of size n is $n_1 = 4nN/(N + 3n)$ when this relation is satisfied by integers. (Otherwise the relation is an approximation.)

27. (Continuation.) Find the value of n_1 for the formula of Exercise 26 when $n = N$. Interpret the result.

28. (Continuation.) In the formula of Exercise 26, show that if n is quite small compared with N, then n_1 is approximately $4n$. Thus halving the standard deviation requires a quadrupling of the sample size, roughly. Comment on the relation to the comparable result for sampling with replacement.

From Exercise 8 of Section 6–1 the mean and standard deviation of the set of N numbers $1, 2, \ldots, N$ are

$$\mu = \frac{N + 1}{2}, \qquad \sigma = \sqrt{\frac{N^2 - 1}{12}}.$$

Use this information and Eq. (16) in Exercises 29 through 31.

29. Show that if n numbers are drawn without replacement from this set, then

$$\sigma_{\bar{X}}^2 = \frac{(N + 1)(N - n)}{12n}.$$

30. (Continuation.) If, in Exercise 29, $N = 1000$, how large must n be for $\sigma_{\bar{X}}$ to be approximately 50?

31. (Continuation.) Regard three-digit numbers in the table of random numbers as the 1000 integers 001, 002, . . . , 999, and 1000 (let 000 stand for 1000). Draw a sample of 32 three-digit numbers *without repetitions*, obtain their average, and see how many units of $\sigma_{\bar{X}}$ that average is from $\mu = 500.5$, the mean of the population.

32. In a lot of 1000 steel bars, the standard deviation of diameters is $\sigma = 0.002$ inch. A sample of 25 of these bars is measured, and the sample average is computed as

an estimate of the mean diameter μ for the lot. Compute the standard deviation of the distribution of possible sample averages. Note that the factor

$$\sqrt{(N - n)/(N - 1)}$$

is very close to 1 in this calculation.

33. For the purpose of estimating mean height, consider drawing a sample of 100 adult males from (a) a fraternal group of 200 adult males, (b) a college of 1000 adult males, (c) a town with 10,000 adult males, (d) a state with 1,000,000 adult males. Suppose the standard deviation for each population, measured in inches, is $\sigma = 3$. Compare the standard deviations of the distributions of sample averages for the several populations. Explain why sampling experts often say with approximate truth, "It's not the size of the population that matters, it's the size of the sample."

34. From a population of measurements of size $N = 100$, a sample of size $n = 10$ is drawn, with replacement, and the average of the measurements in the sample is computed. How large a sample, from the same population, drawn without replacement, would provide the same accuracy, as measured by $\sigma_{\overline{X}}^2$?

35. From a population of size $N = 500$, a sample of size n is to be drawn, without replacement. With each element of the population there is associated a value of a random variable X, and $\overline{X}$ is the average of X's in a sample. It is desired to use the observed value of $\overline{X}$ to estimate the population mean μ, with 95% confidence that the error is no larger than 0.1. From past experience with similar populations, the experimenter is willing to assume that $\overline{X}$ is normally distributed, and that the population standard deviation σ is equal to 3. What percent of the population should he take into his sample? What would your answer have been for a sample drawn *with* replacement?

36. From a population of size $N = 40$, a sample is to be drawn, and the sample average $\bar{x}$ is to be used to estimate the population mean μ. Which provides a more accurate estimate, a sample of 8 drawn without replacement or a sample of 10 drawn with replacement? Why?

37. A population consists of M 1's and $(N - M)$ 0's. A random sample of size n is drawn, without replacement, and X is the number of 1's in the sample, while $\overline{X} = X/n$. Find $\mu_{\overline{X}}$ and $\sigma_{\overline{X}}^2$ in terms of n, M, and N.

38. In Exercise 37, if $M = \frac{1}{2}N$, while N is large and n is of moderate size, how large a sample n is needed to be 75% sure that the number of 1's in the sample is between $0.4n$ and $0.6n$ (a) using Chebyshev's Theorem, (b) assuming that X is normally distributed?

LUCK DOES NOT AVERAGE OUT

For Exercises 39 through 42 use the following information. A bridge deck has 4 suits of 13 cards each, including jack, queen, king, and ace, which are sometimes scored 1, 2, 3, and 4 respectively for bidding purposes. The rest of the cards have zero point count.

39. Compute μ and σ for the probability distribution of point counts per card in a bridge deck. [*Ans:* $\mu = 0.769$, $\sigma = 1.310$]

40. A bridge hand consists of 13 cards dealt without replacement from the deck of 52 described above. What are the mean and standard deviation of the probability distribution of the total point count arising from the sample space of all bridge hands? [*Ans:* $\mu_{\text{sum}} = 10$, $\sigma_{\text{sum}} = 4.13$]

41. Each member of a large bridge club plays 300 hands per season in a certain competition. What are the mean and standard deviation of the sum of the point counts for the probability distribution of the sums yielded by samples of 300 hands? Note that single hands are dealt without replacement of cards, but that successive hands are dealt with replacement of hands. [*Ans:* $\mu_{\text{season}} = 3000$, $\sigma_{\text{season}} = 71.5$]

42. Use the results of Exercise 41 and the normal approximation to suggest that at the end of the season about $\frac{1}{3}$ of the club members are *not* within 72 points of the population average. If we assume that half these people score above and half below average, this means that about $\frac{1}{6}$ of the members have in point count the equivalent of at least an extra 18 aces or 72 jacks above average. Another $\frac{1}{6}$ are similarly below average. If the club is very large, use the normal approximation to show that about 5% of the members are not within 143 points of the mean. Thus 2.5% of the members are lucky enough to have the equivalent in point count of an extra jack or more above average on about half the hands.

43. Let m be the size of a sample drawn with replacement and n the size of a sample drawn without replacement from the same population, of size N. Find a formula for m in terms of n and N so that the variance of averages for the two kinds of samples are equal.

11 Least squares, curve-fitting, and regression

11–1 CURVE-FITTING

Wherever data are processed, the occasion arises for fitting lines, curves, or surfaces to represent relations between two or more variables. The astronomer predicts the path of a satellite; the businessman projects a trend; the biologist relates the size of a growing population to time; the educator compares students' college grades with their high-school grades; the meteorologist forecasts the path of a storm; and the doctor compares the rate of relief from pain with the size of the analgesic dose. All such tasks involve the fitting of curves. For some problems the work is easy and can be done by hand; for others the calculations are horrendous and would not even be considered without modern high-speed computers. Nevertheless, the central ideas behind work in curve-fitting are relatively simple, and we present them in this chapter.

Example 1 *Forgetting.* To discover how fast people forget, the psychologist Strong slowly read lists of 20 words to 5 people and asked them later to recognize the words when mixed with 20 other words. As soon as the list was read, the subjects were kept busy at other intellectual tasks, so that they could not rehearse. Strong, the investigator, imposed penalties for false recognition, and computed the score for retention, R, on the following basis:

$$R = 100 \times \frac{\text{number correct} - \text{number falsely recognized}}{20}.$$

Thus a subject with 14 words correct and 2 falsely recognized had a score of 60%.

The time interval before the subject was asked to recognize a list of words was varied from one minute to one week. The data are given in Table 11–1, where T is the number of minutes between the reading of the list and the testing, and $t = \log T$.

Table 11–1 Data for Strong's experiment on retention

T, number of minutes till testing	t, logarithm of time in minutes	R, average retention score (percent)
1	0	84
5	0.7	71
15	1.2	61
30	1.5	56
60 (1 hr)	1.8	54
120 (2 hr)	2.1	47
240 (4 hr)	2.4	45
480 (8 hr)	2.7	38
720 (12 hr)	2.9	36
1,440 (1 day)	3.2	26
2,880 (2 days)	3.5	20
5,760 (4 days)	3.8	16
10,080 (7 days)	4.0	8

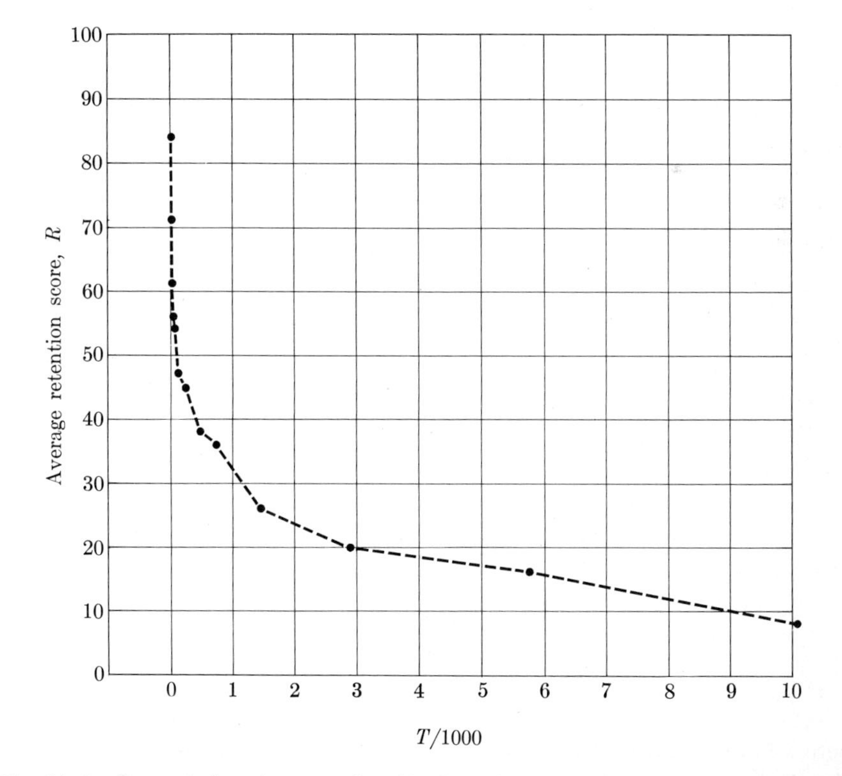

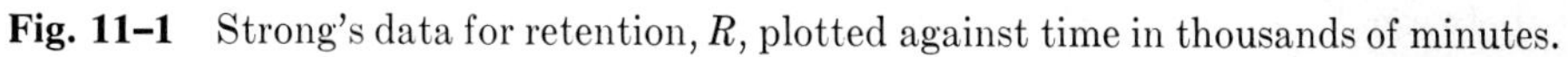

Fig. 11–1 Strong's data for retention, R, plotted against time in thousands of minutes.

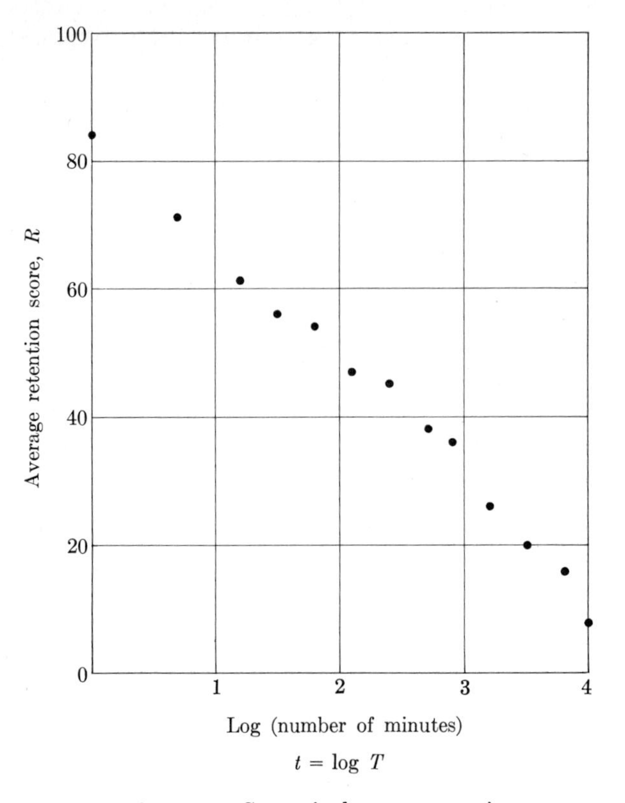

Fig. 11-2 Strong's data on retention.

In Fig. 11-1, the average R for the five subjects is plotted against T. The data are quite curvilinear when plotted this way. The dashed lines merely lead the eye from one point to the next.

An alternative plot of these data is given in Fig. 11-2, where the vertical axis is the average retention score for the five subjects and the horizontal scale is t, the logarithm of T.

Strong, who plotted his data on *ordinary* graph paper, as in Fig. 11-1, noted that, initially, retention scores went down quickly as time increased between reading and testing, but that at about the 8-hour interval scores began to decay more slowly. This suggested that the data be plotted on logarithmic paper (R against t), as in Fig. 11-2. You will observe that while these data fluctuate slightly, they seem to lie quite near a straight line, which means that retention is approximately linearly related to t, or

$$R = mt + b,$$

where b and m are constants. How do we find b and m to determine the relation between R and t, and thus reduce the data to a simple formula?

Black-thread method An old and simple method of fitting a straight line is to stretch a black thread across a graph, and rearrange its position until one's eye is satisfied that the thread fits the points well. Then the coordinates are read at convenient points at the left and right edges of the figure. In our figure, it is convenient to read the intercepts at $t = 0$ and $t = 4$, and then to work out an equation for the line as follows.

The slope-intercept formula for the equation of a straight line is

$$y = mx + b,$$

where b is the y-intercept and m is the slope of the line:

$$m = \frac{\text{rise}}{\text{run}} = \frac{y_2 - y_1}{x_2 - x_1},$$

where (x_1, y_1) and (x_2, y_2) are the coordinates of two points on the line.

In our problem, R corresponds to y, and t to x. We set $t = 0$ and get the R-intercept, R_0. Let the R-reading at $t = 4$ be R_4. Then the slope

$$m = \frac{R_4 - R_0}{4 - 0} = \frac{R_4 - R_0}{4},$$

and $b = R_0$. Thus, since our variables are t and R, the fitted equation is

$$R = \left(\frac{R_4 - R_0}{4}\right) t + R_0. \tag{1}$$

The authors' black thread gave the points (0, 85) and (4, 10). They estimate $m = (10 - 85)/4 = -18.75$ and $R_0 = 85$, and their fitted equation is

$$R = -18.75t + 85.$$

You are asked to perform this same exercise. Your results may differ from these.

For many purposes, the black-thread method is adequate, and its simplicity recommends it. Why search further? There are a number of reasons:

1. *Reliability.* When data are as clean and beautiful as Strong's, there is little problem about the fitting, but when we have a swarm of points with considerable variation (see Fig. 11–8), our eye is not a sure guide in placing the line.

2. *Objectivity.* Scientists prize repeatable objective methods, because they fear that unconscious bias on the part of an observer may distort the results.

3. *Generalization.* It is desirable to have a general method of curve-fitting that can be extended to new and more complicated situations. For example, we may wish to fit a quadratic $y = ax^2 + bx + c$, and the thread method is not satisfactory for this work. Or we may wish to introduce more variables, e.g., how does height of seedlings depend on amount of water *and* amount of fertilizer?

Once we see the need for more general methods, our task is to find some and to make a choice among them.

Why fit curves? Before proceeding with the search for general methods, let us consider further why we want to fit lines and curves. Here are some reasons:

1) *To get a formula.* We want a concise way to express the relation between variables.

2) *To forecast.* We may wish to use the relation for forecasting results.

3) *To assess predictions.* We may wish to assess the reliability of predictions by looking at the scatter of observed points about the fitted curve.

4) *To measure goodness-of-fit.* We may have a theory that we wish to test. For example, Strong may have felt from earlier work that retention was a linear function of log T, and his results support that hypothesis.

5) *To evaluate constants.* An established theory may suggest the form of a curve to be fitted; then, fitting the curve to data assesses the constants. For example, if s is distance in feet, t is time in seconds, and g is the gravitational constant, then a body dropped from rest in a vacuum near the surface of the earth falls according to the law $s = \frac{1}{2}gt^2$. We can drop a body and measure the time t required to traverse a distance s. These data can be used to estimate g.

In this chapter, we especially treat the fitting of straight lines, and this needs some justification; after all, many relations are curvilinear. Some reasons for studying the fitting of straight lines are:

a) *Generality of method.* The technique we develop for fitting straight lines extends to other methods, so the foundation for generalization is laid.

b) *Actual linearity.* Many relations are linear or nearly so.

c) *Transformations.* Even when a relation is not linear, often an easy transformation of one or both of the variables produces a nearly linear relation. For example, Strong used the logarithm of time rather than time itself, a common transformation.

d) *Limited range.* Often researchers study a limited range of a variable. The functional relation over the limited range may be nearly linear even though the total functional relation is far from linear. Examples: (a) The average height of males in the United States is obviously nonlinear when plotted against age. Yet average heights from age 12 to age 14 *are* nearly linear when plotted against age. (b) On the circumference of a circle 2 inches in radius, look at a segment about $\frac{1}{2}$ inch long; it is nearly a straight line. Yet what curve have you studied that is less linear than a circle?

e) *Convenience in reporting.* Even when a function is not a straight-line relation, a researcher may find it convenient to report the linear approximation, for ease in communication. For convenience in interpolation, the straight line has no equal. That is one reason why researchers use transformations, as mentioned in (c) above.

In the next sections we provide a general method for fitting straight lines.

EXERCISES FOR SECTION 11–1

1. Apply the black-thread method to obtain an equation for the retention data of Fig. 11–2.
2. (a) Use the equation fitted in Exercise 1 to find the value of t, and then T, when $R = 0$. (b) The work in part (a) is an *extrapolation* from the data; why must it be regarded with caution, or even distrust?
3. Use the black-thread method to find an equation for a line fitting the points $(0, 0)$, $(1, 0)$, $(2, 1)$, $(3, 1)$, $(4, 2)$, $(5, 2)$.
4. Use the black-thread method to find an equation for a line fitting the points $(-2, 3)$, $(-1, 2)$, $(0, 1)$, $(0, 0)$, $(1, -1)$, $(3, -3)$.
5. Plot the 10 points with coordinates $(-1, -1)$, $(0, -2)$, $(1, 0)$, $(1, 2)$, $(2, 2)$, $(3, -3)$, $(3, 1)$, $(4, 3)$, $(5, 0)$, $(5, 1)$ on ordinary graph paper and use the black-thread method to find an equation of the fitted straight line.
6. (Continuation.) Collect the values of m obtained by members of the class for Exercise 5. Get the median (middle value) of these m's. Did anyone have a value grossly different from the median? What would the mean of the m's have been?
7. Plot the points $(0, 0)$, $(0, 1)$, $(1, 0)$, $(1, 1)$ on ordinary graph paper and use the black-thread method to find an equation of a fitted straight line.
8. (Continuation.) Is there a line perpendicular to the one you fitted in Exercise 7 that fits the data equally well? If so, find its equation.
9. Study the four points of Exercise 7 and give a list of four equations that a reasonable person might suggest for close-fitting lines.
10. (a) Suggest an objective criterion for choosing unambiguously among straight lines that can be fitted to a set of data. (b) Apply your criterion to the fitting of the four points of Exercise 7. (c) Apply your criterion to the fitting of the three points $(0, 0)$, $(1, 1)$, $(2, 0)$.
11. An apparatus measures falling time, t, to the hundredth part of a second when an object is dropped a distance s, in feet. The following data are gathered:

s	4	3	2	1
t	0.50	0.43	0.35	0.25

Find g by fitting a straight line through the origin with the black-thread method. (To obtain linearity, plot $T = t^2$ as the horizontal axis.)

11–2 THE METHOD OF LEAST SQUARES

When we use a black thread to fit a straight line, we try to make sure the thread is close to the points. Mentally, we compromise, making some points farther from and some closer to the line. To measure the closeness of a point to a fitted line, we use the *vertical* deviation, since we are trying to predict y from x. Strong didn't want to predict t from retention, he wanted to predict retention from t.

If we have fitted the line $y = x + 1$ to a set of points, then the y-coordinate of the point P on the line with $x = 3$ is $y = 3 + 1 = 4$. Therefore, if the point Q with coordinates (3, 5) belongs to the fitted set, its vertical deviation from the line is $PQ = 5 - 4 = 1$. More generally, the vertical deviation d of a point with coordinates (x_1, y_1) from the line $y = mx + b$ is

$$d = y_1 - (mx_1 + b). \tag{1}$$

Example 1 Figure 11–3 shows three points (the small table within the figure gives coordinates) that have been fitted by eye with the equation

$$y = \tfrac{3}{2}x - 1. \tag{2}$$

Since we plan to use the line to predict y from x, find the vertical deviations from the line to the three given points.

Solution Upon substitution in Eq. (2), the fitted values of y for $x = 1, 2, 3$ are found to be $y = \frac{1}{2}, 2, 3\frac{1}{2}$, respectively. We summarize the work in Table 11–2.

Table 11–2 Deviations for Fig. 11–3

x	Observed y's	Fitted y's	Deviations for y (observed − fitted)
1	1	$\frac{1}{2}$	$1 - \frac{1}{2} = \frac{1}{2}$
2	$1\frac{1}{2}$	2	$1\frac{1}{2} - 2 = -\frac{1}{2}$
3	4	$3\frac{1}{2}$	$4 - 3\frac{1}{2} = \frac{1}{2}$

The deviations seem rather small, and the visual fit satisfactory. But without a criterion, we can scarcely discuss the possibility of finding a better-fitting line.

What sorts of criteria might we use? We could ask that the sum of the deviations be zero. But there are many lines that have this property, including ill-fitting ones such as the horizontal line with equation $y = \frac{13}{6}$. We could ask that the sum of the absolute vertical deviations be minimized. But the intractability that ruled out absolute deviations as a measure of variability in Chapter 6 is even more of a roadblock in fitting lines. The program is extremely laborious.

The next suggestion usually is to *minimize the sum of the squared vertical deviations.* It turns out that this method is mathematically tractable, generalizes easily, and is not difficult to carry out. It is part of the general *method of least squares.* We shall examine this method in connection with Example 1.

In looking at Fig. 11–3, we intuitively feel that the slope of the line is right, but possibly the y-intercept could be improved.

Example 1 (Continued.) Let us retain the slope of the line of Fig. 11–3 and, by numerical trial and error, test various values of the y-intercept, to find the

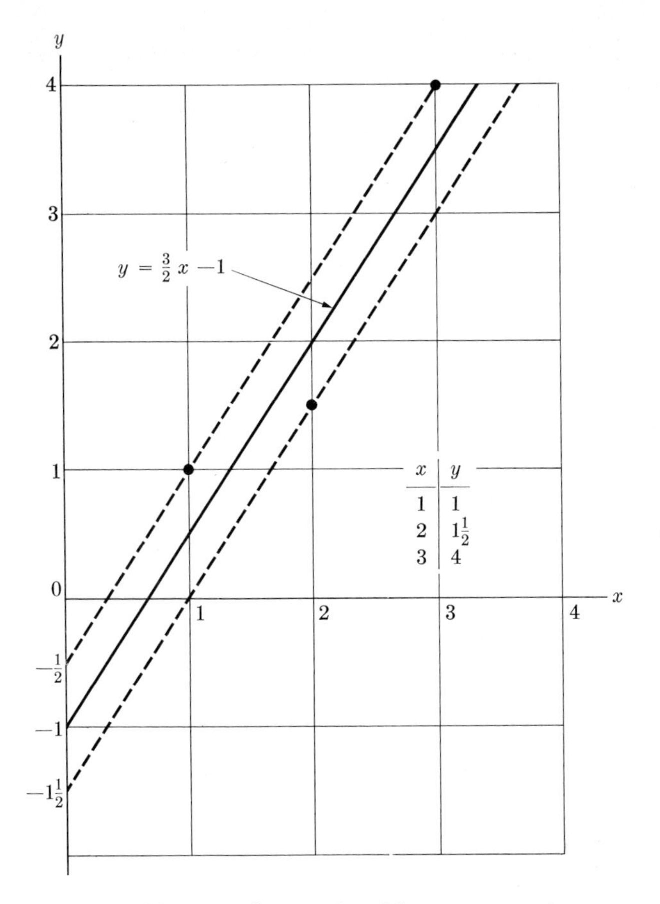

Fig. 11–3 Line $y = \frac{3}{2}x - 1$ fitted by eye to 3 points.

y-intercept that minimizes the sum of squares of vertical deviations (we abbreviate "vertical deviations" to "deviations").

Solution We surely do not want all points on one side of the line, so we restrict attention to values of the y-intercept between $-1\frac{1}{2}$ and $-\frac{1}{2}$ (see the dashed lines of Fig. 11–3). It seems reasonable, too, that if an adjustment is made, the new line should move toward the two points *above* the present line. Therefore let us try as y-intercepts the values $-.5$, $-.6$, $-.7$, $-.8$, $-.9$, -1.0 (our present value). In this range, the amount we add to our present y-intercept, -1, can be subtracted from the deviations for the two endpoints, and from the deviation for the middle point. For example, when the intercept is $b = -.9$, we have added .1 to our old intercept. Therefore, the deviations obtained by adjusting the old observations at the endpoints are $.5 - .1 = .4$, and at the middle point the deviation is $-.5 - .1 = -.6$. Results are given in Table 11–3.

To see what value of b approximately minimizes the sum of squared deviations, in Fig. 11–4 we plot values of these sums from Table 11–3.

Table 11–3 Sums of squares of vertical deviations

x \ b	−1.0		−.9		−.8		−.7		−.6		−.5	
	dev	dev²	dev	dev²	dev	dev²	dev	dev²	dev	dev²	dev	dev²
1	.5	.25	.4	.16	.3	.09	.2	.04	.1	.01	0	0
2	−.5	.25	−.6	.36	−.7	.49	−.8	.64	−.9	.81	−1.0	1.0
3	.5	.25	.4	.16	.3	.09	.2	.04	.1	.01	0	0
Sum		.75		.68		.67		.72		.83		1.0

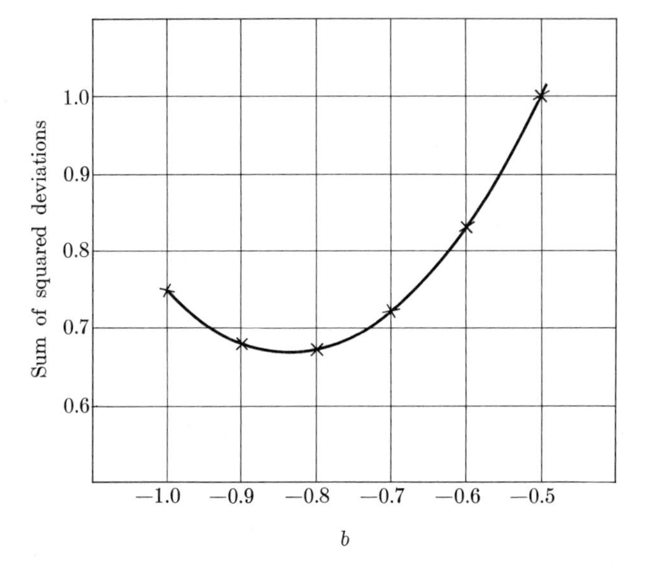

Fig. 11–4 Plot of sums of squared deviations for various intercept values, from data of Table 11–3.

The graph of Fig. 11–4 suggests a parabola. Both Fig. 11–4 and Table 11–3 suggest that the minimizing value of b is about -0.8.

This outcome satisfactorily resolves the present problem. But we want to find a general formula, to speed future work. To this end, we review some facts about the quadratic function.

Minimizing a quadratic function The general quadratic function of x,

$$y = Ax^2 + Bx + C, \qquad A > 0,$$

has a minimum value. To find the minimum value of y, we complete the square as follows:

$$y = A\left(x^2 + \frac{B}{A}x + \frac{B^2}{4A^2}\right) - \frac{B^2}{4A} + C \quad \text{or} \quad y = A\left(x + \frac{B}{2A}\right)^2 - \frac{B^2}{4A} + C.$$

The only term on the right that can be adjusted by a choice of x is

$$A[x + B/(2A)]^2.$$

To minimize y we want that term to be small, and zero is the least we can make it, because both A and $[x + B/(2A)]^2$ are nonnegative. Therefore we have

$$\textit{minimizing value of } x: \quad \boxed{x = -\frac{B}{2A}\cdot} \tag{3}$$

Thus, if $x = -B/2A$, the quadratic is minimized, and we have

$$\textit{minimum value of } y: \quad \boxed{y = C - \frac{B^2}{4A}\cdot} \tag{4}$$

For future use let us note that

$$B = \text{coefficient of first-degree term in the quadratic,}$$
$$A = \text{coefficient of second-degree term in the quadratic.}$$

We mention this now to avoid confusion later when other letters are used as coefficients.

We have proved what we shall call the fundamental lemma for the method of least squares.

11–1 Lemma: Fundamental lemma for least squares

If the coefficient of the second-degree term of a quadratic function is positive, then the function has a minimum when the variable is set equal to

$$\boxed{-\frac{B}{2A} = -\frac{\text{coefficient of 1st-degree term}}{2\text{ (coefficient of 2nd-degree term)}},} \tag{5}$$

and the minimum value of the quadratic function is

$$\boxed{C - \frac{B^2}{4A} = \text{constant term} - \frac{(\text{coefficient of 1st-degree term})^2}{4\text{ (coefficient of 2nd-degree term)}}\cdot} \tag{6}$$

Example 2 Find the value of x that minimizes

$$y = 3x^2 - 2x + 1,$$

and find the minimum value of the function.

Solution Formulas (5) and (6) give

$$x = -\frac{B}{2A} = \frac{2}{6} = \frac{1}{3}$$

for the minimizing value of x, and

$$y = C - \frac{B^2}{4A} = 1 - \frac{(-2)^2}{4(3)} = \frac{2}{3}$$

for the minimum value of y.

Notation In fitting least-squares lines, it helps to have standard symbols for the values of the slope m and the intercept b that yield the minimum sum of squared deviations, and for the minimum value of the sum of squared deviations, S. We call these the *least-squares values* and use $\hat{m}$, $\hat{b}$, and $\hat{S}$ (read m-hat, b-hat, and S-hat) to distinguish these least-squares values from the general values m, b, and S.

Example 3 *Three-point example.* In the example of Fig. 11-3, choose the intercept b in $y = b + 3x/2$ so as to minimize the sum of squares of the deviations. In other words, find $\hat{b}$.

Solution Let us set up the table of observed and predicted values, using b as an undetermined constant.

Table 11-4 Deviations for the three-point example

x	Observed y	Fitted y	Deviation	(Deviation)^2
1	1	$b + \frac{3}{2}$	$-\frac{1}{2} - b$	$(-\frac{1}{2} - b)^2 = \frac{1}{4} + b + b^2$
2	$1\frac{1}{2}$	$b + 3$	$-\frac{3}{2} - b$	$(-\frac{3}{2} - b)^2 = \frac{9}{4} + 3b + b^2$
3	4	$b + 4\frac{1}{2}$	$-\frac{1}{2} - b$	$(-\frac{1}{2} - b)^2 = \frac{1}{4} + b + b^2$
				Sum $= \frac{11}{4} + 5b + 3b^2$

To find $\hat{b}$, we need to minimize the quadratic function of b obtained as the sum in the last column of Table 11-4: $\frac{11}{4} + 5b + 3b^2$. Applying formula (5) to this quadratic, we find

$$\hat{b} = -\frac{5}{2(3)} = -\frac{5}{6} \approx -0.83.$$

We call $\hat{b}$ "the least-squares value of b"; $\hat{b}$ is close to -0.8, as we deduced from Table 11-3. The value of the sum of squares of deviations at this minimum is, from formula (6),

$$\hat{S} = \frac{11}{4} - \frac{25}{4(3)} = \frac{2}{3} \approx 0.667,$$

where we carry three decimal places to show that the true minimum is only slightly less than the value 0.67 obtained in Table 11-3 for $b = -0.8$.

Our method also works for finding a number b that minimizes

$$S = \sum_{i=1}^{n} (y_i - b)^2,$$

where the y_i's are numbers.

It turns out that the least-squares value of b is $\hat{b} = \bar{y}$, the average of the measurements. Consequently, when we compute the sum of squares in the numerator of a sample variance

$$s_y^2 = \frac{\Sigma(y_i - \bar{y})^2}{n},$$

we are computing a minimum sum of squares. We develop this fact as a theorem through the following examples.

Example 4 Find the number b that minimizes the sum of the squares of the deviations of the numbers 4, -1, 0 from b.

Solution Let S be the sum of squares (sum of squared deviations):

$$S = (4 - b)^2 + (-1 - b)^2 + (0 - b)^2 = 17 - 6b + 3b^2.$$

Applying formula (5) to this quadratic function of b, we get

$$\hat{b} = -\frac{-6}{2(3)} = 1.$$

Observe that 1 is the average of the three numbers 4, -1, 0.

Example 5 Find $\hat{b}$ for the three values y_1, y_2, y_3 of y.

Solution For an arbitrary value of b, let the sum of squares be S as before:

$$S = \sum_{i=1}^{3} (y_i - b)^2 = (y_1 - b)^2 + (y_2 - b)^2 + (y_3 - b)^2.$$

When we expand the squares and collect like powers of b, we get

$$S = y_1^2 + y_2^2 + y_3^2 - 2(y_1 + y_2 + y_3)b + 3b^2.$$

From formula (5), we find

$$\hat{b} = -\frac{-2(y_1 + y_2 + y_3)}{2(3)} = \frac{y_1 + y_2 + y_3}{3} = \bar{y}.$$

From formula (6), the minimum value of S is

$$\hat{S} = y_1^2 + y_2^2 + y_3^2 - \frac{4(y_1 + y_2 + y_3)^2}{4(3)} = \Sigma y_i^2 - 3\bar{y}^2.$$

Note that if $\hat{S}$ were divided by 3 we would have the variance of the y-measurements:

$$\frac{\hat{S}}{3} = s_y^2 = \frac{\sum y_i^2}{3} - \bar{y}^2.$$

Thus $\hat{S}$ is just $3s_y^2$. Actually, the moment we find out that $\hat{b} = \bar{y}$, substitution in the original equation gives

$$\hat{S} = (y_1 - \bar{y})^2 + (y_2 - \bar{y})^2 + (y_3 - \bar{y})^2,$$

the numerator of the definitional form for the variance.

The methods of Example 5 can be generalized at once to give a theorem:

11-2 Theorem: Minimum sum of squares

Let $y_1, y_2, \ldots, y_n$ be measurements, let b be a number, and let

$$\begin{aligned} S &= \sum_{i=1}^{n} (y_i - b)^2 \\ &= (y_1 - b)^2 + (y_2 - b)^2 + \cdots + (y_n - b)^2. \end{aligned}$$

Then $\hat{b}$, the value of b that minimizes S, is

$$\boxed{\hat{b} = \bar{y},} \tag{7}$$

and the minimum value of S is

$$\boxed{\hat{S} = ns_y^2,} \tag{8}$$

where s_y^2 is the variance of the y's.

11-3 Corollary: Sum of deviations

Given the numbers $y_1, y_2, \ldots, y_n$, and $\hat{b}$, the least-squares value of b, then

$$\boxed{\sum_{i=1}^{n} (y_i - \hat{b}) = 0.}$$

Proof From the theorem, $\hat{b} = \bar{y}$. Furthermore,

$$\sum_{i=1}^{n} (y_i - \hat{b}) = \sum y_i - n\hat{b} = \sum y_i - n\bar{y} = \sum y_i - n\left(\frac{\sum y_i}{n}\right) = 0. \quad \square$$

Example 6 Find $\hat{b}$, the value of b that minimizes the squared deviations from 10, -5, 6, 4, 0.

Solution $\hat{b} = \bar{y} = \frac{15}{5} = 3.$

We now have all the equipment that we need for fitting straight lines. In the next sections we complete the task in two steps. First we learn to fit lines through the origin, and then to fit a general line.

EXERCISES FOR SECTION 11-2

1. For the data of Fig. 11-3, find the deviations from the line $y = \frac{13}{6}$ mentioned in the text. Show that their sum is zero. Compare them with the deviations of Table 11-2.
2. Find the value of b that minimizes the sum of the squared deviations from the numbers 8, 6, 9, 3, 5, and evaluate the minimum sum of squares, $\hat{S}$.
3. Check by direct calculation that the sum of the deviations from the mean in Exercise 2 is zero.
4. Find the value of b that minimizes the sum of squares for the numbers 14, -10, -20, 0, and evaluate the minimum sum of squares, $\hat{S}$.
5. Check by direct calculation that the sum of the deviations from the mean in Exercise 4 is zero.
6. If S is the sum of squares of deviations in Theorem 11-2, find the least-squares value of b for the numbers -9, -1, 6, 14, 13, -5, and find $\hat{S}$.
7. If S is the sum of squares of deviations in Theorem 11-2, find the least-squares value of b for the numbers 9, 2, 0, and find $\hat{S}$.
8. For the numbers 4, 3, 2, compute the sum of squares of their deviations from $b = 4, 3.5, 3, 2.5, 2$, and graph the results, indicating $\hat{b}$ and $\hat{S}$ on the graph.
9. Find the value of x that minimizes the quadratic $y = 3x^2 + 2x + 1$, and find the minimum value of y.
10. Find the value of x that minimizes the quadratic $y = x^2 - 2x + 1$, and find the minimum value of y.
11. Why can't we find the minimum value of the quadratic $y = -2x^2 + x - 1$?
12. Find the value of x that minimizes the quadratic in x:
$$a + ax + a^2x^2,$$
where a is a constant, and find the value of the quadratic for that value of x.
13. For the quadratic in b:
$$x^3 + x^2b + xb^2, \qquad x > 0,$$
find the minimizing value of b, and the value of the quadratic for that b. Check the latter value by substituting the minimizing b into the quadratic.
14. a) Write out the steps of the proof of Theorem 11-2 on minimum sum of squares for two numbers, y_1 and y_2.
    b) Write out the steps for $n = 4$.
15. Write out the steps of the proof of Theorem 11-2 on minimum sum of squares for a general value of n.
16. Show that $\sum(y_i - c)^2 = \sum(y_i - \bar{y})^2 + n(\bar{y} - c)^2$, where c is a constant and n is the number of y-measurements. From this result, deduce that $\hat{c} = \bar{y}$.

11-3 FITTING A LINE THROUGH THE ORIGIN

Before beginning the more difficult task of fitting a general line, let us organize the method by doing the easier problem of fitting a line through the origin.

Example 1 *Temperatures below the earth's surface.* Mellor reports temperatures taken at various depths in an artesian well at Grenoble, France. The following table, derived from his data, uses a point about 28 meters below the earth's surface as an origin for both temperature and depth. The number of meters below the origin is x, and y is the number of centigrade degrees above the temperature at the origin.

Number of meters below origin, x	40	150	220	270
Number of degrees above the temperature at the origin, y	1.2	4.7	9.3	10.5

These data are plotted in Fig. 11-5. Use the method of least squares (a) to fit a straight line through the origin, (b) to find from its slope the rate of increase in temperature per additional 100 meters of depth.

Solution An equation of a line through the origin is

$$y = mx, \tag{1}$$

where m is the slope of the line. Let the pairs of measurements be designated

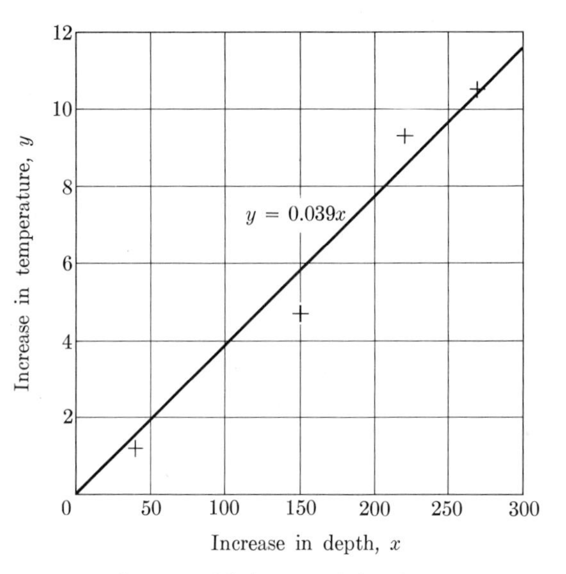

Fig. 11-5 Temperature changes with increased depth in an artesian well at Grenoble, France.

(x_i, y_i), $i = 1, 2, 3, 4$. When we find $\hat{m}$, the least-squares value of m, the estimate of y_i from Eq. (1) is $\hat{m}x_i$. The sum of squares of deviations, then, is

$$\begin{aligned} S &= \sum_{i=1}^{4} (y_i - mx_i)^2 \\ &= (y_1 - mx_1)^2 + (y_2 - mx_2)^2 + (y_3 - mx_3)^2 + (y_4 - mx_4)^2. \end{aligned} \tag{2}$$

Do not lose sight of the fact that the x_i's and the y_i's are known numbers in the table above. It is m that is to be determined so as to minimize S.

When we expand the four squares on the right-hand side of Eq. (2) and collect terms in like powers of m, we get

$$\begin{aligned} S = &(y_1^2 + y_2^2 + y_3^2 + y_4^2) \\ &- 2(x_1y_1 + x_2y_2 + x_3y_3 + x_4y_4)m \\ &+ (x_1^2 + x_2^2 + x_3^2 + x_4^2)m^2. \end{aligned} \tag{3}$$

Note that S is a quadratic in m. We now apply the fundamental lemma of least squares, Lemma 11–1, to get the minimizing value of m:

$$\begin{aligned} \hat{m} = -\frac{B}{2A} &= -\frac{-2(x_1y_1 + x_2y_2 + x_3y_3 + x_4y_4)}{2(x_1^2 + x_2^2 + x_3^2 + x_4^2)} \\ &= \frac{x_1y_1 + x_2y_2 + x_3y_3 + x_4y_4}{x_1^2 + x_2^2 + x_3^2 + x_4^2}, \end{aligned}$$

or, written more compactly,

$$\hat{m} = \frac{\sum x_iy_i}{\sum x_i^2}. \tag{4}$$

Now we are ready to calculate $\sum x_i^2$ and $\sum x_iy_i$, and so complete the numerical problem. Table 11–5 shows the details.

Table 11–5 Calculations for the artesian well example

x_i	y_i	x_i^2	x_iy_i
40	1.2	1,600	48
150	4.7	22,500	705
220	9.3	48,400	2046
270	10.5	72,900	2835
		$\sum x_i^2 = 145{,}400$	$\sum x_iy_i = 5634$

We substitute the numerical results of the bottom line of Table 11–5 into Eq. (4) to get $\hat{m}$:

$$\hat{m} = \frac{5{,}634}{145{,}400} \approx 0.03875.$$

Therefore our fitted line, after rounding to three decimals, is

$$y = 0.039x.$$

Since the slope $\hat{m}$ of our fitted line gives the change in y per unit change in x, we conclude that the temperature increases about 0.039°C per meter, or by about 3.9°C per hundred meters.

The algebraic method of the artesian well example generalizes at once to n points, and we have the theorem:

11-4 Theorem: Fitting $y = mx$

Given the measurement pairs (x_i, y_i), $i = 1, \ldots, n$, then the least-squares line through the origin, whose equation is

$$y = \hat{m}x,$$

has

$$\boxed{\hat{m} = \frac{\sum x_i y_i}{\sum x_i^2},} \qquad i = 1, 2, \ldots, n. \tag{5}$$

Example 2 Given the (x, y) pairs in the first two lines of the following table, fit the least-squares line through the origin.

x	1	1	2	2	3	4	4	
y	1	2	1	3	5	3	4	
x^2	1	1	4	4	9	16	16	$51 = \sum x_i^2$
xy	1	2	2	6	15	12	16	$54 = \sum x_i y_i$

Solution The required squares and cross products are shown in the bottom two lines of the table of data. Dropping subscripts, we have

$$\hat{m} = \frac{\sum xy}{\sum x^2} = \frac{54}{51} \approx 1.06.$$

Therefore the fitted line is $y = 1.06x$.

EXERCISES FOR SECTION 11-3

1. Five points have $\sum x_i y_i = 4$, $\sum x_i^2 = 10$. Find the equation of the least-squares line through the origin.
2. For the four points (0, 0), (0, 1), (1, 0), (1, 1), find the equation of the least-squares line through the origin.
3. Find an equation of the least-squares line through the origin for the following measurement pairs:

$$\begin{array}{lccccccc} x: & 1 & 1 & 2 & 2 & 3 & 4 & 4 \\ y: & 1 & 2 & 2 & 3 & 1 & 3 & 4 \end{array}$$

4. Find an equation of the least-squares line through the origin for the following measurement pairs:

x:	0	1	2	3	4
y:	0	1	4	9	16

5. (Continuation.) Note in Exercise 4 that $y = x^2$, for the given pairs. Substitute $x = 0, 1, 2, 3, 4$ into your fitted equation and compare the results with the true values of y.

6. ($y = \sin x$). For x in degrees, $y = \sin x$ is given to two decimals in the table

x:	0	10	20	30
y:	0	.17	.34	.50

Fit the least-squares line through the origin, and use the resulting equation to estimate sin 5° and sin 25°, checking the values against trigonometric tables.

7. ($y = \tan x$). For x in degrees, $y = \tan x$ is given to two decimals in the table

x:	0	10	20	30
y:	0	.18	.36	.58

Fit the least-squares line through the origin, and use the resulting equation to estimate tan 5° and tan 25°, checking the values against trigonometric tables.

8. [$y = \log(1 + x)$.] Three decimal values of $y = \log(1 + x)$ are given for small values of x in the table (logarithms are to the base e)

x:	−.2	−.1	0	.1	.2
y:	−.223	−.105	0	.095	.182

Fit the least-squares through the origin, and then compare the fitted values of y with the true values for $x = 0.2$ and -0.2.

Remark. Exercises 5, 6, 7, and 8 illustrate point (d) of Section 11–1: over limited ranges many functions are nearly linear.

9. *Pythagorean approximation.* Consider right triangles with sides a, 1, c, where $a \leq 1 \leq c$. We want to develop the approximation $c \approx m(a + 1)$. The Pythagorean theorem yields the one-decimal-place table for selected values of $a + 1$:

$a + 1$:	1	1.2	1.4	1.6	1.8	2.0
c:	1	1.1	1.2	1.3	1.3	1.4

Find the least-squares value of m. It can be used as the multiplier of the sum of the two short sides of a right triangle to estimate roughly the hypotenuse.

10. For the points with coordinates (5, 2), (5, 4), (5, 6), find the least-squares line through the origin.

11. (Continuation.) Invent and prove a theorem about the least-squares line through the origin for the points with coordinates (x, y_1), (x, y_2), (x, y_3).

12. (Continuation.) Extend the theorem of Exercise 11 from 3 to n points.

13. Consider the points with coordinates (2, 4), (3, 6), (5, 10). Find the least-squares line through the origin.

14. Prove that the least-squares line through the origin for the points with coordinates (x_i, kx_i), $i = 1, 2, \ldots, n$, has slope k.

★11-3 SUPPLEMENT. Alternative ways to fit a line through the origin

In Section 11-3, we fitted the line $y = mx$ as a preparation for the least-squares development for the general line. Let us say a little more about fitting a line through the origin as an end in itself.

Each observed point (x_i, y_i) gives an estimate of m, $m_i = y_i/x_i$, $x_i \neq 0$, and we may view our problem as one of putting these several estimates together. Weighted sums suggest themselves: the fitted m is

$$m_f = \sum w_i m_i, \tag{1}$$

where the w_i are the weights. For example, if $w_i = x_i^2/\sum x_i^2$, we find that $m_f = \hat{m}$, the least squares value from Eq. (5) of Section 11-3.

Although we may measure y only once for each x_i, we can readily imagine that for each value of x there is a random variable Y that has a distribution. Repeated measurements at x may give different values of y. In the artesian well example, the temperature at a given depth may differ from one time to another, or even from hour to hour. For our next development, we shall assume that at each x there is a distribution of Y, and that the mean of Y at x is mx. Let us further assume, when needed, that at x the distribution of Y has a variance σ^2, whose value may depend upon x. Let us use Y_i for the random variable in the Y-direction at x_i.

Under our assumption, the distribution of Y_i corresponding to x_i has mean mx_i. Then, assuming fixed weights w_i, the mean value of m_f is

$$\begin{aligned} E(m_f) &= E\left(\sum w_i m_i\right) = \sum w_i E(m_i) \\ &= \sum w_i E(y_i/x_i) \\ &= \sum \frac{w_i m x_i}{x_i} = m \sum w_i. \end{aligned}$$

If $\sum w_i = 1$, then

$$E(m_f) = m,$$

and therefore m_f is an unbiased estimate of m.

While this is a pleasant result, let us note that any weights of the form

$$w_i = x_i^t/\sum x_i^t, \tag{2}$$

for example, would produce an unbiased estimate of m. This means that we have room for choice of weights.

The variability of the Y_i's may depend upon the size of the x_i's. For example, if x is the age and y is the height of a sapling, we expect older saplings to vary more in their heights than younger ones. In many physical measurements among similar objects of different sizes, the larger ones have more variability than the smaller—a distance of 10 feet can usually be measured more accurately than can a mile, using the same methods.

If the Y_i's are independent or even uncorrelated and the variances of their distributions corresponding to x_i are σ_i^2, then the variance of m_f is, from Theorem 10-4 Eq. (1b),

$$\text{Var}(m_f) = \sum w_i^2 \sigma_i^2 / x_i^2, \tag{3}$$

if the weights are fixed.

If the variance of Y for a given x depends on the size of x, we may be able to represent it as proportional to a power of x, kx^r. If we substitute this expression for σ_i^2 in Eq. (3), we find that the value of the variance of m_f is $k\sum w_i^2 x_i^{r-2}$. By calculus methods not given here, we can find the weights w_i that minimize the variance of m_f subject to the condition that $\sum w_i = 1$. We find that the optimum weights are

$$w_i = x_i^{2-r} / \sum x_i^{2-r}. \tag{4}$$

When $r = 0$, $w_i = x_i^2/\sum x_i^2$, and we get

$$r = 0: \qquad m_f = \sum \frac{x_i^2 y_i}{x_i} \Big/ \sum x_i^2 = \frac{\sum x_i y_i}{\sum x_i^2}, \tag{5}$$

which is the least-squares value $\hat{m}$. And so $\hat{m}$ is a particularly good estimate when the scatter about the line is the same for every x, and when the mean of Y_i is mx_i.

When $r = 1$, the variance of Y increases linearly with x, and the minimizing w_i's are $x_i/\sum x_i$, which means that

$$r = 1: \qquad m_f = \sum x_i (y_i/x_i) / \sum x_i = \sum y_i / \sum x_i. \tag{6}$$

In physical measurements, it frequently happens that the variance of Y increases almost linearly with x, and this estimate is attractive under such circumstances.

A second attraction is that if we are working toward a physical law relating x and y, rather than a predictive formula, then the two predictive formulas, namely that for y from x and that for x from y, have the same equation:

$$\begin{aligned} &\text{for predicting } y: \qquad y = \left(\frac{\sum y_i}{\sum x_i}\right) x, \\ &\text{for predicting } x: \qquad x = \left(\frac{\sum x_i}{\sum y_i}\right) y. \end{aligned} \tag{7}$$

This property does not hold for the least-squares line; instead we have

$$\begin{aligned} &\text{for predicting } y: \qquad y = \left(\frac{\sum x_i y_i}{\sum x_i^2}\right) x, \\ &\text{for predicting } x: \qquad x = \left(\frac{\sum x_i y_i}{\sum y_i^2}\right) y. \end{aligned} \tag{8}$$

Example 1 *Artesian well.* In the example relating temperature to depth, the data of Table 11–5 give $\sum y_i = 25.7$, $\sum x_i = 680$. The slope given by Eq. (5) is $m_f = 25.7/680 = 0.038$, which is very near our least-squares slope $\hat{m} = 0.039$. These slopes will agree well when the observed points hug the fitted line.

Example 2 *Craters of Mars.* One theory of craters on planets suggests that the frequency of craters should, for large craters, fall off inversely as the square of the diameter (Marcus, *Science*, June 21, 1968, p. 1334). Pictures taken by Mariner IV show frequencies as follows:

Diameter in km, D	32–45	45–64	64–90	90–128	Sum
$1/D^2$ (for left value of class interval)	.001	.0005	.00025	.000125	.001875
Frequency, F	53	22	14	3	

Let us fit the line of the form $F = m(1/D^2)$.

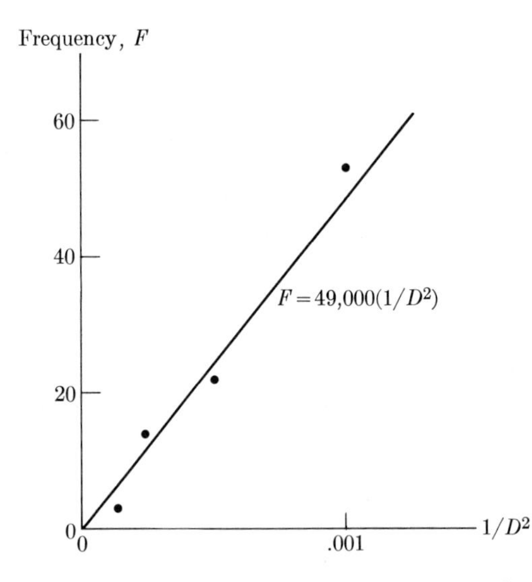

Fig. 11–6 Frequency of Mars crater sizes plotted against the inverse square of D, for large craters.

Solution If the theory is true, then a class interval can be regarded as having a probability p of occurring. The counts then have variance npq, where n is the number of craters counted, and where p is very small and q is near 1 because we are setting aside the vastly more numerous small craters that make n enormous. Consequently, the variance of the frequencies is almost proportional to their size, and we should use the estimate based on $r = 1$. In our problem this gives a slope of about $92/.001875 = 49{,}000$. The plot is shown in Fig. 11–6.

EXERCISES FOR SECTION ★11-3 SUPPLEMENT

1. Let the standard deviation of Y be proportional to x. Find r, the appropriate weights, and the optimum estimate of m.
2. Verify the claim in the text that any weights of the form $w_i = x_i^t/\sum x_i^t$ applied to y_i/x_i give an unbiased estimate of m when the mean of Y at x is mx.

11-4 FITTING A GENERAL LINE

Instead of fitting a general line in the form $y = b + mx$, we use the form

$$y = a + m(x - \bar{x}), \tag{1}$$

where $\bar{x}$ is the average of the x-scores. The use of the deviations from the average, $x - \bar{x}$, instead of x leads to a simplification in algebra. When we have fitted Eq. (1), we have also fitted $y = b + mx$ with $b = a - m\bar{x}$. As usual, m is the slope of the line. And a is the y-coordinate of a point on the line whose x-coordinate is $x = \bar{x}$.

Example 1 Given the points (1, 1), (2, $1\frac{1}{2}$), (3, 4) (see Example 1, Section 11-2, and Fig. 11-3), fit the general least-squares line.

Solution It is convenient to denote the points by (x_1, y_1), (x_2, y_2), (x_3, y_3) until the fitting is completed. For form (1) and given values of a and m, the vertical deviation d_i of the ith point from the line is

$$d_i = y_i - [a + m(x_i - \bar{x})], \qquad i = 1, 2, 3,$$

or

$$d_i = (y_i - a) - m(x_i - \bar{x}), \qquad i = 1, 2, 3.$$

Then

$$S = \sum d_i^2.$$

By choosing a and m, we wish to minimize S. We have

$$d_i^2 = [(y_i - a) - m(x_i - \bar{x})]^2.$$

We expand the square on the right-hand side to get

$$d_i^2 = (y_i - a)^2 - 2(y_i - a)(x_i - \bar{x})m + (x_i - \bar{x})^2 m^2. \tag{2}$$

Next, we break the middle term on the right side of Eq. (2) into two terms:

$$-2y_i(x_i - \bar{x})m + 2a(x_i - \bar{x})m.$$

When these parts are summed, the second vanishes because $\sum(x_i - \bar{x}) = 0$. Thus, summing Eq. (2), we get

$$S = \sum d_i^2 = \sum(y_i - a)^2 - 2\sum y_i(x_i - \bar{x})m + [\sum(x_i - \bar{x})^2]m^2. \tag{3a}$$

Note that a affects only the first sum on the right:

$$\sum(y_i - a)^2. \tag{3b}$$

This sum can be minimized by applying Theorem 11–2 on minimum sums to make

$$\boxed{\hat{a} = \bar{y}.} \tag{4}$$

The last two terms on the right-hand side of Eq. (3a) form a quadratic in m:

$$-2\sum y_i(x_i - \bar{x})m + [\sum(x_i - \bar{x})^2]m^2. \tag{3c}$$

To minimize this quadratic we apply the fundamental lemma, and get

$$\hat{m} = -\frac{B}{2A} = \frac{\sum y_i(x_i - \bar{x})}{\sum(x_i - \bar{x})^2}.$$

The numerator in the rightmost member of the foregoing equation can be rewritten for purposes of calculation as

$$\sum y_i x_i - \bar{x}\sum y_i = \sum x_i y_i - 3\bar{x}(\tfrac{1}{3}\sum y_i) = \sum x_i y_i - 3\bar{x}\bar{y}.$$

Therefore

$$\hat{m} = \frac{\sum x_i y_i - 3\bar{x}\bar{y}}{\sum(x_i - \bar{x})^2}. \tag{5}$$

Finally, the minimum value of S is the minimum of (3b) plus the minimum of (3c). Therefore

$$\hat{S} = \sum(y_i - \bar{y})^2 + C - \frac{B^2}{4A},$$

and $C = 0$. Therefore

$$\hat{S} = \sum(y_i - \bar{y})^2 - \frac{[\sum y_i(x_i - \bar{x})]^2}{\sum(x_i - \bar{x})^2}.$$

Replacing the numerator of the rightmost term yields

$$\hat{S} = \sum(y_i - \bar{y})^2 - \frac{[\sum x_i y_i - 3\bar{x}\bar{y}]^2}{\sum(x_i - \bar{x})^2}. \tag{6}$$

We now replace the x's and y's with their given numerical values, and get

$$\bar{x} = \frac{1 + 2 + 3}{3} = 2, \qquad \bar{y} = \frac{1 + \frac{3}{2} + 4}{3} = \frac{13}{6},$$

$$\sum(x_i - \bar{x})^2 = (1 - 2)^2 + (2 - 2)^2 + (3 - 2)^2 = 2,$$

$$\sum x_i y_i = 1 \times 1 + 2 \times \tfrac{3}{2} + 3 \times 4 = 16.$$

$$3\bar{x}\bar{y} = 13.$$

From Eq. (4b) of Section 6–2, we get

$$\sum(y_i - \bar{y})^2 = \sum y_i^2 - 3\bar{y}^2 = 1 + \tfrac{9}{4} + 16 - 3(\tfrac{13}{6})^2 = 5\tfrac{1}{6}.$$

Finally, we have

$$\hat{a} = \bar{y} = \tfrac{13}{6},$$

$$\hat{m} = \frac{\sum x_i y_i - 3\bar{x}\bar{y}}{\sum(x_i - \bar{x})^2} = \frac{16 - 13}{2} = \frac{3}{2}.$$

Therefore our fitted line is

$$\begin{aligned} y &= \hat{a} + \hat{m}(x - \bar{x}) \\ &= \tfrac{13}{6} + \tfrac{3}{2}(x - 2) \\ &= \tfrac{3}{2}x - \tfrac{5}{6}. \end{aligned}$$

Let us compare these constants with the corresponding values from our visual fit in Section 11–2. The slope is $\frac{3}{2}$, as we guessed in Section 11–2. To check the intercept, we compute $\hat{a} - \hat{m}\bar{x} = \frac{13}{6} - 3 = -\frac{5}{6}$, which also agrees with the minimizing value we obtained by least squares in Section 11–2, when we assumed $m = \frac{3}{2}$.

We substitute numbers into Eq. (6), to get the minimum value of S:

$$\hat{S} = 5\tfrac{1}{6} - \frac{(16 - 13)^2}{2} = \frac{2}{3},$$

as in Section 11–2.

The average squared deviation is

$$\frac{\hat{S}}{n} = \frac{\hat{S}}{3} = \frac{2}{9}.$$

This result is like a sample variance; it is the variance of the deviation scores d_i. Recall that $\hat{S}/n = \sum d_i^2/n$. If $\sum d_i = 0$, $\hat{S}/n$ is the variance of the d_i's. Later, we prove that when $\hat{a}$ and $\hat{m}$ are used in fitting the line, $\sum d_i = 0$.

The derivation for Example 1 is quite general, except that the number of measurements was 3 instead of n. If we extend the derivation to n, we have the following theorem.

11–5 Theorem: The general least-squares line

Given the set of paired measurements (x_i, y_i), $i = 1, 2, \ldots, n$, then the least-squares line

$$y = \hat{a} + \hat{m}(x - \bar{x})$$

is obtained when

$$\boxed{\hat{a} = \bar{y}, \qquad \hat{m} = \frac{\sum x_i y_i - n\bar{x}\bar{y}}{\sum(x_i - \bar{x})^2} = \frac{\sum x_i y_i - n\bar{x}\bar{y}}{\sum x_i^2 - n\bar{x}^2},} \tag{7}$$

where $\sum(x_i - \bar{x})^2 \neq 0$; and the minimum sum of squared deviations is

$$\hat{S} = \sum(y_i - \bar{y})^2 - \frac{[\sum x_i y_i - n\bar{x}\bar{y}]^2}{\sum(x_i - \bar{x})^2}. \tag{8}$$

The line of best fit can be written as

$$y - \bar{y} = \hat{m}(x - \bar{x}). \tag{9}$$

In form (9), it is easy to see that the line passes through the point with coordinates $(\bar{x}, \bar{y})$, which is called the *centroid* of the set of points. Among criteria for a close-fitting line, the requirement that the line pass through the centroid might have been chosen. We are pleased to have satisfied this requirement by least squares, so to speak, at no additional charge. In plotting the least-squares line on a graph, it is often convenient to plot the centroid and draw through it a line with slope $\hat{m}$.

Example 2 *Submarine sinkings.* The data of Table 11–6 show the results of a historical study of the actual number of German submarines sunk each month

Table 11–6 Sinkings of German submarines by U.S. in World War II by months

Month	Actual number, y	Guesses by U.S. (reported sinkings), x
1	3	3
2	2	2
3	6	4
4	3	2
5	4	5
6	3	5
7	11	9
8	9	12
9	10	8
10	16	13
11	13	14
12	5	3
13	6	4
14	19	13
15	15	10
16	15	16
	140	123

$n = 16$

$\sum x = 123$
$\bar{x} = 7.69$
$\sum y = 140$
$\bar{y} = 8.75$
$\sum x^2 = 1287$

$\sum xy = 1431$

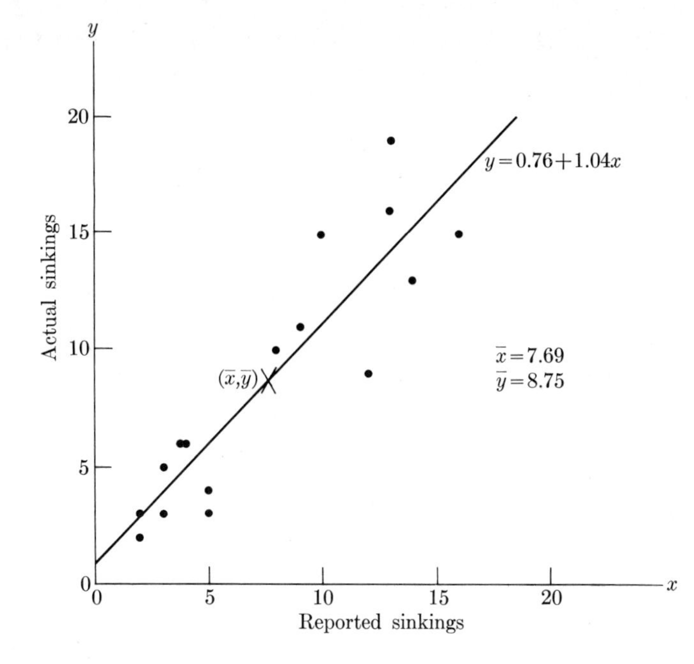

Fig. 11–7 Actual and reported sinkings of German submarines in World War II.

by the U.S. Navy in World War II, together with the U.S. Navy's reports of the number sunk. Fit the general least-squares line for estimating the actual number of sinkings from the U.S. Navy's reports.

Solution Dropping subscripts and using the data from Table 11–6, we find

$$\hat{m} = \frac{\sum xy - n\bar{x}\bar{y}}{\sum x^2 - n\bar{x}^2} = \frac{1431 - 16(123/16)(140/16)}{1287 - 16(123/16)^2} = 1.039,$$

$$\hat{a} = \bar{y} = 140/16 = 8.75,$$

and so the line is

$$y = 8.75 + 1.039(x - 7.69),$$

or approximately

$$y = 0.76 + 1.04x.$$

Thus the actual sinkings were slightly heavier than the Navy's reports implied. In Fig. 11–7 we show the points together with the fitted line.

Remark For such a set of data, it is natural also to consider fitting the line through the origin. Since the variability about the line must increase with increasing x, we might use for our slope $\sum y_i/\sum x_i$, as discussed in Section ★11–3 Supplement. The fitted line then is $y = (140/123)x$, or $y = 1.14x$. The slope is near 1.0, and the line fits about as well as the general line.

Standard error of estimate We want to show that $\hat{S}/n$ is the variance of the vertical deviations from the line, and thus show that $\sqrt{\hat{S}/n}$ is the standard deviation of the d_i's. We shall then have a measure of error in estimation, $\sqrt{\hat{S}/n}$, called the *standard error of estimate*, where $s_d = \sqrt{\hat{S}/n}$.

To prove this fact, we show that $\sum d_i = 0$ for the least-squares line, and thus get $s_d^2 = \hat{S}/n$.

11–6 Theorem: $\sum d_i = 0$

The general least-squares line leads to deviations d_i whose sum is zero.

Proof For a point (x_i, y_i), the fitted line predicts the value of y as

$$y_i' = \bar{y} + \hat{m}(x_i - \bar{x}).$$

Then

$$d_i = y_i - y_i' = (y_i - \bar{y}) - \hat{m}(x_i - \bar{x}),$$

$$\sum d_i = \sum(y_i - \bar{y}) - \hat{m}\sum(x_i - \bar{x}) = 0. \quad \square$$

11–7 Theorem: Variance of deviations

The minimum average squared error, $\hat{S}/n$, is the variance of the deviations from the general least-squares line.

Proof Because $\sum d_i = 0$, we have $\bar{d} = \sum d_i/n = 0$, and therefore

$$\frac{\hat{S}}{n} = \frac{\sum d_i^2}{n} = s_d^2. \quad \square$$

Remark Some authors use the symbol s_e^2 for our s_d^2. Some divide by $n - 1$ instead of by n.

11–8 Definition: Standard error of estimate

The standard error of estimate is

$$\boxed{s_d = \sqrt{\hat{S}/n}.}$$

Thus the standard error of estimate is just the standard deviation of the vertical deviations from the least-squares line. For the submarine sinkings $s_d = 2.35$, which gives a notion of the variability of the points about the line and of the errors in the Navy's reports.

EXERCISES FOR SECTION 11–4

1. Find the general least-squares line for the set of pairs (1, 1), (2, 3), (3, 3) and find the standard error of estimate.
2. (Continuation.) Compute the three vertical deviations in Exercise 1 and compare their absolute magnitudes.
3. Compute the deviation scores for Example 1, compare their magnitudes, and find their sum.

4. Find an equation of the general least-squares line and the standard error of estimate for the points with coordinates

x:	-2	-1	0	0	1	2	2
y:	-2	-2	-2	-1	-1	0	1

5. (Continuation.) Use the equation of Exercise 4 to estimate y when

$$x = -1, 0, 1.$$

6. Find the slope and y-intercept of the general least-squares line fitting the points with coordinates

x:	-1	-1	0	0	0	1	1
y:	0	1	-1	0	1	0	-1

7. (Continuation.) Plot the points and the least-squares line for the data of Exercise 6. Note that the line passes through the mean of the y-values corresponding to each x.
8. In an eight-week course, the final grades A, B, C, D, E were given scores $y = 2, 1, 0, -1, -2$, respectively. Students with more than 3 absences were dropped from the course. The fitted line for predicting grade scores from absences was $y = 1.5 - 0.5x$. What average letter grade is predicted for students with 1 or 3 absences? For a student with 4 absences, if the same equation holds?
9. For the points with coordinates $(0, 0)$, $(0, 1)$, $(1, 0)$, $(1, 1)$, find the general least-squares line for predicting y from x.
10. The general least-squares line for predicting y from x is $y = 3x + 4$. If $\bar{x} = 1$, find $\bar{y}$.
11. Show that if $\bar{x} = 0$, the slope of the general least-squares line and of the least-squares line through the origin are identical.
12. Two points always determine a straight line. But for what kind of pairs of points is it impossible to apply formulas like (7) or (8)?
13. Show that

$$\hat{m} = \frac{(\sum x_i y_i/n) - \bar{x}\bar{y}}{s_x^2}.$$

14. Show that

$$s_d^2 = s_y^2 - \frac{[(\sum x_i y_i/n) - \bar{x}\bar{y}]^2}{s_x^2}.$$

15. Show that $s_d^2 = s_y^2 - \hat{m}^2 s_x^2$.
16. Write out the steps of the proof of Theorem 11–5 on the general least-squares line for two points, (x_1, y_1) and (x_2, y_2).
17. One set of points has x-coordinate x_1, another set has x-coordinate $x_2 \neq x_1$. Use a verbal argument to prove that the general least-squares line passes through the points $(x_1, \bar{y}_1)$, $(x_2, \bar{y}_2)$, where $\bar{y}_i$ is the average y for points with x-coordinate x_i, $i = 1, 2$.
18. For three points with x-coordinates equally spaced, $(x - 1, y_1)$, (x, y_2), $(x + 1, y_3)$, fit the general least-squares line and compare the absolute magnitudes of the deviations. Comment.

19. If k sets of points have x-coordinates $x_1, x_2, \ldots, x_k$ and y-means $\bar{y}_1, \bar{y}_2, \ldots, \bar{y}_k$, and if the ordered pairs $(x_i, \bar{y}_i)$ are collinear, show that the least-squares line passes through these points.
20. Write out the steps of Theorem 11–5 on the general least-squares line for n points (x_i, y_i), $i = 1, 2, \ldots, n$.

11–5 PRECISION OF ESTIMATION AND THE CORRELATION COEFFICIENT

The prediction equation

$$y = \bar{y} + \hat{m}(x - \bar{x}) \tag{1}$$

usually gives us a way to use the relation between x and y to improve the prediction of y. When the slope $\hat{m}$ is zero (or nearly zero), our estimate of y would be

$$y = \bar{y},$$

and the variance of the deviation scores is then just the sample variance of the y's:

$$s_y^2 = \frac{\sum(y_i - \bar{y})^2}{n}. \tag{2}$$

On the other hand, the deviation scores d_i, obtained when we use Eq. (1) to predict y, have variance

$$\begin{aligned} s_d^2 &= \frac{\hat{S}}{n} \\ &= \frac{\sum(y_i - \bar{y})^2}{n} - \frac{(\sum x_i y_i - n\bar{x}\bar{y})^2}{n\sum(x_i - \bar{x})^2}. \end{aligned} \tag{3}$$

We recall from Section 10–4, Eq. (16), that the correlation coefficient r is

$$r = \frac{\sum x_i y_i - n\bar{x}\bar{y}}{n s_x s_y}. \tag{4}$$

We can rewrite the last term of Eq. (3) in terms of the correlation coefficient and s_y^2 as follows:

$$\begin{aligned} \frac{(\sum x_i y_i - n\bar{x}\bar{y})^2}{n\sum(x_i - \bar{x})^2} &= \frac{(\sum x_i y_i - n\bar{x}\bar{y})^2}{n^2 s_x^2} \\ &= \frac{(\sum x_i y_i - n\bar{x}\bar{y})^2 s_y^2}{n^2 s_x^2 s_y^2} \\ &= r^2 s_y^2. \end{aligned} \tag{5}$$

When we substitute the results of Eqs. (2) and (5) into Eq. (3), we get

$$\boxed{s_d^2 = s_y^2 - r^2 s_y^2 = (1 - r^2) s_y^2.} \tag{6}$$

The extent to which s_d^2 is smaller than s_y^2 measures the improvement in prediction, at least in large samples. Let us consider the ratio s_d^2/s_y^2. Using Eq. (6), we have

$$\boxed{\frac{s_d^2}{s_y^2} = 1 - r^2.} \tag{7}$$

Consequently r^2 measures the fraction of s_y^2 that has been removed from the variance of deviations by the fitting of the linear equation, Eq. (1). It is customary to say that r^2 is the fraction of variance, s_y^2, "explained" by the regression line.

What values can r take? Since $0 \le s_d^2 \le s_y^2$ [from Eq. (6)], we have, upon dividing through by s_y^2,

$$0 \le \frac{s_d^2}{s_y^2} \le 1,$$

or

$$0 \le 1 - r^2 \le 1. \tag{8}$$

The values of r that satisfy inequality (8) are

$$-1 \le r \le 1, \tag{9}$$

and this inequality shows that the value of the correlation coefficient can run only from -1 to $+1$.

By rewriting Eq. (1), we can see at once the meaning of various values of r. First we rewrite $\hat{m}$ in terms of r, s_x, and s_y:

$$\begin{aligned}\hat{m} &= \frac{\sum x_i y_i - n\bar{x}\bar{y}}{\sum(x_i - \bar{x})^2} \\ &= \frac{\sum x_i y_i - n\bar{x}\bar{y}}{n s_x^2} \cdot \frac{s_y}{s_y} \\ &= \frac{\sum x_i y_i - n\bar{x}\bar{y}}{n s_x s_y} \cdot \frac{s_y}{s_x},\end{aligned}$$

and finally

$$\boxed{\hat{m} = r\,\frac{s_y}{s_x}.} \tag{10}$$

When we substitute this value into Eq. (1), the forecasting equation becomes

$$\boxed{y - \bar{y} = r\,\frac{s_y}{s_x}(x - \bar{x}).} \tag{11}$$

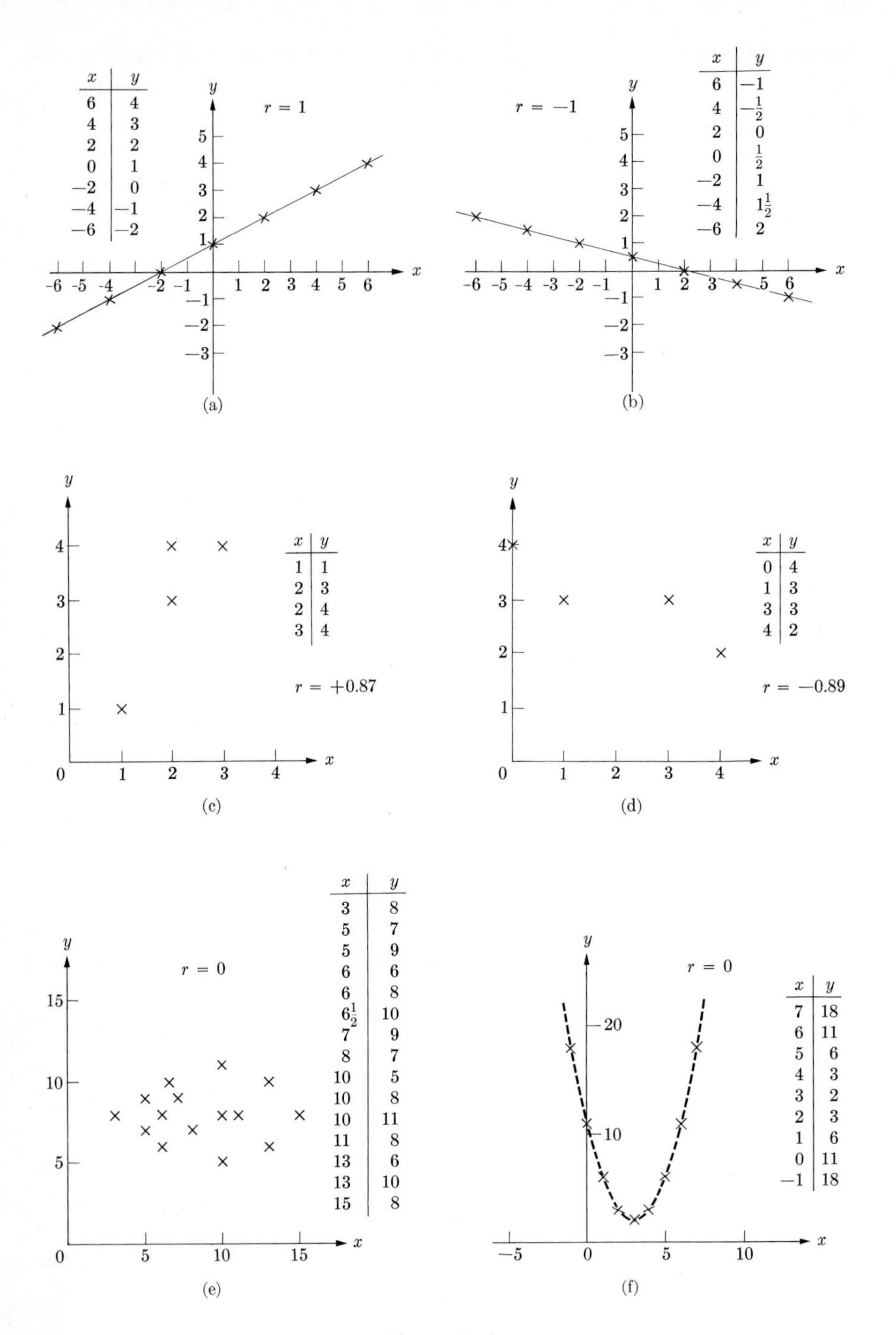
x y
6 4
4 3
2 2
0 1
−2 0
−4 −1
−6 −2
y
r = 1
x
(a)
r = −1
y
x y
6 −1
4 $-\frac{1}{2}$
2 0
0 $\frac{1}{2}$
−2 1
−4 $1\frac{1}{2}$
−6 2
x
(b)
y
x y
1 1
2 3
2 4
3 4
r = +0.87
x
(c)
y
x y
0 4
1 3
3 3
4 2
r = −0.89
x
(d)
x y
3 8
5 7
5 9
6 6
6 8
$6\frac{1}{2}$ 10
7 9
8 7
10 5
10 8
10 11
11 8
13 6
13 10
15 8
y
r = 0
x
(e)
y
r = 0
x y
7 18
6 11
5 6
4 3
3 2
2 3
1 6
0 11
−1 18
20
10
x
(f)

Figure 11-8

The standard deviations s_y and s_x are both positive, so if $r > 0$, then the slope $rs_y/s_x > 0$, and the line rises to the right. Thus as x increases, y increases. If $r < 0$, then the line rises to the left. If $r = 0$, the line is horizontal.

Note that $r = 0$ does *not* imply that no relation exists between y and x, merely that the degree of linear relationship is zero.

Figure 11–8 shows a variety of "scatter diagrams", together with the coordinates of the points and the correlation coefficients between x and y. In practice, there are usually many more points than these, but easy examples are worth study.

Parts (a) and (b) of Fig. 11–8 show that when the points are collinear, then $r = 1$ or -1, as we expect (unless the line is vertical or horizontal).

Parts (c) and (d) show large positive and negative values of r; a line fitted to them would have a value of s_d^2 that is much less than s_y^2.

Parts (e) and (f) show two different examples of $r = 0$. In (e), the points are scattered about rather haphazardly, and we can well imagine that there is little or no relation between y and x. In (f), the points were constructed to fall on the parabola with equation

$$y = (x - 3)^2 + 2;$$

thus there is a mathematical relation between y and x, but it is not measured by r. The reason is that r measures degree of linear relationship, and more advanced methods are required to assess curvilinear relations of this sort.

Calculations For calculations of r made with paper and pencil without the aid of machines, the formula

$$\boxed{r = \frac{(\sum x_i y_i/n) - \bar{x}\bar{y}}{s_x s_y}} \tag{12}$$

is a bit more convenient than Eq. (4), because the numbers in the numerator on the right of Eq. (12) are smaller than those in the corresponding position in Eq. (4).

In these calculations, numbers should not be rounded off too soon, or the information in the data will be lost.

Example 1 *Rounding off too early.* Given the three (x, y)-pairs (1, 1.1), (2, 1.2), (3, 1.4) in the table below, compute r.

x	y	xy
1	1.1	1.1
2	1.2	2.4
3	1.4	4.2
6	3.7	7.7

Solution (with excessive rounding)

$$\bar{x} = \frac{6}{3} = 2, \qquad \bar{y} = \frac{3.7}{3} \approx 1.2,$$

$$\frac{\sum xy}{n} = \frac{7.7}{3} \approx 2.6,$$

$$s_x^2 = \frac{(-1)^2 + 0^2 + (+1)^2}{3} = \frac{2}{3},$$

$$s_y^2 = \frac{\sum(y - \bar{y})^2}{n} \approx \frac{(-0.1)^2 + 0^2 + (0.2)^2}{3}$$

$$= \frac{0.05}{3} \approx 0.02,$$

$$r = \frac{(\sum xy/n) - \bar{x}\bar{y}}{\sqrt{s_x^2 s_y^2}} \approx \frac{2.6 - 2(1.2)}{\sqrt{\frac{2}{3} \times 0.02}} \approx \frac{0.2}{\sqrt{0.01}} = \frac{0.2}{0.1} = 2.$$

The trouble here is not the careless rounding in the calculation of $s_x^2 s_y^2$, though it might have been. Careful calculation of $s_x^2 s_y^2$ gives $0.28/27 \approx 0.01037$, and $\sqrt{0.01037} \approx 0.102$. The rounding in the numerator produced the error. We should have had $\sum xy/n - \bar{x}\bar{y} \approx 2.567 - 2(1.233) = 0.101$ instead of 0.2.

Today, hand and electronic calculating machines are quite widely available, and unless n is small and the numbers of significant figures in the x_i- and y_i-values are small, calculators are used. That is why in our exercises the numbers have been kept modest and easy, even though in practice n may be very large and the numbers cumbersome.

Example 2 *Correlation in the submarine problem.* Since in the submarine example $s_d^2 = 5.526$ and $s_y^2 = 28.56$, we can find r^2 from $1 - s_d^2/s_y^2 = 1 - 0.193 = 0.807$. Thus 81% of the variance of y is accounted for by the reports. Taking the square root gives $r = 0.90$. We would say that the correlation between reported and actual sinkings was very high, even without considering that the final reports had to be based upon statements sent in from the battle zone, where conditions for observation could not have been ideal.

EXERCISES FOR SECTION 11-5

1. From the data of Fig. 11-8(a), calculate directly that $r = 1$.
2. From the data of Fig. 11-8(c), calculate that $r \approx 0.87$.
3. From the data of Fig. 11-8(d), calculate that $r \approx -0.89$.
4. From the data of Fig. 11-8(f), calculate directly that $r = 0$.
5. Show that when $s_x = s_y$, $\hat{m} = r$.
6. When $r = 0$, $s_x > 0$, find the prediction equation of y from x, and explain the use of the expression "from x" in this situation.

7. Given two points not both on either a horizontal or a vertical line, what values can r have? Explain.
8. If $r = 0.3$, what percentage of the variance of y is explained by the regression line? What if $r = -0.3$? What if r is 0.8?
9. The percentage of the variance of y explained by the regression line is 16. What values can r have?
10. If $\bar{x} = \bar{y} = 0$, $s_x = s_y > 0$, find the linear equation for predicting y from x.
11. If $\bar{y} = 2$, $\bar{x} = 3$, $s_x = 5$, $s_y = 10$, $r = -0.4$, find the linear equation for predicting y from x.
12. Show that if x_i and y_i have $\bar{x} = 0$, $\bar{y} = 0$, $s_x = s_y = 1$, then $r = \sum x_i y_i / n$. Note that in this special case x and y are standardized variables, having mean 0 and variance 1.
13. If $s_d^2 = 12$, $s_y^2 = 20$, find the possible values of r.
14. If $r = 0.6$, $s_d = 8$, find s_y.
15. If the variable y is correlated 0.1 with x, what would you say of the effectiveness of x in helping to predict y?
16. If the correlation between x and y is r, show that if $x_i' = x_i + c$, where c is a constant, the correlation between x' and y is also r.
17. Show that if $x_i' = cx_i$, where c is a positive constant, the correlation between x' and y is the same as the correlation between x and y.
18. Show that if c is negative in Exercise 17, the correlation between x' and y is the negative of the correlation between x and y.
19. Show that if x and y are correlated r, the new variables x' and y', where $x_i' = c + dx_i$ and $y_i' = e + fy_i$, are correlated r if $df > 0$, and $-r$ if $df < 0$. This theorem is usually referred to as "the invariance of r under linear transformations in either or both variables".

11-6 MODELS FOR REGRESSION

In addition to the name "curve-fitting" that we have used in this chapter, the theory we treat is also called "regression theory". The origin of this puzzling name will emerge later. Many different mathematical descriptions or models are required to describe the physical situations where regression theory applies. We discuss here two models that arise frequently.

Control-knob model In the control-knob model, one of the variables is more or less under the control of the investigator. When Strong did his retention study (see Section 11-1), he thought retention depended on time, and so he chose to control, as with a control knob on a radio or television set, the set of times in the experiment. These *times* were not randomly drawn. On the other hand, the *retention* of the subjects in the experiments can be regarded as a random variable. If we consider a fixed time until testing, the corresponding variability of retention has at least two important sources.

First, the same person recalling lists repeatedly remembers varying numbers of words. Anyone studying a foreign language has had this experience. Second, as every student knows, individuals differ in their ability to remember. This second source of variability is present in Strong's study, because a population of people was sampled. More generally, in the control-knob model, let us think of picking a single value of one variable, x, and evaluating the other variable Y, repeatedly. The value of Y depends on the outcome of an experiment: Y is a random variable. There is a distribution of the random variable Y for each value of x, but we do not think of x as a random variable.

For any given x, let us suppose that the distribution of Y has a mean $E(Y|x)$ (read "expected value of Y, given x"). Suppose that, as x moves continuously, the points with coordinates $(x, E(Y|x))$ trace out a smooth curve, or a straight line. Suppose, for example, that

$$E(Y|x) = a + m(x - \overline{x}).$$

We have developed methods for fitting a straight-line relation. A control-knob experiment produces paired data (x_i, y_i), $i = 1, 2, \ldots, n$. We fit these data with the line

$$y = \hat{a} + \hat{m}(x - \overline{x}).$$

Then $\hat{a}$ is an estimate of a and $\hat{m}$ is an estimate of m. If we carry out many repetitions of the experiment, then there are many values of $\hat{a}$ and many values of $\hat{m}$. It turns out that $E(\hat{a}) = a$ and $E(\hat{m}) = m$; we shall not prove these equalities, but their demonstration is not difficult.

In the artesian well example (beginning of Section 11-3), the depth below the origin is the independent variable. The experimenter chooses these depths; they do not have a probability distribution. On the other hand, repeated careful measurements of temperatures at a given depth probably would vary from week to week. Thus, for each depth, it is reasonable to think of the temperature difference as a random variable.

It may occur to you that when a temperature-measuring gadget 300 meters long goes into the water, there may be some uncertainty about the depth x. A more complicated model might be needed to handle this source of variation.

Remark In the statistical literature, the control-knob model is known by the unhappy name "the fixed variate model". The word "fixed" refers to the fact that the x's are not values of a random variable.

Example 1 Darian-Smith, Rowe, and Sessle studied the number of electrical impulses per neural response when the skin of a cat was indented to various closely controlled depths. Their figure (shown here as Fig. 11-9) illustrates the control-knob model very well. The 10 observations on each vertical correspond to a single choice of indentation. The distribution of the points gives a notion of the distribution of Y for each chosen x.

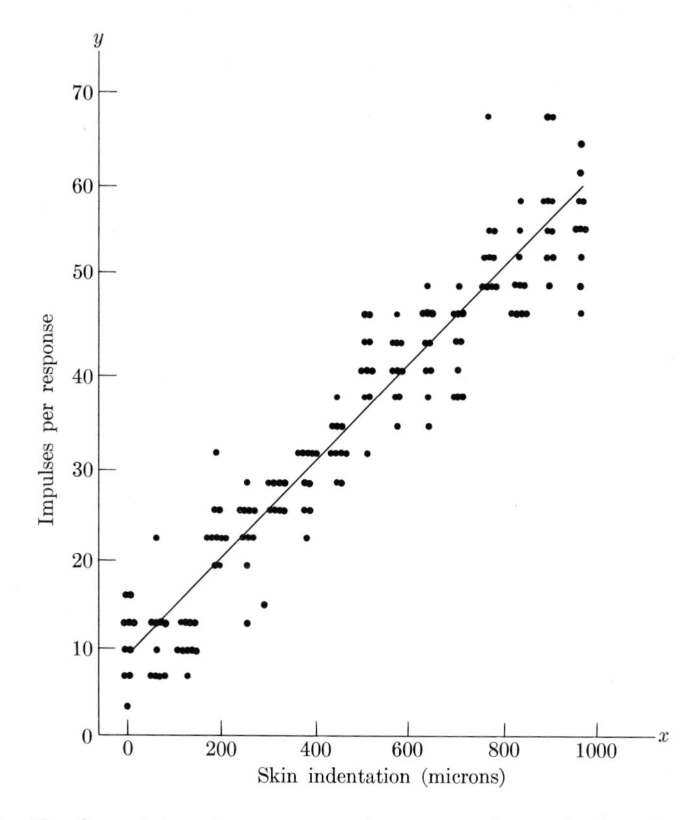

Fig. 11–9 Number of impulses per neural response for a slowly adapting neuron when cat's skin is indented by amounts X (ten responses for each choice of X). From I. Darian-Smith, M. J. Rowe, and B. J. Sessle, " 'Tactile' Stimulus Intensity: Information Transmission by Relay Neurons in Different Trigeminal Nuclei", *Science*, Vol. 160, No. 3829, 17 May 1968, pp. 791–794. Copyright 1968 by the American Association for the Advancement of Science. Reproduced from page 792 with permission of the authors and the American Association for the Advancement of Science.

Bivariate model Suppose that we have a population of individuals, each with two measurable characteristics, such as height and weight. If a person is drawn at random from the population, then the random variable X = height takes a value, and so does Y = weight. In such a sampling procedure, neither variable is controlled by the experimenter. Furthermore, we might wish to predict height from weight or weight from height.

Galton, a biometrician, in the late 1800's probably had heard discussions by the older generation concerning the younger generation: Why was it going to the dogs? In particular, why was it that great fathers didn't have great sons? The sons of great statesmen were often not as impressive as their fathers. Galton was interested in heredity, so he set about to see if he could study such a question in an objective biological situation. He studied tallness rather than greatness.

Table 11–7 Pearson-Lee data on father-son statures

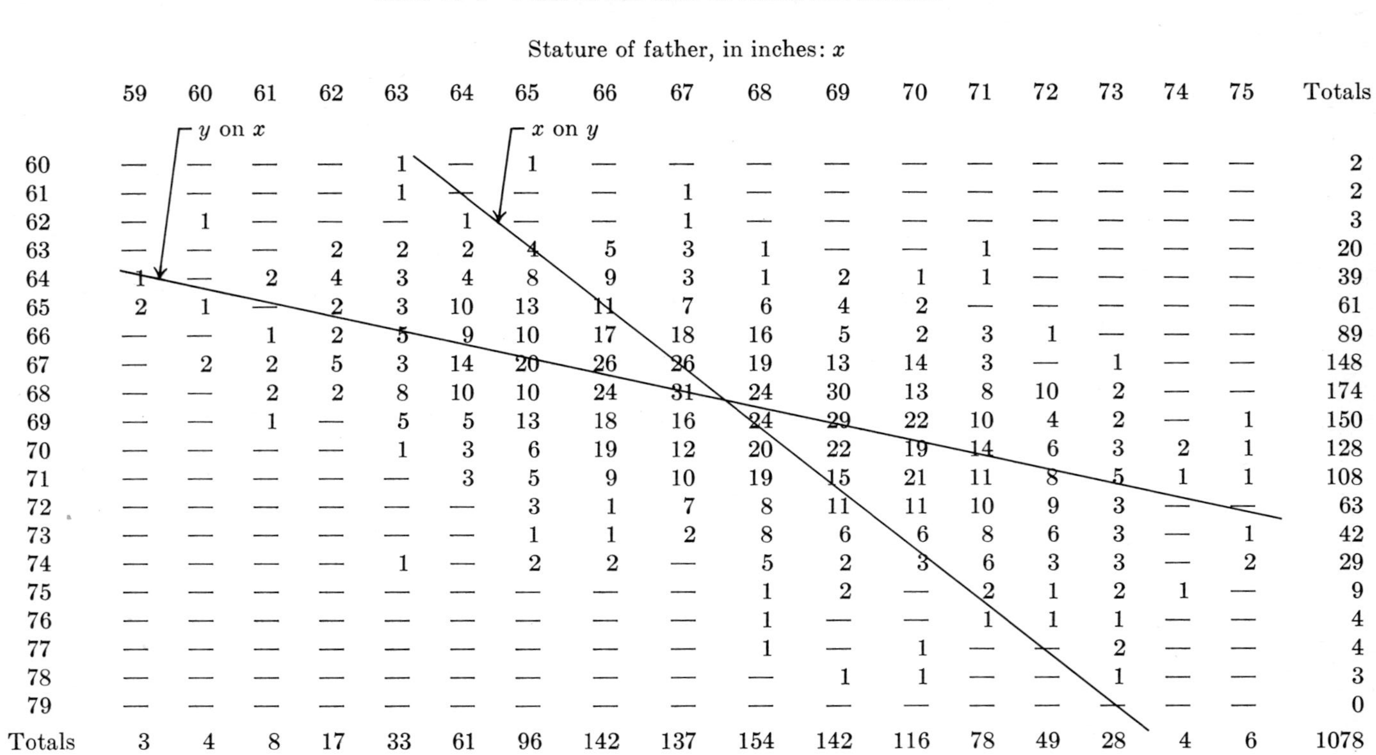

Stature of father, in inches: x

Stature of son, in inches: y	59	60	61	62	63	64	65	66	67	68	69	70	71	72	73	74	75	Totals
60	—	—	—	—	1	—	1	—	—	—	—	—	—	—	—	—	—	2
61	—	—	—	—	1	—	—	—	1	—	—	—	—	—	—	—	—	2
62	—	1	—	—	—	1	—	—	1	—	—	—	—	—	—	—	—	3
63	—	—	—	2	2	2	4	5	3	1	—	—	1	—	—	—	—	20
64	1	—	2	4	3	4	8	9	3	1	2	1	1	—	—	—	—	39
65	2	1	—	2	3	10	13	11	7	6	4	2	—	—	—	—	—	61
66	—	—	1	2	5	9	10	17	18	16	5	2	3	1	—	—	—	89
67	—	2	2	5	3	14	20	26	26	19	13	14	3	—	1	—	—	148
68	—	—	2	2	8	10	10	24	31	24	30	13	8	10	2	—	—	174
69	—	—	1	—	5	5	13	18	16	24	29	22	10	4	2	—	1	150
70	—	—	—	—	1	3	6	19	12	20	22	19	14	6	3	2	1	128
71	—	—	—	—	—	3	5	9	10	19	15	21	11	8	5	1	1	108
72	—	—	—	—	—	—	3	1	7	8	11	11	10	9	3	—	—	63
73	—	—	—	—	—	—	1	1	2	8	6	6	8	6	3	—	1	42
74	—	—	—	—	1	—	2	2	—	5	2	3	6	3	3	—	2	29
75	—	—	—	—	—	—	—	—	—	1	2	—	2	1	2	1	—	9
76	—	—	—	—	—	—	—	—	—	1	—	—	1	1	1	—	—	4
77	—	—	—	—	—	—	—	—	—	1	—	1	—	—	2	—	—	4
78	—	—	—	—	—	—	—	—	—	—	1	1	—	—	1	—	—	3
79	—	—	—	—	—	—	—	—	—	—	—	—	—	—	—	—	—	0
Totals	3	4	8	17	33	61	96	142	137	154	142	116	78	49	28	4	6	1078

He obtained data on heights of fathers and heights of their sons, and thus he had ordered pairs (x_i, y_i), where x_i is the height of the father and y_i the height of the son. Galton's data are not in a form suitable for our purposes, but similar data gathered by Karl Pearson and Alice Lee are shown in Table 11–7.

Galton found that tall fathers, on the average, had sons who were shorter than they, although not as short as the population mean. On the other hand, short fathers had sons on the average taller than they, though not as tall as the population mean. Thus if tallness were regarded as desirable, tall fathers had disappointing sons, but tall sons had disappointing fathers, too. On the other hand, short fathers would have taller sons, and short sons would have taller fathers. (In school a similar effect is familiar. The student at the top of the class on one test can only maintain his position or drop in rank on the next test, while the student at the bottom can only maintain his position or move up.) Galton regarded this result as a "regression" toward the mean of the population, and he called the line fitted to predict a son's height y from a father's height x, the *regression of sons' heights on fathers'*, or the *regression of y on x.*

Then he turned the data around and predicted a father's height x from a son's height y, and obtained the *regression of x on y.* Thus Galton found two regression lines, one for predicting y from x, the other for predicting x from y. And these regression lines are not identical.

The regression of *x* on *y* We know how to get the regression of y on x by fitting $y = \hat{a} + \hat{m}(x - \bar{x})$. The procedure for obtaining the regression of x on y is similar; we fit

$$x = g + h(y - \bar{y}).$$

A deviation score for predicting x_i from y_i is

$$d_i = (x_i - g) - h(y_i - \bar{y}), \qquad i = 1, 2, \ldots, n.$$

We minimize

$$S = \sum_{i=1}^{n} d_i^2 = \sum_{i=1}^{n} [(x_i - g) - h(y_i - \bar{y})]^2.$$

The methods we used in Section 11–4 apply, and we find the minimizing values of g and h to be

$$\boxed{\hat{g} = \bar{x}, \qquad \hat{h} = \frac{\sum x_i y_i - n\bar{x}\bar{y}}{\sum (y_i - \bar{y})^2}.} \tag{1}$$

Recall that the parallel results for the regression of y on x were

$$\hat{a} = \bar{y}, \qquad \hat{m} = \frac{\sum x_i y_i - n\bar{x}\bar{y}}{\sum (x_i - \bar{x})^2}.$$

Example 2 Find the two regression lines for the points plotted in Fig. 11–10.

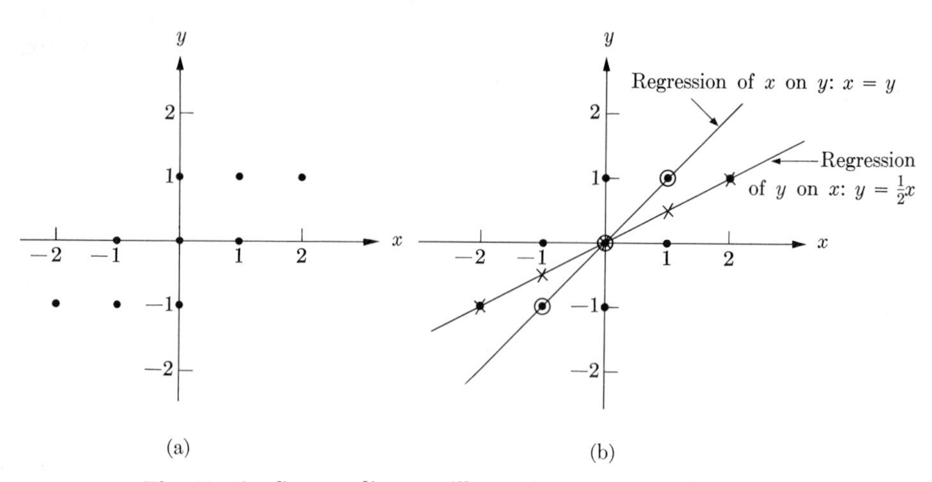

Fig. 11–10 Scatter diagram illustrating two regression lines.

Solution The graphs in Fig. 11–10 have 9 points; the heavy dots are the data. The simplicity of the example makes possible the calculation of the regression lines without using heavy calculations. Exercise 19 of Section 11–4 tells us that when the points plotted at the means of the vertical arrays for each x-value are collinear, the least-squares line passes through the means. In the figure, a cross indicates the column average for one value of x. For example, at $x = 1$, there are two values of y, 1 and 0. These values average 0.5, so the $\times$ in that column is plotted at $y = 0.5$. In this example, all crosses lie on a straight line, and consequently the regression of y on x is the line through the crosses. Because it passes through the origin and has slope $\frac{1}{2}$, its equation is

$$y = \tfrac{1}{2}x \qquad \text{(regression of } y \text{ on } x). \tag{2}$$

Similarly, for a given datum value of y the average of the x's is the same as the x-coordinate of the middle dot in the row for that y. These averages are shown as open circles. They too fall on a straight line, which is the regression line of x on y. It also passes through the origin, but has slope 1, so an equation for it is

$$x = y \qquad \text{(regression of } x \text{ on } y). \tag{3}$$

Thus for predicting y from x we use Eq. (2), but for predicting x from y we use Eq. (3). In this example, these equations are distinct.

The simplicity of the array of points in this example made possible the construction of the equations without the usual calculations. Naturally, such an example would be rare in practice. However, a quick way to get an approximate regression line is to compute column (or row) means, plot them, and pass a visually fitted straight line through them.

The bivariate model differs from the control-knob model in that it has two legitimate regression lines instead of one. In the control-knob model, we would not compute the regression of x on y.

Example 3 *Pearson-Lee data on heights.* Find the two regression lines for heights of fathers and sons.

Solution Heavy computation gives

$$y = 0.51x + 33.9 \qquad \text{(regression of sons' heights on fathers')},$$
$$x = 0.51y + 32.8 \qquad \text{(regression of fathers' heights on sons')}.$$

Lines corresponding to these equations have been printed on Table 11–7 to help give you a feel for the relation between the tabled numbers and the regression lines.

Example 4 *First-ace problem.* In the first-ace problem of Section 1–5, find the regression of the $(r + 1)$st count on the rth. That is, find the regression line for predicting from one count to the next. We know that with perfect shuffling there should be independence between the successive counts, so this calculation provides a check on the model. For example, if high counts are followed by lows, and vice versa, we should get a negative slope for our regression line.

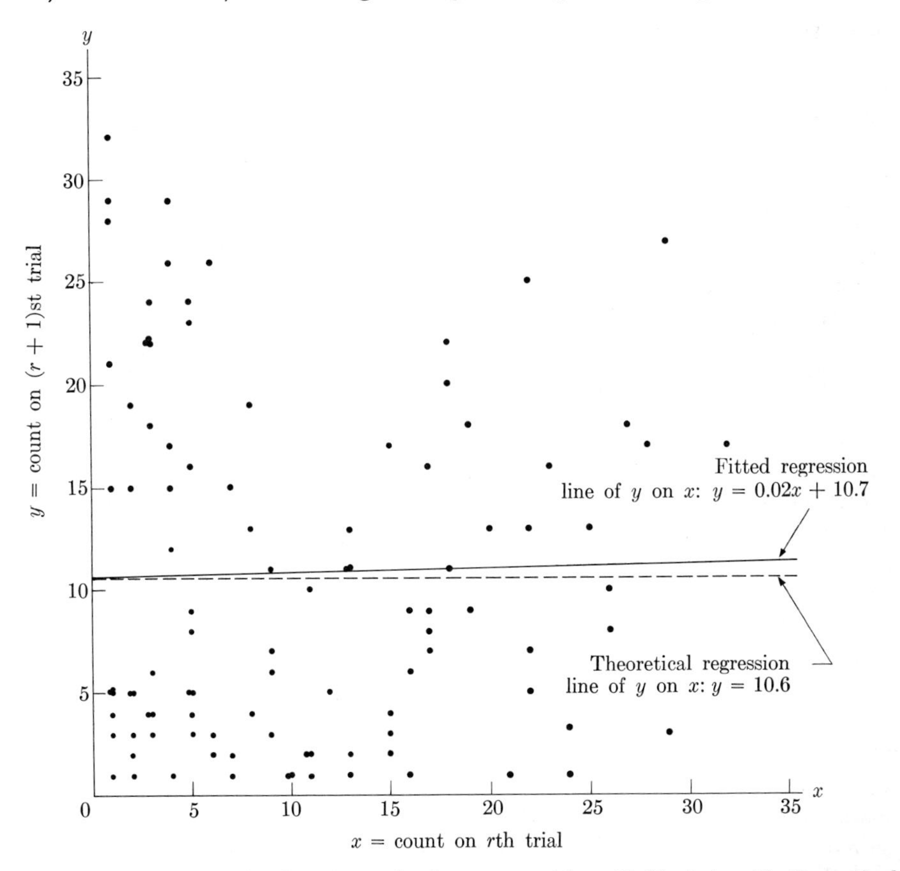

Fig. 11–11 Graph of the data from the first-ace problem, Table 1–1, with theoretical and fitted regression lines for predicting the count on trial $r + 1$ from the count on trial r.

Solution The original data are given in Table 1–1, and from them are derived a set of ordered pairs of successive counts. The original sequence started 5, 4, 29, 3, 24, ... The pairs start (5, 4), (4, 29), (29, 3), (3, 24), ... The data are plotted in Fig. 11–11. (We have added the count for shuffle number 101—it was 5—to get an even 100 pairs.)

If the first and second members of a pair of counts are independent, as we would assume, then the mean of the second count, given the first, is the same as the mean of the second, unconditionally. That is, the expected value of the second count is $\mu_Y = 10.6$ (as we found in Section 1–5) for every possible first count. Therefore the theoretical regression line of y on x is the horizontal line with equation

$$y = 10.6.$$

The use of formulas (7), Theorem 11–5 of Section 11–4, and heavy calculation gives the observed regression for the 100 points of Fig. 11–11 as

$$y = 0.021x + 10.73.$$

Clearly, $\hat{m}$ is near zero, as it should be if we have independence between successive counts, and the constant term 10.73 is close to the theoretical value of 10.6. We continue to be satisfied with the independence assumption for successive counts in the first-ace experiment.

Grouping data and coded values Especially for hand calculation or for tabular presentation, we assign observations to class intervals, usually of equal widths. Height to the nearest inch, ages to the nearest multiple of 10 years, and income to the nearest multiple of \$1000 are common examples. When we use such class intervals, we treat all the measurements in the interval as if they have the middle value. This middle value is sometimes called the *class mark*. By doing this we are able to work with a few numbers instead of a great many. If the intervals are not too coarse, statistics based on these grouped data turn out to be very close to those based on the original data. Most data on continuous variables are rounded in this manner. Even discrete data are grouped in this manner so that we can see the pattern in their frequencies, as we did in the draft-lottery example of Section 3–6.

We used this idea in working with the normal distribution in Section 7–4, when we used class intervals to approximate the variance by forming a discrete distribution from the continuous one.

Real data can have inconveniently large numbers even when they are rounded off or grouped. For example, in a physics experiment one might wind up with values 1122.1, 1122.2, 1122.3, ..., 1122.9. For convenience, we often code these values by making a linear transformation, so that we get small integers to work with. Since we compute means and variances, it is convenient to have 0 near but below the middle of the range. For the numbers just mentioned, one would be inclined to subtract either 1122.4 or 1122.5 from all the numbers. Using 1122.4, his values would be −0.3, −0.2, ..., 0.5. If one wants integers, he multiplies by 10 to get $-3, -2, \ldots, 5$. If x is the original variable, the coded variable is

$$y = 10(x - 1122.4).$$

After computing means and variances of coded variables, the means and variances of the original variables can be recovered using the formulas of Exercises 14, 15, and 16 of Section 6–2. Correlation coefficients are unchanged by coding unless exactly one of the original variables is multiplied by a negative number, and then the sign is changed. (See also Theorem 6–7, Corollary 6–8, and Example 6, Section 6–1.)

Example 5 *Relating extreme temperatures.* Can we predict extreme low temperatures from extreme high ones? Table 11–8 shows the joint frequency distribution of X, the all-time lowest recorded temperature (rounded to the nearest multiple of 10°), and of Y, the all-time highest recorded temperature (rounded to the nearest multiple of 5°), in a state of the U.S.A. Obtain the regression estimates of Y on X and of X on Y.

Table 11–8 Extreme temperatures in the 50 states (Source: *The 1969 World Almanac*, Boston Herald Traveler, 1968)

Y: highest temperature recorded (to nearest multiple of 5)	X: lowest temperature recorded (to nearest multiple of 10)											
	−80	−70	−60	−50	−40	−30	−20	−10	0	10	20	Total f
135					1							1
130												0
125					1							1
120		1	4	4	3	3	1					16
115			2	3	1	3	2					11
110				2	3	3	3	1	1			13
105				3		2						5
100	1						1				1	3
Total f	1	1	6	12	9	11	7	1	1	0	1	50

Solution Table 11–9 expands Table 11–8 to give a very convenient layout for a hand calculation for regression. Note that the central values of the class intervals are shown at the left and the top of the table. The cells show the original frequency from Table 11–8 in bold numerals and two other numbers we explain later. The right-hand and bottom margins show the Y- and X-totals. The next column gives the coded values for Y, called y, and the next row gives the coded values for X, called x. These coded values provide us with decently small numbers for hand work. We chose to subtract 115 from Y because it was near the middle of the Y-distribution and -40 from X because it was near the middle of the X-distribution. This gives us well-centered 0's. Then we divided the new Y's by 5 and the new X's by 10 to get unit steps between the values. The relations between the raw and the coded values are as follows:

$$y = \frac{Y - 115}{5}, \qquad x = \frac{X - (-40)}{10}.$$

When we finish the arithmetic working with x and y, we return to the original units.

Note that entries in the fy column are the product of the entries in the previous two columns for f and y. Similarly the entries in the fy^2 column are the product of the entries in the two previous columns for y and fy.

Let us look now at a cell, say the one for $X = -50$ $(x = -1)$ and $Y = 105$ $(y = -2)$. The numerical and symbolic layouts for the cell are:

	2		xy
numerical: **3**		symbolic: ***f***	.
	6		fxy

The boldface **3** is the cell frequency, f. The number 2 in the upper right corner is xy, the product of the coded values, for the cell $(-1)(-2) = +2$. The lower right-hand number 6 is fxy, the contribution of the cell to the sum of coded cross products, here $3(2) = 6$. We multiply f by xy to get fxy.

Let y' be the regression estimate of y from x, and let x' be the regression estimate of x from y. The regression of y on x is

$$y' - \bar{y} = r\frac{s_y}{s_x}(x - \bar{x}).$$

Substituting the statistics at the bottom of Table 11-9 into the equation gives

$$y' - (-0.20) = -0.266\left(\frac{1.37}{1.79}\right)(x - 0.14),$$

which can be simplified to

$$y' = -0.172 - 0.20x.$$

Replacing the coded values by the original variables gives

$$\frac{Y' - 115}{5} = -0.172 - 0.20\left(\frac{X + 40}{10}\right),$$

or

$$Y' = 110.1 - 0.1X.$$

You are asked to show similarly, in Exercise 16, that

$$X' = 40.6 - 0.70Y.$$

Table 11–9 Extreme temperatures in the 50 states; computational layout

X: lowest temperature recorded (to nearest multiple of 10)

Y: highest temperature recorded (to nearest multiple of 5)	−80	−70	−60	−50	−40	−30	−20	−10	0	10	20	Total f	Coded y	fy	fy^2
135					0 / **1** / 0							**1**	4	4	16
130												**0**	3	0	0
125					0 / **1** / 0							**1**	2	2	4
120		−3 / **1** / −3	−2 / **4** / −8	−1 / **4** / −4	0 / **3** / 0	1 / **3** / 3	2 / **1** / 2					**16**	1	16	16
115			0 / **2** / 0	0 / **3** / 0	0 / **1** / 0	0 / **3** / 0	0 / **2** / 0					**11**	0	0	0
110				1 / **2** / 2	0 / **3** / 0	−1 / **3** / −3	−2 / **3** / −6	−3 / **1** / −3	−4 / **1** / −4			**13**	−1	−13	13
105				2 / **3** / 6		−2 / **2** / −4						**5**	−2	−10	20
100	12 / **1** / 12						−6 / **1** / −6				−18 / **1** / −18	**3**	−3	−9	27
Total f	**1**	**1**	**6**	**12**	**9**	**11**	**7**	**1**	**1**	**0**	**1**	**50** $= n$		$-10 = \sum fy$	$96 = \sum fy^2$
Coded x	−4	−3	−2	−1	0	1	2	3	4	5	6		$\sum fxy = -34$		
fx	−4	−3	−12	−12	0	11	14	3	4	0	6		$7 = \sum fx$		
fx^2	16	9	24	12	0	11	28	9	16	0	36		$161 = \sum fx^2$		

$\bar{x} = \sum fx/n = 7/50 = 0.14$; $\bar{y} = \sum fy/n = -10/50 = -0.20$;
$s_x^2 = \sum fx^2/n - \bar{x}^2 = (161/50) - (0.14)^2 = 3.20$; $s_x = 1.79$;
$s_y^2 = \sum fy^2/n - \bar{y}^2 = (96/50) - (-0.20)^2 = 1.88$; $s_y = 1.37$;
$\text{Cov}(x, y) = \sum fxy/n - \bar{x}\bar{y} = (-34/50) - (0.14)(-0.20) = -0.652$;
$r = \text{Cov}(x, y)/s_x s_y = -0.652/(1.79)(1.37) = -0.266$.

The low correlation shows us that we are not good at predicting one extreme temperature from the other. The 3 outliers in Table 11-8 are Alaska (lower left), Hawaii (lower right), and California (top middle, caused by Death Valley). If these were removed, we would still have a negative correlation which emphasizes that some states are more *temperate* than others, rather than that some states are warmer than others, which is also true. Working with all-time extremes is not the same as working with averages of yearly extremes, or average January lows and July highs. We must expect poorer predictions from these unstable extremes.

You may wonder whether one would compute a regression with such a low correlation. While low correlations may be disappointing, they still may be the best we have, and a small correction may be better than none.

EXERCISES FOR SECTION 11-6

1. Toss 2 coins independently once. If the first coin falls heads, let $x = 1$; if the first coin falls tails, let $x = 0$. The random variable Y is the total number of heads in the 2 tosses.

   Discussion. We have the following table of joint probabilities:

$y \backslash x$	0	1	Totals
0	$\frac{1}{4}$	0	$\frac{1}{4}$
1	$\frac{1}{4}$	$\frac{1}{4}$	$\frac{1}{2}$
2	0	$\frac{1}{4}$	$\frac{1}{4}$
Totals	$\frac{1}{2}$	$\frac{1}{2}$	1

   (a) Compute $E(Y|x)$ for each value of x. (b) Fit the theoretical regression of y on x.

2. Use the tabulations in Exercise 1 to obtain the regression of x on y.
3. Repeat the experiment in Exercise 1, but toss 3 coins independently. As before, let $x = 1$ if the first coin falls heads, $x = 0$ if the first coin falls tails, and Y = total number of heads. (a) Construct the table of joint probabilities for x and y. (b) Compute $E(Y|x)$. (c) Fit the regression of y on x.
4. Use the data in Exercise 3 to obtain the regression of x on y.
5. An engineer's ruler of triangular cross section has faces numbered respectively, 1, 2, 3. Imagine rolling two such rulers, one after another, on the floor. Let X = number on the bottom face of the first ruler and Y = sum of the numbers on the two bottom faces. (a) Compute $E(Y|x)$ for each x. (b) Fit the regression of y on x.

6. Use the data in Exercise 5 to fit the regression of x on y.
7. Metal rods of length x units are inspected for flaws. For a given length x, the fraction with each number of flaws, y, is shown in the following table:

		x, length of rods		
		1	2	3
	0	.6	.3	.2
Y,	1	.3	.5	.3
number of	2	.1	.1	.3
flaws	3	0	.1	.2

(a) Compute $E(Y|x)$ for each x. (b) Fit the regression of y on x.

8. Four coins are tossed and the value of the number of heads x is recorded. Then the coins that showed tails are retossed and the value of the number of total heads Y is recorded. For each x complete $E(Y|x)$, and then find a prediction equation for estimating Y from x.
9. In Exercise 8, the initial number of heads X is a random variable. For each value of Y, compute $E(X|y)$, and find the prediction equation for X from y.
10. Let x be the number of the trial, and let R be a random digit. Let

$$Y = x + R.$$

This is an example of a control-knob model for regression. Show that

$$E(Y|x) = x + 4.5.$$

11. (Continuation.) Refer to Exercise 10 and use the last 5 random digits on the top line of your random-number table to get values of R for trials 0, 1, 2, 3, 4, and find the value of Y for each trial. Now fit the resulting (x, y)-data with a general least-squares line, and compare the slope and y-intercept with those of the theoretical line in Exercise 10.
12. (Continuation.) In Exercise 10 let $Y = x + \overline{R}$, where $\overline{R}$ is the average of 5 random digits. Use the averages of the last 5 digits in the first 5 rows of your random-number tables to get five (x, y)-pairs, and fit the general least-squares line. Compare the slope and y-intercept of the fitted line with those of $y = x + 4.5$.
13. (Continuation.) For Exercise 12 compute s_d and compare it with the theoretical value $\sigma = \sqrt{\frac{1}{5} \cdot \frac{99}{12}} \approx 1.3$.

Note. You may wonder why we do exercises like 11, 12, and 13, when we already know the theoretical regression from Exercise 10. The answer is that we need to gain experience in relating data to a model, and that experience only comes when we know both data results and the model. In practice, it is rare that we know the model and its constants exactly. If we only fit data and compare data results with fitted lines, we may get an exaggerated view of how well the model is being reproduced by the data.

14. The following data are derived by grouping the statures in Table 11–7 by 5-inch intervals.

		x				
		60	65	70	75	Total
y	62	10	49	7	0	66
	67	22	332	262	6	622
	72	0	88	257	25	370
	77	0	0	13	7	20
	Total	32	469	539	38	1078

Use this table to compute $\bar{x}$, $\bar{y}$, s_x, s_y, $\sum xy$, and from these obtain the regression of y on x and the regression of x on y. In calculating, you will find it convenient to replace x by $x' = (x - 70)/5$ and y by $y' = (y - 67)/5$, and then make appropriate adjustments at the end. Compare your equations with those shown in the text.

15. Three cards are numbered 1, 2, and 3. One card is drawn at random from the three and its value is recorded as x, then a second card is drawn at random from the two remaining cards and its value y is recorded. For each x (= 1, 2, 3) find $E(Y|x)$, and fit the regression of y on x to the resulting means.

16. For Example 5 show that the regression of X on Y is

$$X' = 40.6 - 0.70Y.$$

17. Delete the 3 outlying points from Table 11–9 and compute the new correlation.

12 Statistical inference for measured variables

12–1 SIMILARITIES TO OUR PREVIOUS WORK

In Chapter 9, we introduced both confidence limits and testing of hypotheses for binomial probabilities and their differences. That exposition illustrates the general ideas of statistical inference that are applicable to many other problems. Of course, new problems require technical variations to cope with the changes in distributions that arise. In this chapter, we shall treat these same two methods of statistical inference as they apply to means and to differences of means for observations expressed as measurements. The new material has many ties to the work of Chapter 9.

In Chapter 9, we exploited the properties of the binomial distribution for small samples, of the hypergeometric distribution for 2×2 tables having few cases, and in large samples we exploited approximate normality, both of a single observed proportion and of differences in observed proportions. Here we exploit the theorems of Sections 10–1, 10–2, and 10–3 for similar purposes as they apply to means.

Let us compare some examples. In Chapter 9, we used a sample to set confidence limits on the fraction of defective piece parts being manufactured by a process, and we statistically compared the proportions of patients recovering when two kinds of bandages were used. Here we set confidence limits on the mean diameter of a mass-produced piece part, based on the diameters of a small sample of these parts. We look at the difference in lactose tolerance of people from two racial groups, where the tolerance is measured by the peak amount of certain substances found in the blood. We want to see whether the races actually differ on this characteristic.

When we move from observed binomial proportions to sample means, the change in plot is slight because so many features are identical. For example, $\overline{p}$ is an observed mean of a set of 0's and 1's obtained from a sequence of binomial trials, and so we have already been making inferences about means. The normality that simplified the treatment of large-sample problems is employed here too, because we shall usually suppose that our sample means are approximately normally distributed. Just as we wanted to set confidence limits on the unknown value of p, the probability of success on a single trial, we want in this chapter to set confidence limits on μ, the mean of a population. Similarly, whereas we wanted to test whether p could have some stated value, here we want to know whether the data are compatible with a stated value for μ. Where we discuss differences of proportions in Chapter 9, here we discuss differences of means.

Just as we had a special small-sample treatment in the hypergeometric approach of Chapter 9, we shall need a special small-sample approach to means and differences of means based upon a distribution called the t-distribution. We introduce it on an *ad hoc* basis. It is one of the three mathematical devices in this book that we have introduced, for the purpose of applications, without derivation. We shall try to make it understandable.

Two common ideas run throughout the large-sample approach to confidence limits and testing hypotheses. First, we have strong confidence that the interval

$$\text{observed statistic} \pm \text{multiple of standard deviation of the statistic}$$

contains the true value of the quantity being estimated when the multiplier exceeds 2. For example, $\bar{x} \pm 3\sigma_{\bar{X}}$ likely contains μ_X. Second, and nearly equivalently, the ratio

$$\frac{\text{observed statistic} - \text{hypothetical mean}}{\text{standard deviation of the statistic}}$$

should produce a number near zero when the hypothetical mean equals the true mean. When the hypothetical mean departs from the true mean, the ratio may depart substantially from zero. When we say "near zero" we think of a range of numbers that runs, say, from -2 to $+2$, or to be rather extreme, from -3 to $+3$. When ratios fall outside such intervals, we are inclined to believe that the hypothetical mean is not the true mean. Whoever grasps these two ideas has mastered the fundamental devices behind a whole host of statistical techniques, even though he may not have their details in hand. For example, he may not know how to estimate the appropriate variance. He may not know what would be a sensible null hypothesis in a particular problem. But he would have the idea that we often measure the distance between a hypothetical and an observed value in terms of the variability of the statistic.

In this chapter, we are obliged to give a whole sequence of methods tailored to specific problems, but they all use one of the two ideas mentioned. Special difficulties arise when we are hard pressed to get a measure of variability, and then the variety of possibilities lengthens the treatment. The reader may prefer not to study all of these different possibilities at this time, but rather to observe now what cases are treated and to refer to them when he has a practical need for a specific technique. The cases treated here are the ones most frequently occurring in practice.

The reader may especially wish to attend to the changeover from the normal distribution to the t-distribution when the population variances are not known.

Understanding the relation between the normal distribution and the t-distribution, and the origin of the t-distribution, is important for practical work. To avoid having only a cookbook knowledge of the material, one must understand what gives rise to the t-distribution and how one could get estimates of its significance levels by simulation (or by other methods). If he understands this, then he can still handle many new problems where the mathematics may be out of sight. Whereas if he knows only the formulas presented here, he is limited to solving the problems of this chapter.

12-2 CONFIDENCE LIMITS ON OR TESTS OF μ, KNOWN σ

The normality of the distribution of averages of several values, guaranteed by the central limit theorem, and our theorems on the mean and variance of differences of random variables solve many important statistical problems. We use these facts to set confidence limits on population means and on differences between population means, by applying them to data from independent samples. We also test hypotheses about such means and differences. The methods are easily explained by illustrative examples. We shall use two-sided confidence limits and tests throughout. If he needs to, the reader can readily extend these ideas to one-sided limits or tests.

Example 1 *Confidence limits.* Machine setup. An operator of a machine tool sets his machine to make rods of a particular diameter. Different rods produced by this machine have different diameters; they have standard deviation $\sigma = .0020$ inch about their mean μ, and the diameters are approximately normally and independently distributed. The first 9 pieces made after a new machine setup have a sample mean $\bar{x} = 1.2000$ inches. Set 95% confidence limits on the population mean diameter for this setup.

Solution We proceed much as we did in using $\bar{p}$ to set confidence limits on p in Section 9-3. Let $\overline{X}$ be the random variable which is the mean diameter of the sample. Since the items are independent and the measurements are approximately normally distributed, we know from our work in the first three sections of Chapter 10 that $\overline{X}$ is also approximately normally distributed. Then the random variable $Z = (\overline{X} - \mu)/\sigma_{\overline{X}}$ is distributed approximately according to the standard normal and so the probability is .95 that

$$-1.96 \leq Z \leq 1.96 \qquad \text{or} \qquad -1.96 \leq \frac{\overline{X} - \mu}{\sigma_{\overline{X}}} \leq 1.96.$$

We rearrange this inequality, just as we rearranged similar inequalities in Section 9-3, to get upper and lower limits for μ and we find

$$\overline{X} + 1.96\sigma_{\overline{X}} \geq \mu \geq \overline{X} - 1.96\sigma_{\overline{X}}. \qquad (1)$$

You are asked to derive Eq. (1) in Exercise 6.

When we evaluate this result for our particular sample, we get, after recalling that $\sigma_{\overline{X}} = \sigma/\sqrt{n} = .002/3 \approx .00067$,

$$1.2013 \geq \mu \geq 1.1987. \tag{2}$$

This gives the operator a good notion of where his setting actually is. By choosing multipliers other than 1.96, one can vary the confidence level.

The approximate normality assumed for single measurements in this example was not much needed, since the ideas of the central limit theorem promise approximate normality for $\overline{X}$ in most practical cases, and we know the variance. However, the approximate normality for single measurements usually holds in this kind of measurement problem, and we use it later in examples where σ is unknown.

Example 2 *Testing hypotheses.* Machine setup. The operator wanted to set the machine so that $\mu = 1.2008$. Use the data from Example 1 to make a symmetric, two-sided, 5% significance level test of the hypothesis that this is the actual setting.

Solution 1 Since 1.2008 falls inside the 95% confidence interval 1.1987 to 1.2013, we cannot reject the hypothesis that $\mu = 1.2008$ at the 5% level. Nor could we reject any other value of μ in the interval (2) at this level of significance.

Solution 2 The second method of testing is to compute $(\bar{x} - \mu)/\sigma_{\overline{X}}$ and see if it falls between -1.96 and $+1.96$. With $\bar{x} = 1.2000$ and $\mu = 1.2008$, we get

$$(1.2000 - 1.2008)/.00067 = -1.2.$$

Since -1.2 is inside the acceptance region, we accept the null hypothesis.

Remark These two solutions correspond to the two general methods described in Section 12-1. Note that the first solution is a bit more informative because it more readily tells what other values of μ are accepted by the test.

The assumption of known variance In Examples 1 and 2, although we did not know μ, we claimed to know σ.

One might suppose that if the mean μ of a distribution is not known, then its standard deviation σ would not be known either; but, in surprisingly many real-life problems, we may know σ accurately enough for it to be treated as known. By known, we mean in practice that we have an excellent approximation. One reason is that many investigations are made in the presence of a large body of previous experience. In some problems σ is more stable than μ from one population to another, but in others μ and σ are closely related. Hence one must be cautious.

Examples a) *Routine chemical analyses.* Where the same chemical analysis is done routinely and frequently, as in a hospital laboratory, the analysts soon learn the variability of the various analyses—that is, they know what the standard deviation of the measurements is when the same material is analyzed several times.

b) *Annual averages.* We may study changes in a population from one year to the next (say weight in a live adult human population), and we may have a census of last year's population. We plan to sample this year's population looking for a slight change in the mean, but we expect the distribution to have almost the same shape as before. Therefore last year's σ is close enough to the true value for computing $\sigma_{\overline{X}}$.

c) *Standardized tests.* In standardized achievement tests such as College Board examinations, the standard deviations of the scores do not vary much from one part of the country to another, even though the mean does vary a good deal from school to school and region to region. This just seems to be an empirical fact. It may be related to facts in Example (d).

d) *Insensitivity of σ to μ.* Some probability distributions such as the binomial have the feature that the standard deviation is fixed by the mean and depends little on it over a broad range. For the binomial we have $\mu = np$ and $\sigma = \sqrt{np(1-p)}$. If $n = 16$, then for $0.25 \leq p \leq 0.75$, we find that $4 \leq \mu \leq 12$ while $1.73 \leq \sigma \leq 2$. More generally, the key quantity in the standard deviation, $\sqrt{p(1-p)}$, ranges only from 0.433 to 0.500 for $0.25 \leq p \leq 0.75$,

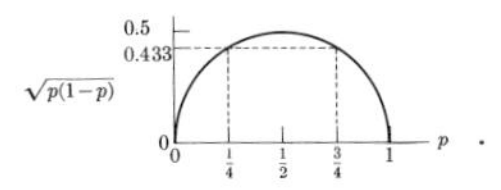

e) *Use of bound.* In some problems, as in the case of the binomial distribution, a useful bound on σ can be obtained. This bound may then be used to give an approximation for a large range of problems. In the binomial case, $\sigma \leq \sqrt{n}/2$ and σ is close to $\sqrt{n}/2$ for a substantial range of values of p.

EXERCISES FOR SECTION 12-2

1. College Entrance Examination Board scores for some tests are approximately normally distributed with standard deviation $\sigma = 100$, and for the whole country the mean is 500. In a random sample of size 25 from the seniors in a particular large school district, the average score was observed to be $\bar{x} = 550$. Set 95% confidence limits on the true mean score for this school district, and use them to decide whether the district's scholars are stronger in the material tested than the national average.

2. *Machine setup.* In the machine setup example, suppose that $\sigma = 0.0030$, $n = 9$, $\bar{x} = 1.2000$. Test the hypothesis $\mu = 1.202$ using the 5% significance level. [*Hint:* Set a 95% confidence interval and see if it includes 1.202.]

3. *One-sided test.* Use the data of Exercise 2 for the machine setup to test the hypothesis that $\mu < 1.202$ against the null hypotheses $\mu \geq 1.202$, using a one-sided 5% significance test.

4. *Power.* Unknown to an investigator, the true mean μ of a normal distribution is 2, but the investigator does know that the standard deviation is 4. If he draws a sample of size $n = 4$ and uses it to set 95% confidence limits on μ, what is the probability that the confidence interval does not include 0? In other words, in repeatedly taking samples of size 4 and setting limits $\bar{x} \pm (1.96)2$ for μ, in what fraction of samples does 0 lie outside the interval?
5. (Continuation.) Find the required probability in Exercise 4 when the sample size is replaced by $n = 16$, and by $n = 64$. Use the data from Exercise 4 together with the new data to draw a curve relating the probability of rejecting the null hypothesis of 0 mean to the sample size. [*Suggestion:* Plot the power against $1/\sqrt{n}$. Note that this is not a power function, but a plot of power against a function of sample size.]
6. Show the steps leading from

$$-1.96 \leq (\overline{X} - \mu)/\sigma_{\overline{X}} \leq 1.96$$

to formula (1) of Example 1.

12–3 CONFIDENCE LIMITS ON OR TESTS OF μ, UNKNOWN σ

What if σ is unknown? If we have approximately normally distributed data and σ is unknown, then in $(\overline{X} - \mu)/\sigma_{\overline{X}}$ it is natural to try to replace $\sigma_{\overline{X}}$ by an estimate of it. One such estimate can be obtained by recalling, first, that $\sigma^2_{\overline{X}} = \sigma^2/n$, where n is the sample size. Second, in Section 6–2, Remark 2, we stated that $\sum(x_i - \bar{x})^2/(n - 1)$ on the average is σ^2 for distributions that have a variance. This implies that $\sum(x_i - \bar{x})^2/(n - 1)$ is an unbiased estimate of σ^2. You are asked to prove this fact in Exercise 4. Using this unbiased estimate for σ^2, we get

$$\text{estimate of } \sigma^2_{\overline{X}} = \frac{\text{estimate of } \sigma^2}{n} = \frac{\sum(x_i - \bar{x})^2}{n(n-1)} = s^2_{\overline{X}}. \tag{1}$$

The extra n in the denominators of the middle two expressions is the "n" of "σ^2/n". We use the notation $s^2_{\overline{X}}$ to parallel $\sigma^2_{\overline{X}}$: the s reminds us that we are dealing with an estimate from the data rather than with a population value, and the $\overline{X}$ reminds us that we are measuring variability of a sample mean rather than of a single measurement.

Our reasoning might be stated thus: for *large* samples $s^2_{\overline{X}}$ is almost sure to be close to $\sigma^2_{\overline{X}}$ and so $z = (\bar{x} - \mu)/\sigma_{\overline{X}}$ and $t = (\bar{x} - \mu)/s_{\overline{X}}$ are likely to have nearly the same value. Let us first see how close we estimate σ using s_X (without worrying about the extra $\sqrt{n}$ of $s_{\overline{X}}$). To give an idea both of the closeness and the variation, we computed s_X in two independent samples of 100 each from the standard normal. We got $s_X = 1.024$ and $s_X = 0.930$, both estimates fairly close to the true value $1(=\sigma)$. Such reasoning is correct as far as it goes, but the issue is this: What is the distribution of

$$T = \frac{\overline{X} - \mu}{s_{\overline{X}}} \tag{2}$$

in *small* samples drawn from the normal distribution? We use a capital letter T here to emphasize that T is a random variable, whereas t is a value of it. (Usually people use t for both the random variable and its value; they are forced in this direction because T is usually reserved for a generalization of the t statistic.) We have no reason to suppose that T is normally distributed, and indeed it is not. Each sample size starting with $n = 2$ produces a different distribution. The symmetry of the normal distribution of the single measurements encourages us to believe that T is symmetrically distributed about 0, and it is. Also, since both the numerator $\overline{X} - \mu$ and the denominator $s_{\overline{X}}$ are essentially measures of distance, we might hope that when the original measurements come from a normal distribution the distribution of T would not depend on the σ of the original normal because the ratio would be unitless, while σ would have units. That happens to be true, as the starred material immediately below shows.

★To change observations from one normal distribution to another, we need only make a linear transformation. If X_i is distributed according to the standard normal distribution, then $aX_i + b$, $a \neq 0$, has mean b and standard deviation $|a|$. If we make this same linear transformation on all the X_i's, $i = 1, 2, \ldots, n$, then the numerator of T becomes $a\overline{X} + b - (a\mu + b) = a(\overline{X} - \mu)$. From Section 6–2, Exercise 16, we also know that this linear transformation multiplies an observed standard deviation by $|a|$. And so T is transformed into $a(\overline{X} - \mu)/|a|s_{\overline{X}} = \pm T$, the sign depending on the sign of a. When a is positive, the new distribution must then be the same as the old; when a is negative, we have the distribution of $-T$, which is identical with that of T because of symmetry. This shows that changing the mean and variance of the observations leaves the distribution of T unchanged as long as the original measurements come from a symmetrical distribution.

Thus even though T has a different distribution for each sample size, these distributions do not depend upon μ or σ. This happy feature makes it easy to proceed to get confidence limits or tests of significance as soon as we have probability tables of these distributions like Table III, which we use for the standard normal. We give such probability levels in Table V.

These distributions are called Student t-distributions, or just t-distributions. We would require calculus methods to derive them. "Student" was the scientific pseudonym of W. S. Gossett, a statistician and chemist who worked for a brewing company. He figured out the formula for the t-distributions by a combination of (1) mathematics, (2) what we call the Monte Carlo method (essentially computing t for many random samples of a given size and analyzing their empirical distributions), and (3) a lucky guess. His methods would have had to give at least a good approximation, but R. A. Fisher later proved that the guess was exactly correct. Student didn't have numbers drawn randomly from a true normal distribution as we have today, but he used instead a great many measurements of lengths of criminals' left middle fingers (and also criminals' heights) obtained from the police files of the time. These lengths were approximately normally distributed. He drew samples from these. Today we could use random normal deviates, which are numbers drawn at random from the standard normal distribution.

Random normal deviates To see some random normal deviates, look at the entries in Table I–B at the back of the book. These are 200 measurements drawn from the standard normal distribution. They can be used to make samples from the normal distribution. Calculations based on these samples could be used to get information about the distribution of statistics derived from samples of normally distributed measurements.

Example 1 *t-distribution;* $n = 2$. Use the table of random normal deviates to get estimates of the probability levels of the t-distribution for samples of size 2 from the normal distribution.

Solution With our 200 measurements we can take 100 independent samples of size 2, compute t for each, and get the observed distribution, shown in Table 12–1. To get an estimate of the central interval containing probability 0.50, we count up 25 numbers starting with the least and count down 25 numbers starting with the largest, and get −1.034 and 0.977. Because of symmetry we average the two absolute values, and estimate 1.01 as the value of t such that the probability is 0.50 between $-t$ and t. Table V shows that 1.00 is the correct answer. If we used 1.01, finer tables than ours show that we would be using probability 0.503 instead of 0.500.

In the same manner we can get estimates for the value of t giving centered probabilities 0.80 and 0.90. For 0.95 our sample is none too large, but we might average the second and third observations at each end and get 9.37. This number is not very close to the correct 12.71 in distance on the t-axis, though

Table 12–1 100 values of t computed from samples of 2 random normal deviates and ordered from least to greatest

−13.6316	−1.2400	−0.3392	0.3991	1.3505
−9.1000	−1.2353	−0.2850	0.4589	1.3810
−8.4545	−1.2099	−0.2837	0.4687	1.4276
−5.8649	−1.0952	−0.2692	0.5359	1.5000
−5.6286	−1.0339	−0.2000	0.5862	1.5185
−5.3750	−1.0315	−0.1633	0.6598	1.6364
−5.3333	−0.9667	−0.0711	0.6774	1.6739
−4.2453	−0.8039	−0.0588	0.6923	1.6755
−3.5333	−0.7987	−0.0145	0.7368	2.0294
−3.4286	−0.7538	−0.0071	0.8155	2.0333
−3.1538	−0.7525	0.0219	0.8173	2.1692
−2.9130	−0.7059	0.0380	0.8301	2.2708
−2.3455	−0.6238	0.0667	0.8605	2.3750
−1.8000	−0.6078	0.1546	0.9055	2.5278
−1.6512	−0.6000	0.1769	0.9176	4.6709
−1.6462	−0.5232	0.2287	0.9773	4.6818
−1.6154	−0.4222	0.2466	1.0488	5.8485
−1.4593	−0.4173	0.3143	1.1320	9.3125
−1.4545	−0.3974	0.3492	1.1322	10.6000
−1.2642	−0.3684	0.3968	1.3256	11.5455

Table 12–2 Summary of Monte Carlo calculations for t, $n = 2$

True value of t	Estimated value of t	Required value of 2-sided probability	Value of probability given by estimated t
1.00	1.01	.50	0.503
3.08	2.80	.80	0.782
6.31	5.16	.90	0.878
12.71	9.37	.95	0.932

the probability is slightly larger than 0.93, and so we are close to our desired goal in probability, even if our t-value is poor. To estimate small tail probabilities accurately we need large samples. The results of our Monte Carlo effort are summarized in Table 12–2. Thus our results closely approximate those of Table V. In a serious research effort, by getting more random normal deviates, say from the RAND tables, and computing more t's, we could get better approximations. We could also extend the method to other values of n. Since Fisher has derived these distributions and others have computed their probabilities as shown in Table V, we do not have to do this Monte Carlo study. But it is comforting to know that we could and that we could solve new problems in this way, as Example 2 illustrates.

★**Example 2** *Sample median.* Is the sample median a better estimate of the population mean for a normal distribution than the sample mean in samples of size $n = 3$?

Solution Symmetry shows that both sample medians and sample means average to μ for the general normal distribution. The question then is not one of bias but of variability. Although we do not present the mathematics appropriate for this problem, we can draw many samples of 3, locate the median (middle number) from each, and compute the variance of these medians. We drew 50 samples of 3 from Table I–B, located the median, $\widetilde{X}$, in each, and computed the sample variance $s^2_{\widetilde{X}} = 0.427$. Since the standard normal has variance $\sigma^2 = 1$, had we sampled from the general normal with arbitrary variance σ^2, our estimate of the variance of the median would have been $0.427\sigma^2$. Since the sample mean has variance $\frac{1}{3}\sigma^2$, and since $0.427 > \frac{1}{3}$, it appears that as far as variances are concerned the sample mean $\overline{X}$ is a better estimator of μ for the normal distribution based on samples of 3 than is the sample median.

The normality in this example matters. For some distributions having higher tails than the normal distribution, the sample median would have performed better than the sample mean.

Figure 12–1 shows examples of Student's t-distribution for samples of size $n = 2$, 3, and infinity, the latter being the standard normal distribution. As

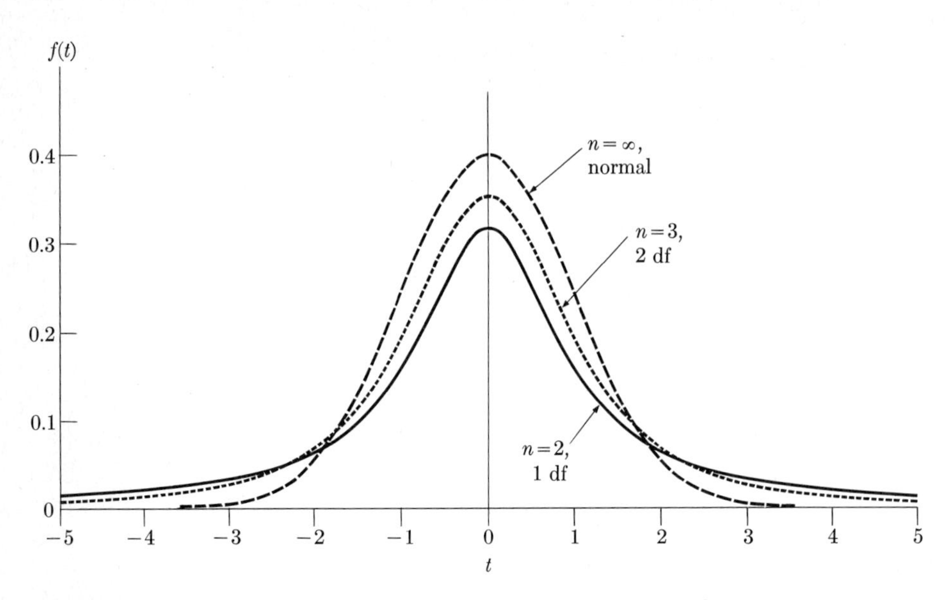

Fig. 12–1 Collection of t-distributions (df = degrees of freedom, see p. 441).

n grows, $s_{\overline{X}}$ is more likely to be very close to $\sigma_{\overline{X}}$, and then the sample value of t becomes indistinguishable from z. Table 12–3 shows that the multiplier 1.96 would have to be replaced by 12.71 for a sample of size 2, essentially multiplying the length of the confidence intervals by 6. When the sample size is as large as 30, the t-value is 2.05 instead of 1.96, and for many purposes both of these values can be comfortably rounded to 2.00. Thus one sometimes hears of "large" samples starting at 30, as if all positive numbers less than 30 were small and the rest were large. This is ridiculous. Of course, the t-distributions apply for each n, and the normal is the limit, as n grows large; but the normal is a very good approximation by the time $n = 30$.

Example 3 *Confidence limits for μ, unknown σ.* In the machine setup example, suppose that $\bar{x} = 1.2000$ as before and that the observed value of $s_{\overline{X}}$ in a sample of 10 was 0.0008. Use these data to set 95% confidence limits on μ.

Solution From Table 12–3 or from Table V at the back of the book, we find that the probability is 0.95 that $-2.26 \le (\overline{X} - \mu)/s_{\overline{X}} \le 2.26$. Solving the

Table 12–3 Values obtained from the t-distributions to replace the multiplier 1.96 used with the normal distribution for setting 95% confidence limits on μ with σ unknown and sample size n

n	2	3	4	5	10	30	100	1000	Normal ∞
t	12.71	4.30	3.18	2.78	2.26	2.05	1.98	1.96	1.96

inequality for upper and lower confidence limits on μ, we get $\overline{X} \pm 2.26 s_{\overline{X}}$. Consequently the numerical limits are $1.2000 \pm 2.26(.0008)$, which gives an upper limit of 1.2018 and a lower limit of 1.1982.

Generalization To set symmetric confidence limits on the mean μ of a normal distribution based on the measurements in a sample of size n, we can use

$$\bar{x} \pm (\text{multiplier})\ s_{\overline{X}},$$

where the multiplier read from Table V depends upon the confidence level chosen and the sample size.

Matched individuals Sometimes we have matched data such that a 2-sample problem can be treated like a one-sample problem. This occurs when the same individual or process is measured twice, or when individuals have been paired as in twin studies to improve control. We illustrated this in our example on methods of memorizing in Section 9–5.

Example 4 *Difference between two means, matched samples.* Random samples of adult males requiring treatment for obesity are assigned to various treatments. Those 10 assigned to one treatment obtained the results shown in Table 12–4. Let μ_B and μ_A be the population mean weights before and after the two-week session of reducing. Set 90% confidence limits on the population mean loss, $\mu_L = \mu_B - \mu_A$, on this regime.

Solution Some thought Mr. B didn't try, but we aren't measuring trying, only reducing. We take differences between the before and after weights to get the loss L for each person and thus treat one sample of 10 weight losses. Setting confidence limits on $\mu_B - \mu_A$ is then the same as setting them on μ_L, the population mean weight loss under this regime. The calculations are shown in Table 12–4.

Table 12–4 Weight changes after two weeks

Patient	Before	After	Loss, L	L^2
A	252	244	8	64
B	274	274	0	0
C	293	288	5	25
D	261	251	10	100
E	243	237	6	36
F	272	267	5	25
G	282	275	7	49
H	248	244	4	16
I	277	272	5	25
J	257	253	4	16
		Total	54	356

$$\overline{L} = 5.4$$

$$\sum(L - \overline{L})^2 = \sum L^2 - n\overline{L}^2 = 356 - 10(5.4)^2 = 64.4$$

$$s_{\overline{L}}^2 = \frac{\sum(L - \overline{L})^2}{n(n-1)} = 64.4/10(9) = 0.716$$

$$s_{\overline{L}} = 0.846$$

From Table V at the back of the book we find that the two-sided symmetrical 90% limits for the t-distribution for a sample of 10 fall at ± 1.83,

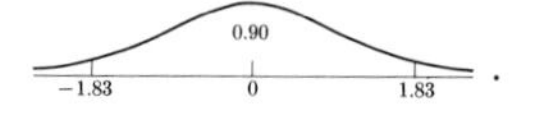

Therefore, since the mean observed loss is $\bar{L} = 5.4$ pounds and $s_{\bar{L}} = 0.846$, the 90% confidence limits are

$$\bar{L} \pm 1.83 s_{\bar{L}} = 5.4 \pm 1.83(0.846) = 3.85, \quad 6.95.$$

And so the mean two-week weight loss based on these data is not well determined by the sample of 10. We are fairly confident, say 90%, that it is between 4 and 7 pounds, and that it is not as small as 0 or as large as 10 pounds, for example.

Remark 1 *Percentage changes.* In problems like this, it is frequently useful to work with percentage changes rather than absolute ones because they may be better behaved. If we had had a 450-pound man, his weight loss might have swamped that of the other individuals, yet his percentage change might have been nearly in line with the others. Our men were all in the 240–300-pound range, and so using percentages would not change the results much.

Remark 2 *Wild observations.* A wild observation—one that is very large compared to the rest, which are near zero—sends the value of t toward unity when $\mu = 0$. To see this, suppose that we have $n - 1$ "small" observations with value 0 and one large value x. Then $\bar{x} = x/n$ and the sum of squares of deviations is

$$\sum x_i^2 - n\bar{x}^2 = x^2 - x^2/n = x^2(n-1)/n.$$

Substituting these values into t (when we assume $\mu = 0$) gives

$$t = \frac{\bar{x} - \mu}{s_{\bar{X}}} = \frac{x/n}{\sqrt{\dfrac{x^2(n-1)}{n^2(n-1)}}} = 1.$$

On the other hand, if $n - 1$ measurements have value u and 1 has value $u + x$, then

$$t = 1 + (nu/x).$$

Had all n values been close to u, so that x was near 0, then t would have become enormous; but when x is large, the value of t is again cut down toward unity.

A wild observation can come from a gross reading, copying or computing error, or from a sudden change in conditions, or it may be part of the natural process to have occasional extreme observations.

Remark 3 *Evidence for normality.* Unless one has a very large number of measurements or extreme evidence of nonnormality (such as a wild observation may give), one must usually depend upon outside information for the adequacy of the normal assumption. In the weight-change problem, we could worry about the 0 and the 10 as possibly including two extreme values. Their net effect, if they are outliers, is to increase the estimated variance and therefore lengthen our confidence limits.

Degrees of freedom The t-distributions are used for many problems, not just those involving one-sample confidence limits or tests. For this reason they are indexed 1, 2, 3, . . . , rather than by sample size 2, 3, 4, . . . , as would be natural if we were dealing only with one-sample problems. For reasons that we shall not go into just now, this index is called *the number of degrees of freedom*. Thus instead of speaking of the t-distribution for a sample of size 2, we speak of the t-distribution with 1 degree of freedom. More generally, for a sample of size n, we have a t-distribution with $n - 1$ degrees of freedom. There is no reason to be upset about subtracting one from each of a set of numbers and then giving the list a new name. The number of degrees of freedom is just a label for the number you use to choose a t-distribution.

To get a hint for the origin of the term, suppose for a moment that we know μ. Then we could use $\sum(X_i - \mu)^2/n$ as an unbiased estimate of σ^2. The point is not so much that the estimate is unbiased as that the n terms $X_i - \mu$ are all independent. This independence makes each term worth one full measurement for estimating σ^2. Knowing μ, we do not "lose any degrees of freedom" when we estimate σ^2. But when we have to replace μ by $\overline{X}$, then $\sum(X_i - \overline{X}) = 0$. This means that the differences are not independent: there is a linear relation among all these deviations. That linear relation, arising because μ is estimated and not known, costs a degree of freedom in estimating σ^2. When we have to estimate more than one mean, we lose more than one degree of freedom, as you will see later. These remarks are not a proof of anything, but they explain to some extent the origin of the name degrees of freedom. Note that if we have only one measurement, we cannot estimate σ^2 at all from our sums of squares if we know nothing of μ, because we have no degrees of freedom left after using our observation to estimate μ.

EXERCISES FOR SECTION 12-3

1. *Hummingbird speeds.* To measure the speed of flight of Blue Sylph hummingbirds, Walter Scheithauer timed a pair on figure-eight courtship flights through a large aviary, over a course of length 73.68 yards. The first 10 flight times in seconds were 3.6, 4.2, 3.2, 4.8, 4.0, 4.2, 3.3, 3.8, 5.1, 3.9. Set 80% confidence limits on the mean flight time over the course. [*Ans:* 3.75, 4.27 in seconds.]
2. Use the answer given in Exercise 1 to set 80% confidence limits on the mean speed in miles per hour. What do you think of the conjecture that over a straight course instead of a figure-eight they might reach 55 miles per hour?
3. *Cloud seeding* (*K. R. Gabriel*). In an experiment intended to increase rain by seeding clouds, a northern region or a central region was seeded every day at random. The net gains in inches of rain owing to seeding in 6 time periods covering 5 years

were as follows:

Period	1	2	3	4	5	6
Total seeding gain	1.5	0.4	−0.1	−0.1	2.0	1.0

Assuming that the 6 periods are independent and have equal weights, test whether the net gain from seeding is positive against the null hypothesis of 0 gain from seeding.

4. Show that $\sum(x_i - \bar{x})^2/(n-1)$ is an unbiased estimate of σ^2. [*Hint:* Work on the numerator. Recall the computational form $\sum(x_i - \bar{x})^2 = \sum x_i^2 - n\bar{x}^2$. Note that $E(X_i^2) = \sigma^2 + \mu^2$ and $E(\overline{X}^2) = \sigma_{\overline{X}}^2 + \mu^2$. Now assemble the information.]
5. Show that in a sample (x_1, x_2) of size $n = 2$ drawn from a normal distribution with mean $\mu = 0$, the t-statistic $(\bar{x} - \mu)/s_{\overline{X}}$ reduces to $(x_1 + x_2)/|x_1 - x_2|$.
6. Use the random normal deviates of Table I–B to draw 19 samples of size 2 from the standard normal distribution. The result of Exercise 5 makes it readily possible to form 19 values drawn from the t-distribution with 1 degree of freedom. Find the value of $|t|$ that includes 9 measurements between $-t$ and t and has one measurement on one boundary. Compare your result with the appropriate value given in Table V. Treat a zero denominator as a very large quotient with the same sign as that of the numerator. Set 0/0 aside. (The authors got 1.12.)

12–4 DIFFERENCES BETWEEN MEANS

Confidence limits on the difference between population means based on two independent normal samples whose populations have known variances If $\overline{X}$ and $\overline{Y}$ are means of samples of sizes n_X and n_Y from normal populations with means μ_X and μ_Y and variances σ_X^2 and σ_Y^2, then the variance of the difference $\overline{X} - \overline{Y}$ is, from Corollary 10–3, for independent observations,

$$\sigma_{\overline{X}-\overline{Y}}^2 = \sigma_{\overline{X}}^2 + \sigma_{\overline{Y}}^2 = \frac{\sigma_X^2}{n_X} + \frac{\sigma_Y^2}{n_Y}. \tag{1}$$

We also use the fact that the distribution of the difference of two independent normally distributed variables is itself normally distributed (Section 10–3). To set 95% symmetrical confidence limits on $\mu_X - \mu_Y$, we compute

$$\bar{x} - \bar{y} \pm 1.96\sigma_{\overline{X}-\overline{Y}}. \tag{2}$$

Example 1 *Lactose tolerance.* To study comparative tolerance of adults of different races to lactose, physicians computed peak rise in blood-reducing substances for 20 Orientals and 20 Caucasians after they had been given a dose of lactose (essentially milk). Larger rises imply more tolerance. The sample means for Caucasians and Orientals are $\bar{x} = 40$ and $\bar{y} = 15$, respectively, and the variances are $\sigma_X^2 = 100$ and $\sigma_Y^2 = 25$. Set 95% confidence limits on the difference between the population means.

Solution We compute

$$\bar{x} - \bar{y} \pm 1.96\sqrt{\frac{\sigma_X^2}{n_X} + \frac{\sigma_Y^2}{n_Y}} = 40 - 15 \pm 1.96\sqrt{\frac{100}{20} + \frac{25}{20}}$$

$$= 25 \pm 1.96(\tfrac{5}{2}) \approx 20, \quad 30.$$

The result shows a substantial difference in tolerance between the two groups of people. Some adults are made quite ill. Further experiments show differences between Caucasians and other racial groups.

Confidence limits on the difference between means of two independent normal populations with unknown, but equal, standard deviations, small samples In the two-sample problem, when the population standard deviations are unknown, there are two main cases: (1) the standard deviations σ_X and σ_Y can be assumed to be equal or, as a practical matter, nearly equal; (2) we cannot assume them equal and know no other relation between them.

Consider Case 1. In Exercise 8, you are asked to prove that an unbiased estimate for the variance of the difference of means is

$$s_{\bar{X}-\bar{Y}}^2 = \frac{\sum(x_i - \bar{x})^2 + \sum(y_i - \bar{y})^2}{n_X + n_Y - 2}\left(\frac{1}{n_X} + \frac{1}{n_Y}\right), \tag{3}$$

and we enter the degrees-of-freedom column of the t-table, Table V, with $n_X + n_Y - 2$ degrees of freedom. We had to estimate μ_X and μ_Y, and so we lost 2 degrees of freedom for estimating the common value of σ^2.

Example 2 *Ambiguity and speech.* To find out if people with less information about a task talk more about it, Hayes, Meltzer, and Lundberg gave 5 groups a clear picture of a complicated Tinkertoy to be built, and 5 groups an ambiguous picture, smaller, and shaded. Each group had to tell a silent builder, who had no picture, how to make the object. The groups were scored for the amount of talking done. The groups with the clear picture had a mean speech score $\bar{x} = 22.8$, with $\sum(x - \bar{x})^2 = 707.6$; and the groups with the ambiguous picture had a mean speech score $\bar{y} = 40.8$, $\sum(y - \bar{y})^2 = 1281.6$. Assuming that the population variances of the speech scores of the two groups are equal, set 90% confidence limits on $\mu_X - \mu_Y$ and test the hypothesis mentioned.

Solution To get the estimated standard deviation of the difference, we substitute in Eq. (3) and get

$$s_{\bar{X}-\bar{Y}}^2 = \frac{707.6 + 1281.6}{8}\left(\frac{1}{5} + \frac{1}{5}\right) = 99.5, \qquad \text{or} \qquad s_{\bar{X}-\bar{Y}} = 9.97.$$

The 90% two-sided level for t with 8 degrees of freedom (Table V) is 1.86, [diagram: 0.90, $-t$, 0, t]. The confidence limits are

$$\bar{x} - \bar{y} \pm 1.86 s_{\bar{X}-\bar{Y}} = 22.8 - 40.8 \pm 1.86(9.97) = +0.6, \quad -36.6.$$

Thus the difference in means, although large and in the hypothesized direction, has quite wide confidence limits. Notice that 0 *is* contained in the confidence interval, and so we would not reject the hypothesis of equality at the 10% level of significance. Nevertheless, we see that the difference encourages us to accept the truth of the hypothesis.

Comments 1) In experiments on another set of subjects, these same investigators got means of 39.2 and 41.3 for the clear and ambiguous groups, respectively, and so the large difference seen in Example 2 was not replicated.

2) The more talking done on the average, the larger we would expect the variance to be; but if the means were actually equal, we should expect the population variances to be approximately equal too.

Unequal unknown standard deviations If both sample sizes are large when the standard deviations are unknown and believed not to be equal, we estimate $\sigma_{\bar{X}}^2$ and $\sigma_{\bar{Y}}^2$ and substitute into $\sqrt{\sigma_{\bar{X}}^2 + \sigma_{\bar{Y}}^2}$ to estimate $\sigma_{\bar{X}-\bar{Y}}$. We then use normal tables to get the multiplier for setting confidence limits or for testing hypotheses. But when sample sizes are small, careful approximations are quite complicated, and so we shall suggest a simple rough-and-ready approximation.

If the sample sizes are small, estimate $\sigma_{\bar{X}-\bar{Y}}^2$ from

$$s'^2_{\bar{X}-\bar{Y}} = \frac{\sum(x-\bar{x})^2}{n_X(n_X-1)} + \frac{\sum(y-\bar{y})^2}{n_Y(n_Y-1)} \tag{4}$$

and take the smaller of $n_X - 1$ and $n_Y - 1$ as the degrees of freedom for the multiplier. When $n_X = n_Y$, Eq. (4) reduces to Eq. (3), and so $s_{\bar{X}-\bar{Y}}^2 = s'^2_{\bar{X}-\bar{Y}}$. But our usage would still be conservative for testing because here we would use $n_X - 1$ as the degrees of freedom instead of $2(n_X - 1)$.

In Eq. (4) we use a prime on s to distinguish this formula from that of Eq. (3).

Example 3 *Rate of use of short words.* Mosteller and Wallace studied the distribution of word lengths in writings of Alexander Hamilton and James Madison. Table 12-5 shows the rate of use of one-letter words per 1000 words of text in

Table 12-5 Rate of use of one-letter words per 1000 words of text

Hamilton	Madison
24	20
21	27
23	19
24	30
33	11
28	17
28	27
37	
$\sum h = 218$	$\sum m = 151$

$$\bar{h} = 218/8 = 27.2, \qquad \bar{m} = 151/7 = 21.6$$

$$s'^2_{\bar{H}-\bar{M}} = \frac{\sum(h-\bar{h})^2}{n_h(n_h-1)} + \frac{\sum(m-\bar{m})^2}{n_m(n_m-1)}$$

$$= \frac{208}{8(7)} + \frac{272}{7(6)} = 10.19$$

$$s'_{\bar{H}-\bar{M}} = 3.19$$

8 essays by Hamilton and 7 by Madison. Set 95% confidence limits on the difference of the rates. Might their rates be equal?

Solution From the t-table (Table V) with 6 degrees of freedom, the multiplier is 2.45. The limits are

$$\bar{h} - \bar{m} \pm 2.45 s'_{\bar{H}-\bar{M}} = 27.2 - 21.6 \pm 2.45(3.19) = -2.2, \quad 13.4.$$

The upper and lower confidence limits for the rates fall on both sides of 0, and so equality of rates of use of one-letter words by the two authors is still a reasonable hypothesis.

EXERCISES FOR SECTION 12-4

1. *Handedness in baseball.* G. R. Lindsey studied the baseball contention that a batter does better when he and the pitcher are opposite-handed than when they are same-handed. He assembled many batting averages based on about 50 successive times at bat, as follows:

Batter and pitcher are:	Number of sets of 50	Grand batting average	Standard deviation of the sets of batting averages
X (opposite-handed)	103	0.263	0.06
Y (same-handed)	106	0.231	0.06

Set 95% confidence limits on the difference between the population batting averages and interpret.

2. *Neuron information.* In comparing the information transmitted by different types of neurons, Darian-Smith, Rowe, and Sessle found the data in Table 12-6. [Note that the sample standard errors ($s_{\bar{X}}$ and $s_{\bar{Y}}$) have been reported in the table rather than the sample standard deviations (s_X and s_Y).]

   Assuming that the two types of neurons have a common σ^2, test at the 5% level the difference in mean channel capacity of receptor afferents and nucleus oralis.

Table 12-6 Sensitivity and channel capacity of neurons

Type of neuron	Channel capacity (bits)		
	Mean	Standard error	n
Receptor afferents, x	2.40	0.06	19
Nucleus oralis, y	2.09	0.07	18
Nucleus caudalis	1.07	0.07	22

3. Use the data of Exercise 2 to test whether the channel capacity of receptor afferents was 0.25 bits larger on the average than that of nucleus oralis.

4. Three types of lobster pots produced the catches given in the following table (T. M. Prudden, after H. Thomas). Set 90% confidence limits on the mean catch of the Fine Mesh Cat Walk.

Catch number:	1	2	3	4	5	6	7	8	9	10
Standard	1	2	0	6	3	2	4	3	0	2
Fine Mesh	4	1	1	5	4	7	6	4	2	2
Non Escape	3	1	4	3	3	2	3	5	1	2

Catch number:	11	12	13	14	15	16	17	18	19
Standard	2	4	2	3	5	2	4	4	3
Fine Mesh	2	3	3	2	10	1	6	7	5
Non Escape	4	2	0	4	4	4	6	3	6

5. Use the data of Exercise 4 to set 95% confidence limits on the difference in the mean catches of the Standard and Fine Mesh pots, and thus test whether the Fine Mesh is a better trap.

6. The data of Exercise 4 are matched for order, but order may not be important. If you treated Exercise 5 as if the data were matched, redo it as if the samples were independent, and vice versa. Compare the conclusions.

7. *Rosier futures.* How much better is the future? Hadley Cantril surveyed samples of people in many countries, asking about what concerned them, how satisfied they were, and how satisfied they expected to be 5 years later. He rated their satisfactions on a scale from 0 to 10. People from all countries thought the future would be better. Table 12–7 compares the improvements in ratings expected in 3 industrialized countries with those expected in a number of not-so-industrialized countries. Set 90% confidence limits on the mean difference in the changes between the two kinds of countries.

Table 12–7

Industrialized, I		Nonindustrialized, N	
United States	1.2	Yugoslavia	1.7
West Germany	0.9	Philippines	1.8
Japan	1.0	Panama	2.2
		Nigeria	2.6
Total	3.1	Brazil	2.7
		Poland	1.1
		India	1.4
		Dominican Republic	4.2
		Total	17.7

$$\bar{\imath} = 3.1/3 = 1.03$$

$$\bar{n} = 17.7/8 = 2.21$$

$$s_{\bar{I}}^2 = \frac{0.0467}{3(2)} = .0078$$

$$s_{\bar{N}}^2 = \frac{6.669}{8(7)} = .1191$$

$$s_{\bar{I}-\bar{N}} = \sqrt{.1269} = 0.356$$

8. Using the fact that

$$\sum (x - \bar{x})^2/(n_X - 1)$$

and

$$\sum (y - \bar{y})^2/(n_Y - 1)$$

are unbiased estimates of σ_X^2 and σ_Y^2, respectively, show that if $\sigma_X^2 = \sigma_Y^2 = \sigma^2$, then

$$\frac{\sum(x-\bar{x})^2 + \sum(y-\bar{y})^2}{n_X + n_Y - 2}$$

is an unbiased estimate of σ^2. Therefore Eq. (3) gives an unbiased estimate of $\sigma^2_{\bar{X}-\bar{Y}}$.

MISCELLANEOUS EXERCISES FOR CHAPTER 12

1. In tossing precision-made dice, Willard Longcor recorded in blocks of 20,000 tosses the number of "evens" (2's, 4's, or 6's). He found the average number of evens to be $\bar{x} = 10{,}009$ for $n = 100$ such blocks, and the sample standard deviation of number of evens $s_X = 78.3$.
   a) Use the sample data to set 95% confidence limits on μ, the theoretical number of evens in 20,000 throws. Is there evidence that the dice are not well balanced?
   b) Use the theoretical variance to set 95% confidence limits on $\bar{x}$ and interpret. [In empirical investigations it is well to gather data in blocks so as to be able to get a measure of the variability of the process directly from real life, even when theory seems to be able to supply it. The reason is that the whole mathematical model may not apply perfectly to the particular process. By wisely gathering his data in blocks, Mr. Longcor can compare not only the theoretical mean with the observed mean, but also the theoretical standard deviation with the observed. As often happens, the observed variability is somewhat higher.]

Data for Exercises 2 and 3. The following times in seconds (W. E. Moore) were required to complete the first 10 subtraction and the first 10 multiplication problems done by a tenth-grader on a standard test. The problems are easy and the tenth-grader is expected to get them all correct.

	Serial order of problem									
Times	1	2	3	4	5	6	7	8	9	10
Subtraction	2.4	1.7	2.3	2.5	2.5	1.3	2.0	3.5	4.4	2.0
Multiplication	6.0	2.6	5.1	3.8	3.6	4.4	5.3	4.9	4.2	5.4
Differences	3.6	0.9	2.8	1.3	1.1	3.1	3.3	1.4	−0.2	3.4

2. Assuming that the average tenth-grader was supposed to do each subtraction problem in 3.5 seconds on the average, test whether this student is significantly faster than the average.
3. Assume that the multiplication and subtraction problems were constructed so as to require equal times of the average child, and then use the differences in times to test whether this student is slower at multiplication than at subtraction.
4. The following times are for the first 10 and second 10 multiplication problems worked by a tenth-grader on a standard test. Assuming that the problems have been equated for time, test whether the student is improving with practice.

Times	Serial order of problem 1	2	3	4	5	6	7	8	9	10
First 10	6.0	2.6	5.1	3.8	3.6	4.4	5.3	4.9	4.2	5.4
Second 10	4.2	3.8	3.3	4.2	3.0	4.4	3.3	2.8	4.6	2.7

5. *Randomness of digits of* π. To test the randomness of early digits in the decimal expansion of $\pi (= 3.14159\ldots)$, patches of 300 digits were divided into 150 nonoverlapping pairs. A pair was scored as a doublet based on the digits (a, b) if both members of the pair were either a or b. So if $a = 0$ and $b = 1$, the pairs (00), (11), (10), (01) would all score as doublets. For each of the 45 (a, b)-pairs, the number of doublets was counted in each of 20 patches. The corresponding counts were made for 20 similar patches of random numbers. For a given (a, b), the expected number of doublets is 6 in 150 nonoverlapping pairs of digits.

   The number of (a, b)-pairs producing 10 or more doublets was computed for each patch. Use the following data to test at the 5% level a difference in mean between the random numbers and the digits of π.

	20 patches of digits of π	20 patches of random numbers
Number of pairs with 10 or more doublets	$\bar{x}_\pi = 3.25$	$\bar{x}_r = 3.65$
Variance of the number of pairs with 10 or more doublets	$s_\pi^2 = 7.88$	$s_r^2 = 10.98$

Projects for high-speed computers

The following projects are supplementary exercises for those who have computers available. Progress on some of them can be made by hand. In many of the problems, the amount of effort or computing time can be adjusted by choosing more or fewer values of the parameters. In the simulations the computer must be able to produce random numbers in some problems, random normal deviates in others. Simulation problems are often helped by judicious application of mathematics. For example, in simulating measurements for a t-distribution, noting that t is distributed symmetrically about 0 means that every observed sample value can be regarded as having occurred twice, once with a $+$ and once with a $-$ sign. Also, many problems can be worked exactly algebraically for small values of the parameters.

No one should try to do all of these projects. Nor is it intended that they be done in order. A lot depends on the strength of the computer and on the experience of those doing the programming. Some projects are fairly short. Others are a good semester's effort. A good idea in figuring the time required for computing is to estimate how long the programming and running should take and then multiply by 5. Until one has experience with the particular problems, it is well to keep the goals modest. The authors believe that one finished problem is more fun and more educational than many started and none finished.

We would be pleased to see copies of solutions.

The notation $n = 1[1]10[5]100$ means that n goes from 1 to 10 in steps of 1 and from 10 to 100 in steps of 5.

COMPUTER PROJECTS BASED UPON TEXTBOOK PROBLEMS

In the following exercises for computer work, the numbering refers the reader to the Section and Exercise number or Example number. Suggestions are then made for revising the textual exercise or example to make it suitable for a computer project. Numbers refer to exercises unless otherwise stated.

Section 1–5. Exercises 1, 2, 7, 8, 10, 12, 13, 14, 15, 16, 17: Write a program using random numbers to simulate the processes in these exercises and carry them out 100 times instead of 5 or 10 as the exercise suggests. Have the computer print out the appropriate quantities.

Exercise 3: Write a program so that the computer makes the frequency distributions and computes and prints out the mean.

Exercises 4, 5, 6, 9: Use the program of Exercise 3.

Section 2–1. Exercise 3, first part: Print out the possible call letters for stations.

Exercise 20: Tabulate for $1 \leq r \leq n$, up to $n = 10$.

Section 2–2. Exercise 17: Tabulate values for the formula for $1 \leq r \leq n$ up to $n = 10$.

Section 2–3. Exercise 21: Verify Table 2–3.

Exercise 22: Extend Table 2–3 to $n = 20$.

Section 2–4. Review Exercises. Exercise 17: Let the grid be $2n$ by $2n$. Tabulate the probability that the fly goes through the middle point (n, n) for $n = 1, 2, 3, \ldots, 50$.

Section 2–5. Exercise 34: Compare exact and approximate answers for $x = -0.5[0.1]0.5$.

Exercise 42: Compare exact and approximate results $x = -0.1[0.01]0.1$ and $n = 1[1]10$.

Section 3–4. Exercise 23: Tabulate for $p = 0.0001, 0.001, 0.01, 0.1$, and $n = 1, 2, 3, 4, 5, 10, 100, 1000$ and compare with np. Discuss the closeness of the approximation.

Section 3–5. Verify Table 3–2.

Tabulate the probabilities for the hypergeometric distribution for all possible tables up to $n = 10$.

Exercise 1: Tabulate the probability for each position.

Exercise 4: Tabulate for $k = 1[1]25$.

Section 3–8. Redo Example 3 using 100 samples for the cases of 4, 5, 6, 7, 8, 9, 10 available randomly selected donors.

Example 4: Program the isolates problem for n individuals choosing r friends each randomly. Then simulate 20 times for each n, r pair, where $r = 1, 2, 3, 4, 5$, and $n = 8, 9, 10, 11, 12$ to estimate the probability that no one is an isolate and the mean number of isolates.

Section 4–1. Exercise 1: Tabulate for $P(U) = 0.1[0.1]0.9$.

Section 4–2. Exercise 11: Adjust the probabilities so as to maximize the minimum value among the three P's.

Section 4–4. Exercise 6: Estimate $P(5 \text{ or } 6)$ from the observed data. Call this value p'. Use it to estimate the expected number of 0's, 1's, 2's, and 3's and compare with the observed values. Is the sum of the errors reduced compared to using $p = \frac{1}{3}$?

Section 4–5. Exercises 5, 6.

Exercise 12: Use the computer to maximize $p^x(1 - p)^{n-x}$ approximately for various values of x, n.

Section 4–6. Exercises 10, 11: By direct calculation.

Section 5–1. Exercise 2.

Exercise 3: Use integers 1 through 100.

Exercise 5: Find the distribution of turning points for 6 points.

Exercise 12: Simulate.

Exercise 20: Make a table of the probabilities for the geometric distribution for $p = 0.1[0.1]0.9$ for values of x running from 1 to a value n, such that

$$\sum_{x=1}^{n} f(x) = 0.999.$$

Of course, n can depend upon p.

Section 5–2. Exercise 1: Write a program for computing means of ungrouped and of grouped data. Test it. Find the mean of 27801 and -0.00426.

Exercises 13, 15.

Exercise 26: Verify the numbers .21, .30, .27, .22.

Section 5–3. Exercise 1: Use your program for computing means from grouped data to compute $E(X)$. If it doesn't work, revise it so that it takes fractional numbers.

Section 6–1. Exercise 1: Write a program for computing variances and standard deviations from theoretical probability distributions. Use it on Exercises 11, 12, 13, 14.

Section 6–2. Write a program that computes $\bar{x}$, s^2, and s for both ungrouped and grouped measurements. Apply it to Exercises 1, 2, 5, 6, 21, 22, 23.

Section 7–4. Compute the ordinates of the normal probability density function for a fine grid and estimate the area under the curve using rectangles.

Section 8–1. Exercise 8: Use your computer to calculate the points directly. If it can, make your computer draw the graph.

Section 8–3. Exercise 7: Compute the exact probability.

Section 9–4. Exercise 4: Let the plan be one of samples of size 50; reject if three or more defectives are found. Compute the operating characteristic.

Exercise 11.

Section 9–6. Example 2: Redo with a prior uniform on the even 0.01's. That is, let $P(p = i/100) = 1/101$, $i = 0, 1, \ldots, 100$.

Section 9–7. Exercise 9: Use the 101-point uniform prior distribution introduced in the computer exercise for Section 9–6 above instead of the 6-point one.

Section 10–1. Compute the exact distribution of the sum of the number of pips when n dice are tossed, $n = 2, 3, 4, 5$. Suggestion: When you find the distribution for the sum for 2, then adjoin another using the independence. To the sum of three, adjoin the fourth, and so on.

Section 10–3. The distribution

$f(x)$:	.1	.2	.3	.4
x:	1	2	3	4

was used to get the distribution of the sum of 5 identically distributed variables shown in Fig. 10–3. Obtain the exact distribution of the sum of 10 measurements independently drawn from this distribution. Compare the results with those obtained from the normal distribution approximation. Compare the agreement with that of Fig. 10–3.

Section 11–4. Write a program for fitting the general least squares line. Be sure that it also provides the standard error of estimate. Apply it to the submarine sinkings data of Example 2, Section 11–4.

Exercise 4: Apply the program to these data.

Section 11–5. Write a program for computing the correlation coefficient,both for ungrouped and for grouped data. Apply it to the submarine sinkings data of Example 2, Section 11–4.

Section 11–6. Apply your regression program to the data of Table 11–7 to get the equations of the two regression lines.

Exercise 14: Obtain the regression equations and the correlation coefficient.

Section 12–3. Use samples of size 3 from Table I-B to get a distribution of observed values of t. Use this distribution to estimate the probability levels of t with 2 degrees of freedom, following the line of Example 1.

Write a program for the t-test for the difference of two means when it is believed that the variances of the two populations being sampled are the same.

MISCELLANEOUS COMPUTER PROJECTS

1. Make a table of $n!$.

2. Make tables of the negative binomial distribution for trials to first success $r = 1, 2, 3, 4, 5$ and probability of success on a single trial $p = .01, .05, .1, .2, .3, .4, .5, .6, .7, .8, .9, .95, .99$.

3. For the binomial distribution, make a table of the mean of $1/(X + 1)$, where X is the number of successes, for various pairs of values of n and p.

4. When n dice are rolled, some face must turn up at least as often as any other. Choose among these most frequent faces and remove all the dice with that face up. Toss the remaining ones and repeat the removal procedure. Find the mean number of trials until all dice have been removed.

5. *Sequential experimentation.* A pair of products A and B are tested by comparing their performance, one trial at a time. The trials are binomial with probability p that product A wins on any given trial. The product winning a trial scores one point, and the investigation continues until one of the products is k points ahead and then that product is declared the winner of the test. Make tables and graphs showing the probability that A wins the test when $k = 1, 2, 3, 4, 5$ and $p = 0.50, 0.51, 0.55, 0.6, 0.75, 0.9, 1.0$.

6. *Estimating the mean of the uniform.* Compare the variances of the sample mean, median, and midrange, which is

$$\tfrac{1}{2}\,(\text{largest observation} + \text{smallest observation}),$$

for samples of size 2, 4, 16, and 64 drawn from a uniform distribution on the interval [0, 1]. Three-digit random numbers should be adequate, but you may find 5 or 8 more convenient. Depending upon the speed of your machine, draw 50, 100, or 1000 samples of each size. Compute the three statistics for each sample and get the variance of each for each sample size. You may also wish to investigate the distribution (percentage points) of these statistics for the various sample sizes. Graphical summaries would be instructive.

7. Draw 50 samples of size 2, 8, 32, 128 from the standard normal and compute

$$s' = \sqrt{\frac{\Sigma(X - \bar{X})^2}{n - 1}}$$

for each; compute the standard deviation of s' for each sample size and plot this against $1/\sqrt{n}$. Provide an empirical formula for $\sigma_{s'}$ for normal distributions showing its dependence on n.

8. *Random walk with one absorbing barrier.* A particle starts at position $x = 1$ on the real axis and walks to the right a step of length one unit with probability p, or to the left a unit step with probability $q(= 1 - p)$. If the particle gets to $x = 0$ it is absorbed, otherwise it continues to walk, according to the same probabilities. It can be shown* that the probability of absorption is 1 when $p \leq \frac{1}{2}$, and that it is q/p when $p \geq \frac{1}{2}$. These results apply after infinite waiting, but we need to know for various values of p the amount absorbed after 1, 3, 5, 7, 9, 19, 39, 99, etc., steps unless the action is essentially already over. Suggestion: try $p = .1, .2, .3, .4, .5, .6, .7, .8, .9$, and explore appropriate numbers of steps. Does the amount absorbed relate approximately linearly to $1/n^a$, where a is some power like $\frac{1}{2}$, 1, $\frac{3}{2}$, or 2, and n is the number of steps?

9. Chemists sometimes make 3 parallel analyses of the same material, compute the average of the three measurements, and discard the one furthest from that average. They then report the average of the remaining 2 measurements.

---

* See Frederick Mosteller, *50 Challenging Problems in Probability with Solutions*, Addison-Wesley, Reading, Mass., 1965.

Assume that "good" observations come from a normal with mean μ and standard deviation σ, but "wild" observations from a normal with mean μ and standard deviation $k\sigma$, $k > 1$. Let p be the probability of a wild observation. If the chemist reports the average $\overline{X}'$ of his remaining 2 measurements, for what values of k and p will $\sigma^2_{\overline{X}'}$ be smaller than he would get from the average of the original 3 measurements?

10. *Epidemic.* For a certain mild disease, individuals are either immune, susceptible, or infectious. If a susceptible catches the disease, he is infectious next day and immune thereafter. Suppose n soldiers line up in random order each day. If one is infectious, he infects any susceptible standing beside him. If both are immune, he infects none. At the start, 1 soldier has the disease, and the rest are susceptible. For various values of n, how long, on the average, will the disease last, how many susceptibles will be left at the end of the epidemic, and what is the largest number of soldiers having the disease at one time? For small n work out exact formulas; for large n, simulate by running the process through a reasonable number of times and analyzing the resulting statistics.

11. *Scanning.* Imagine 100 phone calls started at random times in an interval of, say, 100 minutes, simulated by 100 points dropped randomly on a line of length 1 (better keep at least 5 digits in the random number).

a) What is the distribution of the largest number of calls found in any 5-minute period, when we look at many 100-minute periods? Note that we are to look at every 5-minute period, not just the 20 nonoverlapping ones, and choose one count. Note that 100 points dropped once give only one observation.

b) What is the distribution of the largest interval of time between the first and last of 6 consecutive calls?

c) Compare results for (a) and (b).

d) Generalize the problem by considering dropping other numbers of points and by using scanning lengths other than 5.

Appendixes

appendix 1 | Summations and subscripts

A1–1 SUBSCRIPTS AND THE SUMMATION SYMBOL, $\sum$

We often wish to indicate the sum of several measurements or observations. For example, if 30 students take a test, we may wish to know their average score, which is $\frac{1}{30}$ the sum of their scores. Or we may wish to talk about the sum of the points on the top face of a die thrown many times. It is convenient to be able to express such sums in compact form. The Greek letter $\sum$ (capital *sigma*) is used for this purpose, to denote "summation of".

Suppose, for example, we arrange the names of the 30 students in alphabetical order, and then let x_1 represent the test score of the first student, x_2 the score of the second student, and so on, with x_{30} representing the score of the 30th student. The subscripts 1, 2, . . . , 30 correspond to the positions of the students' names on the alphabetical list. If the first 3 students received scores of 85, 79, and 94, in that order, then

$$x_1 = 85, \qquad x_2 = 79, \qquad x_3 = 94.$$

The sum of the 30 scores could be represented by

$$x_1 + x_2 + \cdots + x_{30}, \tag{1}$$

where the three dots are used to indicate "and so on". Another way of representing the same sum, using the summation symbol $\sum$, is

$$\sum_{i=1}^{30} x_i. \tag{2}$$

We read expression (2): "summation of x-sub-i from $i = 1$ through 30". It has exactly the same meaning as expression (1); both indicate the sum of the 30 scores x_1, x_2, and so on through x_{30}. In other words, the symbol

$$\sum_{i=1}^{30}$$

means that we are to replace i by integers in ascending order, beginning at 1 and ending at 30, and add the results.

The subscript may be any convenient letter, although i, j, k, and n are most frequently used.

Example 1 If $x_1 = -3$, $x_2 = 5$, $x_3 = 7$, and $x_4 = 6$, find

a) $\sum_{i=1}^{4} x_i$, b) $\sum_{i=2}^{4} x_i$, c) $\sum_{j=1}^{3} x_j$,

d) $\sum_{k=1}^{4} 5x_k$, e) $\sum_{n=1}^{3} (x_n + x_{n+1})$, f) $\sum_{i=1}^{4} x_i^2$.

Solutions

a) $\sum_{i=1}^{4} x_i = x_1 + x_2 + x_3 + x_4 = -3 + 5 + 7 + 6 = 15.$

b) $\sum_{i=2}^{4} x_i = x_2 + x_3 + x_4 = 5 + 7 + 6 = 18.$

c) $\sum_{j=1}^{3} x_j = x_1 + x_2 + x_3 = -3 + 5 + 7 = 9.$

d) $$\begin{aligned}\sum_{k=1}^{4} 5x_k &= 5x_1 + 5x_2 + 5x_3 + 5x_4 \\ &= 5(x_1 + x_2 + x_3 + x_4) \\ &= 5(15) = 75.\end{aligned}$$

e) $$\begin{aligned}\sum_{n=1}^{3} (x_n + x_{n+1}) &= (x_1 + x_2) + (x_2 + x_3) + (x_3 + x_4) \\ &= (x_1 + x_2 + x_3) + (x_2 + x_3 + x_4) \\ &= 9 + 18 = 27.\end{aligned}$$

f) $\sum_{i=1}^{4} x_i^2 = x_1^2 + x_2^2 + x_3^2 + x_4^2 = (-3)^2 + 5^2 + 7^2 + 6^2 = 119.$

Remark 1 Part (b) of this example illustrates a sum from $i = 2$ through $i = 4$. The equation "$i = 2$" written beneath the summation sign tells us where the sum starts. The subscript i on x_i is first to be replaced by 2. We then proceed through *the integers* from the starting place (in this case, 2) until we reach the integer corresponding to the symbol written above the summation sign (here, 4). Thus, in x_i, we replace i by 2, 3, and 4, and add the results:

$$x_2 + x_3 + x_4.$$

Remark 2 In part (c) we used the letter j, instead of i, for the subscript on x, and for the corresponding *index of summation.* The notation "$j = 1$" beneath the sigma tells us the first value to substitute for j, and this substitution converts x_j into x_1. We then proceed one-by-one through the integers until we reach the *upper limit of summation*, in this case 3. Then we add the results, and get

$$x_1 + x_2 + x_3.$$

Remark 3 In Example 1(d), we replace the subscript k by 1, 2, 3, and 4, in that order, but we have the common factor 5 in each term. In fact, we see that

$$\sum_{k=1}^{4} 5x_k = 5 \sum_{k=1}^{4} x_k,$$

and this result can easily be generalized. We shall do so in the next section.

Remark 4 In part (e), the index of summation n takes the values 1, 2, and 3. The subscript on x_n takes these same values, but the subscript on x_{n+1} takes the values $n + 1 = 2$, 3, and 4, in that order. By rearranging terms, we also see that

$$\begin{aligned} \sum_{n=1}^{3} (x_n + x_{n+1}) &= \sum_{n=1}^{3} x_n + \sum_{n=1}^{3} x_{n+1} \\ &= \sum_{i=1}^{3} x_i + \sum_{j=2}^{4} x_j. \end{aligned} \tag{3}$$

Remark 5 Equation (3) and Examples 1(c, d, e) illustrate that the letter used for the index of summation is immaterial. This also explains why that index is often called a "dummy index". The only requirement is that when the index is everywhere replaced by the consecutive integers, beginning with the *lower limit of summation* (written beneath the sigma) and extending through the *upper limit of summation* (written above the sigma), we get the desired result by adding these expressions. Thus

$$x_2 + x_3 + x_4 = \sum_{j=2}^{4} x_j = \sum_{k=-1}^{1} x_{k+3} = \sum_{k=-1}^{1} x_{3-k}.$$

A1–1 Definitions:

Summation. With each integer i from m through n, let there be associated a number denoted by x_i. The sum of the numbers

$$x_m, x_{m+1}, \ldots, x_n$$

is represented in summation notation by $\sum_{i=m}^{n} x_i$:

$$\boxed{\sum_{i=m}^{n} x_i = x_m + x_{m+1} + \cdots + x_n.} \tag{4}$$

Limits of summation. In Eq. (4) the *lower limit* of summation is m, the *upper limit* is n.

The omission of limits of summation We sometimes omit the limits and write simply $\sum x_i$. This notation means that the *summation is to extend over all values of x_i under discussion*, unless something is said to the contrary. For instance, if the only values in a particular discussion are x_1, x_2, x_3, x_4, then $\sum x_i$ means $x_1 + x_2 + x_3 + x_4$.

Example 2 If $x_i = i(i - 1)$, evaluate

$$\sum_{i=1}^{5} x_i.$$

Solution

$$\begin{aligned}\sum_{i=1}^{5} x_i &= x_1 + x_2 + x_3 + x_4 + x_5 \\ &= 1(1 - 1) + 2(2 - 1) + 3(3 - 1) + 4(4 - 1) + 5(5 - 1) \\ &= 0 + 2 + 6 + 12 + 20 = 40.\end{aligned}$$

Example 3 Evaluate

$$\sum_{j=0}^{2} \frac{j+1}{j+3}.$$

Solution

$$\sum_{j=0}^{2} \frac{j+1}{j+3} = \frac{0+1}{0+3} + \frac{1+1}{1+3} + \frac{2+1}{2+3} = \frac{1}{3} + \frac{2}{4} + \frac{3}{5} = \frac{43}{30}.$$

Example 4 Express the following sum as a simple function of n:

$$\sum_{k=0}^{n} [(k+1)^2 - k^2]. \tag{5}$$

Solution Replacing k by 0, 1, 2, . . . , n and adding, we get

$$\begin{aligned}\sum_{k=0}^{n} [(k+1)^2 - k^2] &= [1^2 - 0^2] + [2^2 - 1^2] + [3^2 - 2^2] \\ &\quad + \cdots + [(n+1)^2 - n^2].\end{aligned} \tag{6}$$

The positive terms on the right side of Eq. (6) have a sum expressed by

$$1^2 + 2^2 + 3^2 + \cdots + (n+1)^2, \tag{7}$$

and from this we must subtract

$$0^2 + 1^2 + 2^2 + \cdots + n^2. \tag{8}$$

Therefore

$$\begin{aligned}\sum_{k=0}^{n} [(k+1)^2 - k^2] &= [1^2 + 2^2 + \cdots + n^2 + (n+1)^2] \\ &\quad - [0^2 + 1^2 + \cdots + n^2] \\ &= (n+1)^2.\end{aligned} \tag{9}$$

EXERCISES FOR SECTION A1–1

Evaluate the following sums:

1. $\sum_{n=1}^{3} n^2$
2. $\sum_{n=1}^{3} 2^n$
3. $\sum_{n=1}^{3} x^n$
4. $\sum_{k=0}^{2} (2k+1)$
5. $\sum_{k=0}^{3} (2k+1)$
6. $\sum_{k=0}^{4} (2k+1)$
7. $\sum_{i=1}^{3} (i^2+i)$
8. $\sum_{n=-2}^{2} (n^2-4)$
9. $\sum_{n=100}^{102} n$
10. $\sum_{n=0}^{2} (n+100)$
11. Use the result of Eq. (9) and the fact that

$$(k+1)^2 - k^2 = 2k+1$$

is an odd integer to prove that the sum of the first $n+1$ positive odd integers is a perfect square. What square?

12. By expanding the left side of the following equation and rearranging terms, show that

$$\sum_{k=0}^{3} (a_k + b_k) = \sum_{k=0}^{3} a_k + \sum_{k=0}^{3} b_k.$$

Can you generalize this result in two ways?

13. By expanding the left side of the following equation and using the distributive law, show that

$$\sum_{i=1}^{3} 7x_i = 7 \sum_{i=1}^{3} x_i.$$

Generalize the result in as many ways as you can.

14. If all the x_i's are equal to the same constant c, what is the value of $\sum_{i=1}^{n} x_i$?

A1–2 THEOREMS ABOUT SUMMATIONS

In Exercise 12 above, you probably discovered the result stated in the following theorem.

A1–2 Theorem

The summation of the sum of two or more variables is the sum of their summations. Thus,

$$\boxed{\sum_{i=m}^{n} (a_i + b_i) = \sum_{i=m}^{n} a_i + \sum_{i=m}^{n} b_i.} \tag{1}$$

Proof To prove Eq. (1), we need only expand the left side and rearrange terms, as follows:

$$\begin{aligned}\sum_{i=m}^{n} (a_i + b_i) &= (a_m + b_m) + (a_{m+1} + b_{m+1}) + \cdots + (a_n + b_n) \\ &= (a_m + a_{m+1} + \cdots + a_n) + (b_m + b_{m+1} + \cdots + b_n) \\ &= \sum_{i=m}^{n} a_i + \sum_{i=m}^{n} b_i.\end{aligned}$$

The result can be extended to the summation of three or more variables by repeated applications of Eq. (1). For example,

$$\Sigma(a_i + b_i + c_i) = \Sigma a_i + \Sigma(b_i + c_i) = \Sigma a_i + \Sigma b_i + \Sigma c_i,$$

where i goes from m to n in all summations. □

A1–3 Theorem

A constant factor can be moved across the summation sign. Thus, if c is a constant,

$$\boxed{\sum_{i=m}^{n} cx_i = c \sum_{i=m}^{n} x_i.} \tag{2}$$

Proof We expand the left side of Eq. (2) and get

$$cx_m + cx_{m+1} + \cdots + cx_n,$$

which can also be written in the form of the right side of Eq. (2). □

Example 1 $\sum_{i=1}^{3} 2i^2 = 2 \sum_{i=1}^{3} i^2 = 2(1 + 4 + 9) = 28,$

$$\sum_{i=1}^{3} (2i)^2 = 4 \sum_{i=1}^{3} i^2 = 4(1 + 4 + 9) = 56.$$

A1–4 Theorem

The summation of a constant is equal to the product of that constant and the number of integers from the lower limit of summation through the upper limit. Thus

$$\boxed{\sum_{i=1}^{n} c = cn.} \tag{3}$$

Proof If each x_i is equal to c, for $i = 1, 2, \ldots, n$, then

$$\sum_{i=1}^{n} x_i = x_1 + x_2 + \cdots + x_n = c + c + \cdots + c = cn. \quad \square$$

Example 2 Use Eq. (9) of Section A1–1 and Theorems A1–2, A1–3, and A1–4 to evaluate $\sum_{k=0}^{n} k$ as a function of n.

Solution From Eq. (9), we have

$$\sum_{k=0}^{n} [(k+1)^2 - k^2] = (n+1)^2. \tag{4}$$

But also,

$$(k+1)^2 - k^2 = k^2 + 2k + 1 - k^2 = 2k + 1.$$

Reversing the order in Eq. (4) and substituting $2k + 1$ for the difference of squares, we get

$$(n+1)^2 = \sum_{k=0}^{n} (2k+1) = \sum_{k=0}^{n} 2k + \sum_{k=0}^{n} 1 = 2\sum_{k=0}^{n} k + (n+1). \tag{5}$$

To find the value of $\sum k$, we subtract $(n + 1)$ and get, after reversing the sides of the equation,

$$2\sum_{k=0}^{n} k = (n+1)^2 - (n+1) = (n+1)[(n+1) - 1] = (n+1)n.$$

Therefore

$$\boxed{\sum_{k=0}^{n} k = \frac{(n+1)n}{2}.} \tag{6}$$

EXERCISES FOR SECTION A1–2

Write the following summations in expanded form and simplify the results as much as possible.

1. $\sum_{k=0}^{3} ka_k$
2. $\sum_{k=1}^{3} k^2(\frac{1}{2})^k$
3. $\sum_{j=-1}^{2} 10^j$
4. $\sum_{i=2}^{5} (2i - 5)$
5. $\sum_{n=1}^{3} (a_n + b_{n-1})$
6. $\sum_{n=0}^{2} (2x_n - 3y_n)$
7. $\sum_{k=0}^{2} \binom{2}{k} x^k$
8. $\sum_{k=0}^{2} \binom{2}{k} x^{2-k}y^k$
9. $\sum_{x=0}^{3} \binom{3}{x} p^x q^{3-x}$
10. $\sum_{x=0}^{2} \frac{x^2}{x!}$
11. Show that

$$\sum_{k=0}^{n} [(k+1)^3 - k^3]$$

is equal to $(n + 1)^3$.

12. (Continuation.) Use the result of Exercise 11 above and the relation $(k+1)^3 - k^3 = 3k^2 + 3k + 1$ to show that

$$3 \sum_{k=0}^{n} k^2 = (n+1)^3 - 3 \sum_{k=0}^{n} k - \sum_{k=0}^{n} 1.$$

13. (Continuation.) Use the results of Exercise 12 and Eqs. (3) and (6) in the text to show that

$$\boxed{\sum_{k=0}^{n} k^2 = \frac{n(n+1)(2n+1)}{6}.} \tag{7}$$

14. How would the results in Eq. (6) and in Exercise 13 be affected if the lower limits of summation were changed to $k = 1$? Explain.
15. Write each of the following sums in summation form:

    a) $z_1 + z_2 + \cdots + z_{23}$.

    b) $x_1y_1 + x_2y_2 + \cdots + x_8y_8$.

    c) $(x_1 - y_1) + (x_2 - y_2) + \cdots + (x_m - y_m)$.

    d) $x_1^3f_1 + x_2^3f_2 + \cdots + x_9^3f_9$.

16. Prove that $\sum_{i=1}^{n} (x_i - m)^2 = \sum_{i=1}^{n} x_i^2 - 2m \sum_{i=1}^{n} x_i + nm^2$.

17. In Exercise 16, suppose that $m = \bar{x}$, the arithmetic mean of $x_1, x_2, \ldots, x_n$. Prove that

$$\sum_{i=1}^{n} (x_i - \bar{x})^2 = \sum_{i=1}^{n} x_i^2 - n\bar{x}^2.$$

18. Express

$$\sum_{i=1}^{n} (ax_i + by_i)^2$$

as a sum of three summations.

Remark. We sometimes have to deal with sums of the form $\sum x_iy_j$, where the sum is to be extended over certain pairs of values of i and j. Suppose, for example, that i goes from 1 through 3, and j takes the values 1 and 2. Then there are 3×2, or 6, pairs of i, j-values, and the corresponding summation is

$$\sum x_iy_j = x_1y_1 + x_1y_2 + x_2y_1 + x_2y_2 + x_3y_1 + x_3y_2.$$

appendix 2 | A theorem on independence

The purpose of this appendix is to prove the following theorem. The result is used in proving that the variance of the sum of three or more independent random variables is the sum of their variances.

A2–1 Theorem

Let X, Y, and Z be independent random variables whose possible values are

$$x_1, x_2, \ldots, x_m, \qquad y_1, y_2, \ldots, y_n, \qquad z_1, z_2, \ldots, z_t.$$

If $U = X + Y$, then U and Z are independent.

Proof We must show that if u is any of the possible values of $x_i + y_j$, then

$$P(U = u, Z = z_k) = P(U = u) \cdot P(Z = z_k), \tag{1}$$

where z_k is any one of the possible values of Z. Table A2–1 shows the possible ordered pairs (x_i, y_j). By entering one more symbol, say z_1, in each cell, we can represent all the ordered triples (x_i, y_j, z_1), with the value of Z fixed as z_1. Another table like Table A2–1 with z_2 added in each cell can be used to represent the ordered triples (x_i, y_j, z_2) corresponding to the value z_2 for z_k. And so on for the remaining values $z_3, \ldots, z_t$, using a separate table for each value of

Table A2–1 Pairs of values (x_i, y_j)

		Value of y_j			
		y_1	y_2	$\cdots$	y_n
Value of x_i	x_1	(x_1, y_1)	(x_1, y_2)	$\cdots$	(x_1, y_n)
	x_2	(x_2, y_1)	(x_2, y_2)	$\cdots$	(x_2, y_n)
	$\vdots$	$\vdots$	$\vdots$		$\vdots$
	x_m	(x_m, y_1)	(x_m, y_2)	$\cdots$	(x_m, y_n)

z_k. If these t different tables were printed on different sheets of paper, one per sheet, and the sheets of paper were then stacked on top of one another, the three-dimensional stack of tables would represent all possible ordered triples (x_i, y_j, z_k). The cells of Table A2–1 resemble the squares of a chessboard, and the totality of cells in the stack of tables resembles a three-dimensional chessboard.

Now suppose, temporarily, that Table A2–1 has a z_1 written in every cell, so that the entries are converted from (x_i, y_j) into (x_i, y_j, z_1). Since X, Y, and Z are independent, the corresponding probability of the cell is

$$P(x_i) \cdot P(y_j) \cdot P(z_1). \tag{2}$$

To compute the probability $P(U = u, Z = z_1)$, the left member of Eq. (1) for $k = 1$, we must add those probabilities given by the product (2) over all cells having $x_i + y_j = u$:

$$P(U = u, Z = z_1) = \sum P(x_i) \cdot P(y_j) \cdot P(z_1), \tag{3}$$

where the summation runs over those pairs (x_i, y_j) with $x_i + y_j = u$. Since $P(z_1)$ remains constant for all terms in this summation, we get

$$P(U = u, Z = z_1) = P(z_1) \sum P(x_i) \cdot P(y_j), \tag{4}$$

where the summation in Eq. (4) also runs over those pairs with sum $x_i + y_j = u$. This sum is exactly equal to the probability that $X + Y$, or U, is equal to the particular value u. Therefore Eq. (4) gives the result

$$P(U = u, Z = z_1) = P(z_1) \cdot P(u). \tag{5}$$

There is nothing special about z_1 in Eqs. (2) through (5) so far as the proof is concerned. We could equally well repeat the argument for z_2, or z_3, or any other value of z_k. Thus we have

$$P(U = u, Z = z_k) = P(z_k) \cdot P(u),$$

which establishes the result that U and Z are independent.

By a slightly more general argument, we can extend the theorem to more than three independent random variables: any one of them is independent of the sum of the others.

Independence of *aX* and *bY* When X and Y are independent random variables, the new random variables $U = aX$ and $V = bY$ are also independent when a and b are constants. This is readily seen. Suppose that X takes the value x_i with probability $P(X = x_i)$, and that Y takes the value y_i with probability $P(Y = y_i)$. Then the independence promises that

$$P(X = x_i \text{ and } Y = y_i) = P(X = x_i) \cdot P(Y = y_i).$$

Let $u_i = ax_i$ and $v_j = by_j$. If neither a nor b is zero, then, since a given value of x produces one value of u (and similarly for y and v), we have

$$\begin{aligned} P(U = u_i \text{ and } V = v_j) &= P(X = x_i \text{ and } Y = y_j) \\ &= P(X = x_i) \cdot P(Y = y_j) \\ &= P(U = u_i) \cdot P(V = v_j). \end{aligned}$$

If a or b or both are zero, U and V are independent in a degenerate sense. You may wish to study this situation by separately considering the case $a = 0$ and $b = 0$, and the case $a \neq 0$ and $b = 0$.

Independence of functions By a slightly more general argument we can extend the theorem to show that if several random variables $X_1, X_2, \ldots, X_r$ are independent and if the function f_i is a function of X_i only, $i = 1, 2, \ldots, r$, then the functions are independent. We have dealt only with linear functions.

Tables

Table I–A 1000 random digits

00	49487	52802	28667	62058	87822	14704	18519	17889	45869	14454
01	29480	91539	46317	84803	86056	62812	33584	70391	77749	64906
02	25252	97738	23901	11106	86864	55808	22557	23214	15021	54268
03	02431	42193	96960	19620	29188	05863	92900	06836	13433	21709
04	69414	89353	70724	67893	23218	72452	03095	68333	13751	37260
05	77285	35179	92042	67581	67673	68374	71115	98166	43352	06414
06	52852	11444	71868	34534	69124	02760	06406	95234	87995	78560
07	98740	98054	30195	09891	18453	79464	01156	95522	06884	55073
08	85022	58736	12138	35146	62085	36170	25433	80787	96496	40579
09	17778	03840	21636	56269	08149	19001	67367	13138	02400	89515
10	81833	93449	57781	94621	90998	37561	59688	93299	27726	82167
11	63789	54958	33167	10909	40343	81023	61590	44474	39810	10305
12	61840	81740	60986	12498	71546	42249	13812	59902	27864	21809
13	42243	10153	20891	90883	15782	98167	86837	99166	92143	82441
14	45236	09129	53031	12260	01278	14404	40969	33419	14188	69557
15	40338	42477	78804	36272	72053	07958	67158	60979	79891	92409
16	54040	71253	88789	98203	54999	96564	00789	68879	47134	83941
17	49158	20908	44859	29089	76130	51442	34453	98590	37353	61137
18	80958	03808	83655	18415	96563	43582	82207	53322	30419	64435
19	07636	04876	61063	57571	69434	14965	20911	73162	33576	52839

Table I–B 200 random normal deviates, $\mu = 0$, $\sigma = 1$

00	−0.56	+1.51	−0.35	−0.51	−0.27	+0.24	+0.13	+2.10	−0.28	−0.98
01	−0.55	+0.91	−0.55	+1.82	+0.03	+1.26	−1.27	−0.90	−0.77	+0.88
02	+0.74	−1.33	+0.19	−0.33	−0.86	−1.39	+0.07	+0.35	+0.46	−0.24
03	+2.56	−0.11	+1.37	−0.86	−0.40	+0.10	+0.13	−0.32	+2.00	+0.32
04	−1.29	−0.02	+0.17	+1.14	−2.00	−0.19	−0.38	+0.88	−0.34	−0.62
05	+1.25	+0.81	+0.16	−1.43	+1.76	+0.31	−0.38	−0.04	+1.61	−0.33
06	−0.76	+0.82	−1.20	−1.39	−1.23	+2.55	+0.08	+1.29	+2.06	+0.76
07	−1.15	+0.36	+0.57	−2.46	+0.22	+0.54	+1.33	+1.65	−0.35	+0.34
08	−1.77	+0.25	−1.16	−0.81	+1.13	+0.80	+0.61	+1.57	+0.31	+0.91
09	−0.06	+1.21	−0.34	−0.19	+1.03	+0.35	+1.40	−0.13	+0.47	−1.09
10	−1.23	+0.30	−1.13	+0.98	−0.29	+0.05	−0.85	−0.20	+2.00	+0.28
11	−0.14	−0.57	+0.46	−0.12	−1.84	+0.20	+0.41	−0.57	−1.52	+0.75
12	−0.02	−0.44	−0.15	+0.99	+0.29	+0.07	+1.36	+0.28	−0.02	−1.20
13	+1.45	+2.24	−1.01	−0.81	−0.07	−0.60	−0.20	+1.04	−0.57	+0.08
14	−0.64	+1.49	−0.37	−0.92	+0.58	+0.69	−0.19	+1.89	−0.06	−0.56
15	−0.41	+0.56	−0.22	−0.45	−0.09	+1.20	−0.81	−0.42	+0.70	−0.71
16	+0.03	−1.77	−0.52	−0.41	+1.87	−0.19	−1.66	−0.31	+1.27	+0.55
17	+1.41	−0.51	−0.21	−0.86	−0.01	+0.87	+0.58	+0.48	−0.02	+0.11
18	−0.54	−0.10	−0.52	−0.76	+1.53	−1.07	−0.67	+0.70	−1.17	+0.54
19	+1.23	+0.31	−1.97	+0.81	−0.78	+0.52	−2.33	+1.30	+0.51	+2.02

Table II Values of $n!$ and $\log n!$

The values of $n!$ are given to five significant figures, and for $n \geq 9$ these values must be multiplied by a power of ten. This power is the raised number to the right of the five significant figures. For example, $15! \approx 13{,}077 \times 10^8$.

n	$n!$	$\log n!$	n	$n!$	$\log n!$	n	$n!$	$\log n!$
1	1	.00000	26	40,329^{22}	26.60562	51	15,511^{62}	66.19065
2	2	.30103	27	10,889^{24}	28.03698	52	80,658^{63}	67.90665
3	6	.77815	28	30,489^{25}	29.48414	53	42,749^{65}	69.63092
4	24	1.38021	29	88,418^{26}	30.94654	54	23,084^{67}	71.36332
5	120	2.07918	30	26,525^{28}	32.42366	55	12,696^{69}	73.10368
6	720	2.85733	31	82,228^{29}	33.91502	56	71,100^{70}	74.85187
7	5,040	3.70243	32	26,313^{31}	35.42017	57	40,527^{72}	76.60774
8	40,320	4.60552	33	86,833^{32}	36.93869	58	23,506^{74}	78.37117
9	36,288^{1}	5.55976	34	29,523^{34}	38.47016	59	13,868^{76}	80.14202
10	36,288^{2}	6.55976	35	10,333^{36}	40.01423	60	83,210^{77}	81.92017
11	39,917^{3}	7.60116	36	37,199^{37}	41.57054	61	50,758^{79}	83.70550
12	47,900^{4}	8.68034	37	13,764^{39}	43.13874	62	31,470^{81}	85.49790
13	62,270^{5}	9.79428	38	52,302^{40}	44.71852	63	19,826^{83}	87.29724
14	87,178^{6}	10.94041	39	20,398^{42}	46.30959	64	12,689^{85}	89.10342
15	13,077^{8}	12.11650	40	81,592^{43}	47.91165	65	82,477^{86}	90.91633
16	20,923^{9}	13.32062	41	33,453^{45}	49.52443	66	54,434^{88}	92.73587
17	35,569^{10}	14.55107	42	14,050^{47}	51.14768	67	36,471^{90}	94.56195
18	64,024^{11}	15.80634	43	60,415^{48}	52.78115	68	24,800^{92}	96.39446
19	12,165^{13}	17.08509	44	26,583^{50}	54.42460	69	17,112^{94}	98.23331
20	24,329^{14}	18.38612	45	11,962^{52}	56.07781	70	11,979^{96}	100.07841
21	51,091^{15}	19.70834	46	55,026^{53}	57.74057	71	85,048^{97}	101.92966
22	11,240^{17}	21.05077	47	25,862^{55}	59.41267	72	61,234^{99}	103.78700
23	25,852^{18}	22.41249	48	12,414^{57}	61.09391	73	44,701^{101}	105.65032
24	62,045^{19}	23.79271	49	60,828^{58}	62.78410	74	33,079^{103}	107.51955
25	15,511^{21}	25.19065	50	30,414^{60}	64.48307	75	24,809^{105}	109.39461

Table III Normal curve areas

Area under the standard normal curve from 0 to z, shown shaded, is $A(z)$.

Examples. If Z is the standard normal random variable and $z = 1.54$, then

$$\begin{aligned} A(z) = P(0 < Z < z) &= .4382, \\ P(Z > z) &= .0618, \\ P(Z < z) &= .9382, \\ P(|Z| < z) &= .8764. \end{aligned}$$

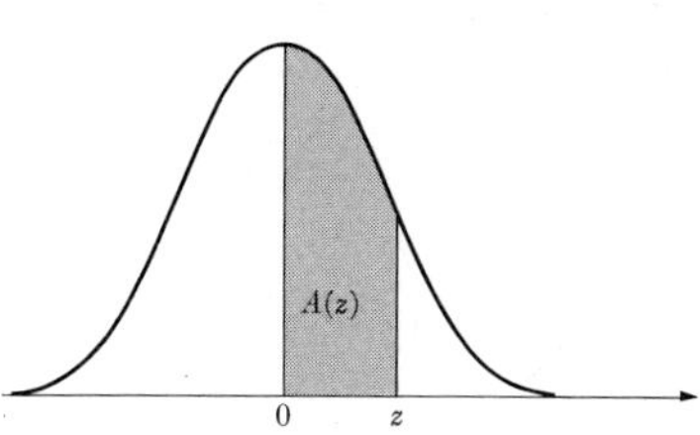

z	.00	.01	.02	.03	.04	.05	.06	.07	.08	.09
0.0	.0000	.0040	.0080	.0120	.0160	.0199	.0239	.0279	.0319	.0359
0.1	.0398	.0438	.0478	.0517	.0557	.0596	.0636	.0675	.0714	.0753
0.2	.0793	.0832	.0871	.0910	.0948	.0987	.1026	.1064	.1103	.1141
0.3	.1179	.1217	.1255	.1293	.1331	.1368	.1406	.1443	.1480	.1517
0.4	.1554	.1591	.1628	.1664	.1700	.1736	.1772	.1808	.1844	.1879
0.5	.1915	.1950	.1985	.2019	.2054	.2088	.2123	.2157	.2190	.2224
0.6	.2257	.2291	.2324	.2357	.2389	.2422	.2454	.2486	.2517	.2549
0.7	.2580	.2611	.2642	.2673	.2704	.2734	.2764	.2794	.2823	.2852
0.8	.2881	.2910	.2939	.2967	.2995	.3023	.3051	.3078	.3106	.3133
0.9	.3159	.3186	.3212	.3238	.3264	.3289	.3315	.3340	.3365	.3389
1.0	.3413	.3438	.3461	.3485	.3508	.3531	.3554	.3577	.3599	.3621
1.1	.3643	.3665	.3686	.3708	.3729	.3749	.3770	.3790	.3810	.3830
1.2	.3849	.3869	.3888	.3907	.3925	.3944	.3962	.3980	.3997	.4015
1.3	.4032	.4049	.4066	.4082	.4099	.4115	.4131	.4147	.4162	.4177
1.4	.4192	.4207	.4222	.4236	.4251	.4265	.4279	.4292	.4306	.4319
1.5	.4332	.4345	.4357	.4370	.4382	.4394	.4406	.4418	.4429	.4441
1.6	.4452	.4463	.4474	.4484	.4495	.4505	.4515	.4525	.4535	.4545
1.7	.4554	.4564	.4573	.4582	.4591	.4599	.4608	.4616	.4625	.4633
1.8	.4641	.4649	.4656	.4664	.4671	.4678	.4686	.4693	.4699	.4706
1.9	.4713	.4719	.4726	.4732	.4738	.4744	.4750	.4756	.4761	.4767
2.0	.4772	.4778	.4783	.4788	.4793	.4798	.4803	.4808	.4812	.4817
2.1	.4821	.4826	.4830	.4834	.4838	.4842	.4846	.4850	.4854	.4857
2.2	.4861	.4864	.4868	.4871	.4875	.4878	.4881	.4884	.4887	.4890
2.3	.4893	.4896	.4898	.4901	.4904	.4906	.4909	.4911	.4913	.4916
2.4	.4918	.4920	.4922	.4925	.4927	.4929	.4931	.4932	.4934	.4936
2.5	.4938	.4940	.4941	.4943	.4945	.4946	.4948	.4949	.4951	.4952
2.6	.4953	.4955	.4956	.4957	.4959	.4960	.4961	.4962	.4963	.4964
2.7	.4965	.4966	.4967	.4968	.4969	.4970	.4971	.4972	.4973	.4974
2.8	.4974	.4975	.4976	.4977	.4977	.4978	.4979	.4979	.4980	.4981
2.9	.4981	.4982	.4982	.4983	.4984	.4984	.4985	.4985	.4986	.4986
3.0	.4987	.4987	.4987	.4988	.4988	.4989	.4989	.4989	.4990	.4990

Table IV Three-place tables of the binomial distribution

Part A of this table gives the values of the function

$$b(x; n, p) = \binom{n}{x} p^x (1-p)^{n-x}$$
$$= \frac{n!}{x!(n-x)!} p^x (1-p)^{n-x}.$$

This is the probability of exactly x successes in n independent binomial trials with probability of success on a single trial equal to p.

Part B gives the values of the cumulative binomial

$$P(X \geq r) = \sum_{x=r}^{n} b(x; n, p)$$
$$= \sum_{x=r}^{n} \binom{n}{x} p^x (1-p)^{n-x}.$$

$P(X \geq r)$ is the probability of r or more successes in n independent binomial trials with probability p of success on a single trial.

In both parts of the table values of the functions are given for x (or r) $= 0, 1, \ldots, n$; $n = 2, 3, \ldots, 25$; and $p = .01, .05, .10, .20, .30, .40, .50, .60, .70, .80, .90, .95,$ and $.99$.

In these tables, each three-digit entry should be read with a decimal preceding it. For entries 1−, the probability is larger than 0.9995 but less than 1. For entries 0+, the probability is less than 0.0005 but greater than 0.

Part A: Individual terms, $b(x; n, p)$

n	x	.01	.05	.10	.20	.30	.40	p .50	.60	.70	.80	.90	.95	.99	x
2	0	980	902	810	640	490	360	250	160	090	040	010	002	0+	0
	1	020	095	180	320	420	480	500	480	420	320	180	095	020	1
	2	0+	002	010	040	090	160	250	360	490	640	810	902	980	2
3	0	970	857	729	512	343	216	125	064	027	008	001	0+	0+	0
	1	029	135	243	384	441	432	375	288	189	096	027	007	0+	1
	2	0+	007	027	096	189	288	375	432	441	384	243	135	029	2
	3	0+	0+	001	008	027	064	125	216	343	512	729	857	970	3
4	0	961	815	656	410	240	130	062	026	008	002	0+	0+	0+	0
	1	039	171	292	410	412	346	250	154	076	026	004	0+	0+	1
	2	001	014	049	154	265	346	375	346	265	154	049	014	001	2
	3	0+	0+	004	026	076	154	250	346	412	410	292	171	039	3
	4	0+	0+	0+	002	008	026	062	130	240	410	656	815	961	4
5	0	951	774	590	328	168	078	031	010	002	0+	0+	0+	0+	0
	1	048	204	328	410	360	259	156	077	028	006	0+	0+	0+	1
	2	001	021	073	205	309	346	312	230	132	051	008	001	0+	2
	3	0+	001	008	051	132	230	312	346	309	205	073	021	001	3
	4	0+	0+	0+	006	028	077	156	259	360	410	328	204	048	4
	5	0+	0+	0+	0+	002	010	031	078	168	328	590	774	951	5
6	0	941	735	531	262	118	047	016	004	001	0+	0+	0+	0+	0
	1	057	232	354	393	303	187	094	037	010	002	0+	0+	0+	1
	2	001	031	098	246	324	311	234	138	060	015	001	0+	0+	2
	3	0+	002	015	082	185	276	312	276	185	082	015	002	0+	3
	4	0+	0+	001	015	060	138	234	311	324	246	098	031	001	4
	5	0+	0+	0+	002	010	037	094	187	303	393	354	232	057	5
	6	0+	0+	0+	0+	001	004	016	047	118	262	531	735	941	6
7	0	932	698	478	210	082	028	008	002	0+	0+	0+	0+	0+	0
	1	066	257	372	367	247	131	055	017	004	0+	0+	0+	0+	1
	2	002	041	124	275	318	261	164	077	025	004	0+	0+	0+	2
	3	0+	004	023	115	227	290	273	194	097	029	003	0+	0+	3
	4	0+	0+	003	029	097	194	273	290	227	115	023	004	0+	4
	5	0+	0+	0+	004	025	077	164	261	318	275	124	041	002	5
	6	0+	0+	0+	0+	004	017	055	131	247	367	372	257	066	6
	7	0+	0+	0+	0+	0+	002	008	028	082	210	478	698	932	7
8	0	923	663	430	168	058	017	004	001	0+	0+	0+	0+	0+	0
	1	075	279	383	336	198	090	031	008	001	0+	0+	0+	0+	1
	2	003	051	149	294	296	209	109	041	010	001	0+	0+	0+	2
	3	0+	005	033	147	254	279	219	124	047	009	0+	0+	0+	3
	4	0+	0+	005	046	136	232	273	232	136	046	005	0+	0+	4
	5	0+	0+	0+	009	047	124	219	279	254	147	033	005	0+	5
	6	0+	0+	0+	001	010	041	109	209	296	294	149	051	003	6
	7	0+	0+	0+	0+	001	008	031	090	198	336	383	279	075	7
	8	0+	0+	0+	0+	0+	001	004	017	058	168	430	663	923	8

Part A: Individual terms, $b(x; n, p)$

n	x	.01	.05	.10	.20	.30	.40	p .50	.60	.70	.80	.90	.95	.99	x
9	0	914	630	387	134	040	010	002	0+	0+	0+	0+	0+	0+	0
	1	083	299	387	302	156	060	018	004	0+	0+	0+	0+	0+	1
	2	003	063	172	302	267	161	070	021	004	0+	0+	0+	0+	2
	3	0+	008	045	176	267	251	164	074	021	003	0+	0+	0+	3
	4	0+	001	007	066	172	251	246	167	074	017	001	0+	0+	4
	5	0+	0+	001	017	074	167	246	251	172	066	007	001	0+	5
	6	0+	0+	0+	003	021	074	164	251	267	176	045	008	0+	6
	7	0+	0+	0+	0+	004	021	070	161	267	302	172	063	003	7
	8	0+	0+	0+	0+	0+	004	018	060	156	302	387	299	083	8
	9	0+	0+	0+	0+	0+	0+	002	010	040	134	387	630	914	9
10	0	904	599	349	107	028	006	001	0+	0+	0+	0+	0+	0+	0
	1	091	315	387	268	121	040	010	002	0+	0+	0+	0+	0+	1
	2	004	075	194	302	233	121	044	011	001	0+	0+	0+	0+	2
	3	0+	010	057	201	267	215	117	042	009	001	0+	0+	0+	3
	4	0+	001	011	088	200	251	205	111	037	006	0+	0+	0+	4
	5	0+	0+	001	026	103	201	246	201	103	026	001	0+	0+	5
	6	0+	0+	0+	006	037	111	205	251	200	088	011	001	0+	6
	7	0+	0+	0+	001	009	042	117	215	267	201	057	010	0+	7
	8	0+	0+	0+	0+	001	011	044	121	233	302	194	075	004	8
	9	0+	0+	0+	0+	0+	002	010	040	121	268	387	315	091	9
	10	0+	0+	0+	0+	0+	0+	001	006	028	107	349	599	904	10
11	0	895	569	314	086	020	004	0+	0+	0+	0+	0+	0+	0+	0
	1	099	329	384	236	093	027	005	001	0+	0+	0+	0+	0+	1
	2	005	087	213	295	200	089	027	005	001	0+	0+	0+	0+	2
	3	0+	014	071	221	257	177	081	023	004	0+	0+	0+	0+	3
	4	0+	001	016	111	220	236	161	070	017	002	0+	0+	0+	4
	5	0+	0+	002	039	132	221	226	147	057	010	0+	0+	0+	5
	6	0+	0+	0+	010	057	147	226	221	132	039	002	0+	0+	6
	7	0+	0+	0+	002	017	070	161	236	220	111	016	001	0+	7
	8	0+	0+	0+	0+	004	023	081	177	257	221	071	014	0+	8
	9	0+	0+	0+	0+	001	005	027	089	200	295	213	087	005	9
	10	0+	0+	0+	0+	0+	001	005	027	093	236	384	329	099	10
	11	0+	0+	0+	0+	0+	0+	0+	004	020	086	314	569	895	11
12	0	886	540	282	069	014	002	0+	0+	0+	0+	0+	0+	0+	0
	1	107	341	377	206	071	017	003	0+	0+	0+	0+	0+	0+	1
	2	006	099	230	283	168	064	016	002	0+	0+	0+	0+	0+	2
	3	0+	017	085	236	240	142	054	012	001	0+	0+	0+	0+	3
	4	0+	002	021	133	231	213	121	042	008	001	0+	0+	0+	4
	5	0+	0+	004	053	158	227	193	101	029	003	0+	0+	0+	5
	6	0+	0+	0+	016	079	177	226	177	079	016	0+	0+	0+	6
	7	0+	0+	0+	003	029	101	193	227	158	053	004	0+	0+	7
	8	0+	0+	0+	001	008	042	121	213	231	133	021	002	0+	8
	9	0+	0+	0+	0+	001	012	054	142	240	236	085	017	0+	9

Part A: Individual terms, $b(x; n, p)$

n	x	p = .01	.05	.10	.20	.30	.40	.50	.60	.70	.80	.90	.95	.99	x
12	10	0+	0+	0+	0+	0+	002	016	064	168	283	230	099	006	10
	11	0+	0+	0+	0+	0+	0+	003	017	071	206	377	341	107	11
	12	0+	0+	0+	0+	0+	0+	0+	002	014	069	282	540	886	12
13	0	878	513	254	055	010	001	0+	0+	0+	0+	0+	0+	0+	0
	1	115	351	367	179	054	011	002	0+	0+	0+	0+	0+	0+	1
	2	007	111	245	268	139	045	010	001	0+	0+	0+	0+	0+	2
	3	0+	021	100	246	218	111	035	006	001	0+	0+	0+	0+	3
	4	0+	003	028	154	234	184	087	024	003	0+	0+	0+	0+	4
	5	0+	0+	006	069	180	221	157	066	014	001	0+	0+	0+	5
	6	0+	0+	001	023	103	197	209	131	044	006	0+	0+	0+	6
	7	0+	0+	0+	006	044	131	209	197	103	023	001	0+	0+	7
	8	0+	0+	0+	001	014	066	157	221	180	069	006	0+	0+	8
	9	0+	0+	0+	0+	003	024	087	184	234	154	028	003	0+	9
	10	0+	0+	0+	0+	001	006	035	111	218	246	100	021	0+	10
	11	0+	0+	0+	0+	0+	001	010	045	139	268	245	111	007	11
	12	0+	0+	0+	0+	0+	0+	002	011	054	179	367	351	115	12
	13	0+	0+	0+	0+	0+	0+	0+	001	010	055	254	513	878	13
14	0	869	488	229	044	007	001	0+	0+	0+	0+	0+	0+	0+	0
	1	123	359	356	154	041	007	001	0+	0+	0+	0+	0+	0+	1
	2	008	123	257	250	113	032	006	001	0+	0+	0+	0+	0+	2
	3	0+	026	114	250	194	085	022	003	0+	0+	0+	0+	0+	3
	4	0+	004	035	172	229	155	061	014	001	0+	0+	0+	0+	4
	5	0+	0+	008	086	196	207	122	041	007	0+	0+	0+	0+	5
	6	0+	0+	001	032	126	207	183	092	023	002	0+	0+	0+	6
	7	0+	0+	0+	009	062	157	209	157	062	009	0+	0+	0+	7
	8	0+	0+	0+	002	023	092	183	207	126	032	001	0+	0+	8
	9	0+	0+	0+	0+	007	041	122	207	196	086	008	0+	0+	9
	10	0+	0+	0+	0+	001	014	061	155	229	172	035	004	0+	10
	11	0+	0+	0+	0+	0+	003	022	085	194	250	114	026	0+	11
	12	0+	0+	0+	0+	0+	001	006	032	113	250	257	123	008	12
	13	0+	0+	0+	0+	0+	0+	001	007	041	154	356	359	123	13
	14	0+	0+	0+	0+	0+	0+	0+	001	007	044	229	488	869	14
15	0	860	463	206	035	005	0+	0+	0+	0+	0+	0+	0+	0+	0
	1	130	366	343	132	031	005	0+	0+	0+	0+	0+	0+	0+	1
	2	009	135	267	231	092	022	003	0+	0+	0+	0+	0+	0+	2
	3	0+	031	129	250	170	063	014	002	0+	0+	0+	0+	0+	3
	4	0+	005	043	188	219	127	042	007	001	0+	0+	0+	0+	4
	5	0+	001	010	103	206	186	092	024	003	0+	0+	0+	0+	5
	6	0+	0+	002	043	147	207	153	061	012	001	0+	0+	0+	6
	7	0+	0+	0+	014	081	177	196	118	035	003	0+	0+	0+	7
	8	0+	0+	0+	003	035	118	196	177	081	014	0+	0+	0+	8
	9	0+	0+	0+	001	012	061	153	207	147	043	002	0+	0+	9

Part A: Individual terms, $b(x; n, p)$

n	x	.01	.05	.10	.20	.30	.40	p .50	.60	.70	.80	.90	.95	.99	x
15	10	0+	0+	0+	0+	003	024	092	186	206	103	010	001	0+	10
	11	0+	0+	0+	0+	001	007	042	127	219	188	043	005	0+	11
	12	0+	0+	0+	0+	0+	002	014	063	170	250	129	031	0+	12
	13	0+	0+	0+	0+	0+	0+	003	022	092	231	267	135	009	13
	14	0+	0+	0+	0+	0+	0+	0+	005	031	132	343	366	130	14
	15	0+	0+	0+	0+	0+	0+	0+	0+	005	035	206	463	860	15
16	0	851	440	185	028	003	0+	0+	0+	0+	0+	0+	0+	0+	0
	1	138	371	329	113	023	003	0+	0+	0+	0+	0+	0+	0+	1
	2	010	146	275	211	073	015	002	0+	0+	0+	0+	0+	0+	2
	3	0+	036	142	246	146	047	009	001	0+	0+	0+	0+	0+	3
	4	0+	006	051	200	204	101	028	004	0+	0+	0+	0+	0+	4
	5	0+	001	014	120	210	162	067	014	001	0+	0+	0+	0+	5
	6	0+	0+	003	055	165	198	122	039	006	0+	0+	0+	0+	6
	7	0+	0+	0+	020	101	189	175	084	019	001	0+	0+	0+	7
	8	0+	0+	0+	006	049	142	196	142	049	006	0+	0+	0+	8
	9	0+	0+	0+	001	019	084	175	189	101	020	0+	0+	0+	9
	10	0+	0+	0+	0+	006	039	122	198	165	055	003	0+	0+	10
	11	0+	0+	0+	0+	001	014	067	162	210	120	014	001	0+	11
	12	0+	0+	0+	0+	0+	004	028	101	204	200	051	006	0+	12
	13	0+	0+	0+	0+	0+	001	009	047	146	246	142	036	0+	13
	14	0+	0+	0+	0+	0+	0+	002	015	073	211	275	146	010	14
	15	0+	0+	0+	0+	0+	0+	0+	003	023	113	329	371	138	15
	16	0+	0+	0+	0+	0+	0+	0+	0+	003	028	185	440	851	16
17	0	843	418	167	023	002	0+	0+	0+	0+	0+	0+	0+	0+	0
	1	145	374	315	096	017	002	0+	0+	0+	0+	0+	0+	0+	1
	2	012	158	280	191	058	010	001	0+	0+	0+	0+	0+	0+	2
	3	001	041	156	239	125	034	005	0+	0+	0+	0+	0+	0+	3
	4	0+	008	060	209	187	080	018	002	0+	0+	0+	0+	0+	4
	5	0+	001	017	136	208	138	047	008	001	0+	0+	0+	0+	5
	6	0+	0+	004	068	178	184	094	024	003	0+	0+	0+	0+	6
	7	0+	0+	001	027	120	193	148	057	009	0+	0+	0+	0+	7
	8	0+	0+	0+	008	064	161	185	107	028	002	0+	0+	0+	8
	9	0+	0+	0+	002	028	107	185	161	064	008	0+	0+	0+	9
	10	0+	0+	0+	0+	009	057	148	193	120	027	001	0+	0+	10
	11	0+	0+	0+	0+	003	024	094	184	178	068	004	0+	0+	11
	12	0+	0+	0+	0+	001	008	047	138	208	136	017	001	0+	12
	13	0+	0+	0+	0+	0+	002	018	080	187	209	060	008	0+	13
	14	0+	0+	0+	0+	0+	0+	005	034	125	239	156	041	001	14
	15	0+	0+	0+	0+	0+	0+	001	010	058	191	280	158	012	15
	16	0+	0+	0+	0+	0+	0+	0+	002	017	096	315	374	145	16
	17	0+	0+	0+	0+	0+	0+	0+	0+	002	023	167	418	843	17

Part A: Individual terms, $b(x; n, p)$

n	x	p .01	.05	.10	.20	.30	.40	.50	.60	.70	.80	.90	.95	.99	x
18	0	835	397	150	018	002	0+	0+	0+	0+	0+	0+	0+	0+	0
	1	152	376	300	081	013	001	0+	0+	0+	0+	0+	0+	0+	1
	2	013	168	284	172	046	007	001	0+	0+	0+	0+	0+	0+	2
	3	001	047	168	230	105	025	003	0+	0+	0+	0+	0+	0+	3
	4	0+	009	070	215	168	061	012	001	0+	0+	0+	0+	0+	4
	5	0+	001	022	151	202	115	033	004	0+	0+	0+	0+	0+	5
	6	0+	0+	005	082	187	166	071	015	001	0+	0+	0+	0+	6
	7	0+	0+	001	035	138	189	121	037	005	0+	0+	0+	0+	7
	8	0+	0+	0+	012	081	173	167	077	015	001	0+	0+	0+	8
	9	0+	0+	0+	003	039	128	185	128	039	003	0+	0+	0+	9
	10	0+	0+	0+	001	015	077	167	173	081	012	0+	0+	0+	10
	11	0+	0+	0+	0+	005	037	121	189	138	035	001	0+	0+	11
	12	0+	0+	0+	0+	001	015	071	166	187	082	005	0+	0+	12
	13	0+	0+	0+	0+	0+	004	033	115	202	151	022	001	0+	13
	14	0+	0+	0+	0+	0+	001	012	061	168	215	070	009	0+	14
	15	0+	0+	0+	0+	0+	0+	003	025	105	230	168	047	001	15
	16	0+	0+	0+	0+	0+	0+	001	007	046	172	284	168	013	16
	17	0+	0+	0+	0+	0+	0+	0+	001	013	081	300	376	152	17
	18	0+	0+	0+	0+	0+	0+	0+	0+	002	018	150	397	835	18
19	0	826	377	135	014	001	0+	0+	0+	0+	0+	0+	0+	0+	0
	1	159	377	285	068	009	001	0+	0+	0+	0+	0+	0+	0+	1
	2	014	179	285	154	036	005	0+	0+	0+	0+	0+	0+	0+	2
	3	001	053	180	218	087	017	002	0+	0+	0+	0+	0+	0+	3
	4	0+	011	080	218	149	047	007	001	0+	0+	0+	0+	0+	4
	5	0+	002	027	164	192	093	022	002	0+	0+	0+	0+	0+	5
	6	0+	0+	007	095	192	145	052	008	001	0+	0+	0+	0+	6
	7	0+	0+	001	044	153	180	096	024	002	0+	0+	0+	0+	7
	8	0+	0+	0+	017	098	180	144	053	008	0+	0+	0+	0+	8
	9	0+	0+	0+	005	051	146	176	098	022	001	0+	0+	0+	9
	10	0+	0+	0+	001	022	098	176	146	051	005	0+	0+	0+	10
	11	0+	0+	0+	0+	008	053	144	180	098	017	0+	0+	0+	11
	12	0+	0+	0+	0+	002	024	096	180	153	044	001	0+	0+	12
	13	0+	0+	0+	0+	001	008	052	145	192	095	007	0+	0+	13
	14	0+	0+	0+	0+	0+	002	022	093	192	164	027	002	0+	14
	15	0+	0+	0+	0+	0+	001	007	047	149	218	080	011	0+	15
	16	0+	0+	0+	0+	0+	0+	002	017	087	218	180	053	001	16
	17	0+	0+	0+	0+	0+	0+	0+	005	036	154	285	179	014	17
	18	0+	0+	0+	0+	0+	0+	0+	001	009	068	285	377	159	18
	19	0+	0+	0+	0+	0+	0+	0+	0+	001	014	135	377	826	19
20	0	818	358	122	012	001	0+	0+	0+	0+	0+	0+	0+	0+	0
	1	165	377	270	058	007	0+	0+	0+	0+	0+	0+	0+	0+	1
	2	016	189	285	137	028	003	0+	0+	0+	0+	0+	0+	0+	2
	3	001	060	190	205	072	012	001	0+	0+	0+	0+	0+	0+	3
	4	0+	013	090	218	130	035	005	0+	0+	0+	0+	0+	0+	4

Part A: Individual terms, $b(x; n, p)$

n	x	.01	.05	.10	.20	.30	.40	p .50	.60	.70	.80	.90	.95	.99	x
20	5	0+	002	032	175	179	075	015	001	0+	0+	0+	0+	0+	5
	6	0+	0+	009	109	192	124	037	005	0+	0+	0+	0+	0+	6
	7	0+	0+	002	055	164	166	074	015	001	0+	0+	0+	0+	7
	8	0+	0+	0+	022	114	180	120	035	004	0+	0+	0+	0+	8
	9	0+	0+	0+	007	065	160	160	071	012	0+	0+	0+	0+	9
	10	0+	0+	0+	002	031	117	176	117	031	002	0+	0+	0+	10
	11	0+	0+	0+	0+	012	071	160	160	065	007	0+	0+	0+	11
	12	0+	0+	0+	0+	004	035	120	180	114	022	0+	0+	0+	12
	13	0+	0+	0+	0+	001	015	074	166	164	055	002	0+	0+	13
	14	0+	0+	0+	0+	0+	005	037	124	192	109	009	0+	0+	14
	15	0+	0+	0+	0+	0+	001	015	075	179	175	032	002	0+	15
	16	0+	0+	0+	0+	0+	0+	005	035	130	218	090	013	0+	16
	17	0+	0+	0+	0+	0+	0+	001	012	072	205	190	060	001	17
	18	0+	0+	0+	0+	0+	0+	0+	003	028	137	285	189	016	18
	19	0+	0+	0+	0+	0+	0+	0+	0+	007	058	270	377	165	19
	20	0+	0+	0+	0+	0+	0+	0+	0+	001	012	122	358	818	20
21	0	810	341	109	009	001	0+	0+	0+	0+	0+	0+	0+	0+	0
	1	172	376	255	048	005	0+	0+	0+	0+	0+	0+	0+	0+	1
	2	017	198	284	121	022	002	0+	0+	0+	0+	0+	0+	0+	2
	3	001	066	200	192	058	009	001	0+	0+	0+	0+	0+	0+	3
	4	0+	016	100	216	113	026	003	0+	0+	0+	0+	0+	0+	4
	5	0+	003	038	183	164	059	010	001	0+	0+	0+	0+	0+	5
	6	0+	0+	011	122	188	105	026	003	0+	0+	0+	0+	0+	6
	7	0+	0+	003	065	172	149	055	009	0+	0+	0+	0+	0+	7
	8	0+	0+	001	029	129	174	097	023	002	0+	0+	0+	0+	8
	9	0+	0+	0+	010	080	168	140	050	006	0+	0+	0+	0+	9
	10	0+	0+	0+	003	041	134	168	089	018	001	0+	0+	0+	10
	11	0+	0+	0+	001	018	089	168	134	041	003	0+	0+	0+	11
	12	0+	0+	0+	0+	006	050	140	168	080	010	0+	0+	0+	12
	13	0+	0+	0+	0+	002	023	097	174	129	029	001	0+	0+	13
	14	0+	0+	0+	0+	0+	009	055	149	172	065	003	0+	0+	14
	15	0+	0+	0+	0+	0+	003	026	105	188	122	011	0+	0+	15
	16	0+	0+	0+	0+	0+	001	010	059	164	183	038	003	0+	16
	17	0+	0+	0+	0+	0+	0+	003	026	113	216	100	016	0+	17
	18	0+	0+	0+	0+	0+	0+	001	009	058	192	200	066	001	18
	19	0+	0+	0+	0+	0+	0+	0+	002	022	121	284	198	017	19
	20	0+	0+	0+	0+	0+	0+	0+	0+	005	048	255	376	172	20
	21	0+	0+	0+	0+	0+	0+	0+	0+	001	009	109	341	810	21
22	0	802	324	098	007	0+	0+	0+	0+	0+	0+	0+	0+	0+	0
	1	178	375	241	041	004	0+	0+	0+	0+	0+	0+	0+	0+	1
	2	019	207	281	107	017	001	0+	0+	0+	0+	0+	0+	0+	2
	3	001	073	208	178	047	006	0+	0+	0+	0+	0+	0+	0+	3
	4	0+	018	110	211	096	019	002	0+	0+	0+	0+	0+	0+	4

Part A: Individual terms, $b(x; n, p)$

n	x	.01	.05	.10	.20	.30	.40	p .50	.60	.70	.80	.90	.95	.99	x
22	5	0+	003	044	190	149	046	006	0+	0+	0+	0+	0+	0+	5
	6	0+	001	014	134	181	086	018	001	0+	0+	0+	0+	0+	6
	7	0+	0+	004	077	177	131	041	005	0+	0+	0+	0+	0+	7
	8	0+	0+	001	036	142	164	076	014	001	0+	0+	0+	0+	8
	9	0+	0+	0+	014	095	170	119	034	003	0+	0+	0+	0+	9
	10	0+	0+	0+	005	053	148	154	066	010	0+	0+	0+	0+	10
	11	0+	0+	0+	001	025	107	168	107	025	001	0+	0+	0+	11
	12	0+	0+	0+	0+	010	066	154	148	053	005	0+	0+	0+	12
	13	0+	0+	0+	0+	003	034	119	170	095	014	0+	0+	0+	13
	14	0+	0+	0+	0+	001	014	076	164	142	036	001	0+	0+	14
	15	0+	0+	0+	0+	0+	005	041	131	177	077	004	0+	0+	15
	16	0+	0+	0+	0+	0+	001	018	086	181	134	014	001	0+	16
	17	0+	0+	0+	0+	0+	0+	006	046	149	190	044	003	0+	17
	18	0+	0+	0+	0+	0+	0+	002	019	096	211	110	018	0+	18
	19	0+	0+	0+	0+	0+	0+	0+	006	047	178	208	073	001	19
	20	0+	0+	0+	0+	0+	0+	0+	001	017	107	281	207	019	20
	21	0+	0+	0+	0+	0+	0+	0+	0+	004	041	241	375	178	21
	22	0+	0+	0+	0+	0+	0+	0+	0+	0+	007	098	324	802	22
23	0	794	307	089	006	0+	0+	0+	0+	0+	0+	0+	0+	0+	0
	1	184	372	226	034	003	0+	0+	0+	0+	0+	0+	0+	0+	1
	2	020	215	277	093	013	001	0+	0+	0+	0+	0+	0+	0+	2
	3	001	079	215	163	038	004	0+	0+	0+	0+	0+	0+	0+	3
	4	0+	021	120	204	082	014	001	0+	0+	0+	0+	0+	0+	4
	5	0+	004	051	194	133	035	004	0+	0+	0+	0+	0+	0+	5
	6	0+	001	017	145	171	070	012	001	0+	0+	0+	0+	0+	6
	7	0+	0+	005	088	178	113	029	003	0+	0+	0+	0+	0+	7
	8	0+	0+	001	044	153	151	058	009	0+	0+	0+	0+	0+	8
	9	0+	0+	0+	018	109	168	097	022	002	0+	0+	0+	0+	9
	10	0+	0+	0+	006	065	157	136	046	005	0+	0+	0+	0+	10
	11	0+	0+	0+	002	033	123	161	082	014	0+	0+	0+	0+	11
	12	0+	0+	0+	0+	014	082	161	123	033	002	0+	0+	0+	12
	13	0+	0+	0+	0+	005	046	136	157	065	006	0+	0+	0+	13
	14	0+	0+	0+	0+	002	022	097	168	109	018	0+	0+	0+	14
	15	0+	0+	0+	0+	0+	009	058	151	153	044	001	0+	0+	15
	16	0+	0+	0+	0+	0+	003	029	113	178	088	005	0+	0+	16
	17	0+	0+	0+	0+	0+	001	012	070	171	145	017	001	0+	17
	18	0+	0+	0+	0+	0+	0+	004	035	133	194	051	004	0+	18
	19	0+	0+	0+	0+	0+	0+	001	014	082	204	120	021	0+	19
	20	0+	0+	0+	0+	0+	0+	0+	004	038	163	215	079	001	20
	21	0+	0+	0+	0+	0+	0+	0+	001	013	093	277	215	020	21
	22	0+	0+	0+	0+	0+	0+	0+	0+	003	034	226	372	184	22
	23	0+	0+	0+	0+	0+	0+	0+	0+	0+	006	089	307	794	23

Part A: Individual terms, $b(x; n, p)$

n	x	.01	.05	.10	.20	.30	.40	p .50	.60	.70	.80	.90	.95	.99	x
24	0	786	292	080	005	0+	0+	0+	0+	0+	0+	0+	0+	0+	0
	1	190	369	213	028	002	0+	0+	0+	0+	0+	0+	0+	0+	1
	2	022	223	272	081	010	001	0+	0+	0+	0+	0+	0+	0+	2
	3	002	086	221	149	031	003	0+	0+	0+	0+	0+	0+	0+	3
	4	0+	024	129	196	069	010	001	0+	0+	0+	0+	0+	0+	4
	5	0+	005	057	196	118	027	003	0+	0+	0+	0+	0+	0+	5
	6	0+	001	020	155	160	056	008	0+	0+	0+	0+	0+	0+	6
	7	0+	0+	006	100	176	096	021	002	0+	0+	0+	0+	0+	7
	8	0+	0+	001	053	160	136	044	005	0+	0+	0+	0+	0+	8
	9	0+	0+	0+	024	122	161	078	014	001	0+	0+	0+	0+	9
	10	0+	0+	0+	009	079	161	117	032	003	0+	0+	0+	0+	10
	11	0+	0+	0+	003	043	137	149	061	008	0+	0+	0+	0+	11
	12	0+	0+	0+	001	020	099	161	099	020	001	0+	0+	0+	12
	13	0+	0+	0+	0+	008	061	149	137	043	003	0+	0+	0+	13
	14	0+	0+	0+	0+	003	032	117	161	079	009	0+	0+	0+	14
	15	0+	0+	0+	0+	001	014	078	161	122	024	0+	0+	0+	15
	16	0+	0+	0+	0+	0+	005	044	136	160	053	001	0+	0+	16
	17	0+	0+	0+	0+	0+	002	021	096	176	100	006	0+	0+	17
	18	0+	0+	0+	0+	0+	0+	008	056	160	155	020	001	0+	18
	19	0+	0+	0+	0+	0+	0+	003	027	118	196	057	005	0+	19
	20	0+	0+	0+	0+	0+	0+	001	010	069	196	129	024	0+	20
	21	0+	0+	0+	0+	0+	0+	0+	003	031	149	221	086	002	21
	22	0+	0+	0+	0+	0+	0+	0+	001	010	081	272	223	022	22
	23	0+	0+	0+	0+	0+	0+	0+	0+	002	028	213	369	190	23
	24	0+	0+	0+	0+	0+	0+	0+	0+	0+	005	080	292	786	24
25	0	778	277	072	004	0+	0+	0+	0+	0+	0+	0+	0+	0+	0
	1	196	365	199	024	001	0+	0+	0+	0+	0+	0+	0+	0+	1
	2	024	231	266	071	007	0+	0+	0+	0+	0+	0+	0+	0+	2
	3	002	093	226	136	024	002	0+	0+	0+	0+	0+	0+	0+	3
	4	0+	027	138	187	057	007	0+	0+	0+	0+	0+	0+	0+	4
	5	0+	006	065	196	103	020	002	0+	0+	0+	0+	0+	0+	5
	6	0+	001	024	163	147	044	005	0+	0+	0+	0+	0+	0+	6
	7	0+	0+	007	111	171	080	014	001	0+	0+	0+	0+	0+	7
	8	0+	0+	002	062	165	120	032	003	0+	0+	0+	0+	0+	8
	9	0+	0+	0+	029	134	151	061	009	0+	0+	0+	0+	0+	9
	10	0+	0+	0+	012	092	161	097	021	001	0+	0+	0+	0+	10
	11	0+	0+	0+	004	054	147	133	043	004	0+	0+	0+	0+	11
	12	0+	0+	0+	001	027	114	155	076	011	0+	0+	0+	0+	12
	13	0+	0+	0+	0+	011	076	155	114	027	001	0+	0+	0+	13
	14	0+	0+	0+	0+	004	043	133	147	054	004	0+	0+	0+	14
	15	0+	0+	0+	0+	001	021	097	161	092	012	0+	0+	0+	15
	16	0+	0+	0+	0+	0+	009	061	151	134	029	0+	0+	0+	16
	17	0+	0+	0+	0+	0+	003	032	120	165	062	002	0+	0+	17
	18	0+	0+	0+	0+	0+	001	014	080	171	111	007	0+	0+	18
	19	0+	0+	0+	0+	0+	0+	005	044	147	163	024	001	0+	19

Part A: Individual terms, $b(x; n, p)$

n	x	.01	.05	.10	.20	.30	.40	p .50	.60	.70	.80	.90	.95	.99	x
25	20	0+	0+	0+	0+	0+	0+	002	020	103	196	065	006	0+	20
	21	0+	0+	0+	0+	0+	0+	0+	007	057	187	138	027	0+	21
	22	0+	0+	0+	0+	0+	0+	0+	002	024	136	226	093	002	22
	23	0+	0+	0+	0+	0+	0+	0+	0+	007	071	266	231	024	23
	24	0+	0+	0+	0+	0+	0+	0+	0+	001	024	199	365	196	24
	25	0+	0+	0+	0+	0+	0+	0+	0+	0+	004	072	277	778	25

Part B: Cumulative terms, $\sum_{x=r}^{n} b(x; n, p)$

n	r	.01	.05	.10	.20	.30	.40	p .50	.60	.70	.80	.90	.95	.99	r
2	0	1	1	1	1	1	1	1	1	1	1	1	1	1	0
	1	020	098	190	360	510	640	750	840	910	960	990	998	1−	1
	2	0+	002	010	040	090	160	250	360	490	640	810	902	980	2
3	0	1	1	1	1	1	1	1	1	1	1	1	1	1	0
	1	030	143	271	488	657	784	875	936	973	992	999	1−	1−	1
	2	0+	007	028	104	216	352	500	648	784	896	972	993	1−	2
	3	0+	0+	001	008	027	064	125	216	343	512	729	857	970	3
4	0	1	1	1	1	1	1	1	1	1	1	1	1	1	0
	1	039	185	344	590	760	870	938	974	992	998	1−	1−	1−	1
	2	001	014	052	181	348	525	688	821	916	973	996	1−	1−	2
	3	0+	0+	004	027	084	179	312	475	652	819	948	986	999	3
	4	0+	0+	0+	002	008	026	062	130	240	410	656	815	961	4
5	0	1	1	1	1	1	1	1	1	1	1	1	1	1	0
	1	049	226	410	672	832	922	969	990	998	1−	1−	1−	1−	1
	2	001	023	081	263	472	663	812	913	969	993	1−	1−	1−	2
	3	0+	001	009	058	163	317	500	683	837	942	991	999	1−	3
	4	0+	0+	0+	007	031	087	188	337	528	737	919	977	999	4
	5	0+	0+	0+	0+	002	010	031	078	168	328	590	774	951	5
6	0	1	1	1	1	1	1	1	1	1	1	1	1	1	0
	1	059	265	469	738	882	953	984	996	999	1−	1−	1−	1−	1
	2	001	033	114	345	580	767	891	959	989	998	1−	1−	1−	2
	3	0+	002	016	099	256	456	656	821	930	983	999	1−	1−	3
	4	0+	0+	001	017	070	179	344	544	744	901	984	998	1−	4
	5	0+	0+	0+	002	011	041	109	233	420	655	886	967	999	5
	6	0+	0+	0+	0+	001	004	016	047	118	262	531	735	941	6
7	0	1	1	1	1	1	1	1	1	1	1	1	1	1	0
	1	068	302	522	790	918	972	992	998	1−	1−	1−	1−	1−	1
	2	002	044	150	423	671	841	938	981	996	1−	1−	1−	1−	2
	3	0+	004	026	148	353	580	773	904	971	995	1−	1−	1−	3
	4	0+	0+	003	033	126	290	500	710	874	967	997	1−	1−	4
	5	0+	0+	0+	005	029	096	227	420	647	852	974	996	1−	5
	6	0+	0+	0+	0+	004	019	062	159	329	577	850	956	998	6
	7	0+	0+	0+	0+	0+	002	008	028	082	210	478	698	932	7
8	0	1	1	1	1	1	1	1	1	1	1	1	1	1	0
	1	077	337	570	832	942	983	996	999	1−	1−	1−	1−	1−	1
	2	003	057	187	497	745	894	965	991	999	1−	1−	1−	1−	2
	3	0+	006	038	203	448	685	855	950	989	999	1−	1−	1−	3
	4	0+	0+	005	056	194	406	637	826	942	990	1−	1−	1−	4
	5	0+	0+	0+	010	058	174	363	594	806	944	995	1−	1−	5
	6	0+	0+	0+	001	011	050	145	315	552	797	962	994	1−	6
	7	0+	0+	0+	0+	001	009	035	106	255	503	813	943	997	7
	8	0+	0+	0+	0+	0+	001	004	017	058	168	430	663	923	8

Part B: Cumulative terms, $\sum_{x=r}^{n} b(x; n, p)$

n	r	.01	.05	.10	.20	.30	.40	p .50	.60	.70	.80	.90	.95	.99	r
9	0	1	1	1	1	1	1	1	1	1	1	1	1	1	0
	1	086	370	613	866	960	990	998	1−	1−	1−	1−	1−	1−	1
	2	003	071	225	564	804	929	980	996	1−	1−	1−	1−	1−	2
	3	0+	008	053	262	537	768	910	975	996	1−	1−	1−	1−	3
	4	0+	001	008	086	270	517	746	901	975	997	1−	1−	1−	4
	5	0+	0+	001	020	099	267	500	733	901	980	999	1−	1−	5
	6	0+	0+	0+	003	025	099	254	483	730	914	992	999	1−	6
	7	0+	0+	0+	0+	004	025	090	232	463	738	947	992	1−	7
	8	0+	0+	0+	0+	0+	004	020	071	196	436	775	929	997	8
	9	0+	0+	0+	0+	0+	0+	002	010	040	134	387	630	914	9
10	0	1	1	1	1	1	1	1	1	1	1	1	1	1	0
	1	096	401	651	893	972	994	999	1−	1−	1−	1−	1−	1−	1
	2	004	086	264	624	851	954	989	998	1−	1−	1−	1−	1−	2
	3	0+	012	070	322	617	833	945	988	998	1−	1−	1−	1−	3
	4	0+	001	013	121	350	618	828	945	989	999	1−	1−	1−	4
	5	0+	0+	002	033	150	367	623	834	953	994	1−	1−	1−	5
	6	0+	0+	0+	006	047	166	377	633	850	967	998	1−	1−	6
	7	0+	0+	0+	001	011	055	172	382	650	879	987	999	1−	7
	8	0+	0+	0+	0+	002	012	055	167	383	678	930	988	1−	8
	9	0+	0+	0+	0+	0+	002	011	046	149	376	736	914	996	9
	10	0+	0+	0+	0+	0+	0+	001	006	028	107	349	599	904	10
11	0	1	1	1	1	1	1	1	1	1	1	1	1	1	0
	1	105	431	686	914	980	996	1−	1−	1−	1−	1−	1−	1−	1
	2	005	102	303	678	887	970	994	999	1−	1−	1−	1−	1−	2
	3	0+	015	090	383	687	881	967	994	999	1−	1−	1−	1−	3
	4	0+	002	019	161	430	704	887	971	996	1−	1−	1−	1−	4
	5	0+	0+	003	050	210	467	726	901	978	998	1−	1−	1−	5
	6	0+	0+	0+	012	078	247	500	753	922	988	1−	1−	1−	6
	7	0+	0+	0+	002	022	099	274	533	790	950	997	1−	1−	7
	8	0+	0+	0+	0+	004	029	113	296	570	839	981	998	1−	8
	9	0+	0+	0+	0+	001	006	033	119	313	617	910	985	1−	9
	10	0+	0+	0+	0+	0+	001	006	030	113	322	697	898	995	10
	11	0+	0+	0+	0+	0+	0+	0+	004	020	086	314	569	895	11
12	0	1	1	1	1	1	1	1	1	1	1	1	1	1	0
	1	114	460	718	931	986	998	1−	1−	1−	1−	1−	1−	1−	1
	2	006	118	341	725	915	980	997	1−	1−	1−	1−	1−	1−	2
	3	0+	020	111	442	747	917	981	997	1−	1−	1−	1−	1−	3
	4	0+	002	026	205	507	775	927	985	998	1−	1−	1−	1−	4
	5	0+	0+	004	073	276	562	806	943	991	999	1−	1−	1−	5
	6	0+	0+	001	019	118	335	613	842	961	996	1−	1−	1−	6
	7	0+	0+	0+	004	039	158	387	665	882	981	999	1−	1−	7
	8	0+	0+	0+	001	009	057	194	438	724	927	996	1−	1−	8
	9	0+	0+	0+	0+	002	015	073	225	493	795	974	998	1−	9

Part B: Cumulative terms, $\sum_{x=r}^{n} b(x; n, p)$

n	r	.01	.05	.10	.20	.30	.40	p .50	.60	.70	.80	.90	.95	.99	r
12	10	0+	0+	0+	0+	0+	003	019	083	253	558	889	980	1—	10
	11	0+	0+	0+	0+	0+	0+	003	020	085	275	659	882	994	11
	12	0+	0+	0+	0+	0+	0+	0+	002	014	069	282	540	886	12
13	0	1	1	1	1	1	1	1	1	1	1	1	1	1	0
	1	122	487	746	945	990	999	1—	1—	1—	1—	1—	1—	1—	1
	2	007	135	379	766	936	987	998	1—	1—	1—	1—	1—	1—	2
	3	0+	025	134	498	798	942	989	999	1—	1—	1—	1—	1—	3
	4	0+	003	034	253	579	831	954	992	999	1—	1—	1—	1—	4
	5	0+	0+	006	099	346	647	867	968	996	1—	1—	1—	1—	5
	6	0+	0+	001	030	165	426	709	902	982	999	1—	1—	1—	6
	7	0+	0+	0+	007	062	229	500	771	938	993	1—	1—	1—	7
	8	0+	0+	0+	001	018	098	291	574	835	970	999	1—	1—	8
	9	0+	0+	0+	0+	004	032	133	353	654	901	994	1—	1—	9
	10	0+	0+	0+	0+	001	008	046	169	421	747	966	997	1—	10
	11	0+	0+	0+	0+	0+	001	011	058	202	502	866	975	1—	11
	12	0+	0+	0+	0+	0+	0+	002	013	064	234	621	865	993	12
	13	0+	0+	0+	0+	0+	0+	0+	001	010	055	254	513	878	13
14	0	1	1	1	1	1	1	1	1	1	1	1	1	1	0
	1	131	512	771	956	993	999	1—	1—	1—	1—	1—	1—	1—	1
	2	008	153	415	802	953	992	999	1—	1—	1—	1—	1—	1—	2
	3	0+	030	158	552	839	960	994	999	1—	1—	1—	1—	1—	3
	4	0+	004	044	302	645	876	971	996	1—	1—	1—	1—	1—	4
	5	0+	0+	009	130	416	721	910	982	998	1—	1—	1—	1—	5
	6	0+	0+	001	044	219	514	788	942	992	1—	1—	1—	1—	6
	7	0+	0+	0+	012	093	308	605	850	969	998	1—	1—	1—	7
	8	0+	0+	0+	002	031	150	395	692	907	988	1—	1—	1—	8
	9	0+	0+	0+	0+	008	058	212	486	781	956	999	1—	1—	9
	10	0+	0+	0+	0+	002	018	090	279	584	870	991	1—	1—	10
	11	0+	0+	0+	0+	0+	004	029	124	355	698	956	996	1—	11
	12	0+	0+	0+	0+	0+	001	006	040	161	448	842	970	1—	12
	13	0+	0+	0+	0+	0+	0+	001	008	047	198	585	847	992	13
	14	0+	0+	0+	0+	0+	0+	0+	001	007	044	229	488	869	14
15	0	1	1	1	1	1	1	1	1	1	1	1	1	1	0
	1	140	537	794	965	995	1—	1—	1—	1—	1—	1—	1—	1—	1
	2	010	171	451	833	965	995	1—	1—	1—	1—	1—	1—	1—	2
	3	0+	036	184	602	873	973	996	1—	1—	1—	1—	1—	1—	3
	4	0+	005	056	352	703	909	982	998	1—	1—	1—	1—	1—	4
	5	0+	001	013	164	485	783	941	991	999	1—	1—	1—	1—	5
	6	0+	0+	002	061	278	597	849	966	996	1—	1—	1—	1—	6
	7	0+	0+	0+	018	131	390	696	905	985	999	1—	1—	1—	7
	8	0+	0+	0+	004	050	213	500	787	950	996	1—	1—	1—	8
	9	0+	0+	0+	001	015	095	304	610	869	982	1—	1—	1—	9

Part B: Cumulative terms, $\sum_{x=r}^{n} b(x; n, p)$

n	r	.01	.05	.10	.20	.30	.40	p .50	.60	.70	.80	.90	.95	.99	r
15	10	0+	0+	0+	0+	004	034	151	403	722	939	998	1−	1−	10
	11	0+	0+	0+	0+	001	009	059	217	515	836	987	999	1−	11
	12	0+	0+	0+	0+	0+	002	018	091	297	648	944	995	1−	12
	13	0+	0+	0+	0+	0+	0+	004	027	127	398	816	964	1−	13
	14	0+	0+	0+	0+	0+	0+	0+	005	035	167	549	829	990	14
	15	0+	0+	0+	0+	0+	0+	0+	0+	005	035	206	463	860	15
16	0	1	1	1	1	1	1	1	1	1	1	1	1	1	0
	1	149	560	815	972	997	1−	1−	1−	1−	1−	1−	1−	1−	1
	2	011	189	485	859	974	997	1−	1−	1−	1−	1−	1−	1−	2
	3	001	043	211	648	901	982	998	1−	1−	1−	1−	1−	1−	3
	4	0+	007	068	402	754	935	989	999	1−	1−	1−	1−	1−	4
	5	0+	001	017	202	550	833	962	995	1−	1−	1−	1−	1−	5
	6	0+	0+	003	082	340	671	895	981	998	1−	1−	1−	1−	6
	7	0+	0+	001	027	175	473	773	942	993	1−	1−	1−	1−	7
	8	0+	0+	0+	007	074	284	598	858	974	999	1−	1−	1−	8
	9	0+	0+	0+	001	026	142	402	716	926	993	1−	1−	1−	9
	10	0+	0+	0+	0+	007	058	227	527	825	973	999	1−	1−	10
	11	0+	0+	0+	0+	002	019	105	329	660	918	997	1−	1−	11
	12	0+	0+	0+	0+	0+	005	038	167	450	798	983	999	1−	12
	13	0+	0+	0+	0+	0+	001	011	065	246	598	932	993	1−	13
	14	0+	0+	0+	0+	0+	0+	002	018	099	352	789	957	999	14
	15	0+	0+	0+	0+	0+	0+	0+	003	026	141	515	811	989	15
	16	0+	0+	0+	0+	0+	0+	0+	0+	003	028	185	440	851	16
17	0	1	1	1	1	1	1	1	1	1	1	1	1	1	0
	1	157	582	833	977	998	1−	1−	1−	1−	1−	1−	1−	1−	1
	2	012	208	518	882	981	998	1−	1−	1−	1−	1−	1−	1−	2
	3	001	050	238	690	923	988	999	1−	1−	1−	1−	1−	1−	3
	4	0+	009	083	451	798	954	994	1−	1−	1−	1−	1−	1−	4
	5	0+	001	022	242	611	874	975	997	1−	1−	1−	1−	1−	5
	6	0+	0+	005	106	403	736	928	989	999	1−	1−	1−	1−	6
	7	0+	0+	001	038	225	552	834	965	997	1−	1−	1−	1−	7
	8	0+	0+	0+	011	105	359	685	908	987	1−	1−	1−	1−	8
	9	0+	0+	0+	003	040	199	500	801	960	997	1−	1−	1−	9
	10	0+	0+	0+	0+	013	092	315	641	895	989	1−	1−	1−	10
	11	0+	0+	0+	0+	003	035	166	448	775	962	999	1−	1−	11
	12	0+	0+	0+	0+	001	011	072	264	597	894	995	1−	1−	12
	13	0+	0+	0+	0+	0+	003	025	126	389	758	978	999	1−	13
	14	0+	0+	0+	0+	0+	0+	006	046	202	549	917	991	1−	14
	15	0+	0+	0+	0+	0+	0+	001	012	077	310	762	950	999	15
	16	0+	0+	0+	0+	0+	0+	0+	002	019	118	482	792	988	16
	17	0+	0+	0+	0+	0+	0+	0+	0+	002	023	167	418	843	17

Part B: Cumulative terms, $\sum_{x=r}^{n} b(x; n, p)$

n	r	.01	.05	.10	.20	.30	.40	.50	.60	.70	.80	.90	.95	.99	r
18	0	1	1	1	1	1	1	1	1	1	1	1	1	1	0
	1	165	603	850	982	998	1−	1−	1−	1−	1−	1−	1−	1−	1
	2	014	226	550	901	986	999	1−	1−	1−	1−	1−	1−	1−	2
	3	001	058	266	729	940	992	999	1−	1−	1−	1−	1−	1−	3
	4	0+	011	098	499	835	967	996	1−	1−	1−	1−	1−	1−	4
	5	0+	002	028	284	667	906	985	999	1−	1−	1−	1−	1−	5
	6	0+	0+	006	133	466	791	952	994	1−	1−	1−	1−	1−	6
	7	0+	0+	001	051	278	626	881	980	999	1−	1−	1−	1−	7
	8	0+	0+	0+	016	141	437	760	942	994	1−	1−	1−	1−	8
	9	0+	0+	0+	004	060	263	593	865	979	999	1−	1−	1−	9
	10	0+	0+	0+	001	021	135	407	737	940	996	1−	1−	1−	10
	11	0+	0+	0+	0+	006	058	240	563	859	984	1−	1−	1−	11
	12	0+	0+	0+	0+	001	020	119	374	722	949	999	1−	1−	12
	13	0+	0+	0+	0+	0+	006	048	209	534	867	994	1−	1−	13
	14	0+	0+	0+	0+	0+	001	015	094	333	716	972	998	1−	14
	15	0+	0+	0+	0+	0+	0+	004	033	165	501	902	989	1−	15
	16	0+	0+	0+	0+	0+	0+	001	008	060	271	734	942	999	16
	17	0+	0+	0+	0+	0+	0+	0+	001	014	099	450	774	986	17
	18	0+	0+	0+	0+	0+	0+	0+	0+	002	018	150	397	835	18
19	0	1	1	1	1	1	1	1	1	1	1	1	1	1	0
	1	174	623	865	986	999	1−	1−	1−	1−	1−	1−	1−	1−	1
	2	015	245	580	917	990	999	1−	1−	1−	1−	1−	1−	1−	2
	3	001	067	295	763	954	995	1−	1−	1−	1−	1−	1−	1−	3
	4	0+	013	115	545	867	977	998	1−	1−	1−	1−	1−	1−	4
	5	0+	002	035	327	718	930	990	999	1−	1−	1−	1−	1−	5
	6	0+	0+	009	163	526	837	968	997	1−	1−	1−	1−	1−	6
	7	0+	0+	002	068	334	692	916	988	999	1−	1−	1−	1−	7
	8	0+	0+	0+	023	182	512	820	965	997	1−	1−	1−	1−	8
	9	0+	0+	0+	007	084	333	676	912	989	1−	1−	1−	1−	9
	10	0+	0+	0+	002	033	186	500	814	967	998	1−	1−	1−	10
	11	0+	0+	0+	0+	011	088	324	667	916	993	1−	1−	1−	11
	12	0+	0+	0+	0+	003	035	180	488	818	977	1−	1−	1−	12
	13	0+	0+	0+	0+	001	012	084	308	666	932	998	1−	1−	13
	14	0+	0+	0+	0+	0+	003	032	163	474	837	991	1−	1−	14
	15	0+	0+	0+	0+	0+	001	010	070	282	673	965	998	1−	15
	16	0+	0+	0+	0+	0+	0+	002	023	133	455	885	987	1−	16
	17	0+	0+	0+	0+	0+	0+	0+	005	046	237	705	933	999	17
	18	0+	0+	0+	0+	0+	0+	0+	001	010	083	420	755	985	18
	19	0+	0+	0+	0+	0+	0+	0+	0+	001	014	135	377	826	19
20	0	1	1	1	1	1	1	1	1	1	1	1	1	1	0
	1	182	642	878	988	999	1−	1−	1−	1−	1−	1−	1−	1−	1
	2	017	264	608	931	992	999	1−	1−	1−	1−	1−	1−	1−	2
	3	001	075	323	794	965	996	1−	1−	1−	1−	1−	1−	1−	3
	4	0+	016	133	589	893	984	999	1−	1−	1−	1−	1−	1−	4

Part B: Cumulative terms, $\sum_{x=r}^{n} b(x; n, p)$

n	r	.01	.05	.10	.20	.30	.40	p .50	.60	.70	.80	.90	.95	.99	r
20	5	0+	003	043	370	762	949	994	1−	1−	1−	1−	1−	1−	5
	6	0+	0+	011	196	584	874	979	998	1−	1−	1−	1−	1−	6
	7	0+	0+	002	087	392	750	942	994	1−	1−	1−	1−	1−	7
	8	0+	0+	0+	032	228	584	868	979	999	1−	1−	1−	1−	8
	9	0+	0+	0+	010	113	404	748	943	995	1−	1−	1−	1−	9
	10	0+	0+	0+	003	048	245	588	872	983	999	1−	1−	1−	10
	11	0+	0+	0+	001	017	128	412	755	952	997	1−	1−	1−	11
	12	0+	0+	0+	0+	005	057	252	596	887	990	1−	1−	1−	12
	13	0+	0+	0+	0+	001	021	132	416	772	968	1−	1−	1−	13
	14	0+	0+	0+	0+	0+	006	058	250	608	913	998	1−	1−	14
	15	0+	0+	0+	0+	0+	002	021	126	416	804	989	1−	1−	15
	16	0+	0+	0+	0+	0+	0+	006	051	238	630	957	997	1−	16
	17	0+	0+	0+	0+	0+	0+	001	016	107	411	867	984	1−	17
	18	0+	0+	0+	0+	0+	0+	0+	004	035	206	677	925	999	18
	19	0+	0+	0+	0+	0+	0+	0+	001	008	069	392	736	983	19
	20	0+	0+	0+	0+	0+	0+	0+	0+	001	012	122	358	818	20
21	0	1	1	1	1	1	1	1	1	1	1	1	1	1	0
	1	190	659	891	991	999	1−	1−	1−	1−	1−	1−	1−	1−	1
	2	019	283	635	942	994	1−	1−	1−	1−	1−	1−	1−	1−	2
	3	001	085	352	821	973	998	1−	1−	1−	1−	1−	1−	1−	3
	4	0+	019	152	630	914	989	999	1−	1−	1−	1−	1−	1−	4
	5	0+	003	052	414	802	963	996	1−	1−	1−	1−	1−	1−	5
	6	0+	0+	014	231	637	904	987	999	1−	1−	1−	1−	1−	6
	7	0+	0+	003	109	449	800	961	996	1−	1−	1−	1−	1−	7
	8	0+	0+	001	043	277	650	905	988	999	1−	1−	1−	1−	8
	9	0+	0+	0+	014	148	476	808	965	998	1−	1−	1−	1−	9
	10	0+	0+	0+	004	068	309	668	915	991	1−	1−	1−	1−	10
	11	0+	0+	0+	001	026	174	500	826	974	999	1−	1−	1−	11
	12	0+	0+	0+	0+	009	085	332	691	932	996	1−	1−	1−	12
	13	0+	0+	0+	0+	002	035	192	524	852	986	1−	1−	1−	13
	14	0+	0+	0+	0+	001	012	095	350	723	957	999	1−	1−	14
	15	0+	0+	0+	0+	0+	004	039	200	551	891	997	1−	1−	15
	16	0+	0+	0+	0+	0+	001	013	096	363	769	986	1−	1−	16
	17	0+	0+	0+	0+	0+	0+	004	037	198	586	948	997	1−	17
	18	0+	0+	0+	0+	0+	0+	001	011	086	370	848	981	1−	18
	19	0+	0+	0+	0+	0+	0+	0+	002	027	179	648	915	999	19
	20	0+	0+	0+	0+	0+	0+	0+	0+	006	058	365	717	981	20
	21	0+	0+	0+	0+	0+	0+	0+	0+	001	009	109	341	810	21
22	0	1	1	1	1	1	1	1	1	1	1	1	1	1	0
	1	198	676	902	993	1−	1−	1−	1−	1−	1−	1−	1−	1−	1
	2	020	302	661	952	996	1−	1−	1−	1−	1−	1−	1−	1−	2
	3	001	095	380	846	979	998	1−	1−	1−	1−	1−	1−	1−	3
	4	0+	022	172	668	932	992	1−	1−	1−	1−	1−	1−	1−	4

Part B: Cumulative terms, $\sum_{x=r}^{n} b(x; n, p)$

n	r	.01	.05	.10	.20	.30	.40	p .50	.60	.70	.80	.90	.95	.99	r
22	5	0+	004	062	457	835	973	998	1−	1−	1−	1−	1−	1−	5
	6	0+	001	018	267	687	928	992	1−	1−	1−	1−	1−	1−	6
	7	0+	0+	004	133	506	842	974	998	1−	1−	1−	1−	1−	7
	8	0+	0+	001	056	329	710	933	993	1−	1−	1−	1−	1−	8
	9	0+	0+	0+	020	186	546	857	979	999	1−	1−	1−	1−	9
	10	0+	0+	0+	006	092	376	738	945	996	1−	1−	1−	1−	10
	11	0+	0+	0+	002	039	228	584	879	986	1−	1−	1−	1−	11
	12	0+	0+	0+	0+	014	121	416	772	961	998	1−	1−	1−	12
	13	0+	0+	0+	0+	004	055	262	624	908	994	1−	1−	1−	13
	14	0+	0+	0+	0+	001	021	143	454	814	980	1−	1−	1−	14
	15	0+	0+	0+	0+	0+	007	067	290	671	944	999	1−	1−	15
	16	0+	0+	0+	0+	0+	002	026	158	494	867	996	1−	1−	16
	17	0+	0+	0+	0+	0+	0+	008	072	313	733	982	999	1−	17
	18	0+	0+	0+	0+	0+	0+	002	027	165	543	938	996	1−	18
	19	0+	0+	0+	0+	0+	0+	0+	008	068	332	828	978	1−	19
	20	0+	0+	0+	0+	0+	0+	0+	002	021	154	620	905	999	20
	21	0+	0+	0+	0+	0+	0+	0+	0+	004	048	339	698	980	21
	22	0+	0+	0+	0+	0+	0+	0+	0+	0+	007	098	324	802	22
23	0	1	1	1	1	1	1	1	1	1	1	1	1	1	0
	1	206	693	911	994	1−	1−	1−	1−	1−	1−	1−	1−	1−	1
	2	022	321	685	960	997	1−	1−	1−	1−	1−	1−	1−	1−	2
	3	002	105	408	867	984	999	1−	1−	1−	1−	1−	1−	1−	3
	4	0+	026	193	703	946	995	1−	1−	1−	1−	1−	1−	1−	4
	5	0+	005	073	499	864	981	999	1−	1−	1−	1−	1−	1−	5
	6	0+	001	023	305	731	946	995	1−	1−	1−	1−	1−	1−	6
	7	0+	0+	006	160	560	876	983	999	1−	1−	1−	1−	1−	7
	8	0+	0+	001	072	382	763	953	996	1−	1−	1−	1−	1−	8
	9	0+	0+	0+	027	229	612	895	987	999	1−	1−	1−	1−	9
	10	0+	0+	0+	009	120	444	798	965	998	1−	1−	1−	1−	10
	11	0+	0+	0+	003	055	287	661	919	993	1−	1−	1−	1−	11
	12	0+	0+	0+	001	021	164	500	836	979	999	1−	1−	1−	12
	13	0+	0+	0+	0+	007	081	339	713	945	997	1−	1−	1−	13
	14	0+	0+	0+	0+	002	035	202	556	880	991	1−	1−	1−	14
	15	0+	0+	0+	0+	001	013	105	388	771	973	1−	1−	1−	15
	16	0+	0+	0+	0+	0+	004	047	237	618	928	999	1−	1−	16
	17	0+	0+	0+	0+	0+	001	017	124	440	840	994	1−	1−	17
	18	0+	0+	0+	0+	0+	0+	005	054	269	695	977	999	1−	18
	19	0+	0+	0+	0+	0+	0+	001	019	136	501	927	995	1−	19
	20	0+	0+	0+	0+	0+	0+	0+	005	054	297	807	974	1−	20
	21	0+	0+	0+	0+	0+	0+	0+	001	016	133	592	895	998	21
	22	0+	0+	0+	0+	0+	0+	0+	0+	003	040	315	679	978	22
	23	0+	0+	0+	0+	0+	0+	0+	0+	0+	006	089	307	794	23

Part B: Cumulative terms, $\sum_{x=r}^{n} b(x; n, p)$

n	r	.01	.05	.10	.20	.30	.40	.50	.60	.70	.80	.90	.95	.99	r
24	0	1	1	1	1	1	1	1	1	1	1	1	1	1	0
	1	214	708	920	995	1−	1−	1−	1−	1−	1−	1−	1−	1−	1
	2	024	339	708	967	998	1−	1−	1−	1−	1−	1−	1−	1−	2
	3	002	116	436	885	988	999	1−	1−	1−	1−	1−	1−	1−	3
	4	0+	030	214	736	958	996	1−	1−	1−	1−	1−	1−	1−	4
	5	0+	006	085	540	889	987	999	1−	1−	1−	1−	1−	1−	5
	6	0+	001	028	344	771	960	997	1−	1−	1−	1−	1−	1−	6
	7	0+	0+	007	189	611	904	989	999	1−	1−	1−	1−	1−	7
	8	0+	0+	002	089	435	808	968	998	1−	1−	1−	1−	1−	8
	9	0+	0+	0+	036	275	672	924	992	1−	1−	1−	1−	1−	9
	10	0+	0+	0+	013	153	511	846	978	999	1−	1−	1−	1−	10
	11	0+	0+	0+	004	074	350	729	947	996	1−	1−	1−	1−	11
	12	0+	0+	0+	001	031	213	581	886	988	1−	1−	1−	1−	12
	13	0+	0+	0+	0+	012	114	419	787	969	999	1−	1−	1−	13
	14	0+	0+	0+	0+	004	053	271	650	926	996	1−	1−	1−	14
	15	0+	0+	0+	0+	001	022	154	489	847	987	1−	1−	1−	15
	16	0+	0+	0+	0+	0+	008	076	328	725	964	1−	1−	1−	16
	17	0+	0+	0+	0+	0+	002	032	192	565	911	998	1−	1−	17
	18	0+	0+	0+	0+	0+	001	011	096	389	811	993	1−	1−	18
	19	0+	0+	0+	0+	0+	0+	003	040	229	656	972	999	1−	19
	20	0+	0+	0+	0+	0+	0+	001	013	111	460	915	994	1−	20
	21	0+	0+	0+	0+	0+	0+	0+	004	042	264	786	970	1−	21
	22	0+	0+	0+	0+	0+	0+	0+	001	012	115	564	884	998	22
	23	0+	0+	0+	0+	0+	0+	0+	0+	002	033	292	661	976	23
	24	0+	0+	0+	0+	0+	0+	0+	0+	0+	005	080	292	786	24
25	0	1	1	1	1	1	1	1	1	1	1	1	1	1	0
	1	222	723	928	996	1−	1−	1−	1−	1−	1−	1−	1−	1−	1
	2	026	358	729	973	998	1−	1−	1−	1−	1−	1−	1−	1−	2
	3	002	127	463	902	991	1−	1−	1−	1−	1−	1−	1−	1−	3
	4	0+	034	236	766	967	998	1−	1−	1−	1−	1−	1−	1−	4
	5	0+	007	098	579	910	991	1−	1−	1−	1−	1−	1−	1−	5
	6	0+	001	033	383	807	971	998	1−	1−	1−	1−	1−	1−	6
	7	0+	0+	009	220	659	926	993	1−	1−	1−	1−	1−	1−	7
	8	0+	0+	002	109	488	846	978	999	1−	1−	1−	1−	1−	8
	9	0+	0+	0+	047	323	726	946	996	1−	1−	1−	1−	1−	9
	10	0+	0+	0+	017	189	575	885	987	1−	1−	1−	1−	1−	10
	11	0+	0+	0+	006	098	414	788	966	998	1−	1−	1−	1−	11
	12	0+	0+	0+	002	044	268	655	922	994	1−	1−	1−	1−	12
	13	0+	0+	0+	0+	017	154	500	846	983	1−	1−	1−	1−	13
	14	0+	0+	0+	0+	006	078	345	732	956	998	1−	1−	1−	14
	15	0+	0+	0+	0+	002	034	212	586	902	994	1−	1−	1−	15
	16	0+	0+	0+	0+	0+	013	115	425	811	983	1−	1−	1−	16
	17	0+	0+	0+	0+	0+	004	054	274	677	953	1−	1−	1−	17
	18	0+	0+	0+	0+	0+	001	022	154	512	891	998	1−	1−	18
	19	0+	0+	0+	0+	0+	0+	007	074	341	780	991	1−	1−	19

Part B: Cumulative terms, $\sum_{x=r}^{n} b(x; n, p)$

n	r	.01	.05	.10	.20	.30	.40	p .50	.60	.70	.80	.90	.95	.99	r
25	20	0+	0+	0+	0+	0+	0+	002	029	193	617	967	999	1−	20
	21	0+	0+	0+	0+	0+	0+	0+	009	090	421	902	993	1−	21
	22	0+	0+	0+	0+	0+	0+	0+	002	033	234	764	966	1−	22
	23	0+	0+	0+	0+	0+	0+	0+	0+	009	098	537	873	998	23
	24	0+	0+	0+	0+	0+	0+	0+	0+	002	027	271	642	974	24
	25	0+	0+	0+	0+	0+	0+	0+	0+	0+	004	072	277	778	25

Table V Probability levels for Student t-distributions*

Degrees of freedom	One-sample size†	Two-sided probability level .50	.80	.90	.95	.98
		One-sided probability level .75	.90	.95	.975	.99
1	2	1.00	3.08	6.31	12.71	31.82
2	3	.82	1.89	2.92	4.30	6.96
3	4	.76	1.64	2.35	3.18	4.54
4	5	.74	1.53	2.13	2.78	3.75
5	6	.73	1.48	2.02	2.57	3.36
6	7	.72	1.44	1.94	2.45	3.14
7	8	.71	1.41	1.89	2.36	3.00
8	9	.71	1.40	1.86	2.31	2.90
9	10	.70	1.38	1.83	2.26	2.82
10	11	.70	1.37	1.81	2.23	2.76
15	16	.69	1.34	1.75	2.13	2.60
30	31	.68	1.31	1.70	2.04	2.46
50	51	.68	1.30	1.68	2.01	2.40
100	101	.68	1.29	1.66	1.98	2.36
1000	1001	.67	1.28	1.65	1.96	2.33
∞‡	∞‡	.67	1.28	1.64	1.96	2.33

* Rounded from D. B. Owen, *Handbook of Statistical Tables* (Addison-Wesley, 1962), with the permission of the publisher.

† For setting confidence limits on the mean of a single sample.

‡ Standard normal distribution.

Table VI Cumulative normal distribution

Cell entry is the probability p that a standard normal random variable Z takes values less than z, where the units digit of z is given in the left-hand column, and the tenths digit is given in the top row.

Examples

1) $p = P(Z < -2.7) = 0.0035$

2) $p = P(Z < 1.4) = 0.9192$

z	.0	.1	.2	.3	.4	.5	.6	.7	.8	.9
−3	.0013	.0010	.0007	.0005	.0003	.0002	.0002	.0001	.0001	.0000+
−2	.0228	.0179	.0139	.0107	.0082	.0062	.0047	.0035	.0026	.0019
−1	.1587	.1357	.1151	.0968	.0808	.0668	.0548	.0446	.0359	.0287
−0	.5000	.4602	.4207	.3821	.3446	.3085	.2743	.2420	.2119	.1841
0	.5000	.5398	.5793	.6179	.6554	.6915	.7257	.7580	.7881	.8159
1	.8413	.8643	.8849	.9032	.9192	.9332	.9452	.9554	.9641	.9713
2	.9772	.9821	.9861	.9893	.9918	.9938	.9953	.9965	.9974	.9981
3	.9987	.9990	.9993	.9995	.9997	.9998	.9998	.9999	.9999	1.0000−

Table VII Inverse of the cumulative normal distribution

Cell entry is the value of z such that the area to the left of z, under the standard normal curve, is p, where the tenths digit of p is given in the left-hand column and the hundredths digit is given along the top.

Examples

1) $p = .16, \quad z = -0.994$

2) $p = .63, \quad z = +0.332$

p	.00	.01	.02	.03	.04	.05	.06	.07	.08	.09
.0	−∞	−2.326	−2.054	−1.881	−1.751	−1.645	−1.555	−1.476	−1.405	−1.341
.1	−1.282	−1.227	−1.175	−1.126	−1.080	−1.036	−0.994	−0.954	−0.915	−0.878
.2	−0.842	−0.806	−0.772	−0.739	−0.706	−0.674	−0.643	−0.613	−0.583	−0.553
.3	−0.524	−0.496	−0.468	−0.440	−0.412	−0.385	−0.358	−0.332	−0.305	−0.279
.4	−0.253	−0.228	−0.202	−0.176	−0.151	−0.126	−0.100	−0.075	−0.050	−0.025
.5	0.000	0.025	0.050	0.075	0.100	0.126	0.151	0.176	0.202	0.228
.6	0.253	0.279	0.305	0.332	0.358	0.385	0.412	0.440	0.468	0.496
.7	0.524	0.553	0.583	0.613	0.643	0.674	0.706	0.739	0.772	0.806
.8	0.842	0.878	0.915	0.954	0.994	1.036	1.080	1.126	1.175	1.227
.9	1.282	1.341	1.405	1.476	1.555	1.645	1.751	1.881	2.054	2.326

EXPLANATION OF CHART I

The curves for Chart I give 95% confidence limits for p, the probability of success on a single binomial trial. These curves also have a Bayesian interpretation based upon a uniform prior distribution for p, which requires further comment.

The curves for $n = 10$ give the shortest interval in the posterior distribution for p that contains probability 0.98; for $n = 15$, the posterior probability is 0.98; for $n = 20$, 0.975; and for $n = 30$, 0.975. The curve for $n = 5$ is also Bayesian at 0.98, except for $\overline{p} = 0$, where the upper limit was changed from 0.479 to 0.500 in order to bring the confidence level up to 0.95, with a corresponding change at $\overline{p} = 1$. For larger values of n—50, 100, 250, 1000—the confidence curves are visually indistinguishable from those of Clopper and Pearson, whose curves for these n's are reproduced with the kind permission of the Biometrika Trustees from "The Use of Confidence or Fiducial Limits Illustrated in the Case of the Binomial", *Biometrika*, Vol. 26 (1934), p. 410.

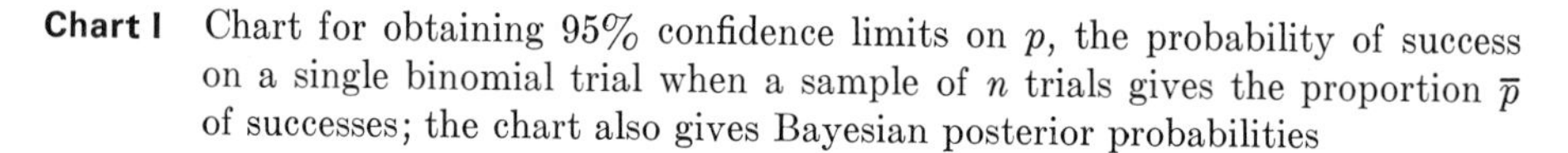

Chart I Chart for obtaining 95% confidence limits on p, the probability of success on a single binomial trial when a sample of n trials gives the proportion $\overline{p}$ of successes; the chart also gives Bayesian posterior probabilities

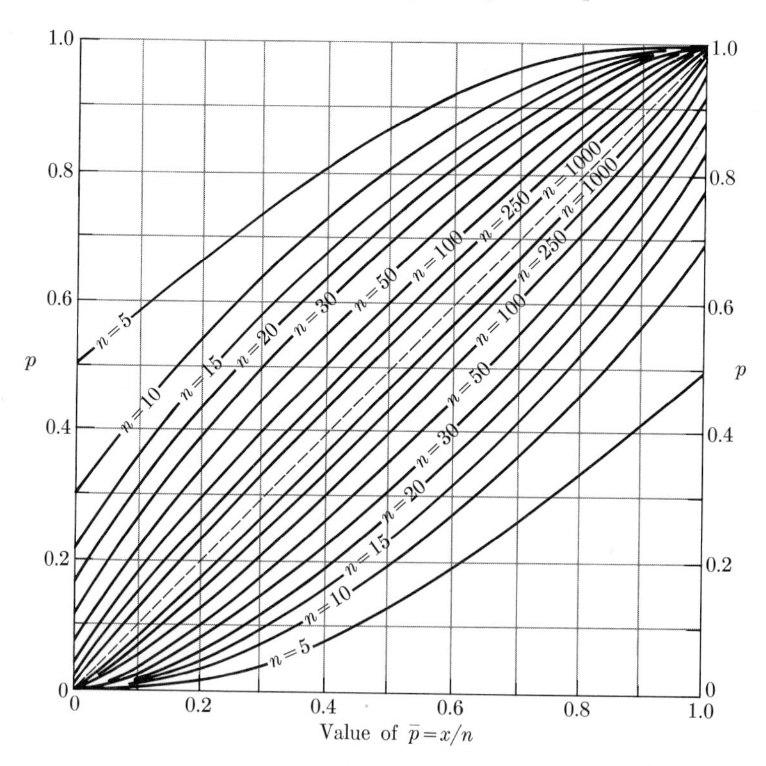

To obtain confidence limits for p, enter the horizontal axis at the observed value of $\overline{p}$. Read the vertical axis at the two points where the two curves for n cross the vertical line erected from $\overline{p}$. *Example:* $\overline{p} = 0.4$, $n = 30$, lower confidence limit 0.22, upper confidence limit, 0.60. The Bayesian posterior probability for this same interval, based upon a uniform prior, is 0.975. See Explanation of Chart I beginning on page 493.

Bibliography

Bibliography

For further titles, see *Subject Guide to Books in Print:* Probabilities, Statistics, and related portions.

I. The following book offers a mathematics refresher:

Walker, H. M., *Mathematics Essential for Elementary Statistics.* New York: Henry Holt and Company, 1951.

II. The books in the following list require approximately the same level of mathematical preparation as this book. Any of these books is suited for supplementary reading.

Berman, S. M., *The Elements of Probability.* Reading, Mass.: Addison-Wesley Publishing Co., Inc., 1969.

Chernoff, Herman, and L. E. Moses, *Elementary Decision Theory.* New York: John Wiley & Sons, Inc., 1959.

Dixon, W. J., and F. J. Massey, Jr., *Introduction to Statistical Analysis.* New York: McGraw-Hill Book Company, Inc., 3rd ed., 1969.

Goldberg, S., *Probability: An Introduction.* Englewood Cliffs, N. J.: Prentice-Hall, Inc., 1960.

Hodges, J. L., Jr., and E. L. Lehmann, *Elements of Finite Probability.* San Francisco: Holden-Day, Inc., 1965.

Johnston, J. B., G. B. Price, and F. S. Van Vleck, *Sets, Functions, and Probability.* Reading, Mass.: Addison-Wesley Publishing Co., Inc., 1968.

Raiffa, Howard, *Decision Analysis: Introductory Lectures on Choices Under Uncertainty.* Reading, Mass.: Addison-Wesley Publishing Co., Inc., 1968.

Schmitt, S. A., *Measuring Uncertainty: An Elementary Introduction to Bayesian Statistics.* Reading, Mass.: Addison-Wesley Publishing Co., Inc., 1969.

Wallis, W. A., and H. V. Roberts, *Statistics: A New Approach.* Glencoe, Ill.: The Free Press, 1956.

Weaver, Warren, *Lady Luck: The Theory of Probability.* Garden City, N. Y.: Doubleday and Co., Inc., 1963.

Wilks, S. S., *Elementary Statistical Analysis.* Princeton, N. J.: Princeton University Press, 1948.

Zeisel, Hans, *Say it With Figures.* New York: Harper and Row, revised 5th ed. 1968.

III. The following books offer a variety of illustrations of uses of probability and statistics:

Bowker, A. H., and G. J. Lieberman, *Engineering Statistics.* Englewood Cliffs, N. J.: Prentice-Hall, Inc., 1959.

Bross, I. D. J., *Design for Decision.* Glencoe, Ill.: The Free Press, 1965.

Cox, D. R., and P. A. Lewis, *Statistical Analysis of Series of Events.* New York: John Wiley & Sons, Inc., 1966.

Edwards, A. L., *Experimental Design in Psychological Research.* New York: Henry Holt and Company, 1960.

Fisher, R. A., *Statistical Methods for Research Workers.* New York: Hafner Publishing Company, Inc., 13th ed., 1964.

Good, I. J., *The Estimation of Probabilities; An Essay on Modern Bayesian Methods.* Cambridge, Mass.: The MIT Press, 1965.

Grant, E. L., *Statistical Quality Control.* New York: McGraw-Hill Book Company, Inc., 3rd ed., 1964.

Levinson, H. C., *Chance, Luck, and Statistics.* New York: Dover, 2nd ed., 1963.

Mosteller, F., and D. L. Wallace, *Inference and Disputed Authorship: The Federalist.* Reading, Mass.: Addison-Wesley Publishing Co., Inc., 1964.

Niven, Ivan, *Mathematics of Choice: How to Count Without Counting.* New York: Random House, Inc. (SMSG Monograph), 1965.

Snedecor, G. W., and W. G. Cochran, *Statistical Methods.* Ames, Iowa: The Iowa State College Press, 6th ed., 1967. (Applications to experiments in agriculture and biology.)

Williams, E. J., *Regression Analysis.* New York: John Wiley & Sons, Inc., 1959.

Williams, J. D., *The Compleat Strategyst.* New York: McGraw-Hill Book Company, Inc., rev. ed., 1965.

Wilson, E. Bright, Jr., *An Introduction to Scientific Research.* New York: McGraw-Hill Book Company, Inc., 1952.

The World of Mathematics. Part VII: The Laws of Chance. Part VIII: Statistics and the Design of Experiments. New York: Simon and Schuster, 1956.

Youden, W. J., *Statistical Methods for Chemists.* New York: John Wiley & Sons, Inc., 1951.

IV. The books in the following list are of a more advanced nature:

Birnbaum, Z. W., *Introduction to Probability and Mathematical Statistics.* New York: Harper, 1962.

Brownlee, K. A., *Statistical Theory and Methodology in Science and Engineering.* New York: John Wiley & Sons, Inc., 1965.

Brunk, H. D., *An Introduction to Mathematical Statistics.* Boston: Blaisdell (Ginn and Company), 2nd ed., 1965.

Freeman, H. A., *Introduction to Statistical Inference.* Reading, Mass.: Addison-Wesley Publishing Co., Inc., 1963.

Freund, J. E., *Mathematical Statistics.* Englewood Cliffs, N. J.: Prentice-Hall, Inc., 1962.

Fry, T. C., *Probability and its Engineering Uses.* Princeton, N. J.: D. Van Nostrand Co., Inc., 2nd ed., 1965.

Hoel, Paul G., *Introduction to Mathematical Statistics.* New York: John Wiley & Sons, Inc., 3rd ed., 1962.

Meyer, P. L., *Introductory Probability and Statistical Applications.* Reading, Mass.: Addison-Wesley Publishing Co., Inc., 1965.

Mood, A. M., and F. A. Graybill, *Introduction to the Theory of Statistics.* New York: McGraw-Hill Book Co., Inc., 2nd ed., 1963.

Schlaifer, R., *Probability and Statistics for Business Decisions.* New York: McGraw-Hill Book Company, Inc., 1959.

V. The books in the following list are more advanced than those in categories II and IV above:

Cramér, H., *Mathematical Methods of Statistics.* Princeton, N. J.: Princeton University Press, 1946.

Drake, A. W., *Fundamentals of Applied Probability Theory.* New York: McGraw-Hill Book Company, Inc., 1967.

Feller, W., *An Introduction to Probability Theory and its Applications.* Vol. 1. New York: John Wiley & Sons, Inc., 3rd ed., 1968.

Ferguson, T. S., *Mathematical Statistics: A Decision-Theoretic Approach.* New York: Academic Press, 1967.

Fisz, M., *Probability Theory and Mathematical Statistics.* New York: John Wiley & Sons, Inc., 3rd ed., 1963.

Lindgren, B. W., *Statistical Theory.* New York: Macmillan, 1962.

Parzen, E., *Modern Probability Theory and its Applications.* New York: John Wiley & Sons, Inc., 1960.

Wilks, S. S., *Mathematical Statistics.* New York: John Wiley & Sons, Inc., 1962.

Answers to even-numbered exercises

Answers to even-numbered exercises

Section 1–6

2. $\frac{1}{6}$; $\frac{1}{2}$; $\frac{1}{3}$; $\frac{1}{2}$; $\frac{1}{2}$

4. $\frac{1}{4}$; $\frac{1}{2}$; $\frac{1}{3}$

6. $\frac{4}{9}$; $\frac{5}{9}$; $\frac{5}{4}$

12. $\frac{1}{12}$; $\frac{11}{12}$; $\frac{1}{11}$

14. $\frac{5}{9}$; $\frac{4}{9}$; $\frac{4}{9}$ (assuming that 1 is not a prime); $\frac{1}{3}$

Section 1–7

2. $\{HHH, HHT, HTH, THH, HTT, THT, TTH, TTT\}$;
   $\{0, 1, 2, 3\}$, where number denotes number of heads
4. $\{pn, pd, pq, np, nd, nq, \ldots\}$
6. $A = \{bbbb, bbbg, bbgb, bgbb, gbbb, bbgg, bgbg, bggb, ggbb, gbgb, gbbg, bggg, gbgg, ggbg, gggb, gggg\}$
   For A: 16; 4; 8
8. See first 3 rows and columns of Table 1–11.
10. $\{ABC, ABD, ABE, ACD, ACE, ADE, BCD, BCE, BDE, CDE\}$
    6; 4; 3; 9
12. $\{(r, r), (r, w), (r, b), (w, r), (w, w), (w, b), (b, r), (b, w), (b, b)\}$ (assuming that at least 2 of each color are in the bag); 3; 6; 2; 8
14. $\frac{1}{3}$; $\frac{1}{3}$; $\frac{1}{3}$

Section 1–8

2. $\frac{1}{6}$
4. $\frac{1}{6}$
6. $r \neq c$; $r = c - 2$; $c \geq r + 2$; $r = 2c$
8. $\frac{1}{6}$; $\frac{1}{2}$; $\frac{1}{3}$; $\frac{1}{2}$; $\frac{1}{6}$; $\frac{1}{6}$
10. $\frac{1}{3}$; $\frac{2}{3}$; $\frac{1}{3}$; $\frac{2}{3}$; $\frac{1}{9}$
12. $\frac{1}{4}$; $\frac{1}{4}$; $\frac{3}{16}$; $\frac{1}{4}$; $\frac{1}{2}$; $\frac{3}{8}$
14. $\frac{3}{5}$; $\frac{3}{10}$; $\frac{9}{10}$; $\frac{1}{10}$

Review Exercises for Sections 1–6, 1–7, and 1–8

2. $\frac{5}{6}$
4. $\frac{13}{35}$; $\frac{3}{7}$; $\frac{4}{7}$; $\frac{1}{2}$; $\frac{3}{14}$; $\frac{5}{14}$
6. Set of real numbers
8. Two are: {1, 2, 3, . . . , 49}, where number denotes position of first ace, numbered from top; and {Ace on top, Ace not on top}
10. $\{2, N2, NN2, NNN2, \ldots\}$. *Note:* This sample space is not finite
12. aa, aA, AA

Section 2–1

2. 720; 480; 240
4. 120; 216; $\frac{1}{36}$
6. 3,024
8. 720
10. 14,400
12. 255
14. $9! - 1$
16. 502
18. 1024; 2^n
20. n^r
22. $\frac{1}{3}$

Section 2–2

2. 1
4. 5040
6. $10(9)8 = 720 = 6!$
8. 90,000; 234,000
10. 1/1260
12. 2/7; 5 to 2
14. 100,000; 30,240; 69,760
16. 42; $8400

Section 2–3

4. 56
6. 399 or 400 counting none
8. 45; 120; 210
10. $20!/(3!5!12!)$
12. $\binom{100}{98} > \binom{20}{4}$
14. $301(201)101 - 2$
16. 21
18. 330; $330 \times 12! \times 8!$
20. 44; 4095
22. 11: 1, 11, 55, 165, 330, 462, 462, 330, 165, 55, 11, 1
    12: 1, 12, 66, 220, 495, 792, 924, 792, 495, 220, 66, 12, 1

Permutation and Combinations: Sets A and B

2. 2450; 1225
4. 190
6. $8!/3!$
8. 360; 120; 240; 12
10. 56
12. 3,168
14. 3,360
16. 90; 18; 72
18. $\binom{8}{4}\binom{10}{5}\binom{9}{3}$
20. 2160
22. $\binom{5}{3}\binom{4}{3}6!$
24. $8(9!)$
26. $8! - 1$
28. 171
30. $\binom{9}{3}\binom{6}{3}\binom{3}{3} = 1680$

Section 2–4

2. $10!/[(2!)^2(3!)^2]$
4. 1716
6. 210
8. $11!/[(3!)^2(2!)^2]$; $9!/[3!2!2!]$; $3(8!)$
10. $\binom{22}{10}$
12. 126
14. $11(11!)/(3!4!5!)$
16. 210; 3/7
18. 126
20. 3876
22. 28; 55
24. $\binom{n}{m} = \dfrac{n!}{m!(n-m)!}$; $\binom{n-1}{m-1} = \dfrac{(n-1)!}{(m-1)!(n-m)!}$;
    $\binom{n-1}{m-1} \Big/ \binom{n}{m} = m/n$
26. $\binom{n+r+1}{r}$
28. $\dfrac{r+1}{n+r+2}$

Review Exercises

2. 72
4. 9204 (including no cutting at some positions, but excluding complete blanks)
6. 16
8. No. He has missed 2

Section 2–5

2. 101; m^{100}, $m^{99}n$, $m^{98}n^2$
4. $\binom{100}{49}a^{51}x^{49}$; $\binom{100}{19}a^{81}x^{19}$; $\binom{100}{60}a^{40}x^{60}$
6. $1 + 5b + 10b^2 + 10b^3 + 5b^4 + b^5$;
   $1 + 5(.01) + 10(.01)^2 + 10(.01)^3 + 5(.01)^4 + (.01)^5$
8. $1 + 7p + 21p^2 + 35p^3 + 35p^4 + 21p^5 + 7p^6 + p^7$
10. $1 - 5x^2 + 10x^4 - 10x^6 + 5x^8 - x^{10}$
12. $x^6 + 6x^5y + 15x^4y^2 + 20x^3y^3 + 15x^2y^4 + 6xy^5 + y^6$
14. $1 + 2a + \frac{3}{2}a^2 + \frac{1}{2}a^3 + \frac{1}{16}a^4$
16. $a^6 + 6a^6x + 15a^6x^2 + 20a^6x^3 + 15a^6x^4 + 6a^6x^5 + a^6x^6$
18. $x^{12} - 6x^{13} + 15x^{14} - 20x^{15} + 15x^{16} - 6x^{17} + x^{18}$
20. $p^{50} + 50p^{49}q + 1225p^{48}q^2$
22. $1 - 40a^2 + 780a^4$
24. 1.02
26. 1.02
28. 3.015
30. 5.1
32. 121.25
34. $1 - x$
36. $2 + 2x$
38. $1 + 3x + 6x^2 + 10x^3$
40. $1 + \dfrac{r}{1!}q + \dfrac{r(r+1)}{2!}q^2 + \dfrac{r(r+1)(r+2)}{3!}q^3$

Section 3–1

2. $\frac{5}{9}$
4. $\frac{1}{12}$
6. 0
8. $\frac{9}{10}$
10. $\frac{1}{2}$
12. $\frac{4}{5}$
14. $\frac{6}{10} + \frac{6}{10} - \frac{3}{10} = \frac{9}{10}$

Section 3–2

2. Yes; yes
4. Less than or equal to 0.4; yes; perhaps
6. $\frac{17}{18}$ 8. $\frac{4}{9}$ 10. $\frac{5}{9}$ 12. $\frac{4}{9}$
14. $\frac{5}{36}$ 16. $\frac{5}{6}$ 18. $\frac{8}{15}$ 20. $\frac{13}{20}$

Section 3–3

2. $(\frac{1}{2})(\frac{1}{4}) = \frac{1}{8}$
   $(\frac{3}{8})(\frac{1}{8}) \neq 0$
4. $\frac{25}{36}$
6. Not independent
8. $\frac{5}{144}$

Section 3–4

2. $\frac{2}{36} = \frac{6}{36} \cdot \frac{2}{6}$
4. $\frac{1}{2}$
8. $\frac{1}{2}$
10. $\frac{1}{5}$
12. $\frac{19}{20}$
14. $\frac{4}{5}$
16. $\frac{1}{2}$
18. $\frac{1}{45}$; $\frac{44}{45}$; $\frac{28}{45}$
20. 0.00075; 0.02425; 0.94575
22. 7
24. $\frac{18}{35}$, $\frac{11}{35}$, $\frac{6}{35}$
26. $\frac{4}{9}$
28. $\frac{3}{20}$, $\frac{17}{20}$
30. $\frac{1}{24}$

Section 3–5

2. $1 - \binom{47}{4}/\binom{52}{4}$; 1
4. $1 - \dfrac{12 \cdot 11 \cdots (12 - k + 1)}{12^k}$
6. $\frac{112}{729}$
8. $\frac{57}{616}$; $\frac{285}{748}$
10. About 0.75; about 3 to 1

Sections 3–6, 3–7, and 3–8

6. $\binom{800}{200}$ points; $1/\binom{800}{200}$

Section 4–1

2. 0.51; 0.70; 0.70; 0.58
4. $\{0U, 1U, 2U, 3U\}$; $\binom{3}{x} p^x(1-p)^{3-x}$, $x = 0, 1, 2, 3$
6. Yes
8. $\frac{1}{2}$; 0

Section 4–2

2. $\frac{1}{2}$; $\frac{1}{2}$; $\frac{1}{4}$; $\frac{3}{4}$; $\frac{1}{2}$; $\frac{1}{2}$; $\frac{3}{4}$; $\frac{1}{4}$
4. 0.025; 0.005; 0.03
6. $1/(n + r)$; $(r + 1)/(n + r)$
8. $\frac{1}{6}$; $\frac{1}{2}$; $\frac{1}{3}$
10. $\frac{4}{21}$
12. 60% against it

Section 4–3

6. 0.000001; about 0.97
8. $\frac{5}{18}$; $\frac{17}{36}$; $\frac{2}{9}$; $\frac{1}{36}$

Section 4–4

4. $\frac{19}{27}$
6. Theoretical: 192, 288, 144, 24
10. Sum of probabilities $= p^4 + 4p^3q + 6p^2q^2 + 4pq^3 + q^4$
$= (p+q)^4 = 1$

Probability	q^4	$4pq^3$	$6p^2q^2$	$4p^3q$	p^4
Number of U's	0	1	2	3	4

Section 4–5

2.

$b(x; 5, .4)$	.078	.259	.346	.230	.077	.010
x	0	1	2	3	4	5

4. When $n = 2$, the probability is 1
6. Two-engine plane preferred unless $p = 0$ or 1
8. $\frac{16}{27}$
10. $\frac{13}{729}$
14. $\frac{1}{2} < q < 1$
16. $\frac{1}{2} < q < 1$

Section 4–6

2. a) 0.617 b) 0.163
4. a) 0.982 b) 0.256 c) 0.024
6. 0.33
8. a) 0.133 b) 0.205 c) 0.927
10. 0.993; 0.196; 0.579
12. 0.050
14. 0.956
16. 0.026
18. 0.358; 0.016
20. Theoretical: 25.6, 128, 256, 256, 128, 25.6
22. a) Both equal 0.412
b) Both equal 0.166
24. a) Both equal 0.919
b) Both equal 0.263
26.

$f(x)$	.031	.156	.312	.312	.156	.031
x	0	1	2	3	4	5

Section 4–7

2. $P(A|B) = 1$
4. $\frac{1}{2}$
6. $\frac{1}{3}$
10. $\frac{2}{5}$; $\frac{3}{8}$
12. About 0.205
14. $\frac{4}{5}$; $\frac{1}{10}$
16. $P(k) = \log k/\log[(2n)!]$

Section 4–8

2. $\frac{15}{17}$; 1; 0; $\frac{4}{9}$; $\frac{5}{9}$
4. $r/(b+r)$; $r/(b+r)$; $r/(b+r)$
6. $\frac{1}{10}$; $\frac{1}{2}$; $\frac{1}{3}$; $\frac{3}{4}$; $\frac{9}{10}$; $\frac{9}{10}$
8. $(\frac{39}{52})(\frac{38}{51})(\frac{37}{50})(\frac{36}{49})(\frac{13}{48}) \approx 0.082$
10. 0.0002228
12. $\frac{7}{10}$; $\frac{3}{10}$
14. ≈ 0.99; ≈ 0.98; No

Section 4–9

2. $\frac{8}{23}$; $\frac{9}{23}$; $\frac{6}{23}$
4. $\frac{8}{233}$; $\frac{225}{233}$
6. $\frac{32}{57}$; $\frac{25}{57}$
8. a) Urn I: $\frac{1}{10}$, $\frac{1}{4}$ Urn II: $\frac{9}{10}$, $\frac{3}{4}$
   b) Urn I: $\frac{1}{10}$, $\frac{1}{28}$ Urn II: $\frac{9}{10}$, $\frac{27}{28}$
10. About $1/10^4$
12. $\frac{12}{17}$
14. Odds proportional to 3×99^n, 5×97^n, 2×90^n
16. $\frac{6}{7}$
18. $P(H_1|E) \approx 1$; $P(H_1|\overline{E}) \approx \frac{1}{2}$

Section 4–10

2. n = number of trials required; $S = \{1, 2, \ldots, n, \ldots\}$; $P(\text{first } H \text{ on trial } n) = (\frac{3}{4})^{n-1}(\frac{1}{4})$
4. n = throws required; $S = \{1, 2, \ldots, n, \ldots\}$; P (first success on trial n) $= (1-p)^{n-1}p$
6. $\dfrac{p^2}{1-2pq}$

Miscellaneous Exercises—Chapter 4

2. $\frac{48}{125}$
4. $\frac{1}{216}$
6. $\frac{5}{72}$
8. $\frac{1}{9}$
10. $\frac{1}{36}$
12. $\frac{25}{396}$
14. $\frac{1}{36}$
16. About 0.1
18. About 0.011
22. About 0.15
24. About 0.33
26. About 0.025
28. $\frac{1}{4}$; $\frac{1}{4}$; $\frac{1}{4}$; $\frac{1}{4}$
30. $\frac{1}{16}$; $(\frac{1}{4})^n$
32. $a > \frac{1}{99}$

Section 5–1

4.

$f(x)$	$\frac{1}{4}$	$\frac{1}{2}$	$\frac{1}{4}$
x	1	2	3

; 2

6. $f(x) = (\frac{1}{36})(6 - |x - 7|)$ for $x = 2, 3, \ldots, 12$

8.

$f(z)$	$\frac{1}{27}$	$\frac{3}{27}$	$\frac{6}{27}$	$\frac{7}{27}$	$\frac{6}{27}$	$\frac{3}{27}$	$\frac{1}{27}$
z	3	4	5	6	7	8	9

14.

$f(x)$	$\frac{1}{16}$	$\frac{4}{16}$	$\frac{6}{16}$	$\frac{4}{16}$	$\frac{1}{16}$
x	-4	-2	0	2	4

16. $1/n!$; 0; $\binom{n}{2} \Big/ n!$

18. a) $q^{x-1}p$ with $p = \frac{1}{6}$, $x \geq 1$
b) $P(X \leq 4) = 671/1296 > \frac{1}{2}$ and $P(X > 4) = 625/1296 < \frac{1}{2}$

Section 5–2

2. 4
4. 3.4
6. 3.66
8. 2.5
10. 30
16. -2.05¢
18. -18¢
20. a) 1 b) 81/256 c) No
22. $-1/19$ dollar
24. 10
26. 5.50
28. $-85/216$¢; -39¢
30. 7/4
32. $n + m/2$
34. $(2n + 1)/3$; $(2n + 1)/3 \approx \frac{2}{3}n$ for large n
36. $(\frac{5}{6})^9$

Section 5–3

2.

Probability	0.2	0.1	0.3	0.3	0.1
Values of $3X - 1$	-7	-4	-1	2	5

$E(3X - 1) = -1 = 3E(X) - 1$

4.

Probability	0.3	0.4	0.3
Values of X^2	4	1	0

$E(X^2) = 1.6$

6. 2.7
8. 8.9
10. 5/3¢
12. 0.4, 0.4, 0.2

St. Petersburg Paradox Exercises

2.

$f(x)$	$\frac{1}{2}$	$\frac{1}{4}$	$\frac{1}{8}$	$\frac{1}{8}$
x	0	1	2	3

4.

$f(x)$	$\frac{1}{2}$	$\frac{1}{4}$	$\frac{1}{8}$	$\cdots$	$1/2^n$	$1/2^n$
x	0	1	2	$\cdots$	$n - 1$	n

6. $\frac{5}{2}$
8. $(n + 2)/2$
10. Infinite
12. \$11

Section 6–1

2. a) 1; 1 b) 1
4. a) 0.2 b) 0.8 c) 0.76 d) 0.872
6. 3.4; 0.84
12. 0.00256; 2.15×10^{-6}; 1.47×10^{-3}
14. 3.1; 0.1275; 0.357
16. 2.7; 0.81; 0.9
18. 720; 1100; 33.2
20. 4; $\frac{4}{3}$
22. $\frac{25}{8}$; $\frac{55}{64}$; 0.927
24. 76; 3.04; 0.19
28. $\frac{1}{2}$; 1; 1
30. $n \geq 82$

Section 6–2

2. 2; $\frac{2}{3}$; 0.816
4. 0; 4; 2
6. $\frac{1}{3}$; 0.042; 0.205
8. 2; 1; 1
10. −7; 2; 1.41
12. Inconsistent because the sample variance is never negative
18. $Y = X + 900$
20. 1.4; 0.49; 0.7
22. a) 7.25; 0.93 b) 72.75; 97.25; 99.875

Section 6–3

2. $\frac{8}{9}$
4. 5
6. a) 1 b) $\frac{4}{9}$
8. 20
12. a) $\frac{1}{4}$ b) $\frac{1}{9}$ c) $\frac{1}{25}$
18. $\frac{1}{18}$
20. $\frac{1}{4}$; 0
22. Yes; $1/(2h^2)$; $1/h^2$

Section 6–4

2. $3n/4$
4. 5
6. 5%
8. a) $n/4$ b) $n/9$
10. $\frac{1}{2^2} < \frac{2}{7}$

Section 7–1

2. $E(XY) = \frac{5}{12}$, $E(X) \cdot E(Y) = \frac{1}{2}$
   $E(X + Y) = \frac{17}{12} = E(X) + E(Y)$
4.

$f(w)$	0	$\frac{7}{12}$	$\frac{5}{12}$
w	0	1	2

   $E(W) = \frac{17}{12}$
6. a) .2, .8, .6, .4
   b) Predict $Y = 1$ if $X = 0$; predict $Y = 0$ if $X = 1$
   c) 70%
8. No
10. a) 0, 3 b) No c) $\frac{3}{8}$ d) 0 e) $\frac{1}{2}$
12. 4
14. $10(\frac{7}{9})^9$

Section 7–2

2. $\frac{1}{2}$; $\frac{1}{4}$; $\frac{1}{40}$; $1 - (\pi/4) \approx 0.215$

4. $\frac{1}{8}$; $\frac{1}{2}$

Section 7–3

2. $\frac{29}{36}$; yes
4. Jump at x is $f(x)$ as given by Table 5–5

Section 7–4

2. a) -0.253 b) -1.645
   c) 1.645 d) 0.675
4. a) 0.0228; 0.7881
   b) 11.85×10^{-6}; 0.9985
8.

h	$A(h)$	$0.4h$	$0.4h - A(h)$
0.01	0.0040	0.004	0+
0.1	0.0398	0.04	0.0002
0.2	0.0793	0.08	0.0007
0.3	0.1179	0.12	0.0021
0.4	0.1554	0.16	0.0046
0.5	0.1915	0.20	0.0085
0.76	0.2764	0.304	0.0276
0.77	0.2794	0.308	0.0286

   Difference exceeds 10% of $A(h)$ when $h = 0.77$

10. b) From Table 7–7, sum is 0.9999
12. .3989; .2420; .0540

Section 8–1

2. 225, $\frac{1}{5}$
10. 0.211 or 0.789
12.

Interval	Actual probability	Chebyshev (at least)	Empirical rule
$10 \leq x \leq 15$	.770	0	.68
$7.5 \leq x \leq 17.5$	.956	.75	.95
$5 \leq x \leq 20$	1−	.89	.997

Section 8–2

2. b) 0.120 c) 0.240
   d) $0.240 = b(0; 4, 0.3)$, and
   $0.120 = \frac{1}{2}b(0; 4, 0.3)$
4. b) $\mu = np = 0.8, \quad \sigma = \sqrt{npq} = 0.8$

Section 8–3

2.

n	2	4	8	16
Maximum error in cumulative	.010	.005	.004	.002

6. 0.0262, 0.0812
8. (0.3989)(0.0632) $\approx$ 0.0252

Section 9–2

2. $P(|\overline{p} - p|) < .05) \approx .95$
4. 96
6. 0.25
8. 0.5
10. $(1 - p)/p$
12. $\frac{1}{4}$; $\frac{1}{20}$; 0.0158
14. 1—
16. 0.7942
18. 0.004
20. 2
22. 0.0337
24. 0.2358

Section 9–3

2. (0.66, 0.94)
4. (0.65, 0.83)
6. (0, 0.14)
8. $n \approx 450$

Section 9–4

2. 0.09, 0.25
6. $n = 10$, $r = 1$, $p = 0.0050$;
   $n = 40$, $r = 3$, $p = 0.0225$;
   $n = 100$, $r = 7$, $p = 0.0375$;
   $n = 1000$, $r = 70$, $p = 0.0550$
8. 220
10. Let $J(p)$ be the probability of accepting lot under original plan; under new plan, the probability is $2 - J(p)$ times as great
12. 1.1%
14. Yes; descriptive level is 1.6%
16. No; descriptive level is 5.23%
18. (0.63, 0.97)
20. a) Coin is unbiased
    b) $\frac{1}{16}$
    c) When the coin is biased in favor of heads
22. Yes
24. a) Chance of getting any pair correct is $p = \frac{1}{2}$
    b) $p \neq \frac{1}{2}$
    c) 5.5%
    d) 0.678
26. a) 0.033 b) 0.623, operating characteristic 0.377

Section 9–5

2. 3
4. Probability is 0.15

6. Yes
8. $z = 2.48$. Probability of table at least this extreme is 0.0066.
10. $z = 0.96$, $P = 1 - 2(.3315) = 0.337$.
12. 0.008, 0.242
14. -0.00075, 0.00139
16. Probability of as good as or better forecasting is 0.343. We would not reject hypothesis of random forecasting.

Section 9–6

2. 0.661, 0.289, 0.049; 2.60
4. 0, 0.370, 0.630; 8.15
6. 0.3

Section 9–7

2. 44 to 1
4. $n = 1.66x$
6. 0.999
8. 1 to 12

Miscellaneous Exercises for Chapter 9

2. 1300
4. Yes
6. Once, if he means p is exactly 1
8. $\frac{89}{110}$
10. $\frac{15}{28}$

Section 10–1

2. a) $\mu_X = \frac{1}{2}, \sigma_X^2 = \frac{1}{4}$ b) $\mu_Y = \frac{1}{3}, \sigma_Y^2 = \frac{2}{9}$
c) $\mu_Z = 2, \sigma_Z^2 = \frac{2}{3}$ d) $\mu_{X+Y} = \frac{5}{6}, \sigma_{X+Y}^2 = \frac{17}{36}$
e) $\mu_{X+Y+Z} = \frac{17}{6}, \sigma_{X+Y+Z}^2 = \frac{41}{36}$

4. a)

Probability	$\frac{2}{18}$	$\frac{5}{18}$	$\frac{6}{18}$	$\frac{4}{18}$	$\frac{1}{18}$
Values of new U	$-\frac{11}{6}$	$-\frac{5}{6}$	$\frac{1}{6}$	$\frac{7}{6}$	$\frac{13}{6}$

b) 0 c) $\frac{41}{36}$; yes

8.

Probability	$\frac{1}{9}$	$\frac{2}{9}$	$\frac{3}{9}$	$\frac{2}{9}$	$\frac{1}{9}$
Values of U	0	1	2	3	4

$\sigma_U^2 = \frac{4}{3} = \sigma_X^2 + \sigma_Y^2$

10.

Probability	$\frac{1}{16}$	$\frac{4}{16}$	$\frac{6}{16}$	$\frac{4}{16}$	$\frac{1}{16}$
Values of U	0	1	2	3	4

$E(U) = 2 = 2E(X)$, Var $(U) = 1 = 2$ Var (X)

12.

Probability	$\frac{7}{27}$	$\frac{13}{27}$	$\frac{7}{27}$
Values of median	1	2	3

Section 10–2

2. a) 5 b) 24 c) 41
   d) 6 e) 15 f) 5
   g) $\sqrt{3}$ h) $\sqrt{2}$
4. Pythagorean theorem with σ_X and σ_Y being the lengths of the sides adjacent to the right angle and σ_U the length of the hypotenuse
6. $0.003\sqrt{2} \approx 0.0042$
8. $-\frac{1}{2}$; 3.23 (dollars)
10. Mean = 1000; standard deviation would be $100\sqrt{2}$ if Mathematics and English scores were independent, but they probably are not
14. a) μ b) $(1 - 2w + 2w^2)\sigma^2$
    c) $\frac{1}{2}$ d) $\sigma^2/2$
18. $\frac{267}{16}$

Section 10–3

2. 5
4. 9; 10
6. 24; 20
8. $\frac{1}{2}$
10. −2.42
12. 20; 2.5
14. 0.0026
16. 0.0228
18. 12
20. 0.0456
24. Choice (a)
26. a) $\frac{1}{16}$ b) 0.0228 c) 0.0014
    d) Yes; that births in a family form a series of binomial trials with regard to sex

Section 10–4

2. Cov $(X, Y) = 0$
6. $a/|a|$
12. $\rho = -1/(n - 1)$
14. a) 1 b) 1
16. Theoretical mean and variance are both 1.
18. $2n\sigma^2$
24. a) $\frac{5}{6}$ b) $\frac{5}{6}$ c) $\frac{5}{36}$
26. 0
28. a) $\frac{25}{36}$ b) $\frac{25}{36}$ c) $\frac{275}{1296}$
30. a) $\frac{91}{36}$ b) $\frac{395}{1296}$
32.

		Second die					
		1	2	3	4	5	6
First die	1	(1, 0)	(2, 1)	(3, 2)	(4, 3)	(5, 4)	(6, 5)
	2	(2, 1)	(2, 0)	(3, 1)	(4, 2)	(5, 3)	(6, 4)
	3	(3, 2)	(3, 1)	(3, 0)	(4, 1)	(5, 2)	(6, 3)
	4	(4, 3)	(4, 2)	(4, 1)	(4, 0)	(5, 1)	(6, 2)
	5	(5, 4)	(5, 3)	(5, 2)	(5, 1)	(5, 0)	(6, 1)
	6	(6, 5)	(6, 4)	(6, 3)	(6, 2)	(6, 1)	(6, 0)

X = maximum, Y = |die 1 − die 2|
Cell entries: (X, Y)

34. $\frac{665}{648}$; 0.51

Section 10–5

2. a) 0; 5

b)

Probability	$\frac{1}{4}$	$\frac{1}{4}$	$\frac{1}{4}$	$\frac{1}{4}$
Sample average	-3	-1	1	3

$\mu = 0, \sigma^2 = 5$

4.

Probability	$\frac{1}{4}$	$\frac{1}{4}$	$\frac{1}{4}$	$\frac{1}{4}$
Sample average	-1	$-\frac{1}{3}$	$\frac{1}{3}$	1

$\mu_{\bar{X}} = 0; \sigma^2_{\bar{X}} = \frac{5}{9}$

6. Means are equal

8. $S = \{(N, D), (N, Q), (N, H), (N, \$), (D, Q), (D, H), (D, \$), (Q, H), (Q, \$), (H, \$)$, and permutations of these$\}$

10. 0.76; 0.1809

12.

Probability	$\frac{1}{10}$	$\frac{1}{10}$	$\frac{1}{10}$	$\frac{1}{10}$	$\frac{1}{10}$	$\frac{1}{10}$	$\frac{1}{10}$	$\frac{1}{10}$	$\frac{1}{10}$	$\frac{1}{10}$
Sample average: $n = 2$	.075	.150	.175	.275	.300	.375	.525	.550	.625	.750

$\mu_{\bar{X}} = 0.38; \sigma^2_{\bar{X}} = 0.045225$

14.

Exercise	*Variance*
10	0.1809 (sum of 2)
11	0.1206 (single observation)
12	0.0452 (average of 2)

16. $S = \{1.20, 1.40, 1.80, 2.00, 2.40, 2.60\}$

$\mu_{\text{sum}} = 1.90 = 2(0.95)$

$\sigma^2_{\text{sum}} = 0.25 = 2(0.1875)(\frac{2}{3})$

18. 1.5; 0.655

20. 0.193

30. 32

32. 0.000395

34. 9 (almost as accurate)

36. 10 with replacement because $\sigma^2/10$ is slightly less than $4\sigma^2/39$

38. a) 100 b) 33

Section 11–1

2. a) 34,100 (minutes)

4. $y = -x + \frac{1}{2}$

Section 11–2

2. 6.2; 22.8

4. -4; 632

6. 3; 454

8. Values of S: 5; 2.75; 2; 2.75; 5

10. $x = 1; y = 0$

12. $x = -1/(2a); y = a - \frac{1}{4}$

Section 11–3

2. $y = x/2$
4. $y = 10x/3$
6. $y \approx 0.0168x$; sin 5°: estimated 0.084, true 0.087; sin 25°: estimated 0.420, true 0.423
8. $y \approx 1.01x$
10. $y = 0.8x$
12. $\hat{m} = \bar{y}/x$

Section 11–4

2. In absolute magnitudes, middle deviation twice an end deviation
4. $y \approx -1 + 0.67(x - \frac{2}{7})$; $s_d \approx 0.53$
6. $\hat{m} = -\frac{1}{2}$; $\hat{a} = 0$
8. B; C; $C-$ or $D+$
10. 7
12. When the x-coordinates are identical, so that $\sum(x_i - \bar{x})^2 = 0$
18. See answer for Exercise 2 and result for Exercise 3

Section 11–5

6. $y = \bar{y}$, for every x
8. 9%; 9%; 64%
10. $y = rx$
14. 10

Section 11–6

2. $x = y/2$
4. $x = y/3$
6. $x = y/2$
8. $y = 2 + x/2$
12. $y = 4.44 + 0.88x$
14. $\bar{x} = 67.70$
    $\bar{y} = 68.60$
    $s_x = 3.075$
    $s_y = 3.069$
    $y = 39.08 + 0.436x$
    $x = 37.66 + 0.438y$

Section 12–2

2. $1.20196 \geq \mu \geq 1.19804$
4. 0.16

Section 12–3

2. $40.2 \geq \mu \geq 35.3$

Section 12–4

2. $0.50 \geq \mu_{\bar{X}-\bar{Y}} \geq 0.12$
4. Confidence interval $\approx$ 3 to 5 lobsters
6. Assuming samples were independent: confidence interval = 0.05 to 2.37
   Assuming data matched for order: confidence interval = 0.45 to 1.97

Miscellaneous Exercises for Chapter 12

2. $t = -3.67$. Reject null hypothesis of no difference in favor of alternative that student is faster than the average, 0.05 significance level.
4. $t = 2.31$. Reject null hypothesis that student did equally well on the 2 sets in favor of the alternative that he improved with practice, 0.05 significance level.

Appendix 1, Section A1–1

2. 14
4. 9
6. 25
8. -10
10. 303
12. $$\sum_{k=0}^{3} (a_k + b_k) = a_0 + b_0 + a_1 + b_1 + a_2 + b_2 + a_3 + b_3$$
$$= (a_0 + a_1 + a_2 + a_3) + (b_0 + b_1 + b_2 + b_3)$$
$$= \sum_{k=0}^{3} a_k + \sum_{k=0}^{3} b_k$$

One generalization is obtained by replacing the upper limit of summation, 3, by n:
$$\sum_{k=0}^{n} (a_k + b_k) = \sum_{k=0}^{n} a_k + \sum_{k=0}^{n} b_k$$
A second generalization is obtained by increasing the number of summands:
$$\sum_{k=0}^{n} (a_k + b_k + c_k) = \sum_{k=0}^{n} a_k + \sum_{k=0}^{n} b_k + \sum_{k=0}^{n} c_k, \text{ and so on}$$

14. nc

Appendix 1, Section A1–2

2. $1^2(\frac{1}{2})^1 + 2^2(\frac{1}{2})^2 + 3^2(\frac{1}{2})^3 = \frac{21}{8}$
4. $(2.2 - 5) + (2.3 - 5) + (2.4 - 5) + (2.5 - 5) = 8$
6. $$(2x_0 - 3y_0) + (2x_1 - 3y_1) + (2x_2 - 3y_2) = 2(x_0 + x_1 + x_2) - 3(y_0 + y_1 + y_2)$$
8. $\binom{2}{0}x^2y^0 + \binom{2}{1}x^1y^1 + \binom{2}{2}x^0y^2 = x^2 + 2xy + y^2 = (x + y)^2$
10. $\dfrac{0^2}{0!} + \dfrac{1^2}{1!} + \dfrac{2^2}{2!} = 1 + 2 = 3$
12. $$(n + 1)^3 = \sum_{k=0}^{n} [(k + 1)^3 - k^3] = \sum_{k=0}^{n} (3k^2 + 3k + 1)$$
$$= 3\sum_{k=0}^{n} k^2 + 3\sum_{k=0}^{n} k + \sum_{k=0}^{n} 1$$

Hence:
$$3\sum_{k=0}^{n} k^2 = (n + 1)^3 - 3\sum_{k=0}^{n} k - \sum_{k=0}^{n} 1$$

14. $\sum_{k=0}^{n} k = \sum_{k=1}^{n} k = \dfrac{(n+1)n}{2},$

because the first term in the summation of Eq. (6) is 0 when $k = 0$. Similarly, the first term of the summation in Eq. (7), Ex. 13, is 0 when $k = 0$, so that

$$\sum_{k=0}^{n} k^2 = \sum_{k=1}^{n} k^2 = \frac{n(n+1)(2n+1)}{6}.$$

16. $$\begin{aligned}\sum_{i=1}^{n} (x_i - m)^2 &= \sum_{i=1}^{n} (x_i^2 - 2mx_i + m^2)\\ &= \sum_{i=1}^{n} x_i^2 + \sum_{i=1}^{n} (-2mx_i) + \sum_{i=1}^{n} m^2\\ &= \sum_{i=1}^{n} x_i^2 - 2m \sum_{i=1}^{n} x_i + nm^2\end{aligned}$$

18. $a^2 \sum_{i=1}^{n} x_i^2 + 2ab \sum_{i=1}^{n} x_i y_i + b^2 \sum_{i=1}^{n} y_i^2$

Index

Index

Numbers in parentheses refer to exercises on the indicated pages.